Photonics for Bioapplications: Sensors and Technology

Photonics for Bioapplications: Sensors and Technology

Guest Editors

Nélia Jordão Alberto
Maria de Fátima Domingues
Nunzio Cennamo
Adriana Borriello

MDPI

Basel • Beijing • Wuhan • Barcelona • Belgrade • Novi Sad • Cluj • Manchester

Guest Editors
Nélia Jordão Alberto
Instituto de Telecomunicações
Aveiro
Portugal

Maria de Fátima Domingues
Biomedical Engineering &
Biotechnology
Khalifa University
Abu Dhabi
United Arab Emirates

Nunzio Cennamo
Department of Engineering
University of Campania
"Luigi Vanvitelli"
Aversa
Italy

Adriana Borriello
Department of Precision
Medicine
University of Campania
"Luigi Vanvitelli"
Naples
Italy

Editorial Office
MDPI AG
Grosspeteranlage 5
4052 Basel, Switzerland

This is a reprint of the Special Issue, published open access by the journal *Biosensors* (ISSN 2079-6374), freely accessible at: https://www.mdpi.com/journal/biosensors/special_issues/photonics_bioapplications.

For citation purposes, cite each article independently as indicated on the article page online and as indicated below:

Lastname, A.A.; Lastname, B.B. Article Title. *Journal Name* **Year**, *Volume Number*, Page Range.

ISBN 978-3-7258-5491-2 (Hbk)
ISBN 978-3-7258-5492-9 (PDF)
https://doi.org/10.3390/books978-3-7258-5492-9

Contents

About the Editors

Nélia Jordão Alberto

Nélia Jordão Alberto completed her PhD in Physics in 2011, from the University of Aveiro (Portugal). Currently, she is an Assistant Researcher at the Instituto de Telecomunicações (Portugal), and a lecturer at the Physics Department of the Aveiro University. She is the author/co-author of more than 120 peer-reviewed papers in journals and conferences, several book chapters, and co-editor of two books. Dr. Alberto was listed in the top 2% of the most-cited scientists according to Stanford University and Elsevier rankings in the years 2020, 2022, 2023 and 2024. Her research interests include the development of photonic-based solutions and their application, with a special focus on the biomedical context. Dr. Alberto has participated as technical program committee member of distinct international events (AOP2024, Globecom2024 SAC-EH, Globecom2025 SAC-EH, IEEE Healthcom 2024 and 2025, IEEE ICC'25 and ICC'26, . . .).

Maria de Fátima Domingues

Maria de Fátima Domingues is an Assistant Professor at the Biomedical Engineering and Biotechnology Department at Khalifa University, Abu Dhabi, UAE and a Research Collaborator at Instituto de Telecomunicações, Aveiro. Maria de Fátima Domingues received her PhD in Physics Engineering from University of Aveiro, Portugal, in December 2014, and in 2015 started a joint Research position at the Instituto de Telecomunicações – Aveiro and the Consejo Superior de Investigaciones Científicas (CSIC)-Madrid, Spain. She is the lead researcher in biomedical applied photonics at the BMEBD-KU, with +130 co authored peer-reviewed publications in this field. In 2024 Dr. Domingues was listed in the top 2% of the most-cited scientists according to Stanford University and Elsevier rankings. She has been actively cooperating in project proposals both as a PI and co-PI, at national (UAE), and international (EU) levels. She is a project reviewer for entities such as the European Research Council (ERC) and Marie Skłodowska Curie Actions HORIZON MSCA DN. She has been part of the organizing committee of various international conferences and scientific meetings, including IEEE HealthCom 2022 and Healthcom 2025, IEEE WF-IoT2024, 2025 and 2026, IEEE ICC 2024 and Globecom 2025. She is an IEEE Senior Member and an active member of several IEEE Societies. Since January 2022, she is the Secretary for the IEEE EMBS Portuguese Chapter.

Nunzio Cennamo

Nunzio Cennamo (NC) was born in Italy in 1975. He received a master's degree in Electronic Engineering in 2002 and the PhD degree in Electronic Engineering in 2005, both from the Second University of Naples, Naples, Italy. He is a full Professor in Electronics at the University of Campania Luigi Vanvitelli, Naples (Italy). His research interests include the design and fabrication of optical sensors, chemical sensors, biosensors and optoelectronic devices. He is the author of more than 230 international journal and conference papers, and 15 patents. He is cofounder of the Spin Off "MORESENSE srl" in Milan (Fondazione Filarete, Milan, Italy). He is an Associate Editor of "*Photonics Research*" (OSA), of "*Applied Sciences*" (MDPI), and "*Sensors*" (MDPI). He is an Organizer and General Chair of several International Conferences, the IEEE "Sensor Applications Symposium" (IEEE SAS 2024), Naples (Italy) July 23-25, "7th International Symposium on Sensor Science" (I3S 2019) 9-11 May 2019, Naples (Italy), and of several editions of the "International Electronic Conference on Applied Sciences" (ASEC 2021, ASEC 2022, ASEC 2023, ASEC 2024, ASEC 2025). He is a Publicity Chair of the last two editions (the 6th and 7th) of the "IEEE International Symposium M&N" (18-20

July 2022 Padua - Italy and 3-5 July 2024 Rome - Italy), and from the 2025 Edition, a member of the IEEE SAS Steering Committee. He has been an invited speaker several times in Webinars, Summer Schools, and International Conferences. He is an Organizer/Chair of several Special Sessions in International Conferences. He is a Principal Investigator of several national projects and Coordinator of the Local Unit of national and international projects (total budget major than 2 M Euros).

Adriana Borriello

Adriana Borriello (AB) obtained her PhD in Cellular Biochemistry in 1999 (Second University of Naples). Since 2015, AB is Associate Professor of Biochemistry at the University of Campania "L. Vanvitelli" and since 2007, Biology Manager at AOU Vanvitelli. She has obtained the National Scientific Qualification for the role of Full Professor, SSD General Biochemistry (2017). AB is Academic Board Member of the "Biochemical and Biotechnological Sciences" PhD Program since 2008 and has intense teaching activities in Medicine and Surgery and Degree Courses for health professionals (Vanvitelli). She is senior member of several Scientific Societies and Associate Editor and Editorial Board Member for international journals, such as *Frontiers in Endocrinology*, *Cells* (MDPI), and *Diseases* (MDPI), as well as Guest Editor for special issues in *Cells* and *Biosensors* (MDPI). AB has been PI/participant in numerous national research projects. Her research interests are related to cell division control, post-translational regulation of the Cip/Kip inhibitors, and characterization of their new roles in controlling cytoskeleton dynamics. Interests also concern genetic/biochemical mechanisms of hematological diseases, including congenital polycythemias due to altered hypoxia response. Recent efforts are devoted to develop, through multidisciplinary collaborative studies, surface plasmon resonance-based point-of-care biosensors for analytes of interest in biomedical applications. She has been invited speaker/Committee member in numerous international conferences and author of more than 80 peer-reviewed papers in international, high-impact journals, including the *New England Journal of Medicine*, *Leukemia*, *Cancer Research*, and *Free Radicals in Biology and Medicine*, and several book chapters, with >3,000 total citations and an H-index of 33 (Scopus).

biosensors

MDPI

Editorial

Photonics for Bioapplications: Sensors and Technology

Nélia Alberto [1,*], Maria Fátima Domingues [1,2,3], Nunzio Cennamo [4] and Adriana Borriello [5]

[1] Instituto de Telecomunicações, University of Aveiro, Campus Universitário de Santiago, 3810-193 Aveiro, Portugal
[2] Department of Biomedical Engineering and Biotechnology, Khalifa University of Science and Technology, Abu Dhabi P.O. Box 127788, United Arab Emirates
[3] Healthcare Engineering Innovation Group, Khalifa University of Science and Technology, Abu Dhabi P.O. Box 127788, United Arab Emirates
[4] Department of Engineering, University of Campania Luigi Vanvitelli, Via Roma 29, 81031 Aversa, Italy
[5] Department of Precision Medicine, University of Campania "Luigi Vanvitelli", Via De Crecchio 7, 80138 Naples, Italy
* Correspondence: nelia@ua.pt

Over the past decade, interest in advancing photonic systems for bioapplications has been steadily growing, and various key factors have driven this trend. One major factor is the rapid progress in photonic technology—effects like miniaturization, increased sensitivity, and improved resolution—which has broadened how these systems can be used.

Another driving force is the rising demand for more precise diagnostics. As the need for non-invasive, real-time, and highly sensitive diagnostic tools increases, photonic systems have become crucial, especially in areas like biosensors and point-of-care devices. Furthermore, new applications have emerged; photonic technologies are now being used in fields such as environmental monitoring, biochemical analysis, biomedical imaging, and personalized medicine, further expanding their reach and significance.

Additionally, the development of new materials and bio-receptors, smart sensing bio-surfaces, innovative sensor designs, and related sensing instrumentation have significantly enhanced the performance of photonics systems. These advancements have made the technology more suitable and adaptable to complex bioapplications. The interdisciplinary nature of this field has also contributed to its development. By merging photonics with biology, chemistry, and medicine, researchers from different scientific backgrounds are finding new opportunities to collaborate and innovate. Overall, this increasing interest in photonic systems reflects their potential to address major challenges in healthcare, environmental science, and biotechnology.

This Special Issue was designed to showcase the latest innovations in photonic sensing and interrogation systems for bioapplications. It addressed recent technological advancements and novel materials to smart bio-surfaces, emerging applications, innovative sensor designs, and cutting-edge instrumentation. Contributions included original research articles, communications, and review papers, aiming to provide a snapshot of the current state of research and the possible future directions for the field. The goal was to create a high-quality collection of papers on the following, or related, key topics:

- Optical fiber biosensing/immunosensors;
- New bio/chemical probes for bioapplications;
- Wearable biomedical sensors;
- Point-of-care devices;
- Molecular diagnosis;
- Biomarker detection;

 https://doi.org/10.3390/bios15070416

- Imaging sensors;
- Optical sensors in e-Health architectures;
- Optical chemical sensors;
- Energy-efficient eHealth architecture;
- Big data analysis for eHealth;
- Plasmonic sensors and interrogation systems;
- Molecularly imprinted polymers for sensing;
- Advanced signal processing techniques;
- Photonic integrated circuits (PICs) for bioapplications;
- Low-cost, miniaturized, selective, and multiparameter photonic devices.

The contributions to this Special Issue resulted in a collection of 11 manuscripts that report advances in photonics for bioapplications, emphasizing the relevance of the topic to the research community and showcasing the diverse applications where these innovative technological developments can enhance both performance and impact.

For instance, Yin et al. developed an enzyme-assisted fluorescent biosensor that utilizes circular single-stranded DNA for miRNA detection (Contribution 1). This device reaches a detection threshold as low as 0.36 nM, outperforming many previous miRNA sensors and demonstrating robust detection even in complex samples like serum, making it a valuable tool for early disease diagnosis.

Trypsin is recognized as a potential biomarker in cancers like colorectal, gastric, and pancreatic. A novel near-infrared sensitive dye–peptide conjugate (SQ-3 PC) for selectively detecting trypsin activity via fluorescence ON/OFF sensing is proposed (Contribution 2). The probe exhibits high sensitivity, a linear response up to 30 nM, and a limit of detection and limit of quantification of 1.07 nM and 3.25 nM, respectively. The device demonstrates strong selectivity for trypsin, even in the presence of other enzymes, making it suitable for complex biological applications.

The communication proposed by Xu et al. reports a new THz metasurface sensor for cinnamoylglycine detection, a urinary biomarker of gestational diabetes mellitus (GDM) (Contribution 3). The device, produced from lithium tantalate prism tetramers on quartz, achieves a high Q-factor of 231, enhancing molecular detection. Simulations show a detection limit of 1.23 $\mu g \times cm^{-2}$, demonstrating excellent sensitivity. Given the risks associated with GDM, like preeclampsia and preterm birth, the sensor offers a faster and more accurate alternative to traditional tests, with potential for detecting other biomarkers and monitoring disease progression.

Park et al. developed an *Asterias forbesi*-inspired surface-enhanced Raman spectroscopy (SERS) with numerous hotspots (Contribution 4). The substrate showed strong SERS performance, minimized oxidation damage, and ensured repeatability for uric acid detection. Under laboratory conditions, uric acid was detected at concentrations as low as 1.16 nM, and selectivity was demonstrated against various metabolites. In a simulated real-world scenario, the sensor successfully detected a wide range of uric acid concentrations, encompassing both deficiency and excess levels, in serum and urine samples.

In the study presented by Jiao et al., the polarimetric aberrations caused by three different illumination schemes in backscattering Mueller Matrix imaging were analyzed (Contribution 5). The optimized schemes established key criteria for selecting spatial illumination patterns in non-collinear backscattering Mueller Matrix measurements, offering valuable insights for advancing quantitative tissue polarimetric imaging and biosensing.

In (Contribution 6), the viscoelastic properties of AuNP-assisted radiofrequency-heated and ablated tissues were analyzed, correlating them with temperature profiles using confocal Brillouin micro-spectroscopy and Mg-silicate-NP-doped temperature-sensing fibers. The results reveal that changes in the Brillouin peak were linked to thermal doses and protein denaturation. The hybrid Brillouin–optical backscattering reflectometry technique shows potential for monitoring tissue deformations during thermal therapies, enabling the real-time tracking of viscoelastic and thermal properties in metal-nanoparticle-embedded tissues for hyperthermal and theranostic applications.

A red-light-driven photoelectrochemical (PEC) sensor based on a g-C_3N_4/CuS/TiO_2 ternary heterojunction for the sensitive detection of the synergistic antioxidant effects of sesamol and other antioxidants is reported in (Contribution 7). Sesamol is a natural antioxidant that reduces free radicals and supports brain function, improving memory and cognition in Alzheimer's patients. By scavenging holes in the valence band, the antioxidants suppress charge carrier recombination, boosting the PEC photocurrent. The sensor exhibited high sensitivity, selectivity, and stability, when applied to detect sesamol in soybean and peanut oils. It holds potential for applications in nutrition analysis, food fraud detection, and medical research, contributing to improved consumer health and product quality.

Zhou et al. developed titanium metal–organic framework (Ti-MOF) microflowers for the detection and removal of Al(III) ions in water (Contribution 8). Ti-MOF offers high water stability, porosity, and strong luminescence, with a detection range of 0.4–15 μM and a limit of 75 nM, well below the World Health Organization's aluminum safety limit. It also allows selective Al(III) detection by shielding interference from Fe(III) using ascorbic acid, making it an effective sensor and adsorbent for water purification.

A nano-slit array sensor based on temperature variation is proposed for the highly sensitive and specific detection of molecular fingerprints (Contribution 9). As the temperature changes, the dielectric properties of the temperature-sensitive semiconductor material InSb dynamically adjust, causing shifts in the transmission resonance angle and producing transmission envelope curves over a broad frequency range. Using this temperature scanning approach, the characteristic fingerprint spectra of hexogen, a high-impact military explosive, were successfully identified, allowing both qualitative and quantitative detection with a sensitivity limit of 1.61 $\mu g/cm^2$.

A distributed optical fiber sensing network was proposed in (Contribution 10) for the two-dimensional in situ thermal mapping of advanced methods for radiofrequency thermal ablation. Radiofrequency ablation was conducted on a bovine phantom using a standard approach, while methods enhanced with agarose and gold nanoparticles were carried out to improve the ablation efficiency. Results showed that agarose-mediated thermal ablation treated the largest area, but with low repeatability, while AuNP-mediated ablation at a 4 mg/mL density offered the best balance between ablation efficacy and repeatability.

The Special Issue concludes with a review paper dedicated to cost-effective optical fiber-based biosensing platforms, exploring various geometries, interrogation techniques, and encapsulation methods (Contribution 11). The potential of these cost-effective optical fiber sensing techniques is highlighted through numerous successful applications, in three key areas: cancer, cardiovascular biomarker detection, and environmental monitoring. Some weaknesses of these systems are also highlighted, emphasizing the need to improve aspects related to repeatability and reproducibility, as well as the development of techniques for sensor reuse.

In conclusion, this Special Issue presents a selection of manuscripts that highlight recent advances in the field of photonics, with a particular focus on bioapplications. The works gathered here reflect the diversity and breadth that characterize both the research domain itself and the applications. Although they represent only a sample of the extensive ongoing developments, the published articles provide a meaningful overview of the current trends and pave the way for future opportunities for research and innovation in the field of photonic sensors, emphasizing their potential contributions to biomedicine and related areas.

We sincerely thank the authors for their valuable contributions to this Special Issue, which brings together original research articles that provide a comprehensive and current overview of a diverse array of topics, all unified by a shared focus on "Photonics for bioapplications: Sensors and Technology". Additionally, we express our gratitude to the academic editors and reviewers for their dedicated efforts in ensuring the scientific rigor and quality of the published manuscripts. Their efforts have significantly enhanced the overall impact and relevance of this Special Issue.

Funding: This work was supported by the FCT—Fundação para a Ciência e Tecnologia, I.P., project reference UIDB/50008: Instituto de Telecomunicações.

Acknowledgments: N.A. is supported by the Scientific Employment Stimulus 2022.00250. CEECIND/CP1716/CT0002, and DOI identifier 10.54499/2022.00250.CEECIND/CP1716/CT0002. MFD work was supported by the Khalifa University grant RC2-2018-022.

Conflicts of Interest: The authors declare no conflicts of interest.

List of Contributions:

1. Yin, X.; Liao, Y.; Li, F.; Li, J.; Du, J. Enzyme-assisted fluorescence biosensor based on circular single-stranded DNA without group modification for microRNA detection. *Biosensors* **2024**, *14*, 527.
2. Balyan, P.; Gupta, S.; Mavileti, S.; Pandey, S.; Kato, T. NIR-sensitive squaraine dye—Peptide conjugate for trypsin fluorogenic detection. *Biosensors* **2024**, *14*, 458.
3. Xu, Q.; Sun, M.; Wang, W.; Shi, Y. All-dielectric metasurface-based terahertz molecular fingerprint sensor for trace cinnamoylglycine detection. *Biosensors* **2024**, *14*, 440.
4. Park, H.; Chai, K.; Kim, W.; Park, J.; Lee, W.; Park, J. Asterias forbesi-inspired SERS substrates for wide-range detection of uric acid. *Biosensors* **2024**, *14*, 8.
5. Jiao, W.; Zhang, Z.; Zeng, N.; Hao, R.; He, H.; He, C.; Ma, H. Complex spatial illumination scheme optimization of backscattering Mueller matrix polarimetry for tissue imaging and biosensing. *Biosensors* **2024**, *14*, 208.
6. Kurbanova, B.; Ashikbayeva, Z.; Amantayeva, A.; Sametova, A.; Blanc, W.; Gaipov, A.; Tosi, D.; Utegulov, Z. Thermo-visco-elastometry of RF-wave-heated and ablated flesh tissues containing Au nanoparticles. *Biosensors* **2023**, *13*, 8.
7. Huang, L.; Yang, J.; Liang, Z.; Liang, R.; Luo, H.; Sun, Z.; Han, D.; Niu, L. Ternary heterojunction graphitic carbon nitride/cupric sulfide/titanium dioxide photoelectrochemical sensor for sesamol quantification and antioxidant synergism. *Biosensors* **2023**, *13*, 859.
8. Zhou, J.; Song, J.; Ma, G.; Li, Y.; Wei, Y.; Liu, F.; Zhou, H. Hierarchical Ti-MOF microflowers for synchronous removal and fluorescent detection of aluminum ions. *Biosensors* **2022**, *12*, 935.
9. Wang, W.; Sun, M.; Lin, J.; Xue, Y.; Shi, Y. Terahertz fingerprint metasurface sensor based on temperature variation for trace molecules. *Biosensors* **2024**, *14*, 318.
10. Sametova, A.; Kurmashev, S.; Ashikbayeva, Z.; Amantayeva, A.; Blanc, W.; Atabaev, T.; Tosi, D. Fiber-optic distributed sensing network for thermal mapping of gold nanoparticles-mediated radiofrequency ablation. *Biosensors* **2022**, *12*, 352.
11. Leitão, C.; Pereira, S.; Marques, C.; Cennamo, N.; Zeni, L.; Shaimerdenova, M.; Ayupova, T.; Tosi, D. Cost-effective fiber optic solutions for biosensing. *Biosensors* **2022**, *12*, 575.

biosensors

MDPI

Article

Enzyme-Assisted Fluorescence Biosensor Based on Circular Single-Stranded DNA Without Group Modification for MicroRNA Detection

Xiaoxue Yin, Yazhen Liao, Feiyu Li, Jianbao Li and Jie Du *

School of Materials Science and Engineering, Hainan University, Haikou 570228, China; 22220856010035@hainanu.edu.cn (X.Y.); 21220856000036@hainanu.edu.cn (Y.L.); 20213007198@hainanu.edu.cn (F.L.); 20213007169@hainanu.edu.cn (J.L.)
* Correspondence: dujie@hainanu.edu.cn

Abstract: Fluorescent biosensor, which has the characteristics of high sensitivity, specificity, and low cost, can be directly detected in physiological fluids such as blood and serum. Therefore, the development of fluorescence sensor platforms for miRNA detection has a positive effect on the prevention and treatment of various diseases. In this paper, miR-34a was selected as a biological indicator of Alzheimer's disease (AD). We designed a circular single-stranded DNA (CSSD) biosensor, which uses two unmodified single-stranded DNA (ssDNA) with complementary ends, DNAa and DNAb, to form CSSD by DNA sequence pairing to improve thermal stability and achieve signal amplification. At the same time, CSSD can react with miR-34a, and then the DNA of the DNA–RNA chain is hydrolyzed by duplex-specific nuclease (DSN enzyme). Finally, miR-34a is released to partake in the subsequent step, thus realizing cycle amplification. By evaluating the change in fluorescence signal under the optimized conditions, we discovered that this approach exhibits impressive sensitivity, with a detection threshold reaching as low as 0.36 nM. This surpasses the performance of numerous preceding miRNA detection biosensors. Furthermore, the system displays excellent detection capabilities even in intricate settings like serum, showcasing a strong ability to differentiate and choose effectively. In summary, this is a signal-off fluorescent biosensor, which realizes the purpose of double amplification of biosensor signal by using CSSD and enzyme assistance so that it can be used as a valuable tool for early diagnosis of diseases.

Keywords: duplex-specific nuclease; fluorescence reduction; GelRed; miR-34a

1. Introduction

Alzheimer's disease (AD) is a secluded, age-related neurodegenerative disease, which is the most common form of dementia [1]. The elderly are the people mainly affected by AD, and the incidence of AD increases with the increase in life expectancy [2]. The incidence of AD is expected to double every five years after the age of 65 [3], and there is currently no cure [4]. In the absence of effective treatment, early diagnosis plays a vital role in improving the subsequent condition, and the diagnosis of Alzheimer's disease is expensive, so it is imperative to create a novel and cost-effective diagnostic approach [5]. MicroRNA (miRNA) are a length of 19 to 25 nucleotides of non-coding RNA [6], and they are involved in many critical biological processes. As a result of miRNA in the important role of gene expression [7], their disorders are associated with the pathology of many diseases [8], such as liver, skin, skeletal muscle, cardiovascular, immune, neurological and inflammatory system diseases, as well as cancer [9]. This also means that the miRNA biomarkers can be used in various diseases [10]. In the past, miR-34a was regarded as a biomarker for malignant tumors and cancers [11–13]. However, recent studies have shown that miR-34a is related to brain neurophysiology and pathology [14], especially Alzheimer's disease. Therefore, in addition to being a cancer marker [13], miR-34a is also expected to be a biomarker for AD [5,14]. Due to the extremely low abundance of biomarkers in cells, the development of sensitive platforms for the determination of biomarkers has broad research prospects [15]. Biosensors, including colorimetric, Raman scattering spectroscopy, electrochemical and fluorescent biosensors, etc., are characterized by high sensitivity, high specificity and high cost effectiveness, and can be directly detected in physiological fluids (blood, serum, etc.) [16]. Many researchers choose to use fluorescence detection biosensors to detect biomarkers [17], and their methods are widely used in medicine, food analysis, environmental and agricultural analysis, and industry [18–20]. DNA fluorescence biosensors transform the chemical energy of the tested biomaterial system into light energy. This categorization divides DNA fluorescence sensors into two groups: those with labeling and those without. Wang et al. designed a unique DNA detection method that involves the assembly of two functional nucleic acids to form a new nucleic acid structure upon exposure to the target DNA [21]. The nucleic acid structure represents a functional catalytic nucleic acid that can undergo substrate cleavage reactions in the presence of Mg^{2+}. This leads to the cleavage of DNA labeled with fluorophore and cataplexy groups, resulting in a transition from low to high fluorescence intensity. Consequently, this method enables the quantitative detection of target DNA, with a sensor sensitivity of 0.01 pM and excellent selectivity.

Double-stranded specific nuclease (DSN) selectively degrades DNA in double-stranded DNA and DNA–RNA hybrids, has little activity against single-stranded nucleic acid molecules and double-stranded RNA, and can distinguish single-base mismatches well. The characteristics of DSN enzymes make them a promising option for signal amplification in DSN-based biosensors [22]. Liao et al. devised a nanochannel biosensor incorporating a DSN enzyme signal amplification technique [23]. In this system, the capture probe DNA is firmly attached to the surface of the nanochannel. Upon interaction with the target miRNA, the nanochannel undergoes hybridization, resulting in the formation of the DNA–RNA duplex. The DSN enzyme cleaved the probe DNA in the DNA–RNA double-stranded structure to extract the target miRNA. The target miRNA can then bind to other DNA probes, leading to signal amplification. GR serves as a secure and reliable intercalating dye, attaching to nucleic acid through electrostatic and charge interactions. It can directly replace the highly toxic ethidium bromide (EB), and its sensitivity is much higher than that of EB.

In this experiment, we designed an enzyme-assisted fluorescence biosensor without group modification. Based on the Watson–Crick base complementary pairing principle, the design had two single-stranded DNA (ssDNA) strands so that their terminals are complementary sequences, and there were two miR-34a binding sites on each DNA strand. The two ssDNA were hybridized to form circular single-stranded DNA (CSSD). GR intercalation dye was then added, which could select the DNA specifically and was fluorescent. The stronger the thermal stability of DNA, the stronger the fluorescence signal [24]. When the DSN enzyme is activated, it can cause the DNA strand in double-stranded DNA-RNA to be hydrolyzed, leading to the loss of the fluorescence signal if there is a specific miRNA that binds to CSSD. The target RNA chain and GR were released and they entered the next cycle. If no target miRNA was present, the fluorescence signal was always present. The sensitivity and specificity of the sensor were measured. This kind of biosensor provides a new idea for RNA detection.

2. Materials and Methods

2.1. Materials

The chemicals and reagents employed in the experiments are of analytical reagent grade and do not need any purification of the received pharmaceutical products. The DSN buffer was made up of a solution containing 50 mM Tris-HCl (pH 8.0), while the DSN master buffer was a mixture of 500 mM Tris-HCl (pH 8.0) and 50 mM $MgCl_2$. DNA annealing buffer (5×, sterile solution, 0.1 M Tris-HC1, 0.1 M NaCl, 0.05 M EDTA) and PCR amplification buffer (10×, pH 8.3, 100 mM Tris-HCl buffer, 15 mM Mg^{2+}, 500 mM KCl.) were purchased from Shanghai Yuanye Biotechnology Co., Ltd. (Shanghai, China); GR nucleic acid dye (10,000×) aqueous solution was purchased from Sangong Bioengineering Co., Ltd. (Shanghai, China); Tris-HCI buffer (pH 8.0) was purchased from Shanghai MacLean Biochemical Technology Co., Ltd. (Shanghai, China); DEPC-treated water was purchased from Beijing Solarbio Science & Technology Co., Ltd. (Beijing, China); glycerol solution (100%, sterile) was purchased from Beijing Ranjeco Technology Co., Ltd. (Beijing, China); DSN enzyme was purchased from Shenzhen Newbond Biotechnology Co., Ltd. (Shenzhen, China); and Fetal bovine serum (FBS) was purchased from Zhejiang Tianhang Biotechnology Co., Ltd. (Deqing, China).

The DNA utilized in this experiment, as shown in Table 1, was supplied by Shanghai Sangong Bioengineering Co., Ltd. (Shanghai, China). The lengths of DNAa and DNAb are 66 nt, and they contain two fragments complementary to miR-34a with a length of 22 nt. In addition, they have a complementary 9 nt fragment at each end. The resulting CSSD has a double chain length of 18 nt and a single chain length of 48 nt.

Table 1. Oligonucleotide sequences used in this experiment.

Name	Sequence (from 5′ to 3′)
miR-34a	5′-UGG CAG UGU CUU AGC UGG UUG U-3′
DNAa	5′-GCTGACTGCACAACCAGCTAAGACACTGCCATT TTACAACCAGCTAAGACACTGCCACGATCTCAC-3′
DNAb	5′-GCTAGAGTGACAACCAGCTAAGACACTGCCATT TTACAACCAGCTAAGACACTGCCACGACTGACG-3′
miR-34a	5′-UAGCUUAUCAGACUGAUGUUGA-3′
miR-192	5′-CUG ACC UAUGAA UUG ACA GCC-3′
miR-34a	5′-UGG AGU GUG ACA AUG GUG UUG-3′
miR-15a	5′-UAG CAG CAC AUA AUG GUU UGU G-3′
miR-660	5′-UAC CCA UUG CAU AUC GGA GGU GU G-3′
miR-155	5′-UUA AUG CUA AUC GUG AUA GGG GU-3′

2.2. Apparatus

Fluorescence signal values were detected using the Shimadzu RF-6000 fluorescence spectrophotometer (Shimadzu, Kyoto, Japan). The experiment uses GR, an intercalated fluorescent dye that selectively binds DNA and RNA strands to produce a fluorescent signal. GR is attached to nucleic acids through electrostatic and charge interactions, and the fluorescence intensity of free GR is negligible. In the presence of ssDNA, GR was embedded into ssDNA to form GelRed-ssDNA complex with significant fluorescence intensity. Strong electrostatic interactions could stabilize the GelRed-ssDNA complex.

2.3. Preparation of Reagents

DNA was diluted to 10 μmol/L and miRNA was diluted to 20 μmol/L with DEPC treated water (enzyme free, sterile). Our choice of buffer for the experiment was the DNA annealing buffer (1×), stored in a refrigerator at 4 °C. The DSN buffer and the DSN master buffer both were stored at a temperature of 4 °C. Lyophilized DSN enzyme was added to DSN buffer at 5 μL per 10 U and an equal volume of 100% glycerol was added. The DSN enzyme, set at 1 U/μL, was stored at −20 °C for utilization.

2.4. Generation of CSSD

According to the previous work of Meng et al. [20], the ends of two ssDNA (DNAa, DNAb) are complementary sequences. First, 1 μM of DNAa and DNAb were added to the annealing buffer with 5 μL each, then heated at 95 °C for 2 min, and then cooled at room temperature for incubation. Finally, circular DNA was formed, and its concentration was 2.5 μmol/L.

2.5. Detection of Target miR-34a

According to the previous work of Sapia et al. [24,25], the experimental steps designed by us are as follows. Add 7.5 μL GR dye (3×) nucleic acid solution to CSSD solution and incubate at 25 °C for 15 min. After adding 2 μL of deionized water for 5 min at 95 °C, the mixture is cooled for 30 min at room temperature. Then, 0.5 μL DSN enzyme (1 U/μL) is added, and the DSN enzyme is inactivated at 45 °C for 1 h and 75 °C for 5 min. The measured fluorescence value is represented by F_0. GR solution was added to CSSD solution and incubated at 25 °C for 15 min. Two μL miR-34a are added, incubated at 95 °C for 5 min, cooled for 30 min, and then the DSN enzyme is added. Then, the fluorescence value F is detected, and finally the fluorescence comparison result (F_0-F) is obtained.

2.6. Testing miR-34a

The fluorescence signal value was measured by a fluorescence spectrophotometer with model RF-6000. The mixture used for the test consisted of 2 mL of ultra-pure water mixed with the experimental solution. The excitation wavelength used in this mixture was 270 nm, while the emission wavelength was 545 nm, and the target scan range was between 510 nm and 580 nm. The excitation frequency band was 5 nm and the emission frequency band was 10 nm.

3. Results and Discussion

3.1. Principle of miRNA Detection

The experimental design of the enzyme-assisted fluorescent biosensor detection principle of miR-34a is shown in Figure 1. In this study, we first designed two DNA single strands with complementary ends, DNAa and DNAb, which contain two complementary sequences of miR-34a, respectively. Therefore, hybridization of DNAa and DNAb produces CSSD with four miRNA binding sites. Since the fluorescence signal intensity of GR is enhanced with the increasing thermal stability of DNA, the fluorescence signal intensity can be improved after the two DNA single strands are combined to form a ring, and thus the fluorescence value F_0 is obtained. Starting from the presence or absence of target detection miRNA, two routes can be obtained. Route one is that when the detection target exists, miR-34a carries out the hybridization reaction with circular DNA. At this time, under the action of the DSN enzyme, the DNA strand in the DNA–RNA hybrid strand is selectively hydrolyzed, miR-34a and GR dye are released, and the fluorescence signal value decreases. The released miR-34a can react with other unreacted circular DNA to achieve cyclic amplification. After the above reaction, the fluorescence value F was obtained after high temperature treatment and the DSN enzyme was inactivated. In route two, no RNA hybridizes with DNA due to the deletion of miR-34a. Even in the presence of DSN enzymes, single-stranded DNA cannot be hydrolyzed, so the fluorescence signal value remains unchanged. Finally, the fluorescence difference F_0-F can be obtained by changing the fluorescence value, thereby determining the presence of miR-34a to be tested. Because the fluorescence values of the two lines are based on the GR dye, F_0 and F are essentially the same.

3.2. Feasibility Study of the miR-34a Assay

According to the excitation spectrum and emission spectrum of the sample to be tested, as shown in Figure 2A, we determined that the excitation wavelength for detection was 270 nm and the emission wavelength was 510–580 nm. In this experiment, in the presence of target miR-34a, two ssDNAs with complementary ends were combined to form CSSD. Oligo is one of the three most popular sequence design software in the world. It has powerful features, and its results are widely recognized. The structure of our designed single-stranded oligonucleotide was manually input into Oligo 7, and the Tm value was calculated using the nearest neighbor method. The Tm of DNAa and DNAb were measured at 77.1 °C and 76.1 °C, respectively. The Tm of PCR product CSSD was 96.9 °C, and the Tm of CSSD/RNA complex was 101.9 °C. The circular DNA strands were subjected

three times. F_0 was the fluorescence signal value for adding only DNA and GR, and F was the fluorescence value for adding RNA and DSN enzymes. As shown in Figure 3A, with the increase in incubation time, the fluorescence difference (F_0-F) first increased and then decreased. The maximum value was reached at 45 min. The reason for it might have been the excessive time of incubation, which likely loosened some bound structures to gradually separate. Finally, the optimal time was chosen to be 45 min.

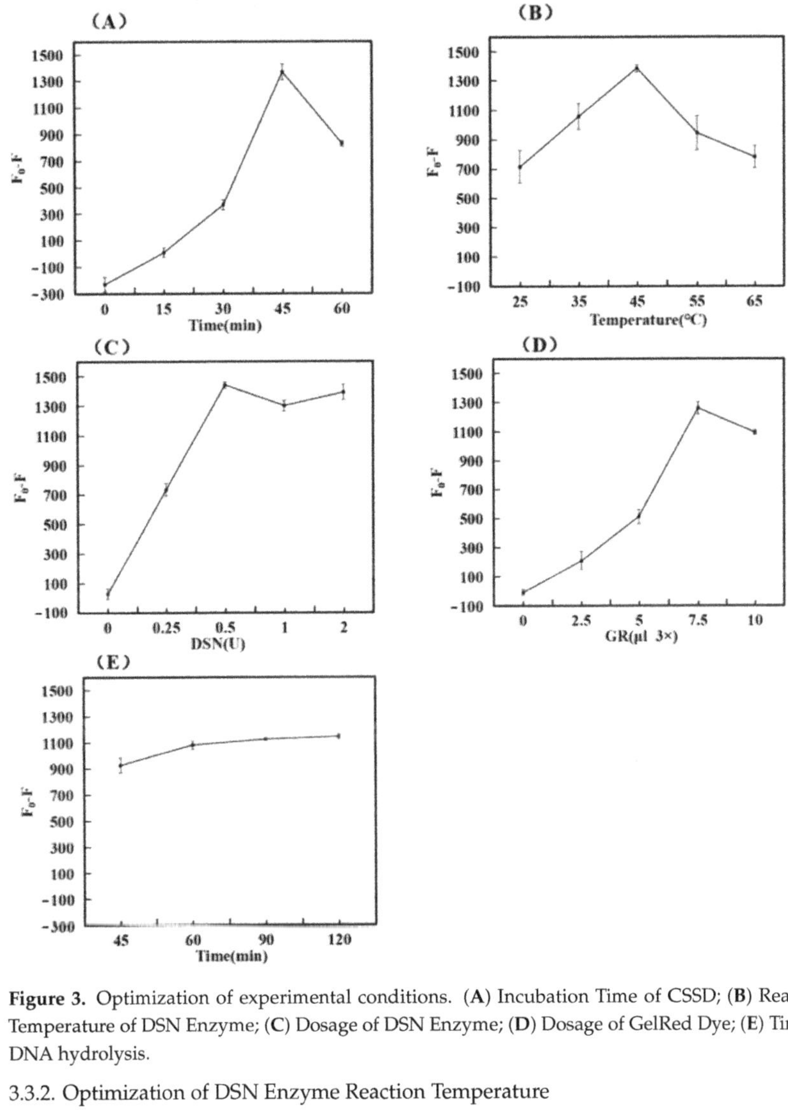

Figure 3. Optimization of experimental conditions. (**A**) Incubation Time of CSSD; (**B**) Reaction Temperature of DSN Enzyme; (**C**) Dosage of DSN Enzyme; (**D**) Dosage of GelRed Dye; (**E**) Time of DNA hydrolysis.

3.3.2. Optimization of DSN Enzyme Reaction Temperature

Temperature is one of the important factors affecting the fluorescence intensity of hybridization. When the temperature drops too low, it can cause the reaction time to stretch or lead to incomplete reactions. Conversely, excessively high temperatures might hinder the activity of the DSN enzyme, thereby impacting the outcomes of the experiment. To address this, we designed five experiments to be tested at different reaction temperatures, and each test was repeated three times. As illustrated in Figure 3B, the difference in fluorescence (F_0-F) shows an increasing trend between 25 °C and 45 °C, while it declines from 45 °C to 65 °C. This latter decrease could be attributed to the elevated temperature impairing enzyme activity. Considering both the cost-effectiveness and practicality of the biosensor, we determined that 45 °C is optimal for the reaction.

3.3.3. Optimization of DSN Enzyme Dosage

The amount of DSN enzyme dosage was related to whether the DNA in the DNA–RNA hybridization chain could be completely hydrolyzed, thus affecting subsequent experiments. The concentration of DSN enzyme configured and used in this experiment was 1 U/μL. To establish the ideal concentration of the DSN enzyme, various amounts were introduced while maintaining consistent experimental conditions, ensuring that the volume of the test solution was standardized with the buffer solution. The outcomes of the tests are displayed in Figure 3C. As the quantity of DSN enzyme rose, the variation in fluorescence (F_0-F) also increased, peaking at 0.5 U. However, beyond this point, additional increases in DSN enzyme dosage did not significantly alter the fluorescence intensity. To optimize resources, the DSN enzyme concentration was therefore maintained at 0.5 U for the following experimental adjustments.

3.3.4. Optimization of GelRed Dye Dosage

The amount of GR affects the intensity of fluorescence signal detection. The concentration of GR dye used in the experiment was 3×. Under the same experimental conditions, GR dye with the same concentration but different amounts was added, and the volume of the tested solution was adjusted with ultra-pure water to be consistent. The experiment was repeated three times. After testing, the results obtained from the experiment are shown in Figure 3D. The difference in fluorescence (F_0-F) rose as the dosage of GR dye escalated, peaking at 7.5 μL before tapering off with additional increments of the dye. This could be attributed to the possibility that at higher GR concentrations, there is less available binding space with DNA, which in turn leads to a reduction in the fluorescence signal. Therefore, the subsequent optimization experiment set the concentration of GR at 7.5 μL.

3.3.5. Optimization of DNA Hydrolysis Time

The hydrolysis time of DNA is also the action time of the DSN enzyme. And the primary cause of the diminished fluorescence signal is the hydrolysis of DNA. When the hydrolysis time is too brief, it results in only a slight attenuation of the signal, ultimately compromising the sensor's sensitivity. Conversely, if the duration is excessively lengthy, it leads to an inefficient use of time. In order to determine the ideal hydrolysis time of DNA, we changed the reaction time of the DSN enzyme and inactivated the enzyme with high temperature after reaching the expected time. Upon analyzing the results, we illustrated our findings in Figure 3E. We observed that the discrepancy in fluorescence grew over time, although the rate of increase began to slow down. Ultimately, we decided that a 60 min hydrolysis period was optimal for DNA.

3.4. *Analysis of Performance of miR-34a Biosensors*

In this study, we designed a DSN enzyme-assisted fluorescent biosensor platform for miR-34a detection. To investigate the detection performance of the sensor, under optimal experimental conditions, different concentrations of miR-34a were detected. Based on the above, the optimized experimental conditions were as follows: 0.5 U DSN enzyme, 1 h DNA incubation time, 45 °C DSN enzyme reaction temperature, and 7.5 μL GR dosage, the fluorescence signal value was the best result. Ten sets of experiments were conducted to measure the fluorescence intensity across various concentrations of miR-34a, with each set repeated thrice for accuracy.

As shown in Figure 4A, this experiment is one in which the fluorescence value decreased with the increase in the target miRNA concentration. The standard detection curve in the sensitivity analysis was derived from the fluorescence value (F_0) of the light blue curve in Figure 4A minus the curve (F) of different concentrations. Ten sets of different data (F_0-F) were obtained, and the linear regression curves F_0-F = 501.19C − 247.04 and R^2 = 0.9816 were obtained through simple fitting, indicating the high feasibility of this experiment. The lowest detection limit (LOD) was determined by tripling the standard deviation (3σ) of the baseline signal. This was calculated using the formula 3σ/S, where σ represents the baseline signal's standard deviation and S stands for the slope of the fitting line in Figure 4B. The detection limit was calculated to be 0.36 nM. The closer R^2 was to one, the higher the accuracy of detecting the target miRNA within the detection range, and the better photobleaching observed in this experiment. As shown in Table 2 and the sensitivity of this work, our work is superior to many previous works on miRNA detection systems.

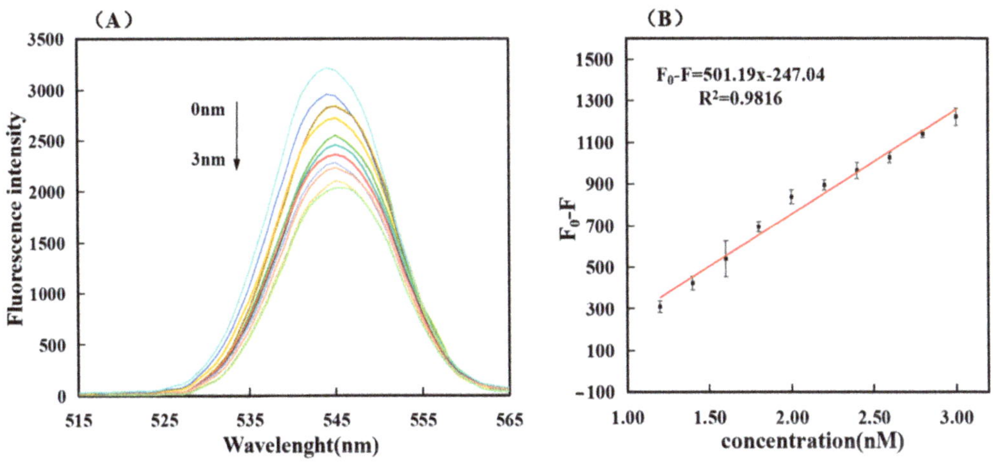

Figure 4. The fluorescence value of miR-34a was measured against the standard detection curve, and the concentrations of top to bottom substances in Figure (**A**) were 0 nM, 1.2 nM, 1.4 nM, 1.6 nM, 1.8 nM, 2.0 nM, 2.2 nM, 2.4 nM, 2.6 nM, 2.8 nM and 3.0 nM, respectively; Figure (**B**) Linear calibration curve of RNA and fluorescence intensity (correlation between RNA concentration and fluorescence signal strength). The error bars represent standard deviations obtained from from triplicate exper.

Table 2. Comparison of some other sensors for miR-34a detection.

Detection Technique	Linear Range	LOD	Reference
PAMAM Dendrimer Modified Electrodes	0–7.5 µg/mL	135 nM	[26]
Ionic Liquid Modified Single-use Electrodes	2–10 µg/mL	125 nM	[27]
Ionic Liquid-Modified Graphite Electrodes	2–10 µg/mL	109 nM	[28]
The Chemical Activation of PGE Surface	0–2.5 µg/mL	72.5 nM	[29]
Graphene Oxide Modified Graphite Electrodes	142–568 nM	41.2 nM	[30]
Electropolymerization	5–80 µg/mL	28.4 nM	[31]
Enzyme-assisted fluorescence biosensor	1.2–3 nM	0.36 nM	This work
Gold Nanoflower @Graphene Quantum Dots Probe	0.4–4 fM	0.1 fM	[32]
Catalytic Hairpin Assembly &DNA Walker	200 pM–10 nM	43.72 pM	[33]
Gold Nanostructures	0.1–1000 nM	39 pM	[5]

3.5. *The Specificity of the Sensor*

The experimental test of resistance to interference from other substances was also an important indicator of sensor performance. In order to see the influence of other miRNA sequences on the detection results, the signal values obtained from the detection of the target miR-34a, and other homologous miRNA were tested under the same optimized conditions. In theory, other non-targeted miRNAs cannot fully react with the designed DNA strand, and the efficiency of enzyme reactions will decrease. In order to verify whether the sensor constructed in this experiment was selective for the target miR-34a, we performed specific detection experiments using different miRNAs. These included miR-155, miR-192, miR-15a, miRNA-122, miR-660, and miRNA-21. Among them, the concentration of miR-34a was 3 nM, and the concentration of other detected substances was 15 nM.

The findings from the experiment are illustrated in Figure 5, where (A) displays the fluorescence spectra for different miRNA detection. Upon examination of the figure, it is clear that the fluorescence values of the remaining miRNA are quite high, whereas the fluorescence level of miR-34a is significantly lower. This suggests that in the experiments, other types of miRNA had difficulty binding to circular DNA and performing enzymatic reactions. The DNA was not hydrolyzed, but other RNAs were stained with GR dye, resulting in enhanced fluorescence intensity. As can be seen from the bar graph in Figure 5B, after three tests, the fluorescence intensity of the target miR-34a decreased, resulting in a positive fluorescence intensity difference, while the staining of other miRNAs increased the fluorescence intensity, resulting in a negative fluorescence intensity difference. Thus, the target RNA was very different from other miRNAs. In summary, this approach exhibits strong discriminatory capacity towards other similar miRNAs, showcasing commendable specificity and selectivity.

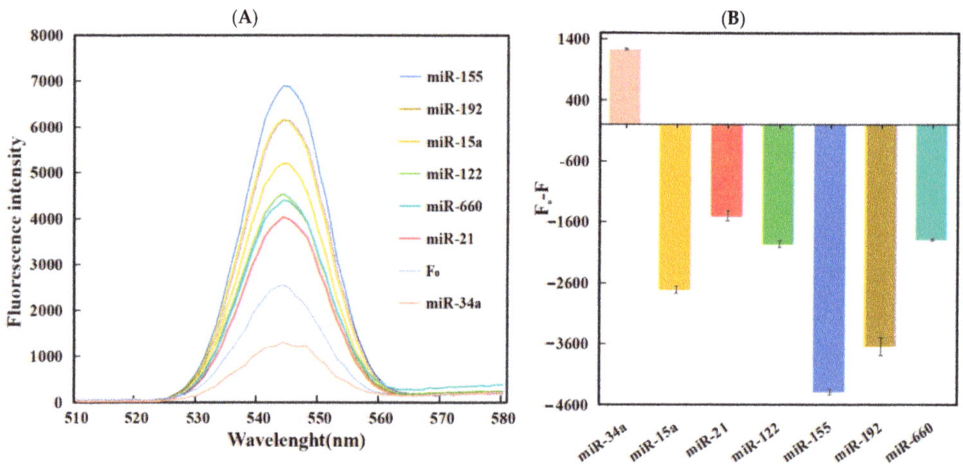

Figure 5. (**A**) Fluorescence spectrum plot of the relationship between miRNA and fluorescence intensity of different species; (**B**) Plot of different miRNAs of fluorescence intensity at the excitation wavelength of 545 nm.

3.6. Detection of miR-34a in Complex Environments

To further verify whether our method was suitable for detection in complex environments, we recovered three different concentrations of the target RNA (1.5 nM, 2.0 nM, 2.5 nM) by adding different concentrations of the target DNA to the reaction to study the reaction in the amplification system. As shown in Table 3, Added (nM) refers to the actual added RNA content. The estimated RNA content, marked as Found (nM), was obtained by calculating the obtained fluorescence value. The ratio between the two is marked as Recovery (%), and RSD is the standard deviation of the recovery rate. The calculated recoveries ranged from 92.8% to 105.6%, and the relative standard deviation recoveries were 5.2%, 6.3%, and 5.5%. The recovery fluctuated within a small range of 100%, meeting the experimental requirements. A lower standard deviation suggests that the test results were more consistent with the expected results, pointing to greater accuracy in the method. The experimental findings indicate that the proposed detection method holds promise for detecting miR-34a in real, intricate biological samples.

Table 3. Recovery of miR-34a in FBS (n = 3).

Sample	Added (nM)	Found (nM)	Recovery (%)	RSD (%)
		1.43	95.2	
1	1.5	1.53	102.0	5.2%
		1.58	105.6	
		2.10	105.0	
2	2.0	1.86	92.8	6.3%
		1.93	96.6	
		2.34	93.7	
3	2.5	2.58	103.0	5.5%
		2.35	94.0	

4. Conclusions

In conclusion, we have constructed an enzyme-assisted fluorescent biosensor, leveraging CSSD for the detection of miRNA. This sensor employs the DSN enzyme to break down the CSSD that is hybridized with the target miR-34a. As a result, miR-34a is freed to engage in the next phase, facilitating cycle amplification. During this process, the DNA's interaction with GR is disrupted, leading to a reduced fluorescence signal. Experimental findings demonstrated a clear distinction between the target miRNA and other RNA, illustrating strong selectivity and suitability for qualitative analysis. Notably, this method is capable of detecting miRNA at remarkably low concentrations, with a limit of detection as low as 0.36 nM and a linear detection range extending from 1.2 nM to 3 nM. Impressive outcomes were achieved even in intricate settings. This approach can also be adapted to identify various RNA biomarkers linked to illnesses by modifying the specified target recognition sequence. It offers considerable practicality and numerous potential applications in disease prevention and treatment, as well as for the quantitative measurement of RNA in medical research, paving the way for innovative studies.

Author Contributions: X.Y. and J.D. performed the experiments and wrote the paper; X.Y., Y.L., F.L. and J.L. analyzed the data, and J.D. conceived and designed the experiments. All authors have read and agreed to the published version of the manuscript.

Funding: This work was financially supported by the Hainan Province Science and Technology Special Fund (ZDYF2023SHFZ110), and the Graduate Students Innovation Research Project of Hainan Province (Qhys2022-77).

Institutional Review Board Statement: Not applicable.

Informed Consent Statement: Not applicable.

Data Availability Statement: The raw data supporting the conclusions of this article will be made available by the authors on request.

Conflicts of Interest: The authors declare no conflicts of interest.

References

1. Valiukas, Z.; Ephraim, R.; Tangalakis, K.; Davidson, M.; Apostolopoulos, V.; Feehan, J. Immunotherapies for Alzheimer's Disease—A Review. *Vaccines* **2022**, *10*, 1527. [CrossRef] [PubMed]
2. Wanionok, N.E.; Morel, G.R.; Fernández, J.M. Osteoporosis and Alzheimer´s disease (or Alzheimer´s disease and Osteoporosis). *Ageing Res. Rev.* **2024**, *99*, 102408. [CrossRef] [PubMed]
3. Ewen, S.T.; Fauzi, A.; Quan, T.Y.; Chamyuang, S.; Yin, A.C.Y. A review on advances of treatment modalities for Alzheimer's disease. *Life Sci.* **2021**, *276*, 119129.
4. Rybak-Wolf, A.; Plass, M. RNA Dynamics in Alzheimer's Disease. *Molecules* **2021**, *26*, 5113. [CrossRef]
5. Pereira, R.L.; Oliveira, D.; Pêgo, A.P.; Santos, S.D.; Moreira, F.T.C. Electrochemical miRNA-34a-based biosensor for the diagnosis of Alzheimer's disease. *Bioelectrochemistry* **2023**, *154*, 108553. [CrossRef]
6. Lu, T.X.; Rothenberg, M.E. MicroRNA. *J. Allergy Clin. Immunol.* **2017**, *141*, 1202. [CrossRef]
7. Takousis, P.; Sadlon, A.; Schulz, J.; Wohlers, I.; Dobricic, V.; Middleton, L.; Lill, C.M.; Perneczky, R.; Bertram, L. Differential expression of microRNAs in Alzheimer's disease brain, blood, and cerebrospinal fluid. *Alzheimer's Dement.* **2019**, *15*, 1468. [CrossRef]

8. Van den Berg, M.M.J.; Krauskopf, J.; Ramaekers, J.G.; Kleinjans, J.C.S.; Prickaerts, J.; Briedé, J.J. Circulating microRNAs as potential biomarkers for psychiatric and neurodegenerative disorders. *Prog. Neurobiol.* **2019**, *185*, 101732. [CrossRef]
9. Tokar, E.J. MicroRNAs: Small molecules with big effects. *Toxicol. Appl. Pharmacol.* **2016**, *312*, 1–2. [CrossRef]
10. Wu, Y.; Li, Q.; Zhang, R.; Dai, X.; Chen, W.; Xing, D. Circulating microRNAs: Biomarkers of disease. *Clin. Chim. Acta* **2021**, *516*, 46. [CrossRef]
11. Chua, C.E.L.; Tang, B.L. miR-34a in Neurophysiology and Neuropathology. *J. Mol. Neurosci.* **2018**, *67*, 235. [CrossRef] [PubMed]
12. Badr, E.A.E.; Ali Assar, M.F.; Gohar, S.F.; Badr, M.H.; Hathout, R.M.; El-kousy, S.M. The clinical impact of miRNA34a and P53 gene expression in colon cancer. *Biochem. Biophys. Rep.* **2018**, *16*, 88. [CrossRef] [PubMed]
13. Krajewska, J.B.; Fichna, J.; Mosińska, P. One step ahead: miRNA-34 in colon cancer-future diagnostic and therapeutic tool? *Crit. Rev. Oncol./Hematol.* **2018**, *132*, 1–8. [CrossRef] [PubMed]
14. Modi, P.K.; Jaiswal, S.; Sharma, P. Regulation of Neuronal Cell Cycle and Apoptosis by MicroRNA 34a. *Mol. Cell. Biol.* **2015**, *17*, 84. [CrossRef] [PubMed]
15. Li, C.-c.; Wang, Z.-y.; Wang, L.-j.; Zhang, C.-y. Biosensors for epigenetic biomarkers detection: A review. *Biosens. Bioelectron.* **2019**, *144*, 111695. [CrossRef] [PubMed]
16. Mittal, S.; Kaur, H.; Gautam, N.; Mantha, A.K. Biosensors for breast cancer diagnosis: A review of bioreceptors, biotransducers and signal amplification strategies. *Biosens. Bioelectron.* **2016**, *36*, 84. [CrossRef]
17. Li, T.; Duan, R.; Duan, Z.; Huang, F.; Xia, F. Fluorescence Signal Amplification Strategies Based on DNA Nanotechnology for miRNA Detection. *Chem. Res. Chin. Univ.* **2019**, *36*, 194. [CrossRef]
18. Deshpande, K.; Kanungo, L. Chemiluminescence and fluorescence biosensors for food application: A review. *Sens. Actuators Rep.* **2022**, *5*, 100137. [CrossRef]
19. Mungroo, N.A.; Neethirajan, S. Biosensors for the Detection of Antibiotics in Poultry Industry—A Review. *Biosensors* **2014**, *4*, 472. [CrossRef]
20. Meng, J.; Chen, S.; Han, J.X.; Tan, Q.; Wang, X.R.; Wang, H.Z.; Zhong, W.L.; Qin, Y.; Qiao, K.L.; Zhang, C.; et al. Derepression of co-silenced tumor suppressor genes by nanoparticle-loaded circular ssDNA reduces tumor malignancy. *Sci. Transl. Med.* **2018**, *10*, eaao6321. [CrossRef]
21. Wang, F.; Elbaz, J.; Orbach, R.; Magen, N.; Willner, I. Amplified Analysis of DNA by the Autonomous Assembly of Polymers Consisting of DNAzyme Wires. *J. Am. Chem. Soc.* **2011**, *133*, 17149. [CrossRef] [PubMed]
22. You, Z.; Yang, Z.; Chen, Y.; Zhang, L. Visual detection of heart failure associated MiRNA with DSN enzyme-based recycling amplification strategy. *RSC Adv.* **2021**, *11*, 18068. [CrossRef] [PubMed]
23. Liao, T.-B.; Luo, K.-X.; Tu, J.-Y.; Zhang, Y.-L.; Zhang, G.-J.; Sun, Z.-Y. DSN signal amplification strategy based nanochannels biosensor for the detection of miRNAs. *Bioelectrochemistry* **2024**, *4*, 108771. [CrossRef] [PubMed]
24. Hao, Y.-Y.; Liu, L.; Zhang, L.-H.; Huang, Q.-L.; Wang, F.; Li, J.; Xu, J.-Q.; Wang, L.-H. Real-time label-free analysis of the thermostability of DNA structures using GelRed. *Nucl. Sci. Tech.* **2018**, *29*, 138. [CrossRef]
25. Sapia, R.J.; Campbell, C.; Reed, A.J.; Tsvetkov, V.B.; Gerasimova, Y.V. Interaction of GelRed™ with single-stranded DNA oligonucleotides: Preferential binding to thymine-rich sequences. *Dyes Pigment.* **2021**, *188*, 109209. [CrossRef]
26. Congur, G.; Erdem, A. PAMAM dendrimer modified screen printed electrodes for impedimetric detection of miRNA-34a. *Microchem. J.* **2019**, *148*, 748. [CrossRef]
27. Yaralı, E.; Kanat, E.; Erac, Y.; Erdem, A. Ionic Liquid Modified Single-use Electrode Developed for Voltammetric Detection of miRNA-34a and its Application to Real Samples. *Electroanalysis* **2019**, *32*, 384. [CrossRef]
28. Kesici, E.; Eksin, E.; Erdem, A. An Impedimetric Biosensor Based on Ionic Liquid-Modified Graphite Electrodes Developed for microRNA-34a Detection. *Sensors* **2018**, *18*, 2868. [CrossRef]
29. Congur, G.; Eksin, E.; Erdem, A. Impedimetric detection of miRNA-34a using graphene oxide modified chemically activated graphite electrodes. *Sens. Actuators A Phys.* **2018**, *279*, 493. [CrossRef]
30. Erdem, A.; Eksin, E.; Isin, D.; Polat, D. Graphene Oxide Modified Chemically Activated Graphite Electrodes for Detection of microRNA. *Electroanalysis* **2017**, *29*, 1350. [CrossRef]
31. Mandli, J.; Amine, A. Impedimetric genosensor for miRNA-34a detection in cell lysates using polypyrrole. *J. Solid State Electrochem.* **2017**, *22*, 1007. [CrossRef]
32. Sun, J.; Cui, F.; Zhang, R.; Gao, Z.; Ji, J.; Ren, Y.; Pi, F.; Zhang, Y.; Sun, X. Comet-like Heterodimers "Gold Nanoflower @Graphene Quantum Dots" Probe with FRET "Off" to DNA Circuit Signal "On" for Sensing and Imaging MicroRNA In Vitro and In Vivo. *Anal. Chem.* **2018**, *90*, 11538. [CrossRef] [PubMed]
33. Feng, X.; Bi, X.; Feng, J.; Hu, S.; Wang, Y.; Zhao, S.; Zhang, L. Proximity-Induced Bipedal DNA Walker for Accurately Visualizing microRNA in Living Cancer Cell. *Anal. Chem.* **2024**, *96*, 10669. [CrossRef] [PubMed]

biosensors

MDPI

Article

NIR-Sensitive Squaraine Dye—Peptide Conjugate for Trypsin Fluorogenic Detection

Priyanka Balyan, Shekhar Gupta, Sai Kiran Mavileti , Shyam S. Pandey * and Tamaki Kato *

Graduate School of Life Science and System Engineering, Kyushu Institute of Technology, 2-4 Hibikino, Wakamatsu-Ku, Kitakyushu-Shi, Fukuoka 808-0196, Japan; priyanka101@mail.kyutech.jp (P.B.); gupta.shekhar932@mail.kyutech.jp (S.G.); mavileti-chilamakuru.kiran-sai678@mail.kyutech.jp (S.K.M.)
* Correspondence: shyam@life.kyutech.ac.jp (S.S.P.); tmkato@life.kyutech.ac.jp (T.K.)

Abstract: Trypsin enzyme has gained recognition as a potential biomarker in several tumors, such as colorectal, gastric, and pancreatic cancer, highlighting its importance in disease diagnosis. In response to the demand for rapid, cost-effective, and real-time detection methods, we present an innovative strategy utilizing the design and synthesis of NIR-sensitive dye–peptide conjugate (**SQ-3 PC**) for the sensitive and selective monitoring of trypsin activity by fluorescence ON/OFF sensing. The current research deals with the design and synthesis of three unsymmetrical squaraine dyes **SQ-1**, **SQ-2**, and **SQ-3** along with a dye–peptide conjugate **SQ-3-PC** as a trypsin-specific probe followed by their photophysical characterizations. The absorption spectral investigation conducted on both the dye alone and its corresponding dye–peptide conjugates in water, utilizing **SQ-3** and **SQ-3 PC** respectively, reveals enhanced dye aggregation and pronounced fluorescence quenching compared to observations in DMSO solution. The absorption spectral investigation conducted on dye only and corresponding dye–peptide conjugates in water utilizing **SQ-3** and **SQ-3 PC**, respectively, reveals not only the enhanced dye aggregation but also pronounced fluorescence quenching compared to that observed in the DMSO solution. The trypsin-specific probe **SQ-3 PC** demonstrated a fluorescence quenching efficiency of 61.8% in water attributed to the combined effect of aggregation-induced quenching (AIQ) and fluorescence resonance energy transfer (FRET). FRET was found to be dominant over AIQ. The trypsin-mediated hydrolysis of **SQ-3 PC** led to a rapid and efficient recovery of quenched fluorescence (5-fold increase in 30 min). Concentration-dependent changes in the fluorescence at the emission maximum of the dyes reveal that **SQ-3 PC** works as a trypsin enzyme-specific fluorescence biosensor with linearity up to 30 nM along with the limit of detection and limit of quantification of 1.07 nM and 3.25 nM, respectively.

Keywords: trypsin detection; aggregation-caused quenching; FRET; NIR detection; squaraine dye; dye–peptide conjugate

1. Introduction

Proteases are enzymes that catalyze the hydrolysis of peptide bonds, playing crucial roles in various physiological and pathological processes [1]. Among these, trypsin, a serine protease produced in the pancreas, is particularly significant due to its role in digestion and its involvement in pancreatic disorders such as pancreatitis and pancreatic cancer [2–4]. Trypsin facilitates the detachment of adherent cells and is therefore used therapeutically in tissue regeneration and as an anti-inflammatory agent [5]. The expression of trypsin is significantly elevated in several human cancer cells found in the stomach, colon, lung, and breast [6]. Recently, trypsin has gained recognition as a potential marker for cancer patients [7,8]. This is because the dysregulation of trypsin has been associated with the development of several tumors, such as colorectal cancer (CRC), gastric cancer, and pancreatic cancer [9,10]. In addition, trypsin is often used in the field of food chemistry for protein identification purposes [11]. In light of its biological significance, there is a significant demand for inventive and accessible assays for trypsin. These assays are essential for the development of effective diagnostic and therapeutic techniques for the treatment of pancreatic diseases and as applications in the proteomics field [12,13]. Various techniques have been developed for trypsin detection, including mass spectroscopy [14,15], gel electrophoresis [16], enzyme-linked immunosorbent assay (ELISA) [17,18], colorimetric [19,20], and fluorometric approaches [21,22]. Among these, fluorogenic sensing technologies have gained prominence due to their simplicity, accessibility, and visual detection capabilities. These methods eliminate the need for sophisticated instrumentation and complex analytical procedures, contributing to their widespread adoption in various fields of study.

Capitalizing on these advantages, researchers have developed a diverse array of sensors employing distinctive fluorescence approaches, specifically designed and validated for the precise and sensitive detection of trypsin activity [23,24]. Recent advances in protease assay methods have focused on developing immunoassays using antibodies to bind target proteases [25,26]. However, while effective for quantifying protease levels, these immunoassays are limited in their ability to map protease activity and its correlation with disease progression. To address this limitation, alternative approaches have been explored, utilizing tailored peptide substrates with optical detection techniques such as absorbance or fluorescence spectroscopy [27–29]. These techniques leverage protease-substrate interactions to measure enzymatic activity directly. One of the challenges in developing fluorescence-based protease assays is the interference from the autofluorescence of biological samples and potential photo-damage to biological species due to high-energy excitation. The use of fluorophores that emit in the near-infrared (NIR) wavelength region offers a solution to these issues, providing increased sensitivity and the potential for deep-tissue bio-imaging [30,31].

In this context, NIR-sensitive squaraine dyes have emerged as promising candidates for biosensing applications due to their exceptional physical and chemical attributes, including strong absorption bands, high molar absorption coefficients, elevated quantum yields, and excellent photostability [32,33]. The absorption and emission properties of squaraine dyes can be tuned from the visible to NIR wavelength region by carefully selecting donor groups with controllable π-conjugation [34,35]. Additionally, SQ dyes exhibit a pronounced aggregation-caused quenching (AIQ) due to the presence of extended π-conjugation leading to a planer molecular framework. This allows for the modulation of their fluorescence properties by controlling their aggregation and dispersion states [36]. This characteristic has been utilized to develop a variety of fluorescent probes, including those for detecting proteins, enzymes, and ions. For instance, a BSA-SQ assembly was designed as a fluorescent "on-off" probe for pepsin detection, where the aggregation of SQ led to fluorescence quenching in the pepsin-catalyzed hydrolysis of BSA [37]. The versatility of SQ dyes in biosensing applications extends to protein detection [38], where their binding to proteins like bovine serum albumin disrupts SQ aggregation, leading to fluorescence enhancement. Furthermore, peptide-conjugated SQ probes based on Förster Resonance Energy Transfer (FRET) have been designed for enzyme detection [39], where enzymatic cleavage terminates the resonance energy transfer process and subsequently increases the fluorescence intensity. Aggregation-induced quenching (AIQ) and FRET can work synergistically to enhance biosensing systems. Hence, there is broad applicability of SQ dyes in developing sensitive and selective fluorescent probes for various biomolecules and ions, capitalizing on their unique aggregation based and energy transfer properties.

In this study, we present the development of an internally quenched homo-labeled fluorescent peptide substrate for the sensitive and selective detection of trypsin enzyme activity. The probe features two NIR-sensitive unsymmetrical squaraine dyes conjugated at each terminal of a trypsin-cleavable peptide sequence. This configuration exhibits fluorescence quenching in its intact state. Upon cleavage by trypsin, the probe demonstrates a significant enhancement of the fluorescence signal in the NIR region, enabling a sensitive detection of trypsin activity. The designed probe offers exceptional sensitivity, capable of detecting trypsin activity at exceedingly low concentrations. Moreover, it demonstrates remarkable selectivity, with minimal interference from other proteases commonly found in complex biological samples. This approach has the potential to improve diagnostics and monitoring of trypsin-related disorders and also has potential applications in proteomics research and drug discovery.

2. Materials and Methods

2.1. Reagents and Materials

All chemicals, solvents, and reagents were of analytical or spectroscopic grade and used as supplied. Fmoc-protected amino acids, Rink amide 4-methylbenzhydrylamine (MBHA)resin,piperidine,1-((Dimethylamino)(dimethyliminio)methyl)-1H-benzo[d][1,2,3] triazole 3-oxide hexafluorophosphate (HBTU), 1H-Benzo[d][1,2,3]triazol-1-ol (HOBt)·H_2O, diisopropylethylamine (DIEA), trifluoroacetic acid (TFA), and 4 M HCl/Dioxane were purchased from Watanabe Chemical Industries, Ltd. (Hiroshima, Japan). Enzymes (α-chymotrypsin, α-trypsin from bovine pancreas, pancreatic elastase, horseradish peroxidase, bovine serum albumin, and papain) were obtained from Sigma-Aldrich Co., LLC (Tokyo, Japan). All other reagents and solvents were procured from Wako Pure Chemical Industries, Ltd. (Osaka, Japan).

2.2. Theoretical Structure Calculation

The theoretical quantum chemical calculation for the structural optimization of squaraine dye–peptide conjugate (**SQ-3 PC**) was conducted on a Dell workstation (8 Core) using the Gaussian G16 program package [40]. The calculation was performed for the isolated molecule in the gaseous state. In the theoretical calculations, judicious selection of a suitable basis set, theory, and functional are required to make a logical balance between the computation cost and accuracy. Considering the huge number of atoms in the molecule, theoretical calculation for structure optimization was conducted on HF/6-31G level of theory to save computation time.

2.3. Instrumentation

Nuclear magnetic resonance (NMR) spectroscopy (500 MHz for 1H NMR) and TOF/FAB-mass spectroscopy in positive ion monitoring mode were used for structural characterization. UV–visible–NIR absorption spectra were recorded on a JASCO V-530 UV/VIS spectrophotometer (JASCO Corporation, Tokyo, Japan). Fluorescence emission spectra were obtained using a JASCO FP-6600 spectrofluorometer (JASCO Corporation, Tokyo, Japan). Analytical HPLC was performed on a Hitachi L-7100 system (Hitachi High-Technologies Corporation, Tokyo, Japan) with an Xterra MS C18-5 μM column (4.6 × 250 mm; Waters Corporation, Milford, MA, USA). The sample was pre-incubated in a EYELA SLI-220 temperature-controlled incubator (Tokyo Rikakikai Co., Ltd., Tokyo, Japan) to ensure thermal equilibration. During spectroscopic measurements, the cuvette holder was equipped with a water-jacketed cell holder connected to a THERMO SUPPLIER FZL-80F circulating water bath system (TAITEC Corporation, Saitama, Japan) maintained at 37 °C.

2.4. Spectroscopic Measurements

Stock solutions of 100 μM trypsin in phosphate buffer saline (PBS) (pH 7.2, 0.1 M) and 100 μM **SQ-3 PC** in dimethyl sulfoxide (DMSO) were prepared and diluted as required. The fluorescence response of **SQ-3 PC** (5 μM) to trypsin was evaluated in H_2O (pH 7.2, 2% DMSO). Fluorescence measurements were taken at 659 nm (excitation and emission slit widths of 5 nm and 6 nm, respectively) at 5 min and then every 10 min for 30 min. Trypsin concentrations ranged from 1 nM to 75 nM. Baseline fluorescence intensity (F_0) was determined using 5 μM **SQ-3 PC** in H_2O with 2% DMSO.

2.5. Determination of Detection Limit

The limit of detection (LOD) and limit of quantification (LOQ) for trypsin-specific fluorescence biosensor was calculated using the following formula:

$$LOD = 3.3\ (SD)/S \quad (1)$$

$$OQ = 10\ (SD)/S \quad (2)$$

where S is the slope of the linear part of the calibration curve and SD is the standard deviation of the sample for consecutive measurements.

2.6. Fluorescence Quenching Efficiency

Fluorescence quenching efficiency was calculated using the following equation:

$$\text{Quenching efficiency } (\%) = [1 - (F_{PC}/F_{dye})] \times 100 \quad (3)$$

where F_{dye} is the emission intensity of SQ-3 dye and F_{PC} is the emission intensity of SQ-3 PC at the same concentration in H_2O (2% DMSO).

3. Results

3.1. Design and Synthesis of Target Molecules

In this study, three unsymmetrical squaraine dyes (**SQ-1**, **SQ-2**, and **SQ-3**) and a dye–peptide conjugate (**SQ-3 PC**) were synthesized and investigated as NIR-sensitive fluorescent molecules for determining trypsin enzyme activity. The dyes are based on a squaraine core with two indole-based terminal groups (Figure 1). The molecular design strategy involved several key modifications to the squaraine core structure to enhance its photophysical properties. Iodine was introduced to investigate the heavy atom effect, potentially improving intersystem crossing. A methoxy group was added as an electron-donating substituent to optimize electronic distribution within the π-conjugated system. Additionally, a propionic acid moiety was incorporated at the N-position of one terminal indole ring to establish an intramolecular charge transfer (ICT) system and provide a reactive carboxylic acid for bioconjugation with peptides via amide bond formation. **SQ-1** features an asymmetric substitution with an iodo group (-I) and a propionated alkyl chain(-CH_2CH_2COOH) on one indole group and a methoxy group and butyl chain(-$CH_2CH_2CH_2CH_3$) on the other. **SQ-2** is similar to **SQ-1** but has butyl chains on both indole groups. **SQ-3** resembles **SQ-1** but lacks the iodine substitution. The synthesis, structural characterization, and analytical data for these compounds and their intermediates are provided in the Supplementary Materials (Schemes S1 and S2).

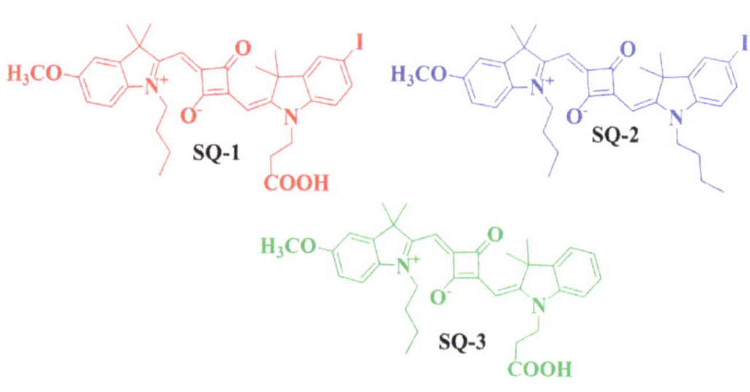

Figure 1. Chemical structure of squaraine dyes (**SQ-1**, **SQ-2**, and **SQ-3**).

SQ-3 was identified as the optimal choice for the preparation of the dye–peptide conjugates due to its distinctive structural attributes, aggregation properties, and optical behavior. Although the substitution of hydrogen with a heavy atom such as iodine in squaraine dyes has been previously reported to enhance fluorescence quantum yield [41], this modification did not prove to be significantly advantageous in the present study. The absence of iodine in **SQ-3** avoids potential heavy metal effects and reduces steric hindrance, potentially facilitating a more efficient coupling to peptides and better interaction with the target enzyme trypsin. In a recent study (Soumya et al.) [42], a symmetrical iodinated squaraine dye is designed to be a photosensitizer for photodynamic therapy, where its stability is crucial for targeted action on tumor tissues. This stability may contribute to its resistance to enzyme hydrolysis. **SQ-3** exhibits the highest aggregation index (AI) in PB of 3.38, attributed to the combination of electron-donating (-OCH_3) and electron-withdrawing (-COOH) groups promoting intermolecular interactions. The amino acid sequence Gln-Arg-Glu was chosen as the substrate for the trypsin enzyme based on the known specificity of trypsin [43]. Specifically, trypsin cleaves peptide bonds between the carboxyl group of arginine or lysine and the amino group of the adjacent amino acid. To optimize the design of our biosensing probe, we incorporated additional structural elements (Figure 2a). Free amino groups with β-Ala and Lys were integrated to serve dual purposes: first, as spacers to provide flexibility towards the substrate, and second, to facilitate the binding of the **SQ-3** squaraine dye to the substrate.

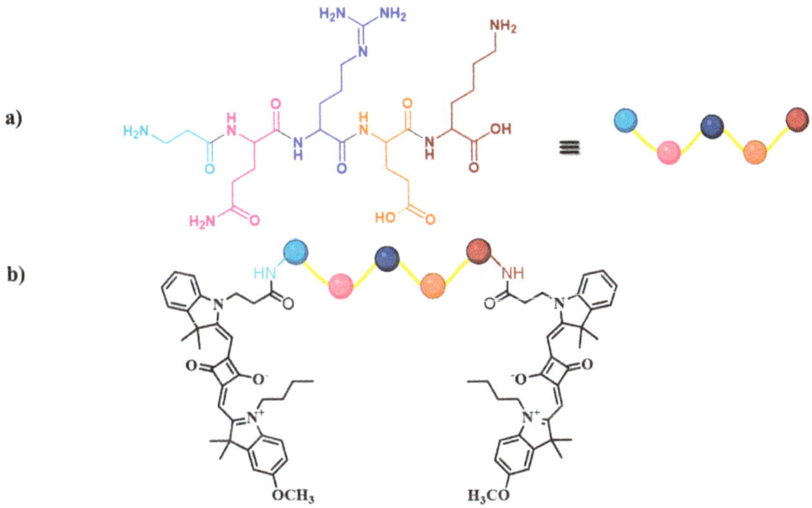

Figure 2. Chemical structure of (**a**) trypsin enzyme-specific peptide and (**b**) dye–peptide conjugate probe (**SQ-3 PC**) for the trypsin enzyme sensing.

The trypsin-specific probe **SQ-3-PC** (Figure 2b) was synthesized using solid-phase peptide synthesis (SPPS). The design incorporated a linear Gln-Arg-Glu tripeptide, chosen for trypsin specificity, with β-Alanine as a spacer and Lysine for the side-chain binding of the terminal squaraine dye. Synthesis began with Fmoc-Lys(Boc)-OH conjugation to Rink Amide MBHA Resin, followed by a stepwise peptide chain construction (Scheme 1).The resin-supported peptide (Fmoc-βAla-Gln(trt)-Arg(Pbf)-Glu(o'tBu)-Lys(Boc)-Resin) underwent Fmoc deprotection with 20% piperidine and Boc removal with HCl/Dioxane (2 M). **SQ-3** was then conjugated at a 2.2:1 molar ratio using HOBt/HBTU as coupling reagents. The final cleavage and deprotection were achieved using a TFA/triisopropylsilane/H_2O (95:2.5:2.5) cocktail. The crude dye–peptide conjugate was precipitated with ether and purified by silica gel column chromatography (chloroform/methanol, 9:1), yielding the final probe. Successful synthesis was confirmed by Time-of-Flight (TOF) mass spectrometric analysis, which revealed a measured m/z of 1702.9112 $[M + H]^+$ (calculated 1701.8901). The detailed synthesis procedure, reaction conditions, and analytical data, including mass spectrometry, are provided in the Supporting Information (Scheme S3).

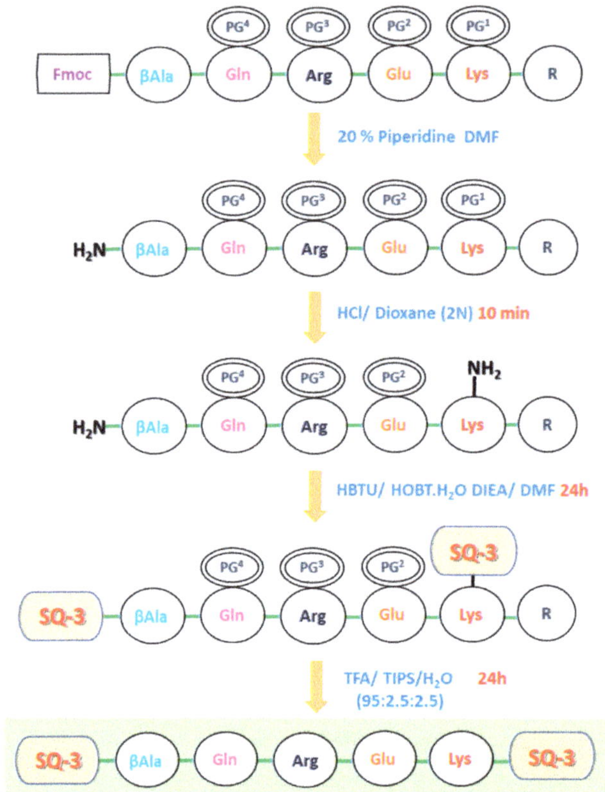

Scheme 1. Schematic representation for the synthesis of dye–peptide conjugate (**SQ-3 PC**); R, Rink Amide MBHA Resin; PG[1], Boc; PG[2], o'tBu; PG[3], Pbf; PG[4], trt.

3.2. Optical Characterization of Squaraine Dyes and Dye–Peptide Conjugate

The photophysical properties of squaraine dyes (**SQ-1**, **SQ-2**, **SQ-3**, and **SQ-3 PC**) were characterized using optical absorption and fluorescence emission spectroscopies in DMSO. These measurements aimed to elucidate the impact of structural modifications on the optical properties. Figure 3a,b present the absorption and emission spectra, respectively, at a dye concentration of 5 μM in DMSO along with the summarization of the deduced optical parameters such as absorption maximum (λ_{abs}), emission maximum (λ_{em}), molar extinction coefficients (ε), and Stokes shifts (Δ) in Table 1. All dyes exhibit strong and sharp optical absorption with the absorption maximum (λ_{abs}) between 652 nm and 659 nm with a weak vibronic shoulder between 600 nm and 620 nm, characteristic of squaraine dyes due to the π-π* transition. **SQ-3**, with -OCH_3 and -COOH groups, has a strong, narrow band in the visible to NIR wavelength region. **SQ-1** and **SQ-2**, containing iodine and -OCH_3 groups on the main ring, show relatively higher bathochromic shifts due to the heavy atom effect, broadening the absorption band and shifting it to longer wavelengths, which is a similar effect as observed by Mayerhöffer et al. [41].

These structural differences not only affect the optical absorption characteristics but also influence the emission properties of the dyes under investigation, as shown in Figure 3b. The enhanced emission of iodine-containing **SQ-1** and **SQ-2** is likely due to the iodine heavy atom effect, which may increase the molecular rigidity or reduce non-radiative decay. **SQ-1** with a single iodine atom achieves an optimal balance between the heavy atom effect and structural asymmetry, resulting in the highest emission intensity. The absence of iodine in **SQ-3** may lead to greater molecular flexibility, increased non-radiative decay, and lower emission intensity. Nevertheless, all of the dyes depict a very small Stokes shift of only 8 nm–15 nm with similar shapes of absorption and emission spectra, which are indicative of the very comparable molecular configurations of the dye in the ground and excited states [44]. Interestingly, the optical behavior observed in the aqueous medium is very different compared to that observed in the DMSO solution. The electronic absorption spectrum of **SQ-3** in H_2O containing 2% DMSO showed a broad absorption spectral feature with λ_{abs} at 653 nm and a highly pronounced vibronic shoulder appearing at 610 nm, as shown in Figure 4a, indicating an enhanced dye aggregation in the aqueous medium.

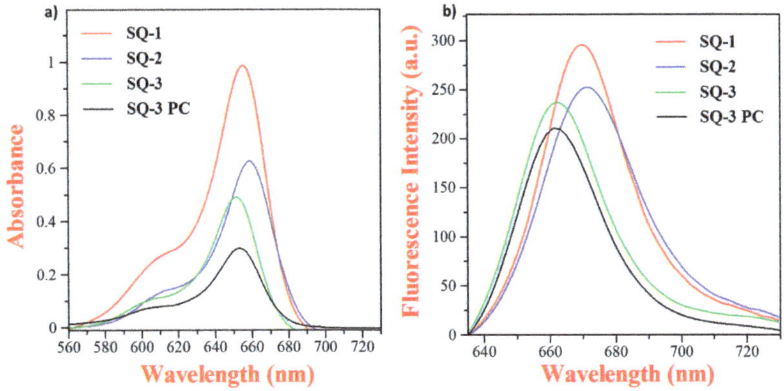

Figure 3. (**a**) Optical absorption and (**b**) fluorescence emission spectra of **SQ-1**, **SQ-2**, **SQ-3**, and **SQ-3 PC** in DMSO (5 μM).

Table 1. Optical parameters for the squaraine dyes and dye–peptide conjugates in DMSO solution.

Squaraine Dye	λ_{abs}	λ_{em}	Δ	ε ($M^{-1}cm^{-1}$)
SQ-1	656 nm	671 nm	15	1.98×10^5
SQ-2	659 nm	672 nm	13	1.26×10^5
SQ-3	652 nm	663 nm	11	1.00×10^5
SQ-3 PC	654 nm	662 nm	8	0.60×10^5

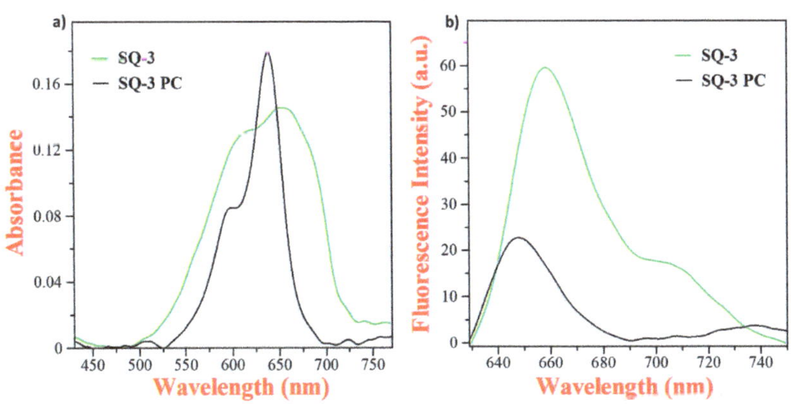

Figure 4. (**a**) Absorption and (**b**) emission Spectra of **SQ-3** and **SQ-3 PC** (5 μM) in H_2O (2% DMSO) solvent.

In contrast, **SQ-3 PC** in the same aqueous medium exhibits a blue-shifted, sharp absorption peak at 639 nm, with a prominent vibronic shoulder at 598 nm and less spectral broadening compared to **SQ-3**. This relatively lesser extent of molecular aggregation in **SQ-3 PC** compared to that of **SQ-3** is attributed to the hampering of aggregation due to the presence of the oligopeptide chain and spacers. Squaraine dyes are prone to dye aggregation forming blue-shifted and co-facial H-aggregates and red-shifted head-to-tail packed J-aggregates (Figure 5) compared to monomeric dye absorption, which is influenced by the structure and molecular environment of the dyes' planar molecular structure [45]. The vibronic shoulder in squaraine dyes is very sensitive to dye aggregation and quite often has been used to index the relative extent of dye aggregation in different molecular environments. In the electronic absorption spectrum of squaraine dyes, the ratio of the absorbance around 600 nm associated with vibronic shoulder and around 650 nm associated with the monomeric dye absorption has been widely utilized to estimate the relative extent of the dye aggregation [46]. A perusal of the normalized absorption spectra of **SQ-3** and **SQ-3 PC** in DMSO and water (Supporting Information Figure S1) reveals that the extent of aggregation of both of the dyes in DMSO is almost similar (0.26), while the behavior is very different in water, with nearly a doubled aggregation index for **SQ-3** dye (0.91) compared to that of **SQ-3 PC** (0.46). At the same time, the nature of the aggregates is primarily H-type [45] in **SQ-3 PC**, while **SQ-3** demonstrated the presence of both the H- and J- aggregates [47]. The unexpectedly low molar absorption coefficient of **SQ-3 PC**, despite containing two squaraine moieties, can be attributed to intramolecular electronic coupling leading to hypochromism, non-optimal conformations due to the flexible linker, and specific solute–solvent interactions in DMSO. These factors collectively contribute to a reduced absorption intensity compared to the expectations for two independent squaraine chromophores.

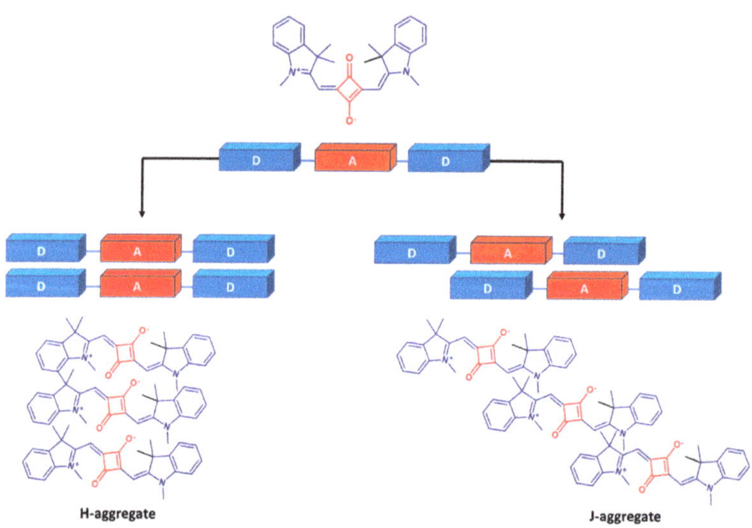

Figure 5. Schematic representation of H- and J-aggregate formation in squaraine dyes.

Notably, in the aqueous solution, **SQ-3-PC** and **SQ-3** exhibited a quenched fluorescence nearly ten times and five times higher (Figure 4b), respectively, than that observed in the DMSO solution. The quantum yield of **SQ-3** and **SQ-3 PC** in water was calculated and found to be 0.006 and 0.0026. This is associated with aggregation-induced quenching (AIQ) due to the enhanced intermolecular interaction in water facilitated by hydrogen bonding between the free -COOH of the dye and water molecules. It is interesting to note that despite the nearly halved aggregation in **SQ-3 PC** in water compared to that of pure dye **SQ-3**, the quenching in fluorescence for **SQ-3 PC** was about three times higher compared to that of **SQ-3**. This clearly corroborates that, apart from AIQ, there might be an alternative pathway for fluorescence quenching in the case of **SQ-3 PC**. A second possibility is the quenching caused by fluorescence resonance energy transfer (FRET). To have FRET-based quenching, two criteria, such as the presence of the two terminal fluorophores within the Forster radius of typically 2 nm–20 nm and a sufficient overlap between the absorption spectrum of one fluorophore with the emission spectrum of another one, need to be fulfilled [41]. Due to a very small Stokes shift of only 11 nm in the case of **SQ-3** dye in DMSO, there is a sufficient overlap between the absorption and emission spectrum of **SQ-3**, as shown in Figure 6a fulfilling the first criterion of FRET in the case of **SQ-3 PC**. Interestingly, the distance between the two faces of **SQ-3** dye in the dye–peptide conjugate **SQ-3 PC** is only 1.92 nm (Figure 6b), which is well within the typical Förster radius, ensuring the efficient energy transfer from one dye molecule to another, fulfilling the second criterion for FRET to occur. By analyzing the fluorescence intensities of the dye (**SQ-3**) and the dye–peptide conjugate (**SQ-3 PC**), we determined that the fluorescence quenching efficiency was 61.8% in H_2O (2% DMSO). Therefore, comparing the AIQ and FRET mechanisms present in **SQ-3 PC**, FRET seems to be more dominant compared to AIQ. The property of conjugation allowed us to investigate the potential application of dye–peptide conjugates for enzyme activity monitoring using fluorescence-based ON/OFF biosensing.

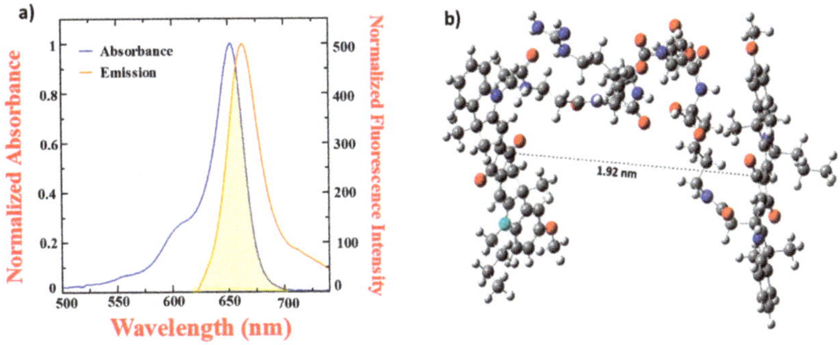

Figure 6. (**a**) Normalized absorption and emission spectra of **SQ-3** in DMSO and (**b**) optimized molecular structure of **SQ-3 PC** calculated using Gaussian G16 program.

3.3. Enzymatic Hydrolysis of SQ-3-PC with Trypsin

The newly designed NIR-sensitive probe **SQ-3-PC** exhibited a highly quenched fluorescence signal caused by the AIQ and FRET, which later played a dominant role. However, after incubation with the trypsin enzyme, there was a substantial increase in fluorescence at 654 nm associated with monomeric absorption. Upon the addition of varying concentrations of this enzyme prepared in the phosphate buffer with a (pH of 7.4, 0.1 mM) in a 5 µM solution of **SQ-3-PC** in H_2O (2% DMSO) followed by incubation for 30 min, there was a significant restoration of quenched fluorescence of the dye, as shown in Figure 7a. A perusal of Figure 7b shows that there was a significant increase in the fluorescence of the dye within 5 min. After 30 min, the fluorescence reached saturation, resulting in a 5-fold increase compared to the fluorescence in the absence of the trypsin enzyme. The observed enhancement in fluorescence intensity subsequent to enzymatic hydrolysis can be attributed to the spatial separation of the **SQ-3** fluorophores beyond their Förster radius. This increased intermolecular distance effectively terminates the resonance energy transfer process, resulting in the cessation of FRET and the concomitant restoration of fluorescence emission. Figure 7c,d show that F_0 indicates the initial fluorescence intensity in the absence of the enzyme, where F represents the fluorescence intensity at a specified point of time after the enzyme is added. Figure 7c demonstrates the fluorescence response of the squaraine-based probe **SQ-3 PC** (5 µM) to varying concentrations of the target enzyme

(0–75 nM). The fluorescence intensity changes as a function of enzyme concentration, with higher enzyme levels inducing more rapid and pronounced spectral shifts. This indicates a direct correlation between the enzyme concentration and probe activation. Concentration-dependent changes in the fluorescence (F/F_o) at the emission maximum, as shown in Figure 7d, reveal that **SQ-3 PC** works as a trypsin enzyme-specific NIR fluorogenic probe with the LOD and LOQ of 1.07 nM and 3.25 nM, respectively. The linear regression equation was y = 0.11327x + 1.05815, R^2 = 0.97043 (Figure S2). Our innovative approach for trypsin detection demonstrates significant advancements over existing methodologies. The developed probe exhibits a remarkably low LOD, which is comparable to or surpasses the most sensitive techniques reported in the recent literature (Table S1). In healthy humans, the concentration of trypsin varies in serum and in the intestine. In serum under fasting conditions, the concentration of trypsin varies from 4 nm to 30 nM [48,49]. Pancreatic diseases such as cystic fibrosis, acute pancreatitis, or the acute phase of chronic pancreatitis are associated with an increased trypsin level of 2.1–71.42 nM in the serum of patients [50]. Thus, the synthesized probe is able to measure the trypsin level in healthy individuals and individuals with a disease associated with the hyper- and hypo-activities of the trypsin enzyme. Furthermore, the detection of trypsin in the NIR wavelength regions allows the utilization of direct body fluids without prior sample processing, owing to the highly diminished autofluorescence.

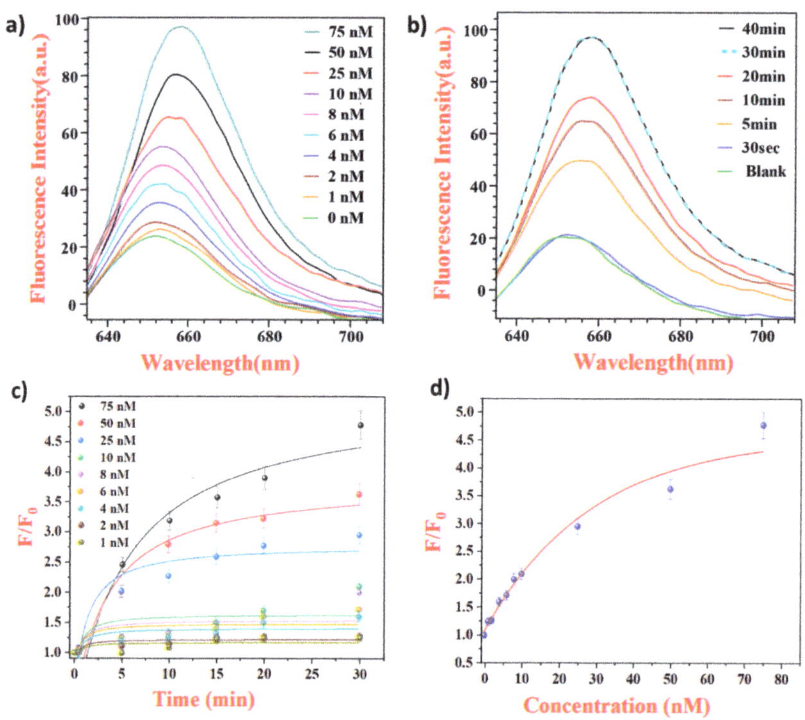

Figure 7. (**a**) Concentration-dependent fluorescence spectra of **SQ-3 PC** (5 μM) in H_2O (2% DMSO) for 30 min with the addition of 0 to 75 nM Trypsin, (**b**) time-dependent fluorescence spectra of **SQ-3PC** (5 μM) in H_2O (2% DMSO) for 30 min with the addition of 75 nM of trypsin enzyme, (**c**) fluorescence spectra of **SQ-3 PC** with change in fluorescence intensity as a function of time with different concentrations of enzyme for a fixed concentration of **SQ-3 PC** (5 μM), and (**d**) plot of ratio of fluorescence intensity (F/F_0) as a function of enzyme concentration.

3.4. Enzyme Selectivity of the Enzyme Probe SQ-3-PC

Apart from the sensitivity and detection limit, the selectivity of a probe is highly desired for targeted disease diagnosis and point-of-care testing devices. The selectivity of the present probe towards trypsin was evaluated by analyzing its response in the presence of potentially interfering enzymes. A 5 µM solution of the newly designed probe **SQ-3 PC** in H_2O (2% DMSO) was incubated with various enzymes, including trypsin, papain, pancreatic elastase, horseradish peroxidase, chymotrypsin, and bovine serum albumin (BSA), each at a concentration of 75 nM, for 30 min. As shown in Figure 8, the emission intensity at 654 nm remained relatively constant in the presence of the competing enzymes. However, a significant increase in emission intensity was observed upon the addition of trypsin. This marked change in fluorescence signal highlights the probe's high selectivity for trypsin, demonstrating its ability to specifically detect trypsin activity while remaining unresponsive to other enzymes present in the sample. These results underscore the potential of this biosensing approach for accurate trypsin detection in complex biological environments.

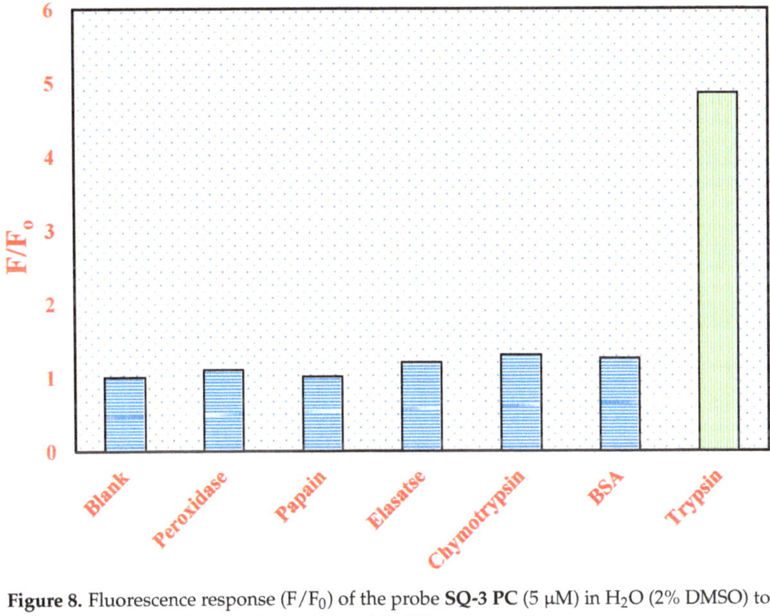

Figure 8. Fluorescence response (F/F_0) of the probe **SQ-3 PC** (5 µM) in H_2O (2% DMSO) towards different potential and competing enzymes (75 nM).

4. Conclusions

In conclusion, we have successfully designed and synthesized novel unsymmetrical squaraine dyes and a potential dye–peptide conjugate for the sensitive and selective detection of the trypsin enzyme. In the aqueous medium, the dye–peptide conjugate **SQ-3 PC** demonstrates reduced dye aggregation but pronounced fluorescence quenching compared to its constituent dye **SQ-3** alone. A fluorescence quenching of 61.8% in the case of **SQ-3 PC** in an aqueous medium compared to its DMSO counterpart is associated with both the AIQ and FRET, where FRET was demonstrated to be the dominant phenomenon for this pronounced fluorescence quenching. The hydrolysis of the designed probe **SQ-3 PC** with trypsin enzyme restores an appreciable amount of the quenched fluorescence within 5 min, and the enhancement in the fluorescence saturates after 30 min. Notably, the probe exhibited high specificity and good selectivity towards trypsin when tested against a panel of potentially interfering enzymes, including BSA, papain, chymotrypsin, horseradish peroxidase, and pancreatic elastase. This selectivity is crucial for the potential application of the probe in complex biological samples, though further validation in such matrices remains an important area for future investigation.

Supplementary Materials: The following supporting information can be downloaded at: https://www.mdpi.com/article/10.3390/bios14100458/s1, Scheme S1: Scheme for synthesis for unsymmetrical squaraine dye **SQ-1**, **SQ-2**; Scheme S2: Scheme for synthesis for unsymmetrical squaraine dye **SQ-3**; Scheme S3. Scheme for the synthesis of peptide sequence; Figure S1: Normalized Absorption Spectra of **SQ-3** and **SQ-3** PC in DMSO (solid line) and H_2O (2% DMSO) (dash line) at a concentration of 5 μM.; Figure S2: Linear correction curve of ratio of fluorescence intensity against trypsin concentration (0 nM to 10 nM). Table S1: Comparison of different methods for the determination of trypsin.

Author Contributions: Conceptualization, P.B., S.S.P. and T.K.; methodology, P.B.; software, P.B.; validation, P.B., T.K. and S.S.P.; formal analysis, P.B.; investigation, P.B.; resources, T.K., S.S.P. and S.K.M.; data curation, P.B. and S.G.; writing—original draft preparation, P.B.; writing—review and editing, P.B. and S.G.; visualization, P.B.; supervision, T.K. and S.S.P. All authors have read and agreed to the published version of the manuscript.

Funding: This research received no external funding.

Institutional Review Board Statement: Not applicable.

Informed Consent Statement: Not applicable.

Data Availability Statement: The data presented in this study are available on request from the corresponding author. The data are not publicly available due to ethical and privacy reasons.

Conflicts of Interest: The authors declare no conflicts of interest.

References

1. López-Otín, C.; Bond, J.S. Proteases: Multifunctional Enzymes in Life and Disease. *J. Biol. Chem.* **2008**, *283*, 30433–30437. [CrossRef] [PubMed]
2. Lasher, D.; Szabó, A.; Masamune, A.; Chen, J.-M.; Xiao, X.; Whitcomb, D.C.; Barmada, M.M.; Ewers, M.; Ruffert, C.; Paliwal, S.; et al. Protease-Sensitive Pancreatic Lipase Variants Are Associated with Early Onset Chronic Pancreatitis. *Am. J. Gastroenterol.* **2019**, *114*, 974–983. [CrossRef]
3. Malla, S.R.; Krueger, B.; Wartmann, T.; Sendler, M.; Mahajan, U.M.; Weiss, F.U.; Thiel, F.G.; De Boni, C.; Gorelick, F.S.; Halangk, W.; et al. Early Trypsin Activation Develops Independently of Autophagy in Caerulein-Induced Pancreatitis in Mice. *Cell. Mol. Life Sci.* **2020**, *77*, 1811–1825. [CrossRef] [PubMed]
4. Atkinson, M.A.; Campbell-Thompson, M.; Kusmartseva, I.; Kaestner, K.H. Organisation of the Human Pancreas in Health and in Diabetes. *Diabetologia* **2020**, *63*, 1966–1973. [CrossRef] [PubMed]
5. Meyer-Hoffert, U.; Rogalski, C.; Seifert, S.; Schmeling, G.; Wingertszahn, J.; Proksch, E.; Wiedow, O. Trypsin Induces Epidermal Proliferation and Inflammation in Murine. Skin. *Exp. Dermatol.* **2004**, *13*, 234–241. [CrossRef]
6. Rakash, S. Role of Proteases in Cancer: A Review. *Biotechnol. Mol. Biol. Rev.* **2012**, *7*, 90–101. [CrossRef]
7. Hadwan, M.H.; Al-Obaidy, S.S.M.; Al-Kawaz, H.S.; Almashhedy, L.A.; Kadhum, M.A.; Khudhair, D.A.; Hadwan, A.M.; Hadwan, M.M. An Optimized Protocol to Assess Trypsin Activity in Biological Samples. *Monatshefte Für Chem. Chem. Mon.* **2023**, *154*, 267–277. [CrossRef]
8. Park, T.; Han, M.; Schanze, K.S.; Lee, S.H. Ultrasensitive Determination of Trypsin in Human Urine Based on Amplified Fluorescence Response. *ACS Sens.* **2023**, *8*, 2591–2597. [CrossRef]
9. Soreide, K.; Janssen, E.; Körner, H.; Baak, J. Trypsin in Colorectal Cancer: Molecular Biological Mechanisms of Proliferation, Invasion, and Metastasis. *J. Pathol.* **2006**, *209*, 147–156. [CrossRef]
10. Gaber, A.; Johansson, M.; Stenman, U.-H.; Hotakainen, K.; Pontén, F.; Glimelius, B.; Bjartell, A.; Jirström, K.; Birgisson, H. High Expression of Tumour-Associated Trypsin Inhibitor Correlates with Liver Metastasis and Poor Prognosis in Colorectal Cancer. *Br. J. Cancer* **2009**, *100*, 1540–1548. [CrossRef]
11. Esimbekova, E.N.; Torgashina, I.G.; Nemtseva, E.V.; Antashkevich, A.A.; Sasova, P.Y.; Kratasyuk, V.A. Trypsin-Based Chemoenzymatic Assay for Detection of Pollutants and Safety Assessment of Food Additives. *Chemosensors* **2023**, *11*, 237. [CrossRef]
12. Woessmann, J.; Petrosius, V.; Üresin, N.; Kotol, D.; Aragon-Fernandez, P.; Hober, A.; Auf Dem Keller, U.; Edfors, F.; Schoof, E.M. Assessing the Role of Trypsin in Quantitative Plasma and Single-Cell Proteomics toward Clinical Application. *Anal. Chem.* **2023**, *95*, 13649–13658. [CrossRef] [PubMed]
13. Rainio, M.; Lindström, O.; Penttilä, A.; Itkonen, O.; Kemppainen, E.; Stenman, U.-H.; Kylänpää, L. Serum Serine Peptidase Inhibitor Kazal-Type 1, Trypsinogens 1 to 3, and Complex of Trypsin 2 and A1-Antitrypsin in the Diagnosis of Severe Acute Pancreatitis. *Pancreas* **2019**, *48*, 374–380. [CrossRef]

14. Boyer, A.E.; Moura, H.; Woolfitt, A.R.; Kalb, S.R.; McWilliams, L.G.; Pavlopoulos, A.; Schmidt, J.G.; Ashley, D.L.; Barr, J.R. From the Mouse to the Mass Spectrometer: Detection and Differentiation of the Endoproteinase Activities of Botulinum Neurotoxins A−G by Mass Spectrometry. *Anal. Chem.* **2005**, *77*, 3916–3924. [CrossRef]
15. Kim, Y.-P.; Lee, B.S.; Kim, E.; Choi, I.S.; Moon, D.W.; Lee, T.G.; Kim, H.-S. Activity-Based Assay of Matrix Metalloproteinase on Nonbiofouling Surfaces Using Time-of-Flight Secondary Ion Mass Spectrometry. *Anal. Chem.* **2008**, *80*, 5094–5102. [CrossRef] [PubMed]
16. Michaud, D. Gel Electrophoresis of Proteolytic Enzymes. *Anal. Chim. Acta* **1998**, *372*, 173–185. [CrossRef]
17. Abuknesha, R.A.; Jeganathan, F.; DeGroot, R.; Wildeboer, D.; Price, R.G. Detection of Proteases Using an Immunochemical Method with Haptenylated–Gelatin as a Solid-Phase Substrate. *Anal. Bioanal. Chem.* **2010**, *396*, 2547–2558. [CrossRef]
18. Steiner, V.J.M.; Williams, D.A.; Moeller, E.M.; Melgarejo, T. Development and Validation of an Enzyme-Linked Immunosorbent Assay for Feline Trypsin-like Immunoreactivity. *Am. J. Vet. Res.* **2000**, *61*, 620–623. [CrossRef]
19. Li, Y.; Bai, H.; Li, C.; Shi, G. Colorimetric Assays for Acetylcholinesterase Activity and Inhibitor Screening Based on the Disassembly−Assembly of a Water-Soluble Polythiophene Derivative. *ACS Appl. Mater. Interfaces* **2011**, *3*, 1306–1310. [CrossRef]
20. Xue, W.; Zhang, G.; Zhang, D. A Sensitive Colorimetric Label-Free Assay for Trypsin and Inhibitor Screening with Gold Nanoparticles. *Analyst* **2011**, *136*, 3136. [CrossRef]
21. Gu, X.; Yang, G.; Zhang, G.; Zhang, D.; Zhu, D. A New Fluorescence Turn-on Assay for Trypsin and Inhibitor Screening Based on Graphene Oxide. *ACS Appl. Mater. Interfaces* **2011**, *3*, 1175–1179. [CrossRef] [PubMed]
22. Xue, W.; Zhang, G.; Zhang, D.; Zhu, D. A New Label-Free Continuous Fluorometric Assay for Trypsin and Inhibitor Screening with Tetraphenylethene Compounds. *Org. Lett.* **2010**, *12*, 2274–2277. [CrossRef] [PubMed]
23. Kaur, J.; Singh, P.K. Trypsin Detection Strategies: A Review. *Crit. Rev. Anal. Chem.* **2022**, *52*, 949–967. [CrossRef] [PubMed]
24. Ong, I.L.H.; Yang, K.-L. Recent Developments in Protease Activity Assays and Sensors. *Analyst* **2017**, *142*, 1867–1881. [CrossRef]
25. Gutiérrez, O.A.; Salas, E.; Hernández, Y.; Lissi, E.A.; Castrillo, G.; Reyes, O.; Garay, H.; Aguilar, A.; Garcia, B.; Otero, A.; et al. An Immunoenzymatic Solid-Phase Assay for Quantitative Determination of HIV-1 Protease Activity. *Anal. Biochem.* **2002**, *307*, 18–24. [CrossRef]
26. Priyanka; Bila, G.; Mavileti, S.K.; Bila, E.; Negrych, N.; Gupta, S.; Tang, L.; Bilyy, R.; Pandey, S.S.; Kato, T. A Biocompatible NIR Squaraine Dye and Dye-Antibody Conjugates for Versatile Long-Term *in Vivo* Fluorescence Bioimaging. *Mater. Adv.* **2024**, *5*, 3940–3949. [CrossRef]
27. Fu, Y.; Liu, L.; Li, X.; Chen, H.; Wang, Z.; Yang, W.; Zhang, H.; Zhang, H. Peptide Modified Manganese-Doped Iron Oxide Nanoparticles as a Sensitive Fluorescence Nanosensor for Non-Invasive Detection of Trypsin Activity in Vitro and in Vivo. *RSC Adv.* **2021**, *11*, 2213–2220. [CrossRef]
28. Cheng, M.; Zhou, J.; Zhou, X.; Xing, D. Peptide Cleavage Induced Assembly Enables Highly Sensitive Electrochemiluminescence Detection of Protease Activity. *Sens. Actuators B Chem.* **2018**, *262*, 516–521. [CrossRef]
29. Zhuo, C.-X.; Wang, L.-H.; Feng, J.-J.; Zhang, Y.-D. Label-Free Fluorescent Detection of Trypsin Activity Based on DNA-Stabilized Silver Nanocluster-Peptide Conjugates. *Sensors* **2016**, *16*, 1477. [CrossRef]
30. Scott, J.I.; Deng, Q.; Vendrell, M. Near-Infrared Fluorescent Probes for the Detection of Cancer-Associated Proteases. *ACS Chem. Biol.* **2021**, *16*, 1304–1317. [CrossRef]
31. Guo, Z.; Park, S.; Yoon, J.; Shin, I. Recent Progress in the Development of Near-Infrared Fluorescent Probes for Bioimaging Applications. *Chem. Soc. Rev.* **2013**, *43*, 16–29. [CrossRef] [PubMed]
32. Khopkar, S.; Shankarling, G. Synthesis, Photophysical Properties and Applications of NIR Absorbing Unsymmetrical Squaraines: A Review. *Dyes Pigment.* **2019**, *170*, 107645. [CrossRef]
33. Ilina, K.; MacCuaig, W.M.; Laramie, M.; Jeouty, J.N.; McNally, L.R.; Henary, M. Squaraine Dyes: Molecular Design for Different Applications and Remaining Challenges. *Bioconjug. Chem.* **2020**, *31*, 194–213. [CrossRef] [PubMed]
34. Li, H.; Tang, Y.; Shen, K.; Lu, J.; Zhang, Z.; Yi, D.; Hao, N.; Fu, Q.; Ye, Z.; Wei, J.; et al. NIR Squaraine Dyes for Dual Colorimetric and Fluorescent Determination of Fe^{3+}, Cu^{2+}, and Hg^{2+} Ions. *RSC Adv.* **2023**, *13*, 17202–17211. [CrossRef]
35. Barcenas, G.; Biaggne, A.; Mass, O.A.; Wilson, C.K.; Obukhova, O.M.; Kolosova, O.S.; Tatarets, A.L.; Terpetschnig, E.; Pensack, R.D.; Lee, J.; et al. First-Principles Studies of Substituent Effects on Squaraine Dyes. *RSC Adv.* **2021**, *11*, 19029–19040. [CrossRef]
36. Chen, H.; Farahat, M.S.; Law, K.-Y.; Whitten, D.G. Aggregation of Surfactant Squaraine Dyes in Aqueous Solution and Microheterogeneous Media: Correlation of Aggregation Behavior with Molecular Structure. *J. Am. Chem. Soc.* **1996**, *118*, 2584–2594. [CrossRef]
37. Jiang, Q.; Chen, Z.; Huang, Y.; Gao, Q.; Luo, C.; Mehdi, M.; Xu, Y.; Li, H.; Sun, S. A Bovine Serum Albumin and Squaraine Dye Assembly Fluorescent Probe for Pepsin Detection. *Microchem. J.* **2023**, *186*, 108361. [CrossRef]
38. Butnarasu, C.; Pontremoli, C.; Moran Plata, M.J.; Barbero, N.; Visentin, S. Squaraine Dyes as Fluorescent Turn-on Probes for Mucins: A Step Toward Selectivity†. *Photochem. Photobiol.* **2023**, *99*, 562–569. [CrossRef]
39. Gupta, S.; Mavileti, S.; Pandey, S.S.; Kato, T. Design and Synthesis of Novel Squaraine-Based Fluorescent Probe for Far-Red Detection of Chymotrypsin Enzyme. *Molecules* **2024**, *29*, 1282. [CrossRef]
40. Gaussian.Com. Expanding the Limits of Computational Chemistry. Available online: https://gaussian.com/ (accessed on 1 August 2024).
41. Mayerhöffer, U.; Fimmel, B.; Würthner, F. Bright Near-Infrared Fluorophores Based on Squaraines by Unexpected Halogen Effects. *Angew. Chem. Int. Ed.* **2012**, *51*, 164–167. [CrossRef]

42. Soumya, M.S.; Abraham, A. Preclinical Evaluation of Symmetrical Iiodinated Squaraine Dye on Experimental Animal Models. *J. Glycobiol. S* **2013**, *1*. [CrossRef]
43. Olsen, J.V.; Ong, S.-E.; Mann, M. Trypsin Cleaves Exclusively C-Terminal to Arginine and Lysine Residues. *Mol. Cell. Proteom.* **2004**, *3*, 608–614. [CrossRef] [PubMed]
44. de Miguel, G.; Marchena, M.; Zitnan, M.; Pandey, S.S.; Hayase, S.; Douhal, A. Femto to Millisecond Observations of Indole-Based Squaraine Molecules Photodynamics in Solution. *Phys. Chem. Chem. Phys.* **2012**, *14*, 1796–1805. [CrossRef] [PubMed]
45. Hu, L.; Yan, Z.; Xu, H. Advances in Synthesis and Application of Near-Infrared Absorbing Squaraine Dyes. *RSC Adv.* **2013**, *3*, 7667. [CrossRef]
46. Yum, J.H.; Moon, S.J.; Humphry-Baker, R.; Walter, P.; Geiger, T.; Nüesch, F.; Grätzel, M.; Nazeeruddin, M.D.K. Effect of Coadsorbent on the Photovoltaic Performance of Squaraine Sensitized Nanocrystalline Solar Cells. *Nanotechnology* **2008**, *19*, 424005. [CrossRef] [PubMed]
47. Chen, G.; Sasabe, H.; Lu, W.; Wang, X.-F.; Kido, J.; Hong, Z.; Yang, Y. J-Aggregation of a Squaraine Dye and Its Application in Organic Photovoltaic Cells. *J. Mater. Chem. C* **2013**, *1*, 6547. [CrossRef]
48. Lake-Bakaar, G.; McKavanagh, S.; Redshaw, M.; Wood, T.; Summerfield, J.A.; Elias, E. Serum Immunoreactive Trypsin Concentration after a Lundh Meal. Its Value in the Diagnosis of Pancreatic Disease. *J. Clin. Pathol.* **1979**, *32*, 1003–1008. [CrossRef]
49. Artigas, J.M.G.; Garcia, M.E.; Faure, M.R.A.; Gimeno, A.M.B. Serum Trypsin Levels in Acute Pancreatic and Non-Pancreatic Abdominal Conditions. *Postgrad. Med. J.* **1981**, *57*, 219–222. [CrossRef]
50. Heinrich, H.C.; Gabbe, E.E.; Ičagić, F. Immunoreaktives Serum Trypsin bei Pankreaserkrankungen. *Klin. Wochenschr.* **1979**, *57*, 1237–1238. [CrossRef]

biosensors MDPI

Communication

All-Dielectric Metasurface-Based Terahertz Molecular Fingerprint Sensor for Trace Cinnamoylglycine Detection

Qiyuan Xu, Mingjun Sun , Weijin Wang and Yanpeng Shi *

School of Integrated Circuits, Shandong University, Jinan 250100, China
* Correspondence: ypshi@sdu.edu.cn

Abstract: Terahertz (THZ) spectroscopy has emerged as a superior label-free sensing technology in the detection, identification, and quantification of biomolecules in various biological samples. However, the limitations in identification and discrimination sensitivity of current methods impede the wider adoption of this technology. In this article, a meticulously designed metasurface is proposed for molecular fingerprint enhancement, consisting of a periodic array of lithium tantalate triangular prism tetramers arranged in a square quartz lattice. The physical mechanism is explained by the finite-difference time-domain (FDTD) method. The metasurface achieves a high quality factor (Q-factor) of 231 and demonstrates excellent THz sensing capabilities with a figure of merit (FoM) of 609. By varying the incident angle of the THz wave, the molecular fingerprint signal is strengthened, enabling the highly sensitive detection of trace amounts of analyte. Consequently, cinnamoylglycine can be detected with a sensitivity limit as low as 1.23 $\mu g \cdot cm^{-2}$. This study offers critical insights into the advanced application of THz waves in biomedicine, particularly for the detection of urinary biomarkers in various diseases, including gestational diabetes mellitus (GDM).

Keywords: terahertz; molecular fingerprint sensor; all-dielectric metasurface; cinnamoylglycine; gestational diabetes mellitus

1. Introduction

Terahertz (THz) spectroscopy, a portion of the electromagnetic spectrum, deals with radiation frequencies from 0.1 THz to 10 THz. This radiation can penetrate a wide variety of non-conducting materials and exhibits non-ionizing properties [1]. Consequently, it is acclaimed as a potent spectroscopic technique due to its label-free and non-destructive attributes. Previous research in THz time-domain spectroscopy has demonstrated it to be a state-of-the-art technology capable of providing distinct molecular fingerprints by detecting vibrational and rotational modes sensitive to molecular structure and environmental factors within the THz frequency range [2,3]. This underscores its substantial potential for applications in biological detection [3,4]. However, traditional THz spectroscopy encounters limitations, notably its reduced sensitivity for detecting trace analytes, which impedes its extensive application. This issue arises because the dimensions of target molecules are significantly smaller than the micrometer scale, whereas the wavelength of THz waves spans approximately from 30 μm to 3000 μm [5], leading to a remarkable mismatch between the absorption cross-section of the analyte and the wavelength. Hence, the interactions between them are too feeble to generate adequate vibrational absorption signals, necessitating a substantial quantity of analyte [6,7]. Recent studies have proposed various approaches to address this issue, encompassing surface plasmon resonances (SPRs) [8], electromagnetic subwavelength structures [9], waveguides [10], metal hole arrays [11], and graphene metamaterials [12]. These sensors can achieve sub-wavelength spatial resolution to enhance detection sensitivity [5]. But they are constrained by substantial ohmic losses and high dispersion of the metal materials [13,14], leading to limited improvements in sensitivity [15,16].

All-dielectric metasurface has been demonstrated as an alternative to traditional metallic metasurfaces. It is based on high-refractive-index dielectrics and leverages Mie-type resonances rather than surface plasmon resonances to obtain strong magnetic resonances as well as electric resonances, thereby enhancing the interaction between electromagnetic waves and materials [17–19]. This innovative design significantly reduces energy losses while elevating the quality factor (Q-factor) of the resonators, thus greatly improving detection sensitivity [17,20]. Furthermore, intrinsic heating can alter the local refractive index and potentially damage biomolecules. A dielectric metasurface can mitigate the plasmonic ohmic loss and heating issues, thereby enabling better repeatability and stability for biosensing [21]. Currently, a precise angle-scanning strategy has been proposed, which is based on guided-mode resonance theory [22]. By monitoring the transmission characteristics of THz waves at different incident angles, richer spectral information can be obtained, thereby improving the detection capability of trace analytes [22,23]. Employing this strategy, a previous study successfully identified lactose and glucose with detection limits of 1.53 $\mu g \cdot cm^{-2}$ and 1.54 $\mu g \cdot cm^{-2}$, respectively [24]. On this basis, maintaining the polarization direction can further refine the enhancement effect and broaden the bandwidth [24].

Cinnamoylglycine, a metabolic byproduct of cinnamic acid metabolism, is derived from dietary sources such as cinnamon [25]. In the human body, it is produced by gut microbes [26] and excreted in urine [27]. Gestational diabetes mellitus (GDM), which involves significant metabolic alterations [28,29], can pose severe health risks to both mother and fetus, such as preeclampsia [30], preterm birth [31], and increased risk of type-2 diabetes later in life [32]. These harms highlight the need for efficient GDM diagnostic methods for regular monitoring and timely intervention. Current methods, including fasting plasma glucose [33,34], oral glucose tolerance testing [35], and the detection of glycated hemoglobin [36], have limitations such as the need for fasting, multiple blood draws, and variability due to external factors. In contrast, studies have suggested a link between urinary cinnamoylglycine levels and diabetes, with diabetic patients exhibiting a clearance rate of 77 mL/min compared to 317 mL/min in healthy controls [37–39]. This proves urinary cinnamoylglycine levels are a practicable biomarker for monitoring metabolic changes associated with GDM and offer a non-invasive alternative. But existing detection methods for cinnamoylglycine, such as high-performance liquid chromatography (HPLC), are time-consuming and costly due to the pretreatment and processing of such highly complex matrices [40]. This underscores the critical need for a rapid, convenient, and accurate detection method for urinary cinnamoylglycine, thereby enhancing the overall management and prognosis of GDM. THz spectroscopy has proven effective for analyzing urine samples and biomarkers [41]. Furthermore, the use of metamaterials has created additional opportunities over the past decade, such as the detection of proteins [42,43] and cancer biomarkers [44]. Therefore, this study introduces a THz sensor to urinary cinnamoylglycine detection, utilizing a sophisticatedly designed all-dielectric metasurface to enhance molecular fingerprints and improve detection sensitivity and specificity.

In this study, we propose an all-dielectric metasurface-based THz sensor, utilizing a periodic symmetric tetramer of lithium tantalate triangular prisms on a quartz substrate. This periodic unit structure, composed of the four high-index tetramer clusters, can induce a strong magnetic dipole (MD) resonance, with a high Q-factor of 231 and figure of merit (FoM) of 609, thus significantly enhancing the molecular fingerprint for detecting trace amounts of analytes. By manipulating the incident angle of a THz wave, a broad envelope curve can be measured due to the shift of resonance frequency, which corresponds to the absorption resonances of the analyte. Consequently, this innovative sensor allows for the detection of cinnamoylglycine, with a detection limit as low as 1.23 $\mu g \cdot cm^{-2}$. Moreover, the proposed metasurface can be efficiently fabricated using micro/nanotechnology, and the testing can be performed with a terahertz time-domain spectrometer [45].

2. Structure and Design

Figure 1 depicts the schematic of the proposed all-dielectric metasurface. It consists of a periodic array of tetramer clusters arranged on a square quartz substrate. Each tetramer cluster is formed by four high-index triangular prisms. The dielectric property is a crucial characteristic of a material, as it fundamentally defines how the material responds to electromagnetic waves. It uniquely determines how radiation propagates through the material by describing its ability to polarize in response to an electric field, thus influencing the propagation of waves within the medium [46]. Lithium tantalate is chosen as the building material for the tetramer clusters due to its minimal imaginary part, resulting in negligible dissipation loss within the relevant frequency range [47]. The complex dielectric permittivity of lithium tantalate can be described as a Lorentz-type dispersion, as is described by the following equation [48]:

$$\varepsilon = \varepsilon_\infty \frac{\omega^2 - {\omega_L}^2 + i\omega\gamma}{\omega^2 - {\omega_T}^2 + i\omega\gamma}, \quad (1)$$

In the equation, the transverse and longitudinal optical phonon frequencies are $\omega_T/2\pi = 26.7$ THz and $\omega_L/2\pi = 46.9$ THz, respectively. The damping factor is $\gamma/2 = 0.94$ THz and the quartz substrate has a refractive index of $n = 2$. For frequencies below the longitudinal optical phonon frequency, the permittivity of lithium tantalate is calculated to be 41.4 [49].

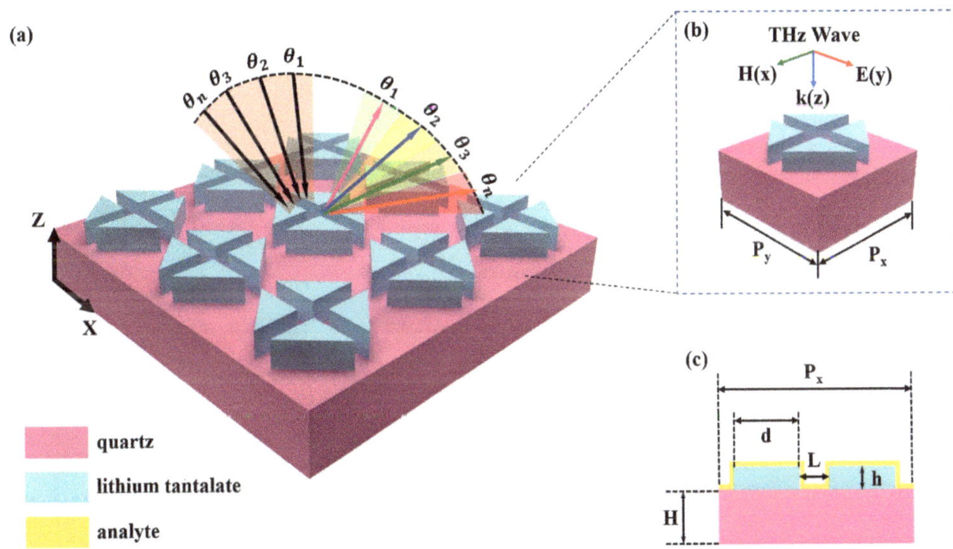

Figure 1. (**a**) The structural diagram of the all-dielectric metasurface, illustrating the periodic arrangement of the high-index triangular prism tetramer based on the quartz substrate; (**b**) a unit cell of the periodic structure with a y-polarized source incident downwards in the z direction; (**c**) the main view of the unit cell (y–z plane) and corresponding parameters.

The geometric parameters of the square unit cell are illustrated in Figure 1b,c, with the side length $P_x = P_y = 135$ μm and the substrate thickness $H = 130$ μm. The prisms are shaped as isosceles right-angled triangles, of which the side length is $d = 90.4$ μm. As shown in the main view of the unit cell in Figure 1c, the distance between adjacent triangular prisms is $L = 33$ μm, and the height of each prism is $h = 49.6$ μm. By directing the incident THz wave to propagate downward along the z direction, the resonance dip within the transmission spectrum can be manipulated, resulting in the formation of an envelope curve that captures the absorption resonances of analytes. This dynamic shifting of the resonance dip is essential for identifying the molecular fingerprints of various substances.

In order to evaluate the performance of the proposed sensor, extensive numerical simulations and spectral response analysis have been carried out with the commercially available three-dimensional finite-difference time-domain (FDTD) software. In the simulation, periodic boundary conditions were applied along the x and y directions to emulate an infinite array, while a perfectly matched layer boundary condition was set along the z-axis to prevent wave reflections. The simulations were conducted with a minimum meshing step of 1 nm to achieve high-resolution accuracy in modeling the electromagnetic fields.

3. Results and Discussion

Figure 2 illustrates the transmission spectra of the proposed metasurface and the electric and magnetic field at the resonance frequency. In Figure 2a, a single resonance at 0.624 THz is observed for both x-polarized and y-polarized waves under normal incidence due to the symmetric arrangement of the metasurface structure. However, as shown in Figure 2b, when the incident angle is adjusted to 37°, the sensor exhibits a strong sensitivity to polarization due to the breaking in symmetry, leading to a frequency shift. The resonance is measured at 0.487 THz for y-polarized waves, showing strong correspondence with the absorption peak of cinnamoylglycine and a high Q-factor of approximately 231. The Q-factor is defined as $Q = f_0/\Delta f$, where the full-width at half-maximum (FWHM) Δf is 2.11 GHz and the resonance frequency f_0 is 0.487 THz. In contrast, the frequency shift for x-polarized waves does not align sufficiently with the targeted absorption peak. Therefore, y-polarized waves were selected for angle-scanning to achieve frequency shifts and obtain the transmission envelope curve.

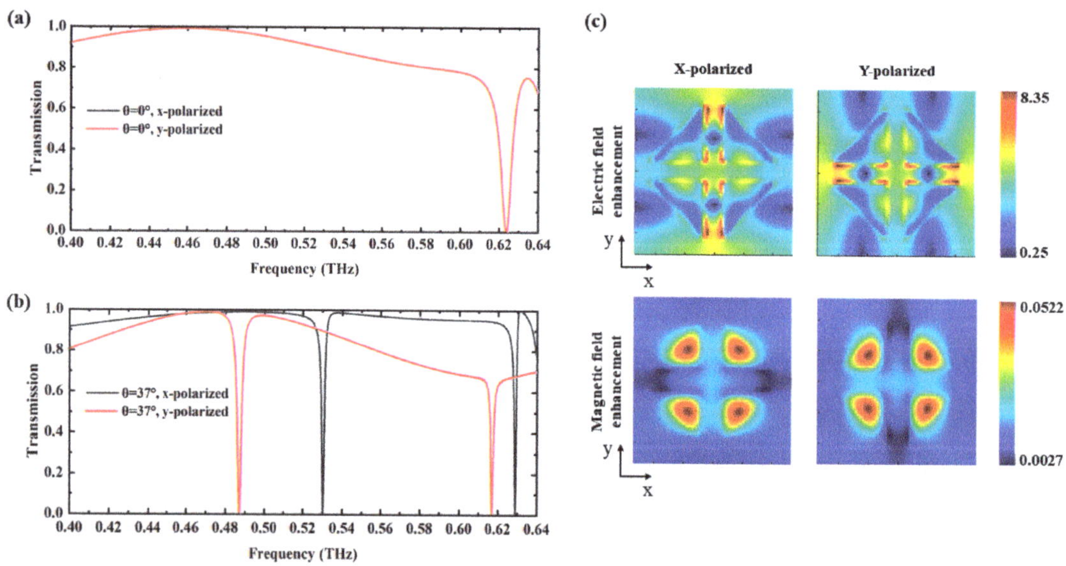

Figure 2. (**a**) Transmission spectra for x-polarized and y-polarized incident waves at 0°; (**b**) transmission spectra for x-polarized and y-polarized incident waves at 37°; (**c**) the electric and magnetic field distribution measured at the surface of the quartz substrate at vertical incidence. The left and right figures correspond to the x-polarized and y-polarized incident wave, respectively.

Figure 2c shows the electric and magnetic field distributions at the substrate surface in the x–y plane. The left and right panels correspond to x-polarized and y-polarized incident waves, respectively. The resonance is identified as an MD mode, characterized by the collective response of four longitudinal MDs [50]. The electric field is predominantly localized in the center region of the cluster, indicating the excitation of the MD resonance in this area, which enhances the interaction between the incident THz waves and analytes. The magnetic field distribution confirms the presence of strong MD resonances, further supporting the sensor's sensitivity to different polarization states and incident angles. By leveraging these distinct field distributions, the sensor can detect analytes more effectively, which improves both the sensitivity and specificity of the detection process.

To investigate the combined spectral response and the relevancy between resonance dip and the incident angle, various angles were selected for analysis. Figure 3 illustrates that each incident angle corresponds to a distinct narrowband unity transmission, with the resonance frequency progressively decreasing as the incident angle increases. This frequency shift results from the alteration of the structure's symmetry due to the change in the incident angle. As this asymmetry intensifies, the frequency shift becomes more observable. Furthermore, modifying the parameters of the tetramer leads to a shift in the resonance frequency of the MD [49]. These influencing factors together create a comprehensive spectral cluster within the transmission spectrum, covering a broad frequency range that aligns closely with the absorption frequency of the analyte. This relevancy is crucial for enhancing the precision of sample identification, as it ensures that the spectral response could precisely mirror the unique absorption characteristics of the analyte.

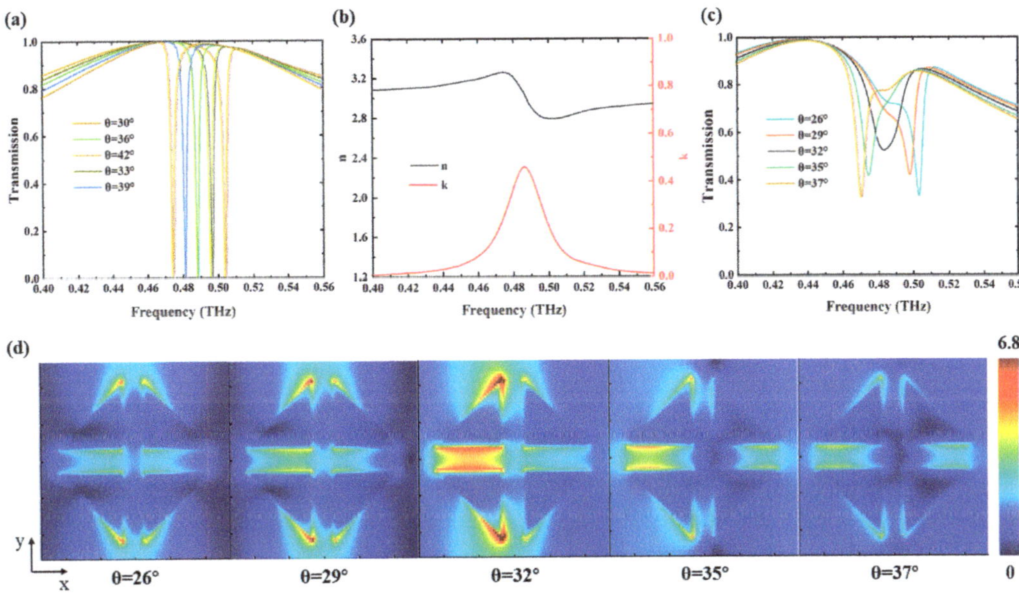

Figure 3. (**a**) Transmission spectra at different incident angles without any analyte; (**b**) the experimentally measured refractive index (n) and extinction coefficient (k) of cinnamoylglycine across the relevant frequency range; (**c**) transmission spectra at different incident angles with a 1 μm thick layer of analyte; (**d**) the electric field distribution measured at the substrate surface in the x–y plane at 0.487 THz for specific incident angles, corresponding to the transmission spectra shown in (**c**), respectively.

When the THz wave interacts directly with the metasurface in the absence of analyte, as shown in Figure 3a, both the minimum transmission value and the linewidth of the transmission spectra remain constant. In Figure 3b, the experimentally measured complex refractive index of cinnamoylglycine is presented, with data extracted using Fresnel formulas [51]. The real part, n, signifies the refractive index, whereas the imaginary part, k, denotes the extinction coefficient. This measurement indicates that the fingerprint absorption peak for cinnamoylglycine is located at 0.487 THz.

The fingerprint detection capability of the metasurface is evaluated by observing the transmission envelope curve with a 1 μm thick cinnamoylglycine layer. Figure 3c displays a unique envelope that peaks at 0.487 THz at an incident angle of 32°, matching the extinction coefficient curve for cinnamoylglycine. This phenomenon is explained by the electric field distributions at specific angles, shown in Figure 3d. At 0.487 THz and an incident angle of 34°, a notable electric field enhancement is observed, with pronounced concentration in the central region of the structure. This enhancement and localization increase the likelihood of wave–matter interactions near the resonant frequency for the analyte on the metamaterial. In contrast, when the incident angle is reduced to 25° or increased to 42°, the electric field strength diminishes.

The sensing sensitivity (S) of the designed metasurface is evaluated based on $S = \Delta f/\sigma$. As is shown in Figure 3c, the resonant frequency at 37° is 0.471 THz, resulting in a frequency shift Δf of 0.0158 THz. With the 1 μm thick cinnamoylglycine layer, the surface concentration is calculated as $\sigma = \rho \times h$, where ρ represents the volume density of cinnamoylglycine, noted as 1.23 g/cm^3. Accordingly, $\sigma = 123\ \mu g/cm^2$. Figure of merit (FoM) is used to assess the comprehensive performance of the sensor, which is defined as $FoM = S/FWHM = S \times Q/f_0$ [52]. Consequently, FoM is obtained as 609.

Figure 4 presents the comprehensive angle-scanning transmission spectra of the metasurface, demonstrating that this strategy effectively distinguishes cinnamoylglycine from other substances due to its distinct absorption resonances. In Figure 4a, the transmission spectra show stable frequency shifts across various incident angles ranging from 13° to 70° in the absence of any analyte, indicating a consistent performance of the metasurface. Conversely, Figure 4b presents the transmission spectra with a 1 μm thick cinnamoylglycine layer on the metasurface. The maximum transmission rate for each incident angle was extracted, and the results were fitted into an envelope curve, plotted as the red line in Figure 4b. The envelope curve peaks at 0.487 THz, which reaches approximately 56.85%, aligning with the characteristic fingerprint spectrum of cinnamoylglycine. The observed phenomenon is attributed to the optical loss of cinnamoylglycine, indicating that the angle-scanning strategy enhances detection sensitivity and accuracy by capturing comprehensive transmission envelopes, thereby facilitating precise analyte identification.

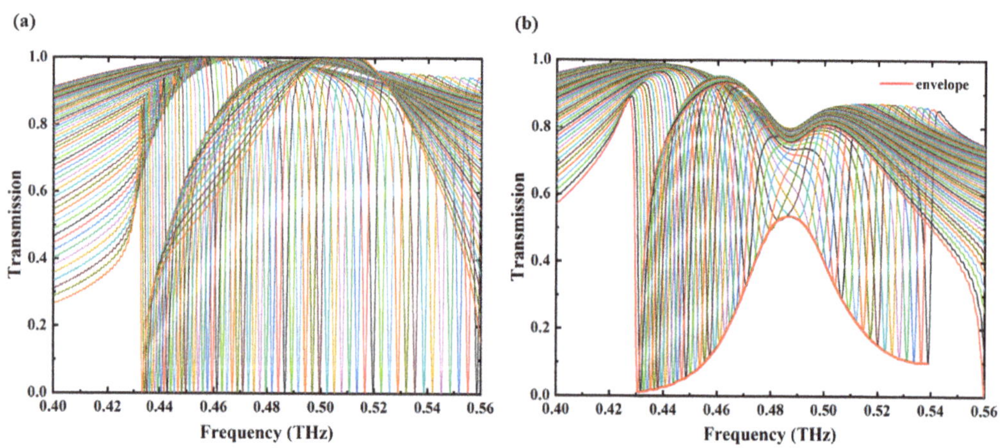

Figure 4. (**a**) Comprehensive transmission spectra without any analyte, with the incident angle ranging from 13° to 70°. Specifically, the rightmost line represents the transmission curve for an angle of 13°, while the leftmost line corresponds to 70°; (**b**) comprehensive transmission spectra with 1 μm thick cinnamoylglycine, with the incident angle ranging from 13° to 62°. The corresponding envelope curve has been plotted by red line in the figure.

23. Meng, D.; Liu, J.; Chen, W.; Cheng, Y.-Y.; You, K.-W.; Fan, Z.-C.; Ye, Q.; Huang, P.-H.; Chen, Y.-S. Study on the enhancement mechanism of terahertz molecular fingerprint sensing. *Results Phys.* **2022**, *39*, 105766. [CrossRef]
24. Zhang, X.; Liu, J.; Qin, J. A terahertz metasurface sensor with fingerprint enhancement in a wide spectrum band for thin film detection. *Nanoscale Adv.* **2023**, *5*, 2210–2215. [CrossRef] [PubMed]
25. Guzman, J.D. Natural cinnamic acids, synthetic derivatives and hybrids with antimicrobial activity. *Molecules* **2014**, *19*, 19292–19349. [CrossRef]
26. Wikoff, W.R.; Anfora, A.T.; Liu, J.; Schultz, P.G.; Lesley, S.A.; Peters, E.C.; Siuzdak, G. Metabolomics analysis reveals large effects of gut microflora on mammalian blood metabolites. *Proc. Natl. Acad. Sci. USA* **2009**, *106*, 3698–3703. [CrossRef]
27. Nutley, B.P.; Farmer, P.; Caldwell, J. Metabolism of trans-cinnamic acid in the rat and the mouse and its variation with dose. *Food Chem. Toxicol.* **1994**, *32*, 877–886. [CrossRef]
28. Butte, N.F. Carbohydrate and lipid metabolism in pregnancy: Normal compared with gestational diabetes mellitus. *Am. J. Clin. Nutr.* **2000**, *71*, 1256S–1261S. [CrossRef]
29. Hadden, D.R.; McLaughlin, C. Normal and abnormal maternal metabolism during pregnancy. *Semin. Fetal Neonatal Med.* **2009**, *14*, 66–71. [CrossRef]
30. Schneider, S.; Freerksen, N.; Röhrig, S.; Hoeft, B.; Maul, H. Gestational diabetes and preeclampsia—Similar risk factor profiles? *Early Hum. Dev.* **2012**, *88*, 179–184. [CrossRef]
31. Hedderson, M.M.; Ferrara, A.; Sacks, D.A. Gestational diabetes mellitus and lesser degrees of pregnancy hyperglycemia: Association with increased risk of spontaneous preterm birth. *Obstet. Gynecol.* **2003**, *102*, 850–856. [CrossRef] [PubMed]
32. Kim, C.; Newton, K.M.; Knopp, R.H. Gestational diabetes and the incidence of type 2 diabetes: A systematic review. *Diabetes Care* **2002**, *25*, 1862–1868. [CrossRef] [PubMed]
33. Nichols, G.A.; Hillier, T.A.; Brown, J.B. Normal fasting plasma glucose and risk of type 2 diabetes diagnosis. *Am. J. Med.* **2008**, *121*, 519–524. [CrossRef] [PubMed]
34. Reichelt, A.J.; Spichler, E.R.; Branchtein, L.; Nucci, L.B.; Franco, L.J.; Schmidt, M.I.; Brazilian Study of Gestational Diabetes Working, G. Fasting Plasma Glucose Is a Useful Test for the Detection of Gestational Diabetes. *Diabetes Care* **1998**, *21*, 1246–1249. [CrossRef]
35. Bogdanet, D.; O'Shea, P.; Lyons, C.; Shafat, A.; Dunne, F. The Oral Glucose Tolerance Test—Is It Time for a Change?—A Literature Review with an Emphasis on Pregnancy. *J. Clin. Med.* **2020**, *9*, 3451. [CrossRef]
36. Kattini, R.; Hummelen, R.; Kelly, L. Early Gestational Diabetes Mellitus Screening With Glycated Hemoglobin: A Systematic Review. *J. Obstet. Gynaecol. Can.* **2020**, *42*, 1379–1384. [CrossRef]
37. Adisakwattana, S. Cinnamic Acid and Its Derivatives: Mechanisms for Prevention and Management of Diabetes and Its Complications. *Nutrients* **2017**, *9*, 163. [CrossRef]
38. de Boer, I.H.; Gao, X.; Bebu, I.; Hoofnagle, A.N.; Lachin, J.M.; Paterson, A.; Perkins, B.A.; Saenger, A.K.; Steffes, M.W.; Zinman, B.; et al. Biomarkers of tubulointerstitial damage and function in type 1 diabetes. *BMJ Open Diabetes Res. Care* **2017**, *5*, e000461. [CrossRef]
39. Menni, C.; Zhu, J.; Le Roy, C.I.; Mompeo, O.; Young, K.; Rebholz, C.M.; Selvin, E.; North, K.E.; Mohney, R.P.; Bell, J.T.; et al. Serum metabolites reflecting gut microbiome alpha diversity predict type 2 diabetes. *Gut Microbes* **2020**, *11*, 1632–1642. [CrossRef]
40. Lu, X.; Zhao, X.; Bai, C.; Zhao, C.; Lu, G.; Xu, G. LC–MS-based metabonomics analysis. *J. Chromatogr. B* **2008**, *866*, 64–76. [CrossRef]
41. Abina, A.; Korošec, T.; Puc, U.; Jazbinšek, M.; Zidanšek, A. Urinary Metabolic Biomarker Profiling for Cancer Diagnosis by Terahertz Spectroscopy: Review and Perspective. *Photonics* **2023**, *10*, 1051. [CrossRef]
42. Hou, X.; Chen, X.; Li, T.; Li, Y.; Tian, Z.; Wang, M. Highly sensitive terahertz metamaterial biosensor for bovine serum albumin (BSA) detection. *Opt. Mater. Express* **2021**, *11*, 2268–2277. [CrossRef]
43. Yan, S.; Xia, L.; Wei, D.; Cui, H.L.; Du, C. Terahertz biosensing of protein based on a metamaterial. In Proceedings of the 2016 IEEE International Conference on Manipulation, Manufacturing and Measurement on the Nanoscale (3M-NANO), Chongqing, China, 18–22 July 2016; pp. 327–330.
44. Geng, Z.; Zhang, X.; Fan, Z.; Lv, X.; Chen, H. A Route to Terahertz Metamaterial Biosensor Integrated with Microfluidics for Liver Cancer Biomarker Testing in Early Stage. *Sci. Rep.* **2017**, *7*, 16378. [CrossRef] [PubMed]
45. Xu, H.; Gao, M.; Tang, X.; Zhang, W.; Luo, D.; Chen, M. Micro/Nano Technology for Next-Generation Diagnostics. *Small Methods* **2019**, *4*, 1900506. [CrossRef]
46. Yang, J.; Long, J.; Yang, L. First-principles investigations of the physical properties of lithium niobate and lithium tantalate. *Phys. B Condens. Matter* **2013**, *425*, 12–16. [CrossRef]
47. Basharin, A.A.; Kafesaki, M.; Economou, E.N.; Soukoulis, C.M.; Fedotov, V.A.; Savinov, V.; Zheludev, N.I. Dielectric Metamaterials with Toroidal Dipolar Response. *Phys. Rev. X* **2015**, *5*, 011036. [CrossRef]
48. Wang, Y.; Han, Z.; Du, Y.; Qin, J. Ultrasensitive terahertz sensing with high-Q toroidal dipole resonance governed by bound states in the continuum in all-dielectric metasurface. *Nanophotonics* **2021**, *10*, 1295–1307. [CrossRef]
49. Sánchez, C.; Agulló-López, F. Transient Effects in the Room-Temperature F-Colouring of NaCl Irradiated with X- or γ-rays. *Phys. Status Solidi (b)* **2006**, *29*, 217–230. [CrossRef]
50. Tuz, V.R.; Khardikov, V.V.; Kivshar, Y.S. All-Dielectric Resonant Metasurfaces with a Strong Toroidal Response. *ACS Photonics* **2018**, *5*, 1871–1876. [CrossRef]

51. Dorney, T.D.; Baraniuk, R.G.; Mittleman, D.M. Material parameter estimation with terahertz time-domain spectroscopy. *J. Opt. Soc. Am. A* **2001**, *18*, 1562–1571. [CrossRef]
52. Liu, B.; Chen, S.; Zhang, J.; Yao, X.; Zhong, J.; Lin, H.; Huang, T.; Yang, Z.; Zhu, J.; Liu, S.; et al. A Plasmonic Sensor Array with Ultrahigh Figures of Merit and Resonance Linewidths down to 3 nm. *Adv. Mater.* **2018**, *30*, e1706031. [CrossRef] [PubMed]

biosensors

MDPI

Article

Asterias forbesi-Inspired SERS Substrates for Wide-Range Detection of Uric Acid

Hyunjun Park [1,†], Kyunghwan Chai [1,†], Woochang Kim [1,†], Joohyung Park [1], Wonseok Lee [2,*] and Jinsung Park [1,*]

1 Department of Biomechatronic Engineering, College of Biotechnology and Bioengineering, Sungkyunkwan University, Suwon 16419, Republic of Korea; guswns1105@g.skku.edu (H.P.); kyungh2012@g.skku.edu (K.C.); wkim91@skku.edu (W.K.); parkjoodori@skku.edu (J.P.)
2 Department of Electrical Engineering, Korea National University of Transportation, Chungju 27469, Republic of Korea
* Correspondence: wslee@ut.ac.kr (W.L.); nanojspark@skku.edu (J.P.)
† These authors contributed equally to this study.

Abstract: Uric acid (UA), the final metabolite of purine, is primarily excreted through urine to maintain an appropriate concentration in the bloodstream. However, any malfunction in this process can lead to complications due to either deficiency or excess amount of UA. Hence, the development of a sensor platform with a wide-range detection is crucial. To realize this, we fabricated a surface-enhanced Raman spectroscopy (SERS) substrate inspired by a type of starfish with numerous protrusions, *Asterias forbesi*. The *Asterias forbesi*-inspired SERS (AF-SERS) substrate utilized an Au@Ag nanostructure and gold nanoparticles to mimic the leg and protrusion morphology of the starfish. This substrate exhibited excellent Raman performance due to numerous hotspots, demonstrating outstanding stability, reproducibility, and repeatability. In laboratory settings, we successfully detected UA down to a concentration of 1.16 nM (limit of detection) and demonstrated selectivity against various metabolites. In the experiments designed for real-world application, the AF-SERS substrate detected a broad range of UA concentrations, covering deficiencies and excesses, in both serum and urine samples. These results underscore the potential of the developed AF-SERS substrate as a practical detection platform for UA in real-world applications.

Keywords: uric acid; wide-range detection; bioinspired; gold nanoparticles; surface-enhanced Raman scattering

1. Introduction

Uric acid (UA), primarily synthesized in the liver, is an organic compound produced endogenously from purine metabolism in the human body. It circulates through the bloodstream and is excreted in urine via kidneys [1]. UA functions as a biomarker for oxidative stress, and the changes in its concentration are associated with various physiological and pathological conditions [2]. Generally, the normal concentration range of UA in the blood is approximately 200 to 500 μM [3,4], and deviations from this range can lead to complications. If UA levels in the blood drop below the normal range (a condition known as hypouricemia), then the risk of acute kidney injury can significantly increase after intense physical activity. Additionally, low UA levels in the blood are associated with renal hypouricemia, which is linked to kidney stones due to an increased level of urinary UA, correlating with the impaired tubular reabsorption of UA [5]. Conversely, heightened UA levels, referred to as hyperuricemia, can result in the formation of urate crystals, which accumulate in joints, cartilage, and adjacent tissues, giving rise to conditions such as gout [6,7]. Additionally, hyperuricemia is associated with inflammatory diseases, kidney issues, and hypertension [8–11]. Therefore, detecting UA concentrations outside the normal range in both blood and urine is crucial for assessing health status and diagnosing diseases.

Due to the critical importance of detecting UA levels, tests for UA in blood and urine are performed via standard procedures in hospitals. However, these tests have limitations, including the need for periodic sample collection and intricate preparation procedures, and these constraints have driven the development of various UA detection methods over the past decades [12–14]. Techniques involving electrochemistry [15,16], fluorescence [17], and spectroscopy [18,19] have been widely employed in UA detection. In particular, electrochemical detection methods are prevalent because of their effectiveness in UA detection. However, UA produces a redox signal similar to dopamine and ascorbic acid, which may cause selectivity problems [20–22]. Recently, spectroscopy based on surface-enhanced Raman scattering (SERS) has garnered significant attention in the biosensing field due to its remarkable sensitivity and exceptional selectivity [23,24]. However, the current SERS-based platforms still face limitations in detecting UA in both deficiency and excess ranges [25]. To overcome this challenge, it is crucial to design nanostructures with high sensitivity and extensive surface areas. Fabricating nanostructures with numerous nanogaps and forming SERS substrates with broad surface areas capable of accommodating high concentrations of analytes are essential for addressing this limitation.

In this study, we drew inspiration from the morphology of *Asterias forbesi*, commonly known as the Forbes Sea Star, to enhance the performance of the SERS substrate. *Asterias forbesi* exhibits numerous protrusions on its torso and legs, resembling structures in which the SERS effect can be maximized by creating nanoscale hotspots. To replicate this structure, we strengthened the SERS effect by fabricating small-scale gold nanoparticles (GNPs) in a gold–silver bimetallic structure, imitating starfish legs and the protuberances of *Asterias forbesi*. This design maximized the SERS effect by increasing the adhesion area with UA and promoting hotspot creation. The *Asterias forbesi*-inspired SERS (AF-SERS) substrate demonstrated an outstanding SERS effect; this result was further confirmed through electromagnetic field simulations. Using the AF-SERS substrate, UA was detected across a wide range of samples, including serum and human urine. Moreover, the substrate enabled the specific identification and detection of UA among various endogenous metabolites. These findings showcase the promising potential of the developed AF-SERS substrate as a sensor platform not only for UA but also for various targets in real-world field settings.

2. Experimental Section

2.1. Chemical Agents

The reagent for SERS substrate fabrication and SERS performance analysis (sodium citrate, chloroauric acid ($HAuCl_4$), potassium dicyanoargentate ($KAg(CN)_2$), Rhodamine 6G(R6G)), UA, selective experimental group (L-ascorbic acid, hydrochloride dopamine, L-cysteine, D-(+)-glucose, creatine), and human serum and sigmatrix urine diluent (artificial urine) were purchased from Sigma Aldrich (St. Louis, MO, USA). A glass slide was purchased from Paul Marienfeld GmbH & Co. KG (Lauda-Königshofen, Germany). UA solution was generated by dissolving it in deionized (DI) water, and the selectivity analysis group was prepared in the same manner. All glassware was washed with piranha solution in concentrated sulfuric acid (98% w/v) and hydrogen peroxide (30% w/v) prior to use. All solutions were prepared using Millipore DI water with a resistivity of at least 18.2 MΩ cm at 25 °C. Additionally, all target samples for final substrate production were reacted with GNP at a 1:1 ratio based on volume.

2.2. Uric Acid Detection Strategy Using AF-SERS Substrate

To develop a sensor capable of simultaneously detecting UA deficiencies and excesses outside the normal range in both blood and urine, a SERS structure with a broad surface area is crucial. This type of structure ensures an effective response even at high UA concentrations [26]. Therefore, we developed a detection strategy based on a SERS substrate inspired by *Asterias forbesi*, commonly known as Forbes Sea Star (Scheme 1). Initially, an electrochemical method was employed to produce a silver nanostructure resembling the legs of a starfish. This was achieved by reducing silver using the $KAg(CN)_2$ solution on the surface of the gold plate through overpotential [27]. However, various substances present in serum and urine can promote silver oxidation, leading to a decline in the intensity of the SERS signal [28]. To address this, we employed galvanic substitution for reducing gold ions using an $HAuCl_4$ solution, thereby coating the surface with gold while simultaneously inducing surface roughening. Subsequently, through the integration of gold-coated silver nanostructures with GNPs, a nanostructure resembling the arms of *Asterias forbesi* was successfully fabricated. This was confirmed through SEM images, demonstrating that the developed structure was similar to the spikes in *Asterias forbesi*. These nanostructures exhibited numerous hotspots, a wide surface area facilitated by high Z-axis and particle characteristics, and a bimetallic enhancement effect, which combined the effective plasmonic properties of silver with the chemical stability of gold, ensuring a strong SERS effect [29]. Through these enhancements, the SERS substrate effectively detected UA in serum and urine samples.

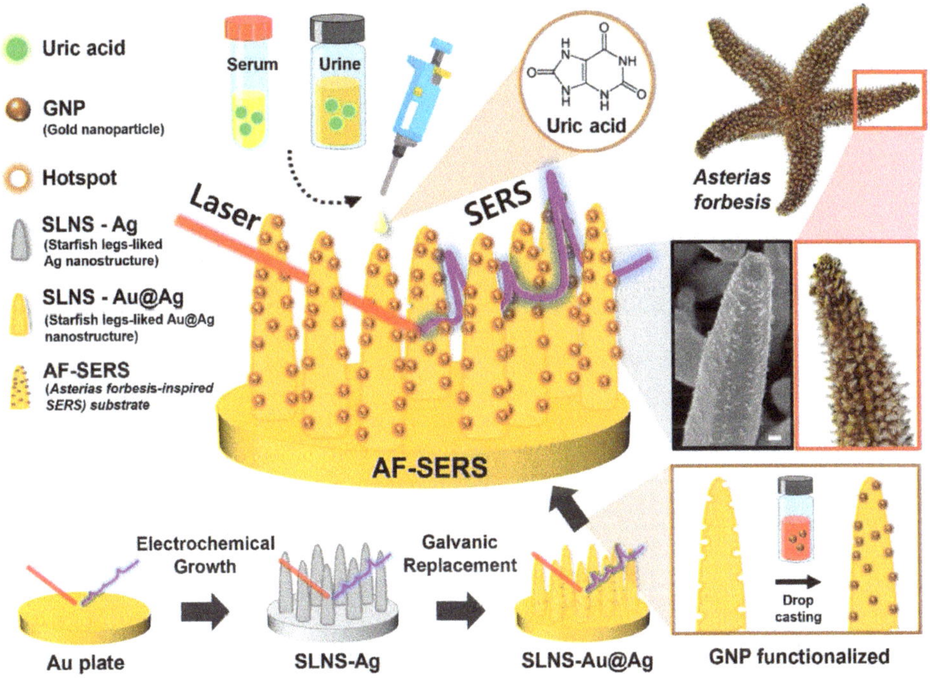

Scheme 1. Schematic of Raman spectroscopic detection of UA derived from serum and urine using an *Asterias forbesi*-inspired SERS substrate.

2.3. Synthesis and Optimization of AF-SERS Substrate

Starfish-leg-like Ag nanostructure (SLNS-Ag) and starfish-leg-like Au@Ag nanostructure (SLNS-Au@Ag) were fabricated using the methods described in previous studies [30,31]. Briefly, bare Au plate was fabricated using an E-beam evaporator, and SLNS-Ag was produced through the reduction in $KAg(CN)_2$ by employing chronoamperometry. Subsequently, 50 μL of 0.2 mM $HAuCl_4$ was placed in a polydimethylsiloxane (PDMS) mold for 30 min to induce galvanic replacement, resulting in the fabrication of SLNS-Au@Ag.

GNPs were synthesized using the Turkevich method [32]. For the synthesis of GNP, 1 mL of 1 wt% $HAuCl_4$ solution and 5 mL of 1 wt% sodium citrate solution were added to 100 mL of DI water and heated at 100 °C for 1 h with vigorous stirring. Afterward, the target substance to be detected was mixed with the synthesized GNP solution in a 1:1 ratio and dropped onto the SLNS-Au@Ag substrate to complete the production of the AF-SERS substrate. For sensitive and precise measurements, each AF-SERS substrate was exclusively used for a single sample, ensuring that none was reused to maintain the highest standards of accuracy.

2.4. SEM- and FEM-Based Simulation Data

The morphologies of SERS substrates were observed via field-emission scanning electron microscopy (FE-SEM; JSM-IT800, JEOL, Tokyo, Japan). SEM images were obtained at 15 kV HV at different magnifications.

Finite elements method (FEM)-based plasmonic simulations of the optimal SERS substrate analysis were performed using the electromagnetic wave module of COMSOL Multiphysics 6.0. The structures of the nanopillar SERS substrate (Au plate, SLNS-Ag, SLNS-Au@Ag) and nanopillar-combined GNP were based on the SEM images. The time harmonic maxwell equation with boundary conditions was calculated for plasmonic SERS substrates. The simulation was vertically projected onto the surface of the SERS substrate with an incident light of 785 nm. The refractive index and simulation function were referenced from a previous study [26], and RGB values were used to analyze the FEM simulation results using Image J software(Image J v.1.53).

2.5. SERS Analysis

SERS spectra were analyzed using Raman spectroscopy (In Via Reflex, Renishaw, Wotton-under-Edge, UK). The laser was focused using a 100× objective lens (Leica DM2700 M, DEU, and Renishaw Centrus Detector, GBR), and the diameter of the spot of the laser beam was approximately 0.4 μm. All SERS spectra were measured using a 785 nm laser with a power of 5.05 mW after 1 s exposure and 10 accumulations.

The Raman spectra of R6G were measured in the range of 567–1675 cm^{-1}, whereas those of UA and the selective experimental group were measured in the range of 497–1622 cm^{-1}.

2.6. Optimization of SERS Substrate

To analyze the optimal SERS substrate, 5 μL droplets of 100 μM R6G were reacted on the following prepared substrates: Au plates, SLNS-Ag, and SLNS-Au@Ag. Additionally, to evaluate the SERS performance of GNP-functionalized substrates, 5 μL droplets of R6G and GNP mixed in a 1:1 volume ratio were applied to each substrate (Au plate, SLNS-Ag, and SLNS-Au@Ag).

To optimize the GNP concentration, the synthesized GNPs were precipitated via centrifugation (Labogene 1524, LABOGENE, Lillerød, Denmark) at 6500× *g* rpm (2669 g) for 30 min, and the supernatant was removed or more DI water was added to prepare GNPs at various concentrations (×0.2–×5). To optimize the quantity of GNPs, various concentrations of GNP solutions were prepared using the synthesized GNP stock solution (×1). Diluted GNP solutions were prepared by diluting the original solution fivefold (×0.2) and twofold (×0.5) using DI water. Concentrated solutions, obtained through centrifugation with concentrations twofold (×2) and fivefold (×5), were prepared for use. The resulting different concentrations of GNPs were mixed with R6G at a 1:1 ratio, and 5 μL droplets were reacted with SLNS-Au@Ag for subsequent SERS measurements.

Initially, a series of SLNS-Ag and SLNS-Au@Ag were prepared and exposed to a challenging 10× PBS (phosphate-buffered saline, 100 mM) environment. Subsequently, these particles were individually retrieved on specified days and thoroughly washed with ethanol (EtOH) and DI water. Afterward, 5 μL of GNP solution combined with 100 μM R6G in a 1:1 ratio were deposited onto the washed substrates containing SLNS-Ag and SLNS-Au@Ag. Subsequently, the variations in the Raman spectra of R6G on each substrate were measured daily.

2.7. Uric Acid Detection with AF-SERS SERS Sensors

All sample preparations involved mixing the sample with 1X GNP at a 1:1 volume ratio and subsequently applying a 5 μL droplet onto the GNP substrate, and samples were allowed to dry prior to SERS measurements.

The reproducibility and uniformity analyses of the fabricated AF-SERS substrate were conducted using a 100 μM UA solution. Reproducibility analysis entailed reacting the UA of the same concentration with six independently prepared substrates to measure the Raman intensity. Uniformity analysis involved measuring the Raman intensity of UA at 50 randomly selected spots on the prepared substrates.

In laboratory conditions, UA detection was performed in the concentration range from 1 nM to 1 mM. The selectivity of the UA sensor was evaluated using various biological fluid materials, including dopamine, L-ascorbic acid, D-(+)-glucose, L-cysteine, and creatine, each at a concentration of 1 mM.

The actual application of the UA sensor involved the use of 10% diluted human serum and artificial urine samples. Various concentrations of UA ranging from 1 μM to 1 mM were prepared using 10% diluted human serum. Additionally, using artificial urine samples, we prepared samples with UA concentrations within the normal range, representing 300 μM, as well as samples indicating UA deficiency with a concentration of 30 μM and excess conditions with a concentration of 3 mM. The Raman measurement conditions were maintained consistent with those used for Raman analysis.

3. Results and Discussion

3.1. Characterization of AF-SERS Substrate

To confirm the characteristics of the AF-SERS substrate, various analyses, including morphology, Raman intensity, and electromagnetic field simulations, were conducted (Figure 1). Initially, SEM was employed for morphological analysis. Figure 1a confirms the successful fabrication of starfish-leg-shaped silver nanostructures (SLNS-Ag) using electrochemical methods. Subsequently, after galvanic replacement, the surface of SLNS-Ag was replaced with gold, resulting in SLNS-Au@Ag. This replacement led to a darker appearance of the substrate surface (Figure S1), and SEM images revealed a slightly roughened surface compared with the smooth SLNS-Ag surface (Figure 1b). This roughening was attributed to the deposition of gold in the areas where silver was displaced. The displacement reaction is described by the following equation [31]:

$$\text{Anode Reaction: } 3Ag \rightarrow 3Ag^{+} + 3e^{-}$$

$$\text{Cathode Reaction: } AuCl4^{-} + 3e^{-} \rightarrow Au + 4Cl^{-}$$

Finally, the SLNS-Au@Ag substrate functionalized with GNP assumed a structure similar to *Asterias forbesi* (Figure 1c,d). Figure S2 further validates the successful attachment of galvanic and gold nanoparticles through EDS analysis, which identifies the components of each Au and Ag. Additionally, the low-magnification SEM images for each condition are shown in Figure S3. Following that, we conducted Raman intensity analysis using R6G, a representative Raman indicator, based on the morphology observed at each step. R6G is a Raman indicator with fluorescent characteristics, and in order to minimize the impact of fluorescence in Raman analysis, we selected a laser with a wavelength of 785 nm [33]. When using the SLNS-Ag substrate as opposed to a gold plate, enhanced Raman signals at the characteristic peak of R6G (1508 cm^{-1}) were observed [34], attributed to the hotspots created by the nanostructures (Figure 1e). Typically, in a Raman substrate utilizing silver, the intrinsic Raman peak of silver appears around 1000 cm^{-1}. However, in Figure 1e, the Raman intensity attributed to the SERS effect of the Raman indicator, R6G, is relatively high, making it challenging to distinguish the Raman peak of silver, which appears at a much lower value. As seen in Figure S4, when the Raman intensity scale is reduced for closer examination, it is confirmed that, unlike the case with the Au plate, a Raman signal around 1000 cm^{-1} is observed in the SLNS-Ag substrate composed of silver [35].

SLNS-Au@Ag exhibited a slight enhancement in Raman intensity compared with SLNS-Ag, attributed to the gold–silver bimetallic effect resulting from the coexistence of gold and silver (Figure 1f) [36]. When evaluating the SERS intensity using the AF-SERS substrate, a significant amplification in the R6G SERS signal was observed (Figure 1g). This outcome might be anticipated solely due to GNP. However, when only GNPs were employed, the performance was inferior compared to when the AF-SERS was employed (Figure S5). These results suggest a synergistic effect between the structures with excellent Z-axis characteristics and nanoparticles. For the analysis of the Raman intensity of each substrate, emphasis was placed on 1508 cm^{-1}, representing the peak value, which manifests the most conspicuous concentration change among the Raman peaks of R6G. Figure 1h presents a bar graph illustrating the average and variance in the Raman intensity at 1508 cm^{-1}, facilitating a quantitative assessment of the Raman intensity for each substrate. The Raman intensity of the AF-SERS substrate surpassed that of the SLNS-Au@Ag substrate by more than 6.2 times, underscoring its outstanding SERS performance. These findings emphasize the highly effective characteristics of *Asterias forbesis*-inspired nanostructures as superior SERS substrates. Notably, the presence of smaller scale nanoparticles, in contrast to a singular metal nanostructure, leads to an increased reaction surface area with the target and the generation of additional hotspots [37,38].

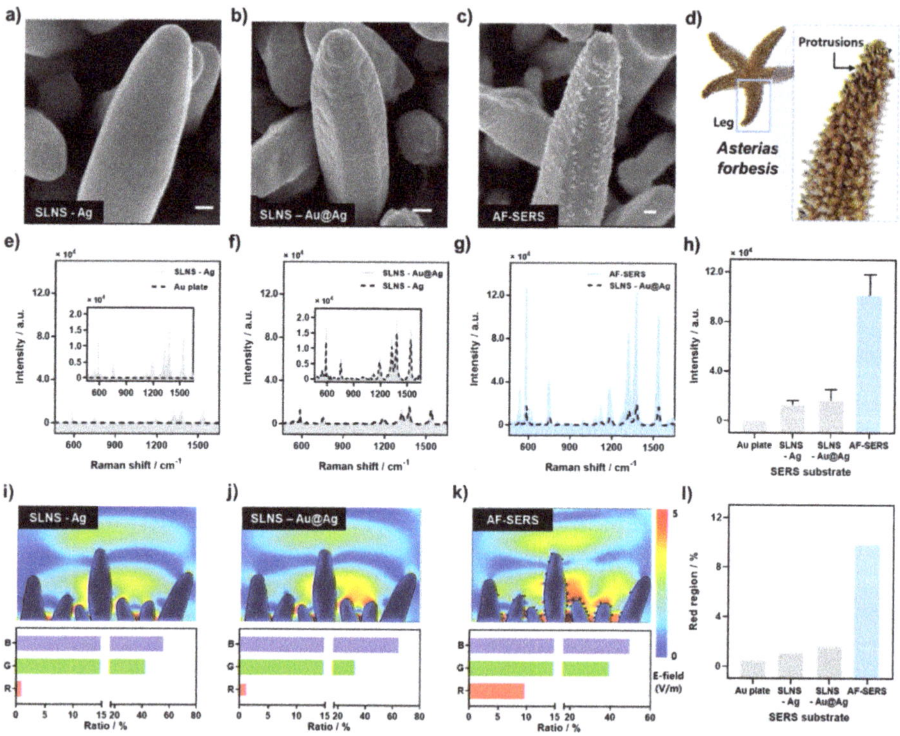

Figure 1. SEM images of (**a**) SLNS-Ag, (**b**) SLNS-Au@Ag, and (**c**) AF-SERS substrate (scale bar: 100 nm). (**d**) Image of the legs and protrusion structures of the *Asterias forbesi*. Raman spectrum comparison data for 100 μM R6G, a Raman emitter: (**e**) Bare Au plate and SLNS-Ag substrate (Gray scale: SLNS-Ag, Black dotted line: Au plate), (**f**) SLNS-Ag and SLNS-Au@Ag substrate (Gray scale: SLNS-Au@Ag, Black dotted line: SLNS-Ag), and (**g**) SLNS-Au@Ag and AF-SERS substrate (Blue scale: AF-SERS, Black dotted line: SLNS-Au@Ag), respectively. (**h**) Raman intensities at 1508 cm^{-1}, a specific peak of R6G on various SERS substrates (Au plate, SLNS-Ag, SLNS-Au@Ag, and AF-SERS,

respectively). FEM-based electromagnetic simulation results and RGB value graph for each area of (**i**) SLNS-Ag, (**j**) SLNS-Au@Ag, and (**k**) AF-SERS substrate. (**l**) Ratio of the red region (in (**i–k**)) for each substrate.

To validate these results, we conducted electromagnetic field simulations, designing the morphologies to resemble the SEM images of each state (Figure 1i–k). The electron density generated through the simulations is depicted using colors, with red indicating high intensity, green representing intermediate intensity, and blue indicating low intensity. Initially, on the Au plate, hotspot formation was minimal, and the red area was scarce (Figure S6). However, as the nanostructures formed at each stage, the electromagnetic field signal was enhanced, leading to an increased presence of the red region (Figure 1i–k). Although SLNS-Ag and SLNS-Au@Ag appeared similar, slightly higher electron density was observed in SLNS-Au@Ag. Remarkably, the AF-SERS structure exhibited a broader red region, particularly on the GNP surface, indicating enhanced hotspots due to the numerous nanogaps formed by particles. This contributed to the reinforcement of SERS intensity. Quantitative analysis revealed that the red region for AF-SERS was 497.32% larger than that for the SLNS-Au@Ag substrates (Figure 1l). Therefore, it can be inferred that AF-SERS demonstrated increased SERS intensity due to the abundant formation of hotspots in both the nanostructures and starfish-leg-shaped structures. Ultimately, through this validation, we confirmed consistency between experimental results and simulation outcomes, providing robust support for our claims.

3.2. Optimization and Sensor Performance of AF-SERS Substrate

The analysis of Raman intensity and simulations underscored the significant impact of the presence or absence of GNPs on the starfish-like nanostructure in determining SERS intensity. Consequently, optimizing the quantity of GNPs corresponding to the protuberances of *Asterias forbesi* was crucial for achieving optimal sensor performance. To explore this, we conducted an analysis of Raman intensity for R6G by varying the amount of GNP to identify the point of optimal effect. The quantity of GNP was adjusted by considering the amount of GNPs initially included in the synthesized GNP solution as 1, allowing for subsequent dilution or concentration to modify the relative GNP quantity. Detailed methods for this adjustment are described in the Experimental Section. Figure 2a presents the Raman spectrum data corresponding to the amount of GNP. As the amount of GNP increases from SLNS-Au@Ag without GNP, the Raman intensity values show an upward trend. To provide a detailed analysis of the Raman intensity of R6G under each condition, a bar graph was generated to represent the Raman intensity at 1508 cm^{-1} (Figure 2b). However, an excessive amount of GNPs can lead to aggregation and overlap, reducing hotspots and consequently diminishing the SERS effect [39,40]. Therefore, the highest Raman intensity was observed at the x1 concentration, confirming the importance of optimizing the amount of GNP for an effective response. Therefore, for subsequent experiments, the AF-SERS substrate fabricated under x1 conditions was used.

Hence, replacing silver with gold through galvanic substitution is a clever strategy to address the oxidation issue, as gold is more resistant to oxidation. This approach can significantly enhance the stability of the sensor, ensuring long-term consistent and reliable performance [41]. It provides a practical solution to overcome the limitations associated with silver structures, particularly when dealing with biomaterials. To assess the stability of the sensor, it was exposed to stringent environmental conditions (immersion in 10× PBS solution for 7 days), and Raman signals (R6G, 100 μM) were compared. The experiment involved using the AF-SERS substrate and GNP + SLNS-Ag substrate, where the galvanic replacement step was omitted. The results indicated that the AF-SERS substrate maintained a consistent Raman intensity over the entire 7-day period, with a relative standard deviation (RSD) value of 7.21% (Figures S7 and 2c). The minimal changes observed over the 7-day period under rigorous environmental conditions imply that AF-SERS is also advantageous for long-term storage. In contrast, the control group, GNP + SLNS-Ag substrate, exhibited a noticeable reduction in Raman intensity at the same peak over time, resulting in a significant

54.76% decrease after 7 days (Figures S8 and 2d). The AF-SERS substrate, coated with gold, demonstrated exceptional resistance to oxidation, effectively preventing a decrease in Raman intensity due to oxidation conditions.

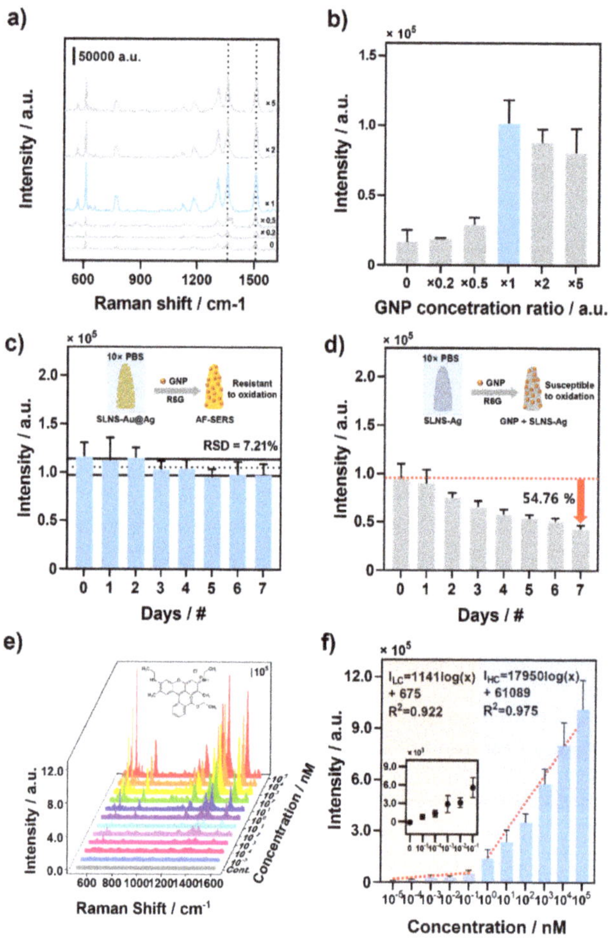

Figure 2. Optimization and SERS performance evaluation of AF-SERS with R6G. (**a**) SERS spectra of R6G depending on GNP concentration. (**b**) Measured values of the SERS peak intensities at 1508 cm^{-1} depending on the GNP concentration. Intensity at specific Raman peaks of R6G in (**c**) AF-SERS and (**d**) GNP + SLNS-Ag substrate measured under extreme conditions (10× PBS solution) for 7 days. (**e**) SERS sensitivity of R6G detection at various concentrations (10^{-14}–10^{-4} M). (**f**) Measured values of SERS peak intensities at 1508 cm^{-1} depending on the R6G concentrations. Inset depicts the magnified SERS peak intensities of R6G at a low concentration (0–10^{-10} M).

To evaluate the sensitivity of the AF-SERS substrate, we utilized various concentrations of R6G, ranging from 100 μM to 10 fM. The specific Raman peak of R6G increased with higher concentrations (Figure 2e). For the quantitative analysis of Raman intensity, the Raman intensity at a specific R6G peak (1508 cm^{-1}) was expressed using mean and variance (Figure 2f). Distinct linear behaviors were observed in both the low concentration (10^{-5}–10^{-1} nM) and high concentration (10^2–10^5 nM). The linear equations for low and high concentration ranges are I_{LC} = 1141 × log(X) + 675 (R^2 = 0.922) and I_{HC} = 17,950X + 61,089 (R^2 = 0.975), respectively, where X represents the concentration of R6G. The limit of detection (LOD) was 1.395 fM, calculated using the formula LOD = 3.3 × standard deviation/slope. Based on these findings, we computed the analytical enhancement factor (AEF) for the AF-SERS substrate using the following formula [42]:

$$AEF = \frac{I_{SERS}/C_{SERS}}{I_{OR}/C_{OR}}$$

Here, I_{SERS} represents the SERS intensity; I_{OR} denotes the SERS intensity observed on the bare substrate; C_{SERS} stands for the concentration of the Raman indicator on the SERS substrate (i.e., LOD); and C_{OR} is the concentration of the Raman indicator on the bare substrate. The computed analytical enhancement factor (AEF) for the AF-SERS was 3.658×10^{11}, signifying enhancement in performance compared to other SERS substrates (Table S1). Consequently, the AF-SERS substrate demonstrated highly sensitive detection of R6G and successfully detected a wide concentration range. These findings suggest a promising potential of the developed substrate for the detection of UA and other analytes in future applications.

3.3. *Performance Evaluation of AF-SERS Substrate with UA*

Subsequently, the suitability of the AF-SERS substrate for UA detection was examined, considering the reproducibility between substrates, random spot reproducibility, and UA detection performance (Figure 3). The normal Raman spectrum of uric acid is depicted in the following figure, with band assignment displayed in Table S2. Initially, six different AF-SERS substrates were fabricated, and each substrate was reacted with a consistent 5 μL of 100 μM UA (Figure 3a). The specific Raman peak of UA occurred at 640 cm^{-1}, and each substrate exhibited a consistent spectral pattern (Figure 3a). Subsequently, the Raman intensity at 640 cm^{-1} was compared for the six substrates (Figure 3d), revealing excellent repeatability with an RSD value of 7.227%.

After reacting UA on a single AF-SERS substrate, we presented the measured Raman spectra for 50 random spots as a heat map image (Figure 3b). The bright line at 640 cm^{-1} in the image indicated the strong SERS intensity of UA on AF-SERS SERS substrates. Similarly, a comparison of Raman intensity at a specific UA peak of 640 cm^{-1} for each spot revealed excellent uniformity with an RSD value of 13.46% (Figure 3e).

Following the examination of various UA concentrations on the AF-SERS substrate, detection experiments were conducted for 10 concentrations, in the range from 1 mM to 1 nM, covering the normal physiological range of 200–500 μM (Figure 3c). As shown in Figure 3c, there is a noticeable tendency for the Raman intensity to increase with higher UA concentrations (10^0–10^{-5} mM) at a specific Raman peak of 640 cm^{-1}, aligning with the molecular structure of UA. The evaluation of the Raman intensity of UA at the specific Raman peak value of 640 cm^{-1} revealed a linear increase in the high concentration range, spanning from the normal physiological range (200–500 μM) to 1 mM (Figure 3f). The linear equation for UA concentration was calculated as I_{HC} = 211.9 X + 31,674 (R^2 = 0.99) in the high concentration range. Additionally, a logarithmic linear increase was observed in the low concentration range (10 nM–100 μM), and the logarithmic linear equation for low concentration UA was calculated as I_{LC} = 9835 × log(X) − 22,636 (R^2 = 0.987; X represents UA concentration (mM)). The LOD for UA was determined to be 1.16 nM using the following formula: LOD = 3.3 × standard deviation/slope of the regression curved line.

Based on these findings, we experimentally validated the ability of our AF-SERS substrate to detect UA concentrations within the normal range, as well as at deficient and excess concentrations. As indicated in Table S3, our results not only exhibit the highest sensitivity of the substrate in terms of LOD but also boast the broadest detection range compared to various UA detection methods, including electrochemical, fluorescence, and spectroscopic methods.

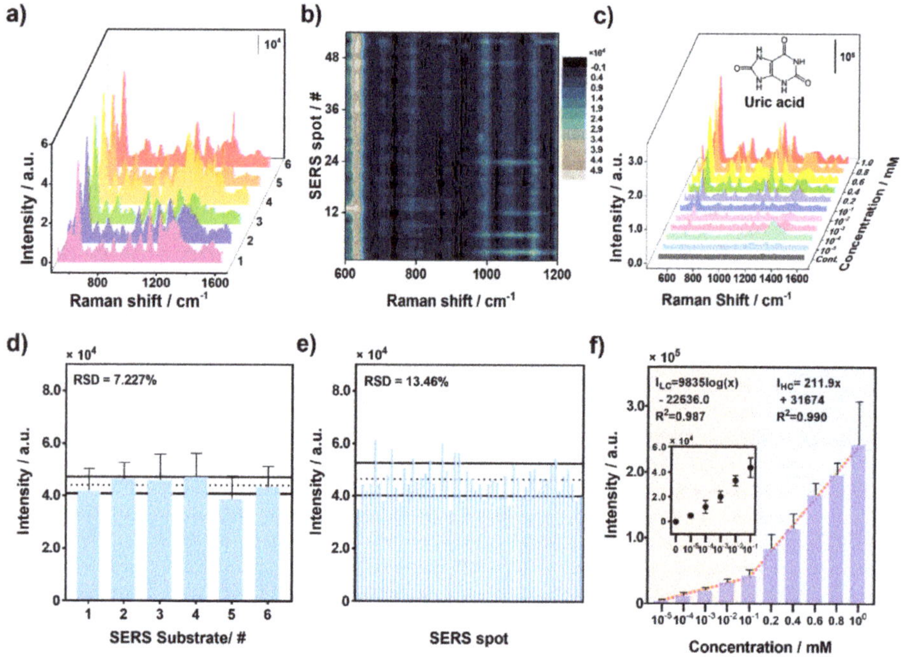

Figure 3. Efficiency and sensitivity analysis of the AF-SERS-based UA sensor. (**a**) R6G Raman spectra of six different AF-SERS substrates and (**b**) Heat map image of R6G Raman spectra at 50 random spots. (**c**) Raman spectra for UA detection at various concentrations on AF-SERS SERS substrate. Inset image is the molecular structure of UA. (**d**) Raman intensity at 1508 cm^{-1}, which is the specific peak of R6G for each substrate. (**e**) Raman intensity measurements at 1508 cm^{-1} for 50 random data points extracted from heat map data. (**f**) Measured value of the SERS peak intensities at 640 cm^{-1} depending on the UA concentration. Inset image shows the results in the low concentration range, presented at an optimal scale.

3.4. Detection Selectivity of UA Sensor with AF-SERS

To evaluate the selective UA detection capability of the fabricated AF-SERS substrate, signals from various metabolites were analyzed (Figure 4). Figure 4a presents the chemical formulations of target molecules used in the selectivity experiments, including UA (I), ascorbic acid (II), creatine (III), dopamine (IV), glucose (V), and L-cysteine (VI), which are representative metabolites [43–46]. UA was tested at a concentration of 800 μM, whereas the remaining substances were experimented with at a high concentration of 1 mM. The Raman spectrum results revealed a distinct and robust intensity at 640 cm^{-1}, the UA-specific Raman peak, exclusively for UA. Other target molecules, even at high concentrations, did not exhibit peaks, which overlapped with UA (Figure 4b). Further analysis focused solely on the Raman intensity at 640 cm^{-1} demonstrated a value of 196,617 $\pm$ 18,102.54 for UA, whereas the remaining molecules displayed significantly lower values, consistently below 15,000 (ascorbic acid: 7559.80 $\pm$ 30,472.24, creatine: 1307.59 $\pm$ 917.42, dopamine: 917.57 $\pm$ 1525.35, glucose: 8298.72 $\pm$ 2247.93, and L-cysteine: 14,426.45 $\pm$ 4684.40) (Figure 4c). This rigorous comparison highlights the specificity of the proposed AF-SERS-based UA sensor, demonstrating its ability to selectively detect UA among other metabolites in the body, despite the relatively high concentrations of other target substances.

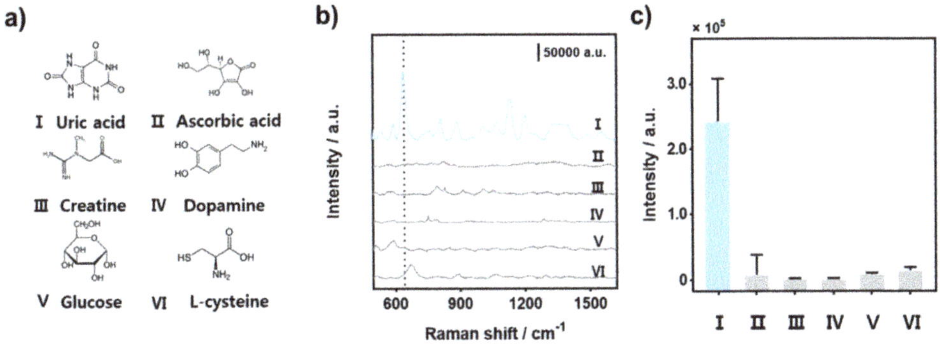

Figure 4. Selectivity analysis of the AF-SERS-based UA sensor. (**a**) Structural formula of selective groups: uric acid (I), ascorbic acid (II), creatine (III), dopamine (IV), glucose (V), and L-cysteine (VI). (**b**) SERS spectra and (**c**) SERS peak intensities at 640 cm^{-1} of UA and the selective groups.

3.5. UA Detection in Real Sample: Human Serum and Urine

Actual samples for detecting UA include human blood and urine. Therefore, to validate the detection capability of UA within the body using the AF-SERS-substrate-based UA sensor, we designed experiments for UA detection in human serum and real urine samples (Figure 5). First, for UA detection in serum, human serum was diluted at a 1:10 ratio (Figure 5a). Subsequently, UA was mixed into the diluted serum at various concentrations to prepare the samples. Samples were prepared from 1 mM to 1 μM, including a concentration of 200–500 μM, which corresponds to the normal range of UA in human serum. As shown in Figure 5b, the Raman signal at 640 cm^{-1} changes significantly as the concentration of UA changes. Specifically, the Raman intensity results by concentration at 640 cm^{-1} and specific Raman peak of UA are shown in Figure 5c. A log-linear increase is observed at low concentrations (1–100 μM), corresponding to conditions in which UA deficiency is anticipated. In contrast, a linear increase is observed at high concentrations (200 μM–1 mM), which are high-risk UA concentrations. The log-linear equation for low concentrations was calculated as $I_{LC} = 2825 \times \log(X) + 3132$ ($R^2 = 0.967$), and the linear equation for high concentrations was calculated as $I_{HC} = 35.59X + 3253$ ($R^2 = 0.9644$). Specifically, X denotes the concentration of UA (mM) in serum. The LOD of UA in serum was determined to be 1.05 μM using the LOD formula. The results indicate a decrease in sensitivity compared to the laboratory environment because various proteins in the serum act as interfering factors. However, despite this, the AF-SERS substrate exhibited detection performance, which far exceeded the normal range.

Following this, urine samples were prepared using synthetic human urine without UA (Figure 5d). Uric acid concentrations in the urine of healthy individuals typically range from 160 to 320 μM [47]. Therefore, we prepared a sample reflecting normal conditions by dissolving 300 μM of uric acid, within the normal range, in artificial urine. Additionally, urine under UA deficiency conditions was diluted to a concentration of 30 μM (1/10 of the normal range), and UA excess conditions were produced by adding UA to 3 mM (10 times beyond the normal range). Figure 5e presents the Raman spectra results for each urine sample. Artificial urine contains not only UA but also various proteins, especially urea. Furthermore, the presence of a Raman-specific peak (1000 cm^{-1}) of urea was confirmed [48]. Among these, a specific peak related to UA appears at 640 cm^{-1}, signifying that it is possible to detect normal UA levels, as well as deficiency and excess states, as shown in Figure 5f. Our UA detection sensor encompasses a broad spectrum within the normal range of uric acid, as illustrated in Figure S9, and in fact boasts a wider coverage compared to conventional detection range [49]. This level of detection capability, which identifies deficiencies and excesses based on UA concentration in urine samples, is also applicable to real urine samples in the future and is deemed suitable for potential application in future real-world field scenarios.

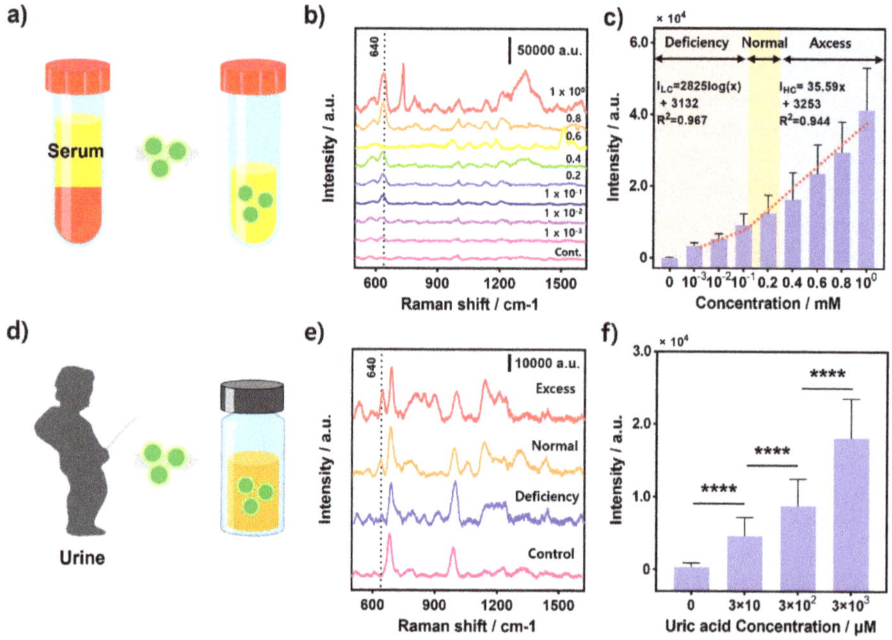

Figure 5. UA detection using the AF-SERS substrate in human serum and artificial urine. (**a**) Schematic of UA sampling in 10% diluted human serum. (**b**) SERS spectra and (**c**) SERS peak intensities at 640 cm^{-1} of 10% diluted human serum in UA in the concentration range of 1 μM–1 mM. (**d**) Schematic of UA sampling in urine. (**e**) SERS spectra and (**f**) SERS peak intensities at 640 cm^{-1} of various UA concentration ranges (deficiency, normal, excess) in urine. Each detection stage can be distinctly differentiated statistically (**** p-value < 0.0001).

4. Conclusions

We developed an AF-SERS substrate for detecting UA by mimicking the ecological structure of a type of starfish. The leg and protrusion morphology of *Asterias forbesi* exhibited a highly enhanced Raman signal, and simulations confirmed that its structure possessed numerous hotspots. The AF-SERS substrate not only exhibited excellent SERS performance but also minimized oxidation-related damage, making it suitable for practical applications. Moreover, we proved the high repeatability and uniformity of the AF-SERS substrate for UA detection. Using the proposed AF-SERS substrate, we successfully detected UA in a laboratory environment down to a concentration of 1.16 nM and demonstrated selectivity against various metabolites. For real-world application of the sensor, we detected UA in human serum and actual urine samples. In serum, we could differentiate UA concentrations beyond the normal range, covering deficiencies and excesses, with a LOD of 1.05 μM. Similarly, we successfully detected UA in artificial urine samples corresponding to deficiency, normal, and excess states. In conclusion, our research confirms that the proposed AF-SERS substrate enables highly sensitive UA detection in real samples. Our findings suggest its potential as a practical detection platform for real-world UA applications.

Supplementary Materials: The following supporting information can be downloaded at: https://www.mdpi.com/article/10.3390/bios14010008/s1. Figure S1: Optical images of a (a) SLNS-Ag and (b) SLNS-Au@Ag substrates; Figure S2: SEM-EDS data of AF-SESR substrate; Figure S3: (a) Raman spectrum of 100 μM R6G on only GNP and AF-SERS substrate respectively. (b) Raman intensity graph at 1508 cm^{-1}, the specific Raman peak of R6G. (Scale bar: 500 nm); Figure S4: Raman spectral data of Au plate and SLNS-Ag substrate for R6G 100 μM. Upon reducing the y-axis scale in a specific region, the Raman signal of silver was identified at around 1000 cm^{-1} (indicated by the red square); Figure S5: (a) Raman spectrum of 100 μM R6G on only GNP and AF-SERS substrate respectively. (b) Raman intensity graph at 1508cm-1, the specific Raman peak of R6G; Figure S6: FEM-based electromagnetic simulation results for Au plate and RGB ratio spectrum of each area; Figure S7: Raman spectra of R6G for each day on AF-SERS substrate when exposed to harsh conditions (PBS buffer solution) for 7 days; Figure S8: Raman spectra of R6G for each day on GNP + SLNS-Ag substrate when exposed to harsh conditions (PBS buffer solution) for 7 days; Figure S9: A graph comparing the excretion concentration range of uric acid in the body using conventional technology, the normal range of uric acid, and the detection range of uric acid using AF-SERS substrate; Table S1: Comparison of enhancement factor performance of SERS substrates using various nanomaterial [37,44,50–53]; Table S2: Vibrational SERS band assignments for Uric acid [54,55]; Table S3: Comparison of the SERS sensing performances of our AF-SERS and other various sensing platform about UA [25,56–63].

Author Contributions: H.P., Visualization, Data Curation, and Writing—Original Draft; K.C., Methodology and Investigation; W.K., Conceptualization and Validation; J.P. (Joohyung Park), Formal Analysis; W.L., Funding Acquisition and Project Administration; J.P. (Jinsung Park), Resources, Supervision, and Writing—Review and Editing. All authors have read and agreed to the published version of the manuscript.

Funding: This research was funded by the National Research Foundation of Korea (NRF) under Grant No. NRF-2023R1A2C2004964, NRF-2022R1I1A1A01066196, NRF-2022R1A2C4001990, and RS-2023-00222737. And Regional Innovation Strategy (RIS) through the NRF By the Ministry of Education (MOE) (2021RIS-001).

Institutional Review Board Statement: Not applicable.

Informed Consent Statement: Not applicable.

Data Availability Statement: Data are contained within the article and Supplementary Materials.

Conflicts of Interest: The authors declare no conflict of interest.

References

1. Maiuolo, J.; Oppedisano, F.; Gratteri, S.; Muscoli, C.; Mollace, V. Regulation of Uric Acid Metabolism and Excretion. *Int. J. Cardiol.* **2015**, *213*, 8–14. [CrossRef] [PubMed]
2. Lytvyn, Y.; Perkins, B.A.; Cherney, D.Z.I. Uric Acid as a Biomarker and a Therapeutic Target in Diabetes. *Can. J. Diabetes* **2015**, *39*, 239–246. [CrossRef] [PubMed]
3. Azmi, N.E.; Ramli, I.; Abdullah, J.; Azmi, M.; Hamid, A.; Sidek, H.; Rahman, S.A.; Ariffin, N.; Yusof, A. A Simple and Sensitive Fluorescence Based Biosensor for the Determination of Uric Acid Using H_2O_2-Sensitive Quantum Dots/Dual Enzymes. *Biosens. Bioelectron.* **2014**, *67*, 129–133. [CrossRef] [PubMed]
4. Kumar, V.; Gill, K.D. To Determine the Uric Acid Concentration in Serum and Urine. In *Basic Concepts in Clinical Biochemistry: A Practical Guide*; Springer: Berlin/Heidelberg, Germany, 2018; pp. 81–84. [CrossRef]
5. Miake, J.; Hisatome, I.; Tomita, K.; Isoyama, T.; Sugihara, S.; Kuwabara, M.; Ogino, K.; Ninomiya, H. Impact of Hyper- and Hypo-Uricemia on Kidney Function. *Biomedicines* **2023**, *11*, 1258. [CrossRef] [PubMed]
6. Mccarty, D.J.; Hollander, J.L. Identification of Urate Crystals in Gouty Synovial Fluid. *Ann. Intern. Med.* **1961**, *54*, 452–460. [CrossRef]
7. Merriman, T.R.; Dalbeth, N. The Genetic Basis of Hyperuricaemia and Gout. *Jt. Bone Spine* **2011**, *78*, 35–40. [CrossRef]
8. Saito, Y.; Tanaka, A.; Node, K.; Kobayashi, Y. Uric Acid and Cardiovascular Disease: A Clinical Review. *J. Cardiol.* **2021**, *78*, 51–57. [CrossRef]
9. Kanbay, M.; Jensen, T.; Solak, Y.; Le, M.; Roncal-Jimenez, C.; Rivard, C.; Lanaspa, M.A.; Nakagawa, T.; Johnson, R.J. Uric Acid in Metabolic Syndrome: From an Innocent Bystander to a Central Player. *Eur. J. Intern. Med.* **2016**, *29*, 3–8. [CrossRef]
10. Bainbridge, S.A.; Roberts, J.M. Uric Acid as a Pathogenic Factor in Preeclampsia. *Placenta* **2008**, *29*, 67–72. [CrossRef]
11. Grassi, D.; Ferri, L.; Desideri, G.; Di Giosia, P.; Cheli, P.; Del Pinto, R.; Properzi, G.; Ferri, C. Chronic Hyperuricemia, Uric Acid Deposit and Cardiovascular Risk. *Curr. Pharm. Des.* **2013**, *19*, 2432–2438. [CrossRef]

12. Mazzara, F.; Patella, B.; Aiello, G.; O'Riordan, A.; Torino, C.; Vilasi, A.; Inguanta, R. Electrochemical Detection of Uric Acid and Ascorbic Acid Using R-GO/NPs Based Sensors. *Electrochim. Acta* **2021**, *388*, 138652. [CrossRef]
13. Martinez-Pérez, D.; Ferrer, M.L.; Mateo, C.R. A Reagent Less Fluorescent Sol–Gel Biosensor for Uric Acid Detection in Biological Fluids. *Anal. Biochem.* **2003**, *322*, 238–242. [CrossRef] [PubMed]
14. Norazmi, N.; Rasad, Z.R.A.; Mohamad, M.; Manap, H. Uric Acid Detection Using Uv-Vis Spectrometer. *IOP Conf. Ser. Mater. Sci. Eng.* **2017**, *257*, 012031. [CrossRef]
15. Chen, J.C.; Chung, H.H.; Hsu, C.T.; Tsai, D.M.; Kumar, A.S.; Zen, J.M. A Disposable Single-Use Electrochemical Sensor for the Detection of Uric Acid in Human Whole Blood. *Sens. Actuators B Chem.* **2005**, *110*, 364–369. [CrossRef]
16. Omar, M.N.; Salleh, A.B.; Lim, H.N.; Ahmad Tajudin, A. Electrochemical Detection of Uric Acid via Uricase-Immobilized Graphene Oxide. *Anal. Biochem.* **2016**, *509*, 135–141. [CrossRef] [PubMed]
17. Zheng, Q.; Xiong, L.; Yu, L.; Wu, D.; Yang, C.; Xiao, Y. An Enzyme-Free Fluorescent Sensing Platform for the Detection of Uric Acid in Human Urine. *J. Lumin.* **2021**, *236*, 118076. [CrossRef]
18. Das, S.C.; Pandey, R.R.; Devkota, T.; Chusuei, C.C. Raman Spectroscopy as an Assay to Disentangle Zinc Oxide Carbon Nanotube Composites for Optimized Uric Acid Detection. *Chemosensors* **2018**, *6*, 65. [CrossRef]
19. Pucetaite, M.; Velicka, M.; Pilipavicius, J.; Beganskiene, A.; Ceponkus, J.; Sablinskas, V. Uric Acid Detection by Means of SERS Spectroscopy on Dried Ag Colloidal Drops. *J. Raman Spectrosc.* **2016**, *47*, 681–686. [CrossRef]
20. Ansah, I.B.; Lee, W.C.; Mun, C.W.; Rha, J.J.; Jung, H.S.; Kang, M.; Park, S.G.; Kim, D.H. In Situ Electrochemical Surface Modification of Au Electrodes for Simultaneous Label-Free SERS Detection of Ascorbic Acid, Dopamine and Uric Acid. *Sens. Actuators B Chem.* **2022**, *353*, 131196. [CrossRef]
21. Manbohi, A.; Ahmadi, S.H. Sensitive and Selective Detection of Dopamine Using Electrochemical Microfluidic Paper-Based Analytical Nanosensor. *Sens. Biosens. Res.* **2019**, *23*, 100270. [CrossRef]
22. Murugan, N.; Jerome, R.; Preethika, M.; Sundaramurthy, A.; Sundramoorthy, A.K. 2D-Titanium Carbide (MXene) Based Selective Electrochemical Sensor for Simultaneous Detection of Ascorbic Acid, Dopamine and Uric Acid. *J. Mater. Sci. Technol.* **2021**, *72*, 122–131. [CrossRef]
23. Tavakkoli Yaraki, M.; Tukova, A.; Wang, Y. Emerging SERS Biosensors for the Analysis of Cells and Extracellular Vesicles. *Nanoscale* **2022**, *14*, 15242–15268. [CrossRef] [PubMed]
24. Weng, S.; Lin, D.; Lai, S.; Tao, H.; Chen, T.; Peng, M.; Qiu, S.; Feng, S. Highly Sensitive and Reliable Detection of MicroRNA for Clinically Disease Surveillance Using SERS Biosensor Integrated with Catalytic Hairpin Assembly Amplification Technology. *Biosens. Bioelectron.* **2022**, *208*, 114236. [CrossRef] [PubMed]
25. Lu, D.; Cai, R.; Liao, Y.; You, R.; Lu, Y. Two-Dimensional Glass/p-ATP/Ag NPs as Multifunctional SERS Substrates for Label-Free Quantification of Uric Acid in Sweat. *Spectrochim. Acta A Mol. Biomol. Spectrosc.* **2023**, *296*, 122631. [CrossRef] [PubMed]
26. Park, H.; Park, J.; Lee, G.; Kim, W.; Park, J. Detection of Chlorpyrifos Using Bio-Inspired Silver Nanograss. *Materials* **2022**, *15*, 3454. [CrossRef] [PubMed]
27. Bang, D.; Chang, Y.W.; Park, J.; Lee, T.; Park, J.; Yeo, J.S.; Kim, E.K.; Yoo, K.H.; Huh, Y.M.; Haam, S. One-Step Electrochemical Fabrication of Vertically Self-Organized Silver Nanograss. *J. Mater. Chem. A Mater.* **2013**, *1*, 4851–4857. [CrossRef]
28. Matikainen, A.; Nuutinen, T.; Itkonen, T.; Heinilehto, S.; Puustinen, J.; Hiltunen, J.; Lappalainen, J.; Karioja, P.; Vahimaa, P. Atmospheric Oxidation and Carbon Contamination of Silver and Its Effect on Surface-Enhanced Raman Spectroscopy (SERS). *Sci. Rep.* **2016**, *6*, 37192. [CrossRef] [PubMed]
29. Fan, M.; Lai, F.J.; Chou, H.L.; Lu, W.T.; Hwang, B.J.; Brolo, A.G. Surface-Enhanced Raman Scattering (SERS) from Au:Ag Bimetallic Nanoparticles: The Effect of the Molecular Probe. *Chem. Sci.* **2012**, *4*, 509–515. [CrossRef]
30. Kim, W.; Park, J.; Kim, W.; Jo, S.; Kim, M.; Kim, C.; Park, H.; Bang, D.; Lee, W.; Park, J. Bio-Inspired Ag Nanovilli-Based Sandwich-Type SERS Aptasensor for Ultrasensitive and Selective Detection of 25-Hydroxy Vitamin D3. *Biosens. Bioelectron.* **2021**, *188*, 113341. [CrossRef]
31. Kim, W.; Lee, W.; Park, H.; Park, J.; Kim, W.; Kang, B.; Choi, E.; Kim, C.S.; Park, J.O.; Lee, G.; et al. Biomimetic Nano-Pine-Pollen Structure-Based Surface-Enhanced Raman Spectroscopy Sensing Platform for the Hypersensitive Detection of Toxicants: Cadmium and Amyloid. *ACS Sustain. Chem. Eng.* **2022**, *10*, 3180–3190. [CrossRef]
32. Turkevich, J.; Stevenson, P.C.; Hillier, J. A Study of the Nucleation and Growth Processes in the Synthesis of Colloidal Gold. *Discuss. Faraday Soc.* **1951**, *11*, 55–75. [CrossRef]
33. Seo, M.; Ha, J.W. Effective Surface-Enhanced Raman Scattering of Randomly Branched Gold Nano-Urchins with Rhodamine 6G as Raman Reporters. *Microchem. J.* **2018**, *140*, 47–51. [CrossRef]
34. Ujihara, M.; Dang, N.M.; Imae, T. Surface-Enhanced Resonance Raman Scattering of Rhodamine 6G in Dispersions and on Films of Confeito-Like Au Nanoparticles. *Sensors* **2017**, *17*, 2563. [CrossRef] [PubMed]
35. Gualerzi, A.; Picciolini, S.; Meloni, M.; Lax, A.; Silani, V.; Bedoni, M. Human Salivary Raman Fingerprint as Biomarker for the Diagnosis of Amyotrophic Lateral Sclerosis. *Sci. Rep.* **2020**, *10*, 10175. [CrossRef]
36. Cheng, Z.Q.; Li, Z.W.; Yao, R.; Xiong, K.W.; Cheng, G.L.; Zhou, Y.H.; Luo, X.; Liu, Z.M. Improved SERS Performance and Catalytic Activity of Dendritic Au/Ag Bimetallic Nanostructures Based on Ag Dendrites. *Nanoscale Res. Lett.* **2020**, *15*, 117. [CrossRef] [PubMed]
37. Lin, X.; Li, L.J.; Guo, H.Y.X.; Li, R.; Feng, J. Preparation of 3D Nano Silver Trees/Sea Urchin-like Gold and SERS Detection of Uric Acid. *Spectrochim. Acta A Mol. Biomol. Spectrosc.* **2024**, *305*, 123464. [CrossRef]

38. Jia, F.; Barber, E.; Turasan, H.; Seo, S.; Dai, R.; Liu, L.; Li, X.; Bhunia, A.K.; Kokini, J.L. Detection of Pyocyanin Using a New Biodegradable SERS Biosensor Fabricated Using Gold Coated Zein Nanostructures Further Decorated with Gold Nanoparticles. *J. Agric. Food Chem.* **2019**, *67*, 4603–4610. [CrossRef] [PubMed]
39. Atta, S.; Vo-Dinh, T. Ultra-Trace SERS Detection of Cocaine and Heroin Using Bimetallic Gold–Silver Nanostars (BGNS-Ag). *Anal. Chim. Acta* **2023**, *1251*, 340956. [CrossRef]
40. Lu, H.; Jin, M.; Ma, Q.; Yan, Z.; Liu, Z.; Wang, X.; Akinoglu, E.M.; van den Berg, A.; Zhou, G.; Shui, L. Ag Nano-Assemblies on Si Surface via CTAB-Assisted Galvanic Reaction for Sensitive and Reliable Surface-Enhanced Raman Scattering Detection. *Sens. Actuators B Chem.* **2020**, *304*, 127224. [CrossRef]
41. Wen, H.; Inose, T.; Hirai, K.; Akashi, T.; Sugioka, S.; Li, J.; Peeters, W.; Fron, E.; Fortuni, B.; Nakata, Y.; et al. Gold-Coated Silver Nanowires for Long Lifetime AFM-TERS Probes. *Nanoscale* **2022**, *14*, 5439–5446. [CrossRef]
42. Le Ru, E.C.; Blackie, E.; Meyer, M.; Etchegoint, P.G. Surface Enhanced Raman Scattering Enhancement Factors: A Comprehensive Study. *J. Phys. Chem. C* **2007**, *111*, 13794–13803. [CrossRef]
43. Yan, Q.; Zhi, N.; Yang, L.; Xu, G.; Feng, Q.; Zhang, Q.; Sun, S. A Highly Sensitive Uric Acid Electrochemical Biosensor Based on a Nano-Cube Cuprous Oxide/Ferrocene/Uricase Modified Glassy Carbon Electrode. *Sci. Rep.* **2020**, *10*, 10607. [CrossRef]
44. Tan, Y.; Qi, M.; Jiang, H.; Wang, B.; Zhang, X. Determination of Uric Acid in Serum by SERS System Based on VO-MnCo2O4/Ag Nanozyme. *Anal. Chim. Acta* **2023**, *1274*, 341584. [CrossRef]
45. Baqi, O.; Shatery, A.; Omer, K.M. Selectivity Enhancement for Uric Acid Detection via In Situ Preparation of Blue Emissive Carbon Dots Entrapped in Chromium Metal—Organic Frameworks. *ACS Omega* **2022**, *7*, 16576–16583. [CrossRef]
46. Putra, B.R.; Nisa, U.; Heryanto, R.; Khalil, M.; Khoerunnisa, F.; Ridhova, A.; Thaha, Y.N.; Marken, F.; Wahyuni, W.T. Selective Non-Enzymatic Uric Acid Sensing in the Presence of Dopamine: Electropolymerized Poly-Pyrrole Modified with a Reduced Graphene Oxide/PEDOT:PSS Composite. *Analyst* **2022**, *147*, 5334–5346. [CrossRef]
47. Pang, X.; Yan, R.; Li, L.; Wang, P.; Zhang, Y.; Liu, Y.; Liu, P.; Dong, W.; Miao, P.; Mei, Q. Non-Doped and Non-Modified Carbon Dots with High Quantum Yield for the Chemosensing of Uric Acid and Living Cell Imaging. *Anal. Chim. Acta* **2022**, *1199*, 339571. [CrossRef]
48. Wang, W.; Chen, Y.; Xiao, C.; Xiao, S.; Wang, C.; Nie, Q.; Xu, P.; Chen, J.; You, R.; Zhang, G.; et al. Flexible SERS Wearable Sensor Based on Nanocomposite Hydrogel for Detection of Metabolites and PH in Sweat. *Chem. Eng. J.* **2023**, *474*, 145953. [CrossRef]
49. Ruggiero, C.; Cherubini, A.; Ble, A.; Bos, A.J.G.; Maggio, M.; Dixit, V.D., Lauretani, F.; Bandinelli, S.; Senin, U.; Ferrucci, L. Uric Acid and Inflammatory Markers. *Eur. Heart J.* **2006**, *27*, 1174–1181. [CrossRef]
50. Huang, C.Y.; Hsiao, H.C. Integrated EC-SERS Chip with Uniform Nanostructured EC-SERS Active Working Electrode for Rapid Detection of Uric Acid. *Sensors* **2020**, *20*, 7066. [CrossRef]
51. Bhattacharjee, G.; Majumder, S.; Senapati, D.; Banerjee, S.; Satpati, B. Core-Shell Gold @silver Hollow Nanocubes for Higher SERS Enhancement and Non-Enzymatic Biosensor. *Mater. Chem. Phys.* **2020**, *239*, 122113. [CrossRef]
52. Li, J.; Cui, X.; Yang, X.; Qiu, Y.; Li, Y.; Cao, H.; Wang, D.; He, W.; Feng, Y.; Yang, Z. Quantification of Uric Acid Concentration in Tears by Using PDMS Inverse Opal Structure Surface-Enhanced Raman Scattering Substrates: Application in Hyperuricemia. *Spectrochim. Acta A Mol. Biomol. Spectrosc.* **2022**, *278*, 121326. [CrossRef]
53. Wang, K.; Chen, Z.; Li, Y.; Zhang, Y. Top-down Produced CdSe Quantum Dots as an Ultrasensitive SERS Platform for the Detection of Uric Acid. *Mater. Chem. Front.* **2023**, *7*, 1624–1632. [CrossRef]
54. Westley, C.; Xu, Y.; Carnell, A.J.; Turner, N.J.; Goodacre, R. Label-Free Surface Enhanced Raman Scattering Approach for High-Throughput Screening of Biocatalysts. *Anal Chem.* **2016**, *88*, 5898–5903. [CrossRef]
55. Goodall, B.L.; Robinson, A.M.; Brosseau, C.L. Electrochemical- Surface Enhanced Raman Spectroscopy (E-SERS) of Uric Acid: A Potential Rapid Diagnostic Method for Early Preeclampsia Detection. *Phys. Chem. Chem. Phys.* **2013**, *15*, 1382–1388. [CrossRef]
56. Kokulnathan, T.; Wang, T.J.; Kumar, E.A.; Duraisamy, N. An-Ting Lee An Electrochemical Platform Based on Yttrium Oxide/Boron Nitride Nanocomposite for the Detection of Dopamine. *Sens. Actuators B Chem.* **2021**, *349*, 130787. [CrossRef]
57. Zhu, X.; Xuan, L.; Gong, J.; Liu, J.; Wang, X.; Xi, F.; Chen, J. Three-Dimensional Macroscopic Graphene Supported Vertically-Ordered Mesoporous Silica-Nanochannel Film for Direct and Ultrasensitive Detection of Uric Acid in Serum. *Talanta* **2022**, *238*, 123027. [CrossRef]
58. Wang, H.; Xie, A.; Li, S.; Wang, J.; Chen, K.; Su, Z.; Song, N.; Luo, S. Three-Dimensional g-C_3N_4/MWNTs/GO Hybrid Electrode as Electrochemical Sensor for Simultaneous Determination of Ascorbic Acid, Dopamine and Uric Acid. *Anal Chim. Acta* **2022**, *1211*, 339907. [CrossRef]
59. Zhao, X.Y.; Yang, Q.S.; Wang, J.; Fu, D.L.; Jiang, D.K. A Novel 3D Coordination Polymer Constructed by Dual-Ligand for Highly Sensitive Detection of Purine Metabolite Uric Acid. *Spectrochim. Acta A Mol Biomol. Spectrosc.* **2021**, *262*, 120065. [CrossRef]
60. Li, F.; Chen, J.; Wen, J.; Peng, Y.; Tang, X.; Qiu, P. Ratiometric Fluorescence and Colorimetric Detection for Uric Acid Using Bifunctional Carbon Dots. *Sens. Actuators B Chem.* **2022**, *369*, 132381. [CrossRef]
61. Sumalatha, V.; Anujya, C.; Balchander, V.; Dhanalaxmi, B.; Pradeep Kumar, M.; Ayodhya, D. Hydrothermal Fabrication of N-CeO_2/p-CuS Heterojunction Nanocomposite for Enhanced Photodegradation of Pharmaceutical Drugs in Wastewater under Visible-Light and Fluorometric Sensor for Detection of Uric Acid. *Inorg. Chem. Commun.* **2023**, *155*, 110962. [CrossRef]

62. Jiang, L.; Wang, L.; Zhan, D.S.; Jiang, W.R.; Fodjo, E.K.; Hafez, M.E.; Zhang, Y.M.; Zhao, H.; Qian, R.C.; Li, D.W. Electrochemically Renewable SERS Sensor: A New Platform for the Detection of Metabolites Involved in Peroxide Production. *Biosens. Bioelectron.* **2021**, *175*, 112918. [CrossRef] [PubMed]
63. Durai, L.; Badhulika, S. A Wearable PVA Film Supported TiO_2 Nanoparticles Decorated $NaNbO_3$ Nanoflakes-Based SERS Sensor for Simultaneous Detection of Metabolites and Biomolecules in Human Sweat Samples. *Adv. Mater. Interfaces* **2022**, *9*, 2200146. [CrossRef]

biosensors MDPI

Article

Complex Spatial Illumination Scheme Optimization of Backscattering Mueller Matrix Polarimetry for Tissue Imaging and Biosensing

Wei Jiao [1,†], Zheng Zhang [1,†], Nan Zeng [1], Rui Hao [1], Honghui He [1,*], Chao He [2,*] and Hui Ma [1]

1 Guangdong Research Center of Polarization Imaging and Measurement Engineering Technology, Shenzhen Key Laboratory for Minimal Invasive Medical Technologies, Institute of Biopharmaceutical and Health Engineering, Tsinghua Shenzhen International Graduate School, Tsinghua University, Shenzhen 518055, China; jiaow23@mails.tsinghua.edu.cn (W.J.); zhangzhe21@mails.tsinghua.edu.cn (Z.Z.); zengnan@sz.tsinghua.edu.cn (N.Z.); haor22@mails.tsinghua.edu.cn (R.H.); mahui@tsinghua.edu.cn (H.M.)

2 Department of Engineering Science, University of Oxford, Parks Road, Oxford OX1 3PJ, UK

* Correspondence: he.honghui@sz.tsinghua.edu.cn (H.H.); chao.he@eng.ox.ac.uk (C.H.)

† These authors contributed equally to this work.

Abstract: Polarization imaging and sensing techniques have shown great potential for biomedical and clinical applications. As a novel optical biosensing technology, Mueller matrix polarimetry can provide abundant microstructural information of tissue samples. However, polarimetric aberrations, which lead to inaccurate characterization of polarization properties, can be induced by uneven biomedical sample surfaces while measuring Mueller matrices with complex spatial illuminations. In this study, we analyze the detailed features of complex spatial illumination-induced aberrations by measuring the backscattering Mueller matrices of experimental phantom and tissue samples. We obtain the aberrations under different spatial illumination schemes in Mueller matrix imaging. Furthermore, we give the corresponding suggestions for selecting appropriate illumination schemes to extract specific polarization properties, and then provide strategies to alleviate polarimetric aberrations by adjusting the incident and detection angles in Mueller matrix imaging. The optimized scheme gives critical criteria for the spatial illumination scheme selection of non-collinear backscattering Mueller matrix measurements, which can be helpful for the further development of quantitative tissue polarimetric imaging and biosensing.

Keywords: polarimetry; Mueller matrix; backscattering imaging; spatial illumination; bio-imaging and sensing

1. Introduction

Polarization techniques have been widely applied to biomedical imaging and sensing with their label-free, non-invasive, and microstructure-sensitive advantages [1–3]. Mueller matrix (MM) polarimetry is more and more prevalently used nowadays, as it encodes numerous polarization properties comprehensively [4–6]. To quantitatively characterize the polarization properties such as diattenuation, retardance, and depolarization, polarization basic parameters (PBPs) are conventionally obtained using Mueller Matrix Polar Decomposition [7], Mueller Matrix Transformation [2], and other MM decomposition methods [8–10]. PBPs can increase the image contrast of specific tissue structures with characteristic properties, such as size, shape, fiber orientation, and alignment, thus enhancing polarimetric biosensing [2]. The primary techniques for MM measurement can be categorized as transmission MM microscopy [11] and backscattering MM polarimetry [12,13]. Thin tissue sample (with a thickness less than 3×10^1 μm) measurements can be taken using transmission microscopy, while bulk tissue sample (with a thickness larger than 3×10^3 μm) measurements require backscattering polarimetry, such as bio-structural and optical properties sensing [14,15], polarimetric endoscopy [16,17], minimally invasive surgery [5], and skin tissue analysis [18–20] for in vivo scenarios. Despite the fact that backscattering MM polarimetry has great potential for various applications, there are still largely unexplored questions. The precision of MM polarimetry can be affected by many factors, including the low signal-to-noise ratio induced by the light field [21], the azimuthal dependence of the data obtained [22], and other measurement factors [2]. One of these is the uncertain incidence and detection angles resulting from uneven biomedical tissue surfaces, namely the surfaces of internal organs with complicated topography [13].

In general, the backscattering MM polarimetric setup consists of a polarization state generator (PSG) and a polarization state analyzer (PSA), which are located on the same side of the sample. The angle formed between the PSG (or PSA) and the normal of the tissue surface is named incidence angle θ (or detection angle ζ), and the angle formed between the PSG and PSA is named absolute spatial angle ψ, which is the sum of θ and ζ. There are four backscattering polarimetry schemes, including the following: (i) normal incidence with normal emergent light; (ii) oblique incidence with normal emergent light; (iii) normal incidence with oblique emergent light; (iv) oblique incidence with oblique emergent light. The complex spatial illumination scheme refers to the spatial relationship between the PSG, PSA, and sample, including the four different schemes above. Ideal collinear reflection MM measurement, as in scheme (i), can reduce the amount of measuring errors by avoiding the orientation effects [23]. However, it requires a non-polarizing beam splitter to ensure the emergent light is colinear with incidence, which increases the cost of the device, induces cumulative errors after frequent use, and asks for an intricate calibration [2]. On the other hand, uneven biomedical tissue surfaces make it extremely difficult to take this way. Conversely, the non-linear reflection approach, although inducing orientation effects, can effectively overcome the above problems [13]. Therefore, non-colinear reflection MM measurement equipment has been increasingly used in biomedical studies and applications.

As an optical method, backscattering polarimetry can be used to detect chemical compounds in tissues or cells by optical signals as part of a biosensor, together with other techniques such as surface plasmon resonance [24,25]. For instance, biosensors based on Mueller matrix measurements have been applied to the detection of miRNA [24], bovine serum albumin [25], dengue virus, and glucose [26–28]. Specifically, a decomposition Mueller matrix polarimetry was proposed for detecting miRNA [24], showing that the polarization parameter could be used as a quantitative sensing index of chemical compounds. A cutting-edge biosensing method that tracks the binding reaction between bovine serum albumin and its antibodies by measuring the phase difference between p- and s-polarization was demonstrated [25]. Moreover, Mueller matrix polarimetric techniques were also used to sense the glucose concentration in aqueous solutions or detect dengue virus [26–28]. However, in the above-mentioned works, we can hardly find a clear criterion of the spatial illumination angle selection, which may have a potential influence on the polarimetric measurement. Both the incidence and detection angles of non-collinear backscattering polarimetric devices are inconsistent in different Mueller matrix polarimetry-based biosensors. For instance, the detection angle was 80° in [24], and the absolute space angle was 60° in [26]. Therefore, we believe that analyzing the impact of complex spatial illumination on backscattering Mueller matrix polarimetry is crucial, which is the focus of this study.

Some recent works have demonstrated that non-colinear reflection can have a complex impact on MM measurement [2]. For instance, the polarimetric aberrations in the MM elements and PBPs at different θ, as in scheme (ii), have been analyzed previously [13]. In this study, we analyze the detailed features of complex spatial illumination-induced aberrations for backscattering MM. We measure the MMs of an anisotropic silk phantom designed previously [13] and porcine liver tissue to explore the influence of the different schemes (ii)−(iv), on MM imaging. The experimental results demonstrate that different polarimetric aberrations can be induced in two-periodic MM elements, as in schemes (ii) and (iii), which are verified using the mean-square error (MSE) and energy spectral density. Furthermore, we find that when measuring MMs with an adjusting distribution of θ and ζ, as in scheme (iv), the aberrations can be significantly alleviated. Additionally, as ζ increases, obvious image distortions can be observed, performing as transverse compressions parallel to the incident plane. The relationship between ζ and the extent of distortions is analyzed using linear regression (LR), and the MM elements can be reconstructed accordingly. Based on the analysis, this study gives corresponding suggestions for selecting appropriate polarimetric schemes to extract specific polarization properties, which can be helpful for the further development of quantitative tissue polarimetric imaging and biosensing.

2. Materials and Methods

2.1. Experimental Setup and Tissue Samples

In this study, we adopt an experimental setup for 3 × 3 backscattering MM imaging based on a division of focal plane (DoFP) camera [29–31]. As shown in Figure 1a, the monochromatic light emitted from the LED (3 W, 633 nm, $\Delta\lambda$ = 20 nm, Daheng Optic, Beijing, China) is modulated by the PSG consisting of a collimating objective lens L1 (Hengyang Optic, Guangzhou, China) and three fixed polarizers P1 (extinction ratio > 1000:1, LBTEK Optic, Changsha, China), and then scattered by a tissue sample. The P1 is driven by a screw linear motor M (FSK30, FUYU Technology, Chengdu, China) to generate 0°, 45°, and 90° linear states of polarization illuminations in series during the measurement. The scattered photons from the tissue samples are captured by the DoFP camera (PHX050S-P, Lucid Vision Labs, Richmond, BC, Canada) after passing through the imaging objective L2 (Hengyang Optic, Guangzhou, China). The DoFP camera contains numerous units capable of autonomously extracting intensity information from four different linear polarizers within different emergent lights. As illustrated in Figure 1b, by utilizing the grayscale images, we can construct the sample's MMs and further analyze them by extracting the corresponding PBPs, which characterize the polarization properties of the tissue sample.

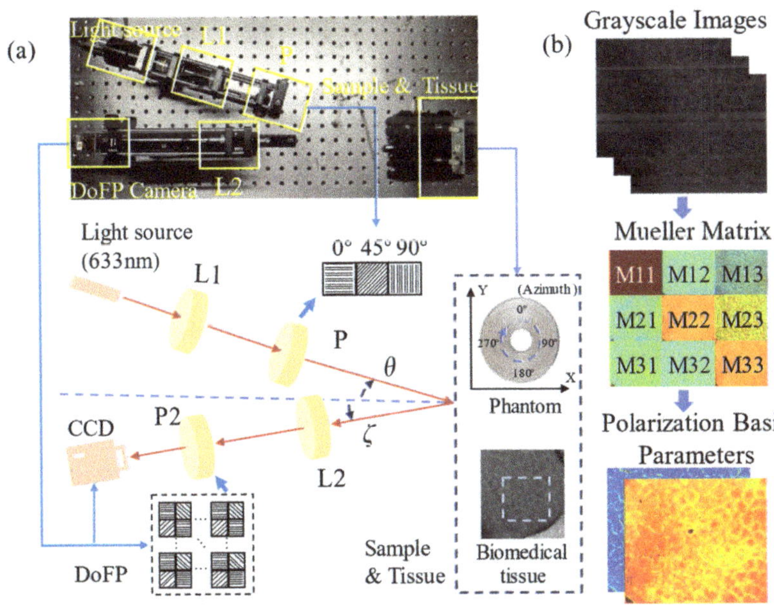

Figure 1. Schematics of the experimental setup and samples: (**a**) Backscattering MM setup using a DoFP camera. L1 and L2, lenses; P1 and P2, polarizers; θ, the oblique incidence angle; ζ, the oblique detection angle; sample and tissue, concentrically aligned silk phantom, and porcine liver tissue; (**b**) flowchart of polarization parameter acquisition based on the 3 × 3 MMs.

Here, we adopt an anisotropic scattering phantom consisting of concentrically aligned silk fibers, as shown in Figure 1a, which can generate the MMs of cylindrical scatterers along all azimuthal directions in the imaging X−Y plane in a single measurement [4]. Additionally, we utilize the porcine liver tissue as an ex vivo experimental sample for validation. This animal experimentation work was approved by the Ethics Committee of Tsinghua Shenzhen International Graduate School, Tsinghua University.

Prior to measurement, calibration of each optical component is conducted using a polarimeter (PAX1000, Thorlabs, Newton, NJ, USA) to ensure that systematic errors are maintained within 1%. During measurement, a constant distance between the PSA (or DoFP camera) and the tissue surface is rigorously maintained to minimize the errors arising from focal length variations. Meanwhile, the angles between the center line and both the PSG and sample arms are systematically adjusted to modulate θ and ζ as indicated in Figure 1a. The 3 × 3 MMs of the sample are reconstructed according to Equations (1) and (2), where $DoFP$, L_2, P, and L_1 correspond to the 3 × 3 MMs of the optical components, S_{in} represents the input Stokes vector, and S_{out} represents the output Stokes vector [13].

$$S_{out} = DoFP \times L_2 \times MM_S \times P \times L_1 \times S_{in}, \tag{1}$$

$$MM_S = DoFP^{-1} \times S_{out} \times S_{in}{}^{-1} \times P^{-1}. \tag{2}$$

2.2. Azimuthal-Dependent Curves of Mueller Matrix Elements

To comprehensively illustrate the aberrations in MM elements induced by complex spatial illumination, the azimuthal-dependent curves of nine MM elements are constructed according to the method described in our previous work [13]. As shown in Figure 2a,b, we maintain ζ (or θ) at 0° and systematically modulate θ (or ζ) from 0 to 40° in steps of 10°, respectively. Subsequently, the acquired curves are shown in Figure 2d,e. Upon comparing the two groups of curves, we can evidently notice that the curves of M12, M21, M13, and M31 exhibit opposite variations with increasing θ or ζ, while the curves of the other MM elements have identical variations with increasing θ or ζ. Additionally, as depicted in Figure 2f, we measure the azimuthal-dependent curves by simultaneously modifying both θ and ζ, shown in Figure 2c. These curves maintain the identical ψ and illustrate the influence of different distributions of θ and ζ on the aberrations. A detailed analysis will be provided in Section 3.

2.3. Analysis Methods

For quantitative analysis, we define the MMs measured when ψ is 10° and θ or ζ are equal to 0° as the referenced MMs. The MSEs between the curves for the other θ or ζ and the corresponding reference curves are calculated as Equation (3) for the two-periodic elements M12, M21, M13, and M31, respectively. In Equation (3), $C(M_j)$ represents the calculated azimuthal-dependent curve, $C(M_{ref})$ represents the corresponding reference curve, and N represents the point number in the curves. In Figure 2d,e, the M12 and M21 curves exhibit a noticeable shift along the vertical axis. Thus, we calculate the direct current components of energy spectral density for different curves.

$$MSE(C(M_j),\ C(M_{ref})) = \frac{1}{N}\sum_{i=1}^{N}\left\|C(M_j)_i - C(M_{ref})_i\right\|^2. \tag{3}$$

Since the anisotropic silk fiber phantom is circularly shaped, we introduce the ellipticity as a parameter to characterize the image distortions induced by the increasing ζ, as shown in Equation (4), where y and x are the longitudinal and transverse distances of the phantom and r is the radius of the phantom. The analytical relationship between ζ and the ellipticity is also shown in Equation (4).

$$Ellipticity = \frac{y-x}{y} = \frac{r - r\cos\varsigma}{r} = 1 - \cos\varsigma. \tag{4}$$

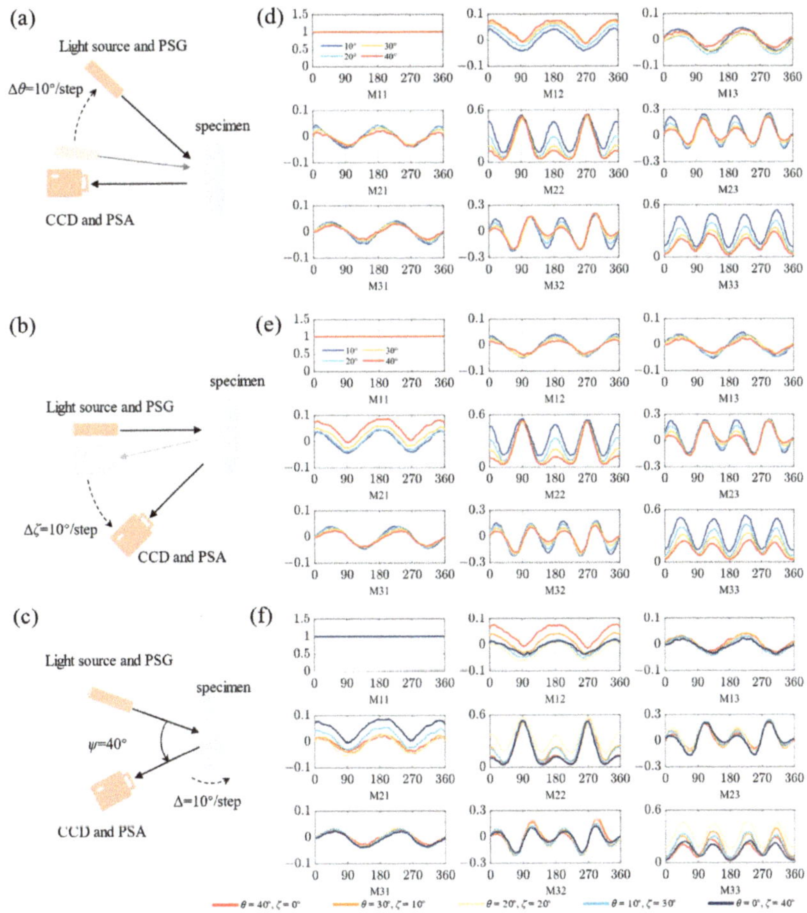

Figure 2. Azimuthal-dependent curves of 3 × 3 MM elements constructed as schemes (ii)−(iv): (**a**) incident angle θ modulated; (**b**) detection angle ζ modulated; (**c**) absolute spatial angle ψ modulated. The corresponding curves: (**d**) incident angle θ modulated; (**e**) detection angle ζ modulated; (**f**) absolute spatial angle ψ modulated. The horizontal axis shows the azimuth angle of the fibers, and the vertical axis represents the MM element values. All the elements are normalized by M11.

The ellipticity results are analyzed through the LR and least squares method in order to validate the correctness of the analytical relationship. In Equations (5) and (6), x is the deg of θ or ζ, y is the ellipticity of corresponding x, $\overline{x}$ and $\overline{y}$ are the means of x and y, and n is the data number.

$$Variable = \frac{\sum xy - n\overline{xy}}{\sum x^2 - n\overline{x}^2}, \tag{5}$$

$$Intercept = \overline{y} - Variable \cdot \overline{x}. \tag{6}$$

To validate the significant alleviation of aberrations in PBPs achieved by measuring MMs with an adjusting distribution of θ and ζ, we characterize the depolarization of the tissue sample by measuring the *LDoP* (linear degree of polarization) and b_2 (linear depolarization and anisotropies) parameters [32], which are sensitive to the oblique incidence [13,33]. They can be calculated as Equations (7)–(9).

$$b = \frac{M22 + M33}{2}, \tag{7}$$

$$b_2 = 1 - b, \tag{8}$$

$$LDoP = \frac{M21 + M22}{M11 + M12}. \tag{9}$$

We construct the frequency distribution histograms (FDHs) of the *LDoP* and b_2 parameters, which can systematically characterize the distribution of the overall polarization properties with PBP images. Several image distance measurements, such as the histogram distance (*HD*), Euclidean norm, and Structure Similarity Index Measure (*SSIM*) [34], are utilized to evaluate the alleviation effects quantitatively. Meanwhile, these measurements can also be used to identify the structural information as different schemes. The formulas for these measurements can be described as Equations (10) and (11),

$$HD(x,y) = MSE(histogram(x), histogram(y)), \tag{10}$$

$$SSIM(x,y) = \frac{(2\mu_x\mu_y + C_1)(2\sigma_{xy} + C_2)}{(\mu_x^2 + \mu_y^2 + C_1)(\sigma_x^2 + \sigma_y^2 + C_2)}, s.t. C_1 = (0.01 \times L)^2,\ C_2 = (0.03 \times L)^2, \tag{11}$$

where μ_x, μ_y, σ_x, σ_y, and σ_{xy} are the local means, standard deviations, and cross-covariance for MM images x, y, and $L = 1$ for MM images.

3. Results

3.1. Aberrations Induced by Complex Spatial Illumination of Two-Periodic MM Elements

It is widely recognized that the D_L derived from M12 and M13 represents the linear diattenuation, while the P_L derived from M21 and M31 represents the linear polarizance of the tissue sample [32]. D_L and P_L can be calculated as Equations (12) and (13).

$$D_L = \sqrt{M12^2 + M13^2}, \tag{12}$$

$$P_L = \sqrt{M21^2 + M31^2}. \tag{13}$$

Previously, we demonstrated that the symmetries of the M12 and M21 pairs, together with the M13 and M31 pairs, can be broken when θ is more than 20° [13]. In this work, we measure MMs and construct the azimuthal-dependent curves of the anisotropic silk phantom, as in schemes (ii) and (iii), as shown in Figures 2d and 2e, respectively. The azimuthal-dependent curve values of M12, M21, M13, and M31 are within the range of -0.1 and 0.1 when θ and ζ are less than 20°. As θ increases, the M12 curve progressively shifts towards positive, while the M21 and M31 curves exhibit amplitude variations but maintain both positive and negative values. Similarly, as ζ increases, the M21 curve shifts towards positive, while the M12 and M13 curves have variations in amplitude but still exist as positive and negative values. Furthermore, these can be more clearly seen in the MM images of the phantom in Figure 3a,c. It should be noted that the curves' variations of other MM elements are identical, as in both schemes (ii) and (iii), like period degeneracy in four-periodic elements. These observations imply that the different polarimetric aberrations in two-periodic elements with identical ψ may be directly correlated to different schemes.

In Figure 3b,d, we quantitatively analyze difference aberrations in MM elements by calculating the MSE of curves using schemes (ii) and (iii), respectively. Different ring areas with various radii for each MM element are collected to calculate the means and standard deviations (SDs) of the MSE. The error bars on the plots represent the SDs of the data. In consideration of the image distortions due to the oblique emergent light, the SDs, which do not exceed 5% of the means, validate the reliability of our experimental results. As shown in Figure 3b,d, there is a significant difference between the two conditions, as in schemes (ii) and (iii). Figure 3b shows significant increases in the MSE between the M12 curves with θ exceeding 20° and the referenced M12 curve. Notably, when θ increases to 40°, the MSE even rises to 1.5×10^{-3}, indicating a significant aberration in M12. Similarly, Figure 3d shows significant increases in the MSE between the M21 curves with ζ exceeding 20° and the referenced M21 curve. Notably, when ζ increases to 40°, the MSE even rises to 2.3×10^{-3}, indicating a significant aberration in M21. Additionally, the curves of M12, M13, M21, and M31 are similar to the trigonometric functions, and the curves of M12 and M21 shift along the vertical axis with an increase in θ or ζ. Therefore, we calculate the direct current components of the energy spectral density at the above eight different angles to quantitatively analyze the aberrations, as shown in Table 1. For most curves, their direct current components tend to be zero, and the variations do not exceed 3×10^{-4}, excluding the curves of M12 or the curves of M21. When θ or ζ attain significant magnitudes, such as achieving 40°, the direct current components of the M12 and M21 curves can escalate to 2×10^{-3}. These results also show that false-positive linear diattenuation is induced as θ increases and false-positive linear polarizance is induced as ζ increases. Through these quantitative analyses, it can be proven that the aberrations are different from schemes (ii) and (iii), which cannot be ignored in the polarization properties analysis.

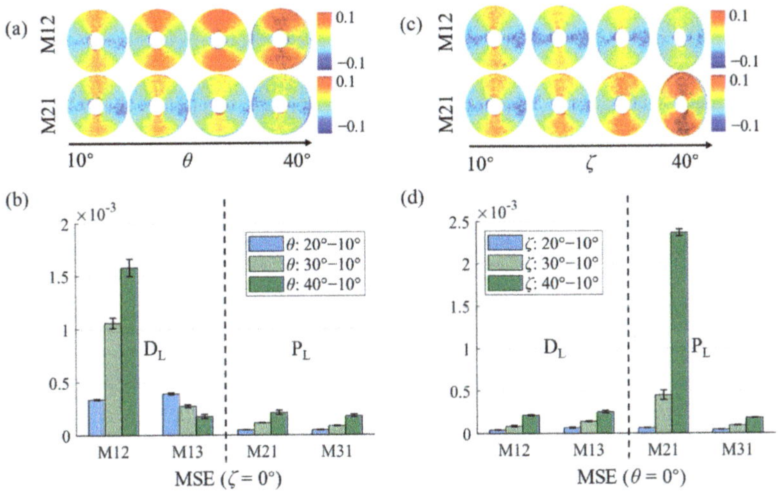

Figure 3. M12 and M21 images and analysis of anisotropic silk phantom under different θ and ζ: (**a**) M12 and M21 images with θ from 10° to 40°; (**b**) MSE with different θ; (**c**) M12 and M21 images with ζ from 10° to 40°; (**d**) MSE with different ζ.

Table 1. Direct current components of the energy spectral density with different MM elements.

Degrees	M12 ($\times 10^{-5}$)	M13 ($\times 10^{-5}$)	M21 ($\times 10^{-5}$)	M31 ($\times 10^{-5}$)
$\theta = 10°, \zeta = 0°$	0.87	0.38	0.30	0.13
$\theta = 20°, \zeta = 0°$	40.83	25.66	0.29	0.0080
$\theta = 30°, \zeta = 0°$	119.83	12.61	0.07	0.03
$\theta = 40°, \zeta = 0°$	173.45	2.49	2.42	0.01
$\theta = 0°, \zeta = 10°$	0.89	1.04	0.68	0.10
$\theta = 0°, \zeta = 20°$	3.11	2.98	4.27	0.19
$\theta = 0°, \zeta = 30°$	4.13	0.29	48.64	1.13
$\theta = 0°, \zeta = 40°$	5.87	3.81	252.78	4.24

In summary, when θ or ζ is beyond 30°, we can obtain more accurate M21 and M31, as in scheme (ii), or obtain more accurate M12 and M13, as in scheme (iii); however, it should be noted that image distortions will be included by the oblique emergent light.

3.2. MM Image Distortions Induced by Oblique Emergent Light

We find that the MM images of the phantom undergo transverse compressions, defining the transverse direction as parallel to the incident plane. The transverse compressions result in the MM image distortions and the destruction of their structural information. To quantitatively analyze the distortions, ellipticity is utilized as a characterization parameter, which represents the distortions of the MM images and can be obtained by Equation (4). As shown in Figure 4a, it is evident that the variations in θ exert minimal influence on the distortions of the MM images, while the shape of the MM images transforms into an ellipse with an increasing ζ. In Figure 4b, the lengths of the columns with the identical color, representing the identical θ, progressively escalate as ζ increases from 10° to 40°. Notably, this phenomenon is most obvious when θ is 0° or 10°. This intuitively illustrates that the increase in ζ aggravates the extent of distortions, while the varying θ has no obvious influence on it.

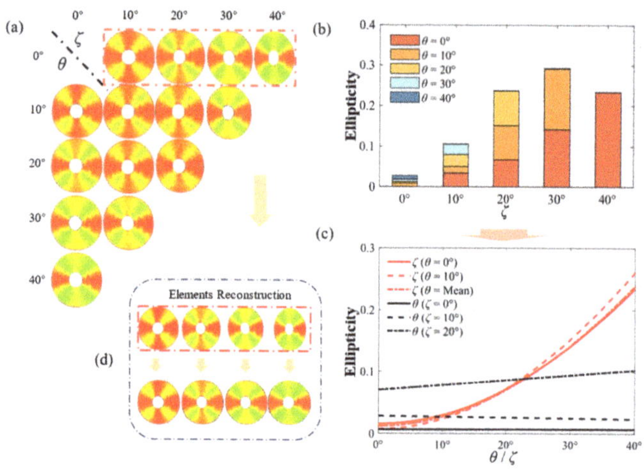

Figure 4. Schematic of ellipticity variation: (**a**) the M22 elements for different θ and ζ; (**b**) histogram of ellipticity for different θ and ζ; (**c**) LR fitting curves obtained by the ellipticity data; (**d**) the reconstruction of the MM elements.

Therefore, according to the experimental setup model, the analytical relationship between ellipticity and ζ is constructed as shown in Equation (4). The analytical relationship illustrates that the cosine value of ζ has a linear relationship with the ellipticity of the MM images. As shown in Figure 4c, we utilize the LR to verify the correctness of the analytical relationship using the above ellipticity data. The fitting curves for the variable θ and the constant ζ are predominantly close to horizontal, indicating a limited association between ellipticity and θ. Additionally, these curves possess different intercepts, indicating that ζ has a predominant influence on ellipticity. Conversely, the curves for the constant θ and the variable ζ are extremely aligned with the curves constructed in Equation (4), affirming the accuracy of the analytical relationship. Moreover, we can introduce a compensation coefficient σ to mitigate the distortions in the transverse direction through Equation (14), where x_{mit} is the distance of the reconstructed MMs and x_{dis} is the distance of the distorted MMs. The reconstructed MM images can be obtained using the bilinear interpolation algorithm. As shown in Figure 4d, this approach effectively mitigates distortions in the MM images.

$$x_{mit} = \sigma x_{dis}, \ \sigma = \frac{1}{1 - Ellipticity}. \tag{14}$$

To systematically verify the reliability of our conclusions, the calculated results of LR based on the least squares method are demonstrated in Table 2. When θ is treated as the variable, the slope of the regression equation converges to approximately zero, while the intercept enlarges with an increasing ζ. Conversely, when ζ is considered as the variable, the slope and the intercept are approximately equal to -1 and 1, aligning closely with the values outlined in Equation (4), respectively. The conspicuous linear correlation between ζ and ellipticity leads to an exceedingly small F below 0.5%, indicating a statistically significant correlation. The exceedingly small p-value, which is also below 0.5%, further reinforces the reliability of the above conclusions. All of these conclusions are applicable to the majority of distortion models caused by oblique emergent light and can be used for MMs' reconstruction, which can recover their structural information. Consequently, we advocate for obtaining MM measurements using scheme (ii) rather than scheme (iii) to acquire undistorted MM images. In instances where measuring MMs using scheme (iii) is deemed indispensable, we suggest restoring the structural integrity of the images using Equation (14).

Table 2. LR results.

Different Degrees	Intercept	X Variable	Significance F	p-Value
θ (ζ = 0°)	0.0066	0.000013	0.92	0.21
θ (ζ = 10°)	0.028	−0.00012	0.80	0.070
θ (ζ = 20°)	0.070	0.00080	0.30	0.052
ζ (θ = 0°)	0.95	−0.93	0.0014	0.0011
ζ (θ = 10°)	1.091	−1.08	0.0058	0.0051
ζ (θ = Mean)	0.98	−0.96	0.000054	0.000040

3.3. *Calibration of Polarization Properties by Adjusting the Distribution of θ and ζ*

It should be noted that MMs measured using schemes (ii) or (iii) with ψ exceeding 20° display significant aberrations, which could potentially induce inaccuracies in characterizing the polarization properties. However, we find that measuring MMs with an adjusting distribution of θ and ζ, as in scheme iv), can remarkably alleviate the aberrations. In Figure 2f, by observing the azimuth dependence curve with θ equal to ζ, some qualitative variations are revealed, as follows: (1) The vertical displacements between the M12, M21, M13, and M31 curves and referenced curves are maintained within 0.03. (2) The M22 curve has a sharp peak at the special azimuth location of 90°. (3) Both the M23 and M32 curves have two sharp peaks at the special azimuth locations of 150° and 210°, respectively. (4) The M33 curve amplitude and minimum approximately increase to 0.4 and 0.2, respectively. These variations make the MMs' curves extremely congruent with the referenced curves. Meanwhile, period degeneracy occurs in the central-block MM elements when θ or ζ exceed 20°, as in schemes (ii) and (iii), which achieves excellent periodic recovery when θ approaches approximate equality to ζ, as in scheme (iv), as shown in Figure 4a. These results reveal that the MMs measured with an adjusting distribution of θ and ζ exhibit a great resemblance to the referenced MMs, which may provide a strategy for obtaining accurate MMs with ψ exceeding 30°.

To further explore the improvement of the polarization properties through this strategy, we measure MMs on porcine liver tissue to validate the alleviation effect on the aberrations in *LDoP* and b_2. To avoid the destruction of the structural information, we adopt scheme (ii) to measure MMs when ζ is 0°. As depicted in Figure 5a, subtle structural distortions are observed in the measured MM images when θ and ζ are 20°. Notably, the polarization properties of the microstructure, including *LDoP* and b_2, are perceived to be more accurate when using scheme (iv). When θ is 40° and ζ is 0°, the aberrations of the PBPs validate the false-positive and false-negative calculated results of depolarization. It should be noted that b_2 is commonly utilized in cancer detection due to its sensitivity to small particles, so the incidence of the above phenomena needs to be reduced [32]. Furthermore, in Figure 5b, when θ is 40° and ζ is 0°, the curves exhibit substantial peak shifts along the horizontal axis compared to the referenced curves, which can be recovered by measurement with an adjusting distribution of θ and ζ. We calculate *HD*, where a smaller distance indicates a higher precision in polarization properties, to verify the reliability of the conclusion. The tabulated results reveal that parameters *LDoP* and b_2 with θ and ζ equal to 20° are closer to the corresponding results of the reference parameters, which merely demonstrate 1.4% and 12% *HD*s in comparison to the parameters measured in scheme (ii). These results illustrate the capability of measuring MMs using scheme (iv) to preserve nearly all accurate depolarization information. Furthermore, measuring polarization properties using scheme (iv) can be more accurate and avoid uncertain aberrations caused by oblique incidence or oblique emergent light in biological tissues.

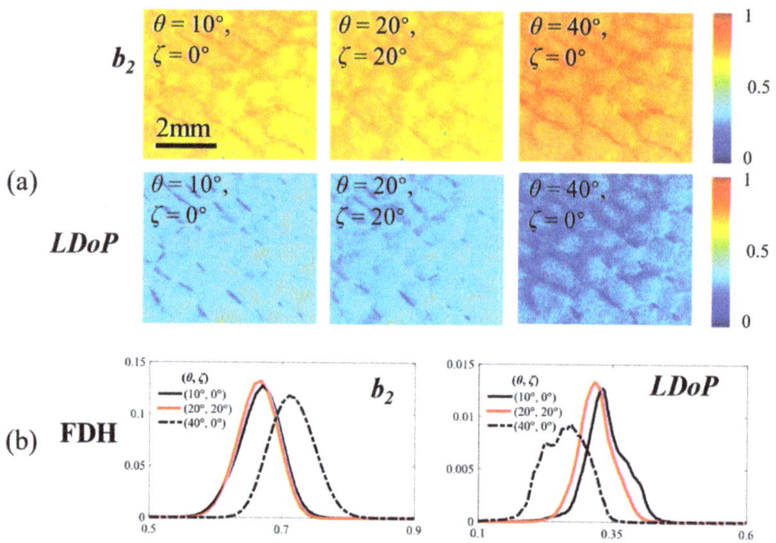

Figure 5. Polarization property recovery: (**a**) parameter images of *LDoP* and b_2 at different θ and ζ; (**b**) corresponding FDHs of the curves.

To assess the precision of the recovery, we calculate the Euclidean norm and *SSIM* between the PBP images using scheme (ii) (or scheme (iv)) and the referenced PBP images. In Table 3, employing the Euclidean norm as a metric reveals that the distances between the b_2 and *LDoP* images using scheme (iv) and the referenced images are only 11.53 and 19.96, respectively, indicating an effective overall recovery. Meanwhile, despite the MM images being distorted, as in scheme (iv), the higher *SSIM* values of b_2 and *LDoP*, which imply greater similarity to the original structure, suggest that the precision of the polarization properties in the microstructure can partly compensate for the loss of macroscopic structural information. Therefore, when ψ exceeds 30°, measuring MMs with an adjusting distribution of θ and ζ is more precise if the overall MMs and PBPs are of the utmost concern.

Table 3. Quantitative results of the recovery by different similarity metrics.

Similarity Metrics	b_2 **(θ, ζ)**		***LDoP*** **(θ, ζ)**	
	(20°, 20°)	**(40°, 0°)**	**(20°, 20°)**	**(40°, 0°)**
HD ($\times 10^{-6}$)	0.16	10.83	1.29	10.67
Euclidean norm	11.53	49.67	19.96	79.15
SSIM	0.90	0.88	0.81	0.75

4. Discussion

MM polarimetry plays a crucial role in optical biosensors due to its significant performance and advantages in mapping the microscopic morphological information of biological tissues. In our previous work, we focused on the influence of oblique incidence with normal emergent light, as in scheme (ii), and proposed specific optimizing strategies [13]. Here, we further explore the detailed characteristics of aberrations induced in all non-collinear conditions, as in schemes (ii)–(iv), which are adaptable to more scenarios in practical polarimetric applications. The conclusions of this study for scheme (ii) are consistent with our previous work, while the analysis and optimization strategies of schemes (iii) and (iv) are the essential aims of this work, providing a thorough complement to the previous conclusions. Furthermore, this study serves as a methodological contribution to offer valuable insights into the design and construction of optical systems for backscattering MM polarimetry in bioimaging and sensing applications, such as bio-structural and optical properties sensing [14,15], polarimetric endoscopy [16,17], and skin tissue evaluation [18–20].

5. Conclusions

In this study, we systematically analyzed the polarimetric aberrations induced by three different illumination schemes in backscattering MM imaging. These schemes were distinguished by incidence and emergent light in non-collinear conditions. Additionally, comprehensive comparisons were performed among the MMs obtained under these schemes, considering ψ as 10°, 20°, 30°, and 40°, respectively. We found that measuring MMs with oblique emergent light can induce specific distortion in MM images, hindering the characterization of tissue structural information. Notably, a linear relationship emerged between the cosine value of ζ and the extent of transverse compression, and this could be used to reconstruct the MM images. Additionally, when measuring MMs as in scheme (ii) with θ exceeding 20°, the aberrations of M12 and M13 exhibited increasing intensities. When it came to scheme (iii) with ζ exceeding 20°, M21 and M31 exhibited significantly increasing intensities. However, the variations in other MM elements remained identical. It should be noted that measuring MMs as different schemes may have different effects on the polarization parameters. Furthermore, we found that scheme (iv) could alleviate the polarimetric aberrations. Quantitative analysis of PBPs using porcine liver tissues revealed a significant improvement in the alleviation of PBP aberrations when MMs were measured with an adjusting distribution of θ and ζ. These findings delivered some crucial guidance for choosing the appropriate spatial illumination for non-collinear MM imaging, as follows: (1) If possible, measure MMs with a normal emergent light to obtain accurate structural information. (2) For linear polarizance, measure MMs as in scheme (ii) to accurately obtain M21 and M31. (3) For linear diattenuation, measure MMs as in scheme (iii) to obtain M12 and M13 accurately. (4) To obtain overall MMs and PBPs, measure MMs as in scheme (iv) with an adjusting distribution of θ and ζ. In summary, the optimized schemes provided critical criteria for the spatial illumination scheme selection of non-collinear backscattering MM measurements, which can be helpful for the further development of quantitative tissue polarimetric imaging and biosensing.

Author Contributions: W.J.: methodology, software, formal analysis, and writing—original draft. Z.Z.: methodology, investigation, validation, and writing—original draft. N.Z.: methodology and investigation. R.H.: methodology and investigation. H.H.: conceptualization, methodology, writing—review and editing, and supervision. C.H.: methodology, resources, and supervision. H.M.: conceptualization, resources, and supervision. All authors have read and agreed to the published version of the manuscript.

Funding: Shenzhen Key Fundamental Research Project (No. JCYJ20210324120012035).

Institutional Review Board Statement: Not applicable.

Informed Consent Statement: Not applicable.

Data Availability Statement: Data underlying the results presented in this paper are not publicly available at this time but may be obtained from the authors upon reasonable request.

Conflicts of Interest: The authors declare that there are no conflicts of interest related to this paper.

References

1. Ramella-Roman, J.C.; Novikova, T. *Polarized Light in Biomedical Imaging and Sensing: Clinical and Preclinical Applications*; Springer International Publishing: Cham, Switzerland, 2023.
2. He, C.; He, H.; Chang, J.; Chen, B.; Ma, H.; Booth, M.J. Polarisation Optics for Biomedical and Clinical Applications: A Review. *Light. Sci. Appl.* **2021**, *10*, 194. [CrossRef] [PubMed]
3. Tukimin, S.N.; Karman, S.B.; Ahmad, M.Y.; Wan Kamarul Zaman, W.S. Polarized Light-Based Cancer Cell Detection Techniques: A Review. *IEEE Sens. J.* **2019**, *19*, 9010. [CrossRef]
4. Chue-Sang, J.; Gonzalez, M.; Pierre, A.; Laughrey, M.; Saytashev, I.; Novikova, T.; Ramella-Roman, J.C. Optical Phantoms for Biomedical Polarimetry: A Review. *J. Biomed. Opt.* **2019**, *24*, 030901. [CrossRef] [PubMed]
5. Qi, J.; Elson, D.S. Mueller Polarimetric Imaging for Surgical and Diagnostic Applications: A Review. *J. Biophotonics* **2017**, *10*, 950. [CrossRef] [PubMed]
6. Alali, S.; Vitkin, A. Polarized Light Imaging in Biomedicine: Emerging Mueller Matrix Methodologies for Bulk Tissue Assessment. *J. Biomed. Opt.* **2015**, *20*, 061104. [CrossRef] [PubMed]
7. Lu, S.Y.; Chipman, R.A. Interpretation of Mueller Matrices Based on Polar Decomposition. *J. Opt. Soc. Am. A* **1996**, *13*, 1106. [CrossRef]
8. Ossikovski, R. Analysis of Depolarizing Mueller Matrices through a Symmetric Decomposition. *J. Opt. Soc. Am. A* **2009**, *26*, 1109. [CrossRef] [PubMed]
9. Ossikovski, R. Differential Matrix Formalism for Depolarizing Anisotropic Media. *Opt. Lett.* **2011**, *36*, 2330. [CrossRef] [PubMed]
10. Ortega-Quijano, N.; Arce-Diego, J.L. Mueller Matrix Differential Decomposition. *Opt. Lett.* **2011**, *36*, 1942. [CrossRef] [PubMed]
11. Ahmad, I.; Ahmad, M.; Khan, K.; Ashraf, S.; Ahmad, S.; Ikram, M. Ex Vivo Characterization of Normal and Adenocarcinoma Colon Samples by Mueller Matrix Polarimetry. *J. Biomed. Opt.* **2015**, *20*, 056012. [CrossRef] [PubMed]
12. Yasuda, M.; Akimoto, T. High-Contrast Fluorescence Microscopy for a Biomolecular Analysis Based on Polarization Techniques Using an Optical Interference Mirror Slide. *Biosensors* **2014**, *4*, 513. [CrossRef] [PubMed]
13. Zhang, Z.; Hao, R.; Shao, C.; Mi, C.; He, H.; He, C.; Du, E.; Liu, S.; Wu, J.; Ma, H. Analysis and Optimization of Aberration Induced by Oblique Incidence for In-Vivo Tissue Polarimetry. *Opt. Lett.* **2023**, *48*, 6136. [CrossRef] [PubMed]
14. Shang, L.; Tang, J.; Wu, J.; Shang, H.; Huang, X.; Bao, Y.; Xu, Z.; Wang, H.; Yin, J. Polarized Micro-Raman Spectroscopy and 2D Convolutional Neural Network Applied to Structural Analysis and Discrimination of Breast Cancer. *Biosensors* **2023**, *13*, 65. [CrossRef] [PubMed]
15. Phan, Q.H.; Han, C.Y.; Lien, C.H.; Pham, T.T.H. Dual-Retarder Mueller Polarimetry System for Extraction of Optical Properties of Serum Albumin Protein Media. *Sensors* **2021**, *21*, 3442. [CrossRef] [PubMed]
16. Manhas, S.; Vizet, J.; Deby, S.; Vanel, J.C.; Boito, P.; Verdier, M.; Martino, A.D.; Pagnoux, D. Demonstration of Full 4×4 Mueller Polarimetry through an Optical Fiber for Endoscopic Applications. *Opt. Express* **2015**, *23*, 3047. [CrossRef] [PubMed]
17. Trout, R.M.; Gnanatheepam, E.; Gado, A.; Reik, C.; Ramella-Roman, J.C.; Hunter, M.; Schnelldorfer, T.; Georgakoudi, I. Polarization Enhanced Laparoscope for Improved Visualization of Tissue Structural Changes Associated with Peritoneal Cancer Metastasis. *Biomed. Opt. Express* **2022**, *13*, 571. [CrossRef] [PubMed]
18. Bargo, P.R.; Kollias, N. Measurement of Skin Texture through Polarization Imaging: Measurement of Skin Texture through Polarization Imaging. *Brit. J. Dermatol.* **2010**, *162*, 724. [CrossRef] [PubMed]
19. Rey-Barroso, L.; Peña-Gutiérrez, S.; Yáñez, C.; Burgos-Fernández, F.J.; Vilaseca, M.; Royo, S. Optical Technologies for the Improvement of Skin Cancer Diagnosis: A Review. *Sensors* **2021**, *21*, 252. [CrossRef] [PubMed]
20. Jütte, L.; Roth, B. Mueller Matrix Microscopy for In Vivo Scar Tissue Diagnostics and Treatment Evaluation. *Sensors* **2022**, *22*, 9349. [CrossRef] [PubMed]
21. Ning, T.; Ma, X.; Li, Y.; Li, Y.; Liu, K. Efficient Acquisition of Mueller Matrix via Spatially Modulated Polarimetry at Low Light Field. *Opt. Express* **2023**, *31*, 14532. [CrossRef] [PubMed]
22. Hu, Z.; Ushenko, Y.A.; Soltys, I.V.; Dubolazov, O.V.; Gorsky, M.P.; Olar, O.V.; Trifonyuk, L.Y. Analytical Review of the Methods of Multifunctional Digital Mueller-Matrix Laser Polarimetry. In *Mueller-Matrix Tomography of Biological Tissues and Fluids*; Springer: Singapore, 2024; pp. 1–11.
23. Chen, Z.; Meng, R.; Zhu, Y.; Ma, H. A Collinear Reflection Mueller Matrix Microscope for Backscattering Mueller Matrix Imaging. *Opt. Lasers Eng.* **2020**, *129*, 106055. [CrossRef]
24. Phan, Q.H.; Dinh, Q.H.; Pan, Y.C.; Huang, Y.T.; Hong, Z.H. Decomposition Mueller Matrix Polarimetry for Enhanced miRNA Detection with Antimonene-Based Surface Plasmon Resonance Sensor and DNA-Linked Gold Nanoparticle Signal Amplification. *Talanta* **2024**, *270*, 125611. [CrossRef] [PubMed]
25. Ho, H.P.; Law, W.C.; Wu, S.Y.; Liu, X.H.; Wong, S.P.; Lin, C.; Kong, S.K. Phase-Sensitive Surface Plasmon Resonance Biosensor Using the Photoelastic Modulation Technique. *Sens. Actuator B-Chem.* **2006**, *114*, 80. [CrossRef]
26. Phan, Q.H.; Lo, Y.L. Stokes-Mueller Matrix Polarimetry System for Glucose Sensing. *Opt. Lasers Eng.* **2017**, *92*, 120. [CrossRef]
27. Phan, Q.-H.; Jian, T.H.; Huang, Y.R.; Lai, Y.R.; Xiao, W.Z.; Chen, S.W. Combination of Surface Plasmon Resonance and Differential Mueller Matrix Formalism for Noninvasive Glucose Sensing. *Opt. Lasers Eng.* **2020**, *134*, 106268. [CrossRef]
28. Le, H.M.; Le, T.H.; Phan, Q.H.; Pham, T.T.H. Mueller Matrix Imaging Polarimetry Technique for Dengue Fever Detection. *Opt. Commun.* **2022**, *502*, 127420. [CrossRef]

29. Qi, J.; Elson, D.S.; Stoyanov, D. Eigenvalue Calibration Method for 3 × 3 Mueller Polarimeters. *Opt. Lett.* **2019**, *44*, 2362. [CrossRef] [PubMed]
30. Khaliq, A.; Ashraf, S.; Gul, B.; Ahmad, I. Comparative Study of 3 × 3 Mueller Matrix Transformation and Polar Decomposition. *Opt. Commun.* **2021**, *485*, 126756. [CrossRef]
31. Nguyen, T.V.; Nguyen, T.H.; Nguyen, N.B.T.; Huynh, C.K.; Le, T.H.; Phan, Q.H.; Pham, T.T.H. Evaluation of Optical Features of Fibronectin Fibrils by Backscattering Polarization Imaging. *Optik* **2023**, *272*, 170304. [CrossRef]
32. Iqbal, M.; Ahmad, I.; Khaliq, A.; Khan, S. Comparative Study of Mueller Matrix Transformation and Polar Decomposition for Optical Characterization of Turbid Media. *Optik* **2020**, *224*, 165508. [CrossRef]
33. Gao, W. Independent Differential Depolarization Parameters and Their Physical Interpretations. *Opt. Laser Technol.* **2023**, *161*, 109156. [CrossRef]
34. Wang, Z.; Bovik, A.C.; Sheikh, H.R.; Simoncelli, E.P. Image Quality Assessment: From Error Visibility to Structural Similarity. *IEEE Trans. Image Process.* **2004**, *13*, 600. [CrossRef] [PubMed]

biosensors

MDPI

Article

Thermo-Visco-Elastometry of RF-Wave-Heated and Ablated Flesh Tissues Containing Au Nanoparticles

Bayan Kurbanova [1], Zhannat Ashikbayeva [2], Aida Amantayeva [2], Akbota Sametova [2], Wilfried Blanc [3], Abduzhappar Gaipov [4], Daniele Tosi [2,5] and Zhandos Utegulov [1,*]

[1] Department of Physics, School of Sciences and Humanities, Nazarbayev University, Astana 010000, Kazakhstan
[2] School of Engineering and Digital Sciences, Nazarbayev University, Astana 010000, Kazakhstan
[3] Université Côte d'Azur, INPHYNI, CNRS UMR7010, Avenue Joseph Vallot, 06108 Nice, France
[4] Department of Medicine, Nazarbayev University School of Medicine, Astana 010000, Kazakhstan
[5] National Laboratory Astana, Laboratory of Biosensors and Bioinstruments, Astana 010000, Kazakhstan
* Correspondence: zhutegulov@nu.edu.kz

Abstract: We report non-contact laser-based Brillouin light-scattering (BLS) spectroscopy measurements of the viscoelastic properties of hyperthermally radiofrequency (RF)-heated and ablated bovine liver and chicken flesh tissues with embedded gold nanoparticles (AuNPs). The spatial lateral profile of the local surface temperature in the flesh samples during their hyperthermia was measured through optical backscattering reflectometry (OBR) using Mg–silica-NP-doped sensing fibers distributed with an RF applicator and correlated with viscoelastic variations in heat-affected and ablated tissues. Substantial changes in the tissue stiffness after heating and ablation were directly related to their heat-induced structural modifications. The main proteins responsible for muscle elasticity were denatured and irreversibly aggregated during the RF ablation. At T > 100 °C, the proteins constituting the flesh further shrank and became disorganized, leading to substantial plastic deformation of biotissues. Their uniform destruction with larger thermal lesions and a more viscoelastic network was attained via AuNP-mediated RF hyperthermal ablation. The results demonstrated here pave the way for simultaneous real-time *hybrid optical sensing of viscoelasticity and local temperature* in biotissues during their denaturation and gelation during hyperthermia for future applications that involve mechanical- and thermal-property-controlled theranostics.

Keywords: radiofrequency; heating; hyperthermia; ablation; fiber-optic; sensors; Brillouin spectroscopy; viscoelastic; gold nanoparticles; thermometry; temperature; protein denaturation; flesh; metal nanoparticle; theranostics

1. Introduction

For localized treatment of cancer in the liver, brain, and breast [1,2] and of Barrett's esophagus syndrome, varicose veins, and uterine fibroids, *radiofrequency ablation (RFA)* at mild powers is often employed as a minimally invasive, safe, and low-risk therapeutic procedure [3,4]. During RFA, the electromagnetic (EM) waves passing through biological tissues via an applied needle cause their local heating due to the propagation of an alternating electrical current (AC) modulated at frequencies ranging from 100 kHz to 300 GHz [5], which causes ionic agitation and cellular necrosis, which can ultimately cause tissue coagulation [6]. However, a lack of control of thermal energy deposition during RFA can lead to such undesirable therapeutic results as malignant transformation [7]. Therefore, the control of the RFA procedure is essential for defining how much tissue is destroyed while avoiding charring and excessive damage to surrounding healthy tissues.

Compared to other EM wave sources, such as visible and near-infrared light, RFA causes much deeper heat penetration into tissues and, therefore, is applicable for the treatment of deeply localized solid tumors [8]. However, it is impossible to generate desirable ablation zones that cover spatially extended tumors, and the employment of a large RF power and a longer time in order to extend the ablation zone will lead to undesirable tissue burns. Adding a mediating absorptive agent, such as *metal nanoparticles (NPs)*, expands the ablation zone during RFA. Hyperthermal heating occurs in the nanometer-scale vicinity of embedded NPs, thus increasing the local temperatures up to hundreds of degrees on a micrometer scale, leading to a more spatially extended temperature distribution within the heated sample, which causes the disruption of cells in the vicinity. More importantly, loading selected NPs with high thermal and electrical conductivities driven by free electrons into tissues promotes wider and deeper ablation zones due to random spatial distributions of localized heating sources attributed to these metal NPs [9].

Gold nanoparticles (AuNPs) are promising for the targeted thermal destruction of cancer due to their abilities of surface functionalization and hyperthermal RF heating. Moreover, AuNPs are less toxic than other metallic NPs and do not chemically interact with most biological substances [10]. The multifunctionality of AuNPs makes them ideal candidates for theranostic applications [11]. Furthermore, with the increase in NP concentration, the ablation efficiency increases. However, NPs with sizes smaller than 5 nm and larger than 100 nm are not biologically active. Very small NPs are easily cleared by the kidney, while NPs larger than 100 nm are sequestered by the white blood cells that defend the immune system [12].

The viscoelastic properties of tissues vary due to tissue dehydration, protein denaturation, and coagulation during hyperthermal treatments, and these properties make them good candidates for dosimetry [13]. Monitoring *viscoelastic variations* in tissues is critical during thermal ablation, since cell death under hyperthermal treatment is driven by AC-induced thermal stresses in the presence of magnetic nanofluids [14]. It was revealed that thermal conduction and tissue volume relaxation parameters play a significant role in tissue bio-thermo-mechanics [15]. A fully nonlinear thermo-visco-hyperelastic finite-element algorithm was developed to describe the nonlinear characteristics of bioheat transfer in deformed soft tissues with thermal expansion/shrinkage during hyperthermal treatment [16].

Turning to the effects of AuNPs on the elastic properties of tissue, the conjugation of AuNPs to fibrilized collagen, as a soft tissue filler, leads to the formulation of AuNP–collagen constructions and increases the collagen's crosslinking longevity. The mechanical properties of tissues are enhanced as a function of crosslinking [17]. However, to the best of our knowledge, no investigations have been performed on the *thermo-viscoelasticity of tissues during and after AuNP-mediated hyperthermal* treatment. Additionally, the viscoelasticity monitoring of RF-heated tissues is very critical if one wants to understand and control the evolution of the mechanical properties of tissues during and after RF heating and ablation. In addition, it is well known that the mechanical properties of the extracellular matrices of biological tissues and their biopolymer components are essential for normal tissue function, and disturbances in these properties frequently occur in disease [18].

The awareness that diseases ranging from osteoarthritis to atherosclerosis, cancer, and diabetes are linked to alterations in the mechanical properties of the affected tissue has stimulated research on the mechanical properties of entire tissues [18]. Since proteins are major components of flesh (e.g., beef or chicken), it is important to emphasize that the *viscoelastic properties of RF-heated and ablated flesh will be dominated by the viscoelasticity of the major RF-heated and ablated proteins* constituting these flesh tissues. In terms of the percentage of mass content, chicken flesh actually has more protein than beef flesh [19]. Heat-ablated tissue is stiffer than untreated tissue and is used to measure spatial stiffness variations and to quantify Young's modulus in tissues after thermal treatment [20]. The elastic and viscous components of a gelated myofibrillar protein network started to increase from 30 to 45 °C [21]. The formation of a viscoelastic gel network in ribbonfish meat was observed during heating, and the maximum rate of increase in the storage and loss of elastic moduli was found to be in the temperature range of 56.8 to 63.3 °C [22]. However, the viscoelastic properties of beef muscle showed a slight decrease for T < 55 °C, then a sharp increase for 55 °C < T < 80 °C, and a saturation at higher temperatures [23]. During the heating of cancerous biological tissues, the cytotoxicity of tumors occurred at 42 °C, while protein coagulation started at 60 °C. The temperature range of 60 to 100 °C is known as an effective parameter for obtaining the largest ablation area. However, temperatures above 100 °C cause tissue evaporation and carbonization, which have undesirable effects [24].

For the assessment of the elastic and viscous properties of tissues, various techniques have been utilized, such as ultrasound elastography [25], dynamic nanoindentation [20], dynamic rheology [21], and Brillouin light-scattering (BLS) spectroscopy [26]. The latter technique is based on the inelastically scattered laser light from thermally excited GHz acoustic waves in a probed condensed medium. BLS spectroscopy is a unique non-destructive and non-contact technique for probing the viscoelastic properties of various materials, and it was traditionally applied in soft matter and materials science studies [27,28], geosciences [29], biomedical applications [30,31], and the detection of biomarkers [32,33]. More importantly, advances in confocal micro-Brillouin light scattering have enabled the rapid noninvasive monitoring of tissue biomechanics down to the micrometer scale, thus opening the way for live cell imaging [34,35]. Turning to cancer detection, Brillouin shifts in non-regressing and regressing melanomas and in healthy tissue samples were found significantly different, with the healthy tissue being the softest and non-regressing melanoma being the stiffest [36]. The elastic moduli of muscle and rat tail tendon collagen were obtained, and hydrogen bond force constants were estimated with BLS [37,38]. In addition, BLS allowed the observation of the propagation of sound waves along a one-dimensional periodic array of sarcomeres in rabbit psoas muscle myofibrils [39]. A biomechanical contrast of muscular, connective, epithelial, and nervous tissues was presented through high-resolution Brillouin microscopy imaging [26]. A temperature-dependent stiffening and gelation process of albumin from chicken eggs was reported through BLS [40].

As for local optical thermometry, the OBR technique stands as the most modern approach to temperature sensing with fiber-optic sensors (FOS) during thermal ablation. In particular, Mg–silicate-NP-doped fibers have greater backscattering signals than those of standard single-mode fibers, thus allowing the extension of OBR to the real-time measurement of the spatial distribution of temperature over the inner plane of thermal ablation, providing a valuable alternative to thermal imaging with a simpler implementation [41]. Thermal modifications in photodamaged biological samples can be experimentally monitored by chaotic attractors, which depend on optical transmittance [42].

In this work, we report local viscoelastic variations through BLS spectroscopy within chicken muscle and bovine liver tissues with embedded AuNPs *after* RF heating and ablation. The distributions of Mg–silicate-NP-doped sensing fibers along an RF applicator during ablation allowed for the planar measurement of the local temperature profile and assessment of the ex vivo RF-ablated zone. BLS combined with Mg–silicate-NP-doped sensing fibers allowed us to obtain *viscoelastic sensing correlated with the detection of the local temperature* of AuNP-mediated hyperthermally RF-heated and ablated flesh tissues. The nature of the high Brillouin peak shift and broadening contrast between normal tissue and the created thermal lesions was studied. In addition, the relationship between the change in viscoelasticity and the obtained thermal dose of the tissues was deeply investigated to better understand the AuNP-assisted hyperthermal RF ablation phenomenon.

2. Materials and Methods

Brillouin spectra were recorded in the 180°-backscattering configuration by using a 6-pass tandem Fabry–Perot Interferometer TFP-2 (Table Stable Ltd., Mettmenstetten, Switzerland) [43]. The free spectral range was set to 25 GHz. Laser light with a 532 nm wavelength from Verdi-G2 (Coherent, Santa Clara, CA, USA) was used with a mild incident beam power kept below 10 mW to prevent sample damage. The laser spot diameter with a 5× microscope objective was 5 μm. The Brillouin spectral changes were measured at different spatial points on RF-heated and ablated mammal tissues corresponding to the temperatures measured with temperature-sensing fibers.

The ex vivo ablation of fresh chicken and bovine liver tissues was conducted by using an RF/MW Hybrid generator (Leanfa s.r.1.) according to the method in [44], and this was aligned with the European Union's "Three Rs" principle. Before use, the meat and livers were stabilized to room temperature at 20–21 °C. The RF power for the ablation procedure was set to 60 W, while the frequency value was 450 kHz. To control the temperature change during radiofrequency ablation, four Mg–silicate-NP-doped fibers spliced to single-mode optical fibers (SMFs) were placed in parallel on the y-axis on meat at a 5 mm distance from each other, as depicted in Figure 1. The generator was set to the safe mode, allowing the termination of the ablation procedure when the impedance of 800 Ω was reached. The active electrode (AE), which was in the form of a research-grade single-tip applicator with a 3 mm diameter and a 160 mm length, was inserted between the 2nd and 3rd fibers to deliver the RF waves, while the meat was positioned on a metallic plate that was used as a passive electrode (PE). The positions of both electrodes and sensing fibers were fixed for all of the experiments. A schematic representation of the experimental setup containing the RF generator, Fabry–Perot interferometer, optical backscatter reflectometer (OBR), and the distributed temperature-sensing optical fiber system is shown in Figure 1.

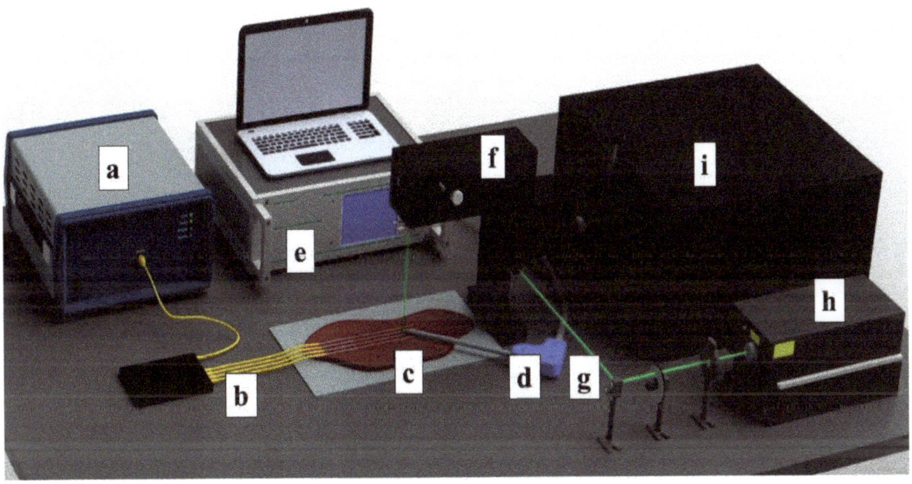

Figure 1. Schematic view of the radiofrequency ablation setup. The setup consisted of: (a) OBR Luna 4600 with a computer used for data acquisition, (b) optical fibers, (c) a bovine liver, (d) the RFA applicator, (e) a hybrid RF/MWA generator used in the RF mode, (f) a confocal microscope, (g) a laser path, (h) a Verdi-G2 532 nm laser source, and (i) a tandem Fabry–Perot interferometer TFP-2.

The synthesis of AuNPs was conducted by using the citrate reduction method, as presented by Turkevic et al. [45]. The average size (d = 20 nm) of the NPs was estimated with a transmission electron microscope (TEM, JEOL JEM—1400 Plus, Indianapolis, IN, USA) image, as shown in Figure 2e. Based on previous studies, Au NPs with a size of >20 nm could take a longer time to be excreted from the body during in vivo experiments, as described in [46]. The optimal density of NPs was selected as 1 mg/mL based on outcomes from previous works [47]. The RF ablation was performed in two conditions: on pristine tissues and on tissues treated with AuNPs with a size of 20 nm dispersed in 0.2% agarose solution. A total of 200 μL of AuNPs dispersed in an agarose solution were injected into the tissue by using a syringe in proximity to the active electrode. The ablation was performed three times under the same ablation conditions to measure the post-heating Brillouin spectral variations with a coupled confocal microscope.

The temperature-sensing setup was based on a commercial OBR (Luna Inc., OBR4600, Roanoke, VA, USA) that worked in the continuously distributed sensing mode. The OBR instrument was connected to the sensing fibers with different lengths by 1 × 8 wideband splitter operating in the third optical window. The OBR delivered a signal via a fiber link and then measured the Rayleigh-scattering-induced reflection. Return losses experienced along the fiber length could be estimated by recording the propagation time of backscattered light. The RFA wavelength of each fiber section's reflection spectrum shifted depending on the change in the temperature. The measurements were taken at 45 mm intervals at the tip of each fiber's spectrum, and these points were grouped into a 2D matrix based on the physical arrangement of the fibers. The obtained 2D thermal maps in the lateral planes of the tissues were arranged in different colors that indicated the temperature changes. The choice of sensing parameters represented a trade-off between the spatial resolution and temperature accuracy; in the experiments, the spatial resolution was set to 2.0 mm, while the overall 'useful' sensing length was set to 45 mm while taking all four fibers into account in this window. All optical connections, including the led-in fibers, splitters, and extenders, were based on a standard single-mode fiber (SMF, corning SMF-28), while the sensing fibers were based on Mg–silicate-NP-doped fibers and had a high spatial resolution (2 mm).

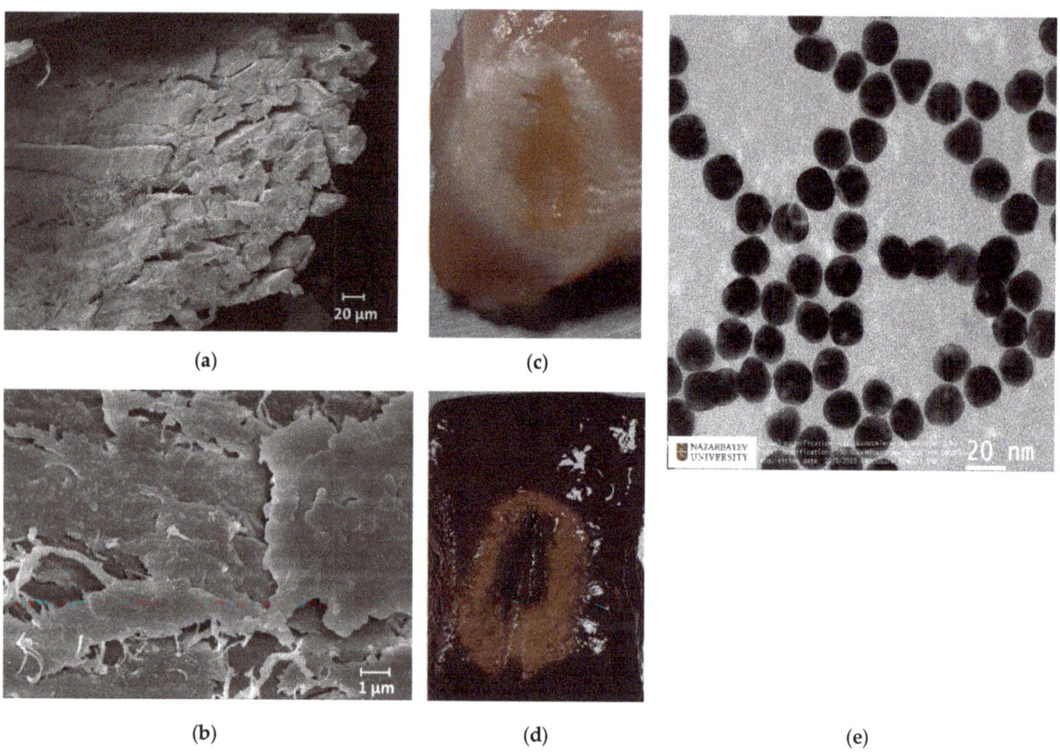

Figure 2. Scanning electron microscopy images of skeletal muscle tissue (**a**) and a closer view of the connective tissues (**b**). Ablated chicken muscle (**c**) and bovine liver (**d**). Transmission electron microscopy image of AuNPs (**e**).

3. Results and Discussion

Generally, both chicken and bovine liver flesh consists of approximately 71–75% moisture, 16–17% protein, 5–8% fat, and 1% ash [19]. Chicken muscle tissue contains muscle fibers and connective tissue, where connective tissue covers the muscle fibers, as can be seen in Figure 2a,b. The *myofibrillar proteins* constitute 50 to 55% of the chicken muscle's total protein content, while *sarcoplasmic proteins* account for approximately 30–34%. The remaining 10–15% of the proteins are *connective tissue proteins*. Myosin and actin are the most important proteins that constitute the myofibrillar structure [48]. The bovine liver is composed of smaller histological structures called lobules, which are roughly hexagonal in shape. In addition, the bovine liver is covered with a connective tissue capsule, which is a thin but tough fibrous supporting connective framework of densely interwoven collagen fibers [49]. The fibers of connective tissue consist of collagen, elastic, and reticular fibers; they are, thus, responsible for the liver's elasticity and tensibility [50].

The heating process had a considerable effect on the biological tissues' properties, as the changes in toughness with denaturation temperatures or indirect interpretations of structural changes in tissue components occurred [51]. Figure 3 demonstrates the chicken muscle's 2D thermal map in the x-y plane, which was processed from the data obtained along the lengths of all four Mg–silicate-NP-doped sensing fibers. The thermal maps of the tissues present the changes in the temperature during the thermal ablation procedure due to the dissipation of RF power in the bovine liver and chicken muscle tissues. The active electrode was positioned at 3.5 cm from the left side, while the sensing system was placed on the right side. The heat pattern clearly showed the maximum temperature increase at the tip of the active electrode and the heat distribution within the tissue, which are depicted in different colors. Images of the RF-ablated bovine liver and chicken meat are shown in Figure 2c,d. Adding AuNPs to the tissues led to a more extensive spatial temperature distribution with a lower maximum temperature; thus, much more uniform temperature distributions along the tissue could be obtained with AuNPs. Hyperthermal RF heating occurred, whereas the AuNPs absorbed the RF waves and produced heat via the limited movement of free electrons inside the NPs. The asymmetric heat pattern was likely due to the heterogeneous properties of the tissue, which deviated from the RF ablation pattern, particularly for fast ablation phenomena.

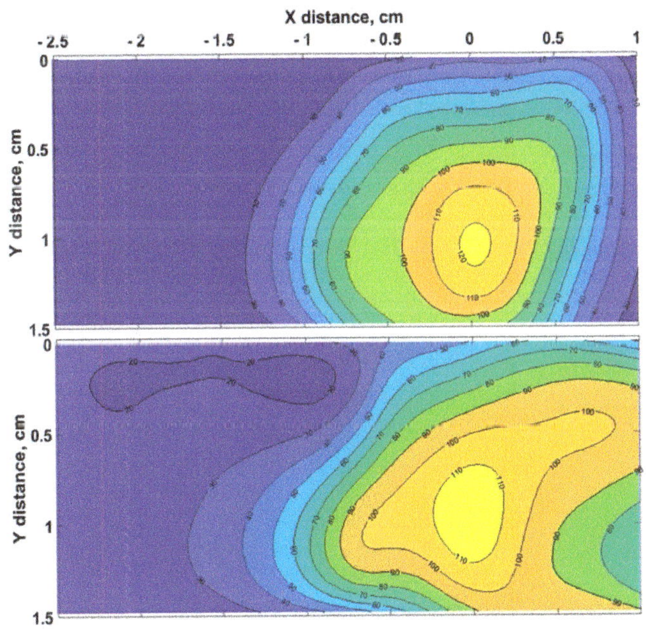

Figure 3. Typical two - dimensional thermal map of pristine chicken muscle (top) and tissues with AuNPs added (down) in the xy lateral plane (x = direction parallel to the RFA applicator and to the sensing fibers).

Figure 4 demonstrates the quantification of the (a) cytotoxic (temperatures > 42 °C) and (b) thermally damaged (temperatures > 60 °C) regions of the chicken muscle tissue for each experimental condition. As can be seen from the bar charts, the AuNP-mediated RFA achieved a larger area of thermal heating and ablation results than it provided in the pristine chicken meat. The same trend was observed for the bovine liver, and both tissues had similar thermal properties.

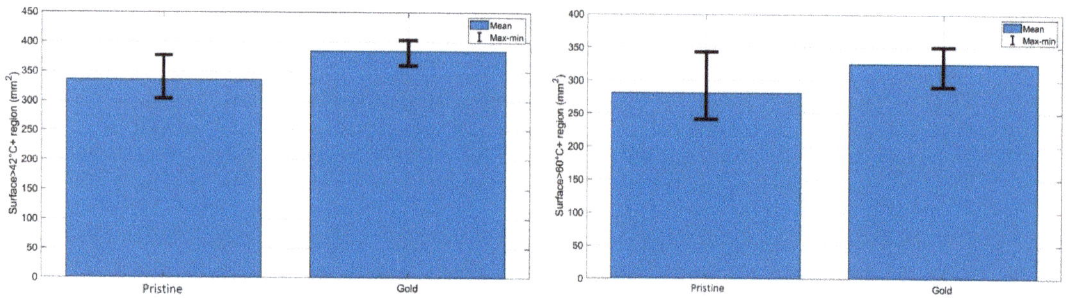

Figure 4. Evaluation of the surface area exposed to temperatures exceeding (**a**) 42 °C and (**b**) 60 °C at the maximum ablation temperature; the bar charts show the maximum, minimum, and mean values of the areas over three experiments for each RFA condition.

Generally, chicken meat was found to enhance its stiffness upon heating in two phases. The first phase took place at T = 40–60 °C, which was likely due to the heat-induced denaturation of myofibrillar proteins, especially myosin. The second stage led to further stiffening at T = 65–80 °C, which could be ascribed to the denaturation of intramuscular collagen, which was associated with the initial breakage of hydrogen bonds, thus loosening the fibrillar structure with the subsequent contraction and dissolution of the collagen molecules, as well as their eventual gelation [52]. Most of the water in living muscle is held within the myofibrils, and longitudinal shrinkage of the myofibrils at high temperatures causes great water loss. Other studies suggested that the *thermal injury threshold* for RFA in chicken tissue is 65 °C [53]. After thermal treatment at temperatures in the range from 53 to 63 °C, the bovine liver's collagen and reticular fibers were denatured, and this led to the breaking of hydrogen bonds and an irreversible transformation of the crosslinked triple-helical structure into a more random coiled structure [54]. At $T > 63$ °C, the cells had shrunk as much as they could, and most of the water in the cells was forced out [55].

Figure 5a demonstrates the typical Brillouin spectra of pristine and AuNPs-treated chicken muscle tissue corresponding to the measured local temperatures. The back-scattering configuration employed in our BLS measurements enabled us to measure only the bulk longitudinal acoustic waves. The Brillouin peaks decayed in intensity with the increase in the RF-induced temperature due to the heat-induced removal of water content from the tissues, thus effectively reducing the elasto-optic coupling. The ablation temperatures correlated well with the Brillouin shift and linewidth changes, with a frequency contrast from 8 and 1.5 GHz, respectively, at the outer transition zone's boundary and of 14.3 and 9 GHz at the condensation boundary. The rate of the Brillouin shift and linewidth initially increased sharply and changed slightly at the condensation boundary. At high temperatures from 65 to 100 °C, low mechanical contrast was observed, as most of the cells were destroyed due to coagulative necrosis taking place in this zone. Similar results were obtained by using shear wave elastography methods [48] in which the velocities and attenuation of propagated shear acoustic waves were measured [56].

Figure 5b illustrates the temperature-dependent Brillouin shift (a) and linewidth (b) of pristine and AuNP-treated bovine liver and chicken muscle tissues. Initially, at T = 20–30 °C, the Brillouin shift and linewidth stayed the same, i.e., the tissues did not undergo denaturation, and the mechanical changes were typically reversible for this temperature elevation range. However, at $T \geq 40$ °C, a slight increase in the Brillouin shift was noticed, which corresponded to the cytotoxic region. A sharp increase in stiffness was observed with further temperature elevation until the onset of ablation. Once ablation occurred, almost constant values of the Brillouin frequency were registered at $T > 60$ °C. The viscous property showed a generally increasing trend as a function of the temperature for both tissues before and after adding AuNPs, as evidenced by the variations in the Brillouin width.

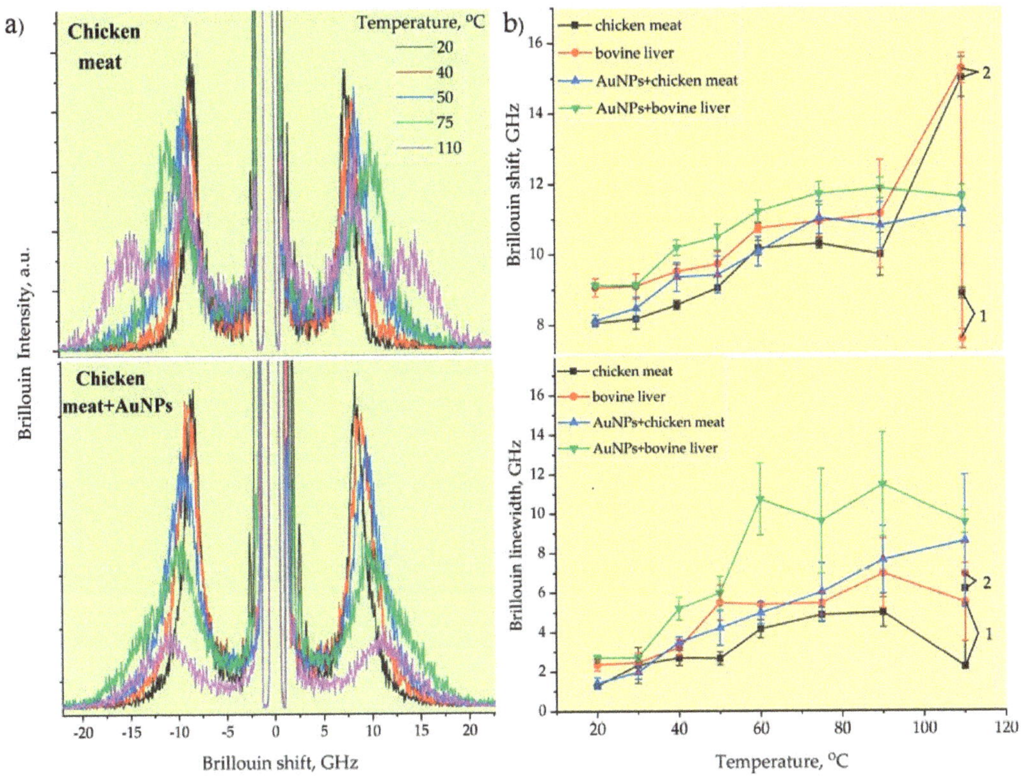

Figure 5. Typical Brillouin spectra of RF - ablated chicken muscle tissues without and with AuNPs (**a**). The Brillouin frequencies and linewidths of chicken meat and bovine liver with and without AuNPs (**b**).

At the initial temperatures, the differences in the Brillouin frequency and linewidth between the muscle and liver tissues were almost 1 GHz. The bovine liver was stiffer and more viscous than the chicken muscle tissue. However, the heating of the pristine tissues up to 110 °C for 40–50 s led to incomplete cell destruction, which was observed in the split Brillouin peaks. The position of the first peak corresponded to the tissue's initial Brillouin frequency value and was related to the non-damaged tissues, while the second peak shifted to a much higher Brillouin frequency of 15 GHz, corresponding to the likely formation of an organic polymeric phase related to completely burned (dead) tissues. Furthermore, the AuNP-mediated hyperthermal RF ablation even enabled the uniform destruction of tissues at the micrometer scale, where no Brillouin peak splitting took place. The AuNPs with a size of 20 nm absorbed RF energy and quickly released heat into the surrounding region due to the increased electron-surface scattering, since the size of the NPs was significantly smaller than the mean free path of the electrons in gold. As can be seen from Figure 5b, the tissues with AuNPs were stiffer and more viscous compared to the pristine ones. The implementation of metallic AuNPs in soft tissue led to the formation of an NP–biotissue composite medium, and further heating of this metal–organic medium caused the formation of a more viscoelastic gel network with NPs inside through protein denaturation.

4. Conclusions

We investigated viscoelastic properties and correlated them with the temperature profiles of AuNP-assisted hyperthermally RF-heated and ablated tissues through confocal Brillouin micro-spectroscopy combined with Mg–silicate-NP-doped temperature-sensing fibers. It was revealed that the significant spectral changes in the Brillouin peak position and linewidth were directly related to the obtained thermal dose and the consequent protein denaturation processes. Increased thermal lesion areas with uniform and complete cell destruction were obtained via the loading of Au NPs into tissues. In addition, compared to pristine tissues without any NPs, further heating of the NP–tissue composite medium led to an increase in the composite tissues' viscoelastic properties at the same temperatures. These results suggest that our hybrid Brillouin–OBR technique can possibly define a tissue's structural deformations during thermal therapies, thus encouraging the use of this technique for the dosimetric control of hyperthermia in future applications. This study opens an avenue for real-time simultaneous localized monitoring of the viscoelastic and thermal properties of metal-nanoparticle-embedded bio-tissues across their heat-driven structural phase transitions in hyperthermal and theranostic applications.

Author Contributions: Conceptualization, Z.U., B.K, D.T. and Z.A.; methodology, B.K., Z.U. and A.S.; software, A.A.; validation, B.K., Z.U., A.G. and W.B.; formal analysis, B.K., Z.U. and D.T.; investigation, B.K., Z.U. and Z.A.; data curation, B.K. and A.A.; writing—original draft preparation, B.K. and Z.U.; writing—review and editing, B.K., Z.U., D.T. and Z.A.; visualization, B.K. and Z.A.; supervision, Z.U. and D.T.; funding acquisition, Z.U. and A.G. All authors have read and agreed to the published version of the manuscript.

Funding: The authors acknowledge the funding from the Nazarbayev University Collaborative Research Program (CRP) for 2020–2022 (Funder Project Reference: 091019CRP2105), EPICGuide (code: 240919FD3908), and SMARTER (code: 091019CRP2117).

Institutional Review Board Statement: Not applicable.

Informed Consent Statement: Not applicable.

Data Availability Statement: The data presented in this work are not publicly available at this time, but can be obtained upon reasonable request from the authors.

Conflicts of Interest: The authors declare no conflict of interest.

References

1. Stauffer, P.R.; Goldberg, S.N. Introduction: Thermal ablation therapy. *Int. J. Hyperth.* **2004**, *20*, 671–677. [CrossRef] [PubMed]
2. Knavel, E.M.; Brace, C.L. Tumor ablation: Common modalities and general practices. *Tech. Vasc. Interv. Radiol.* **2013**, *16*, 192–200. [CrossRef] [PubMed]
3. Almeida, J.I.; Raines, J.K. Radiofrequency ablation and laser ablation in the treatment of varicose veins. *Ann. Vasc. Surg.* **2006**, *20*, 547–552. [CrossRef] [PubMed]
4. Shaheen, N.J.; Sharma, P.; Overholt, B.F.; Wolfsen, H.C.; Sampliner, R.E. Radiofrequency ablation in Barrett's esophagus with dysplasia. *N. Engl. J. Med.* **2009**, *360*, 2277–2288. [CrossRef] [PubMed]
5. D'Arsonval, M.A. Action physiologique des courants alternatifs. *CR Soc. Biol.* **1891**, *43*, 283–286.
6. Rossi, S.; Fornari, F.; Pathies, C.; Buscarini, L. Thermal lesions induced by 480 kHz localized current field in guinea pig and pig liver. *Tumori J.* **1990**, *76*, 54–57. [CrossRef]
7. Obara, K.; Matsumoto, N.; Okamoto, M.; Kobayashi, M.; Ikeda, H.; Takahashi, H.; Katakura, Y.; Matsunaga, K.; Ishii, T.; Okuse, C. Insufficient radiofrequency ablation therapy may induce further malignant transformation of hepatocellular carcinoma. *Hepatol. Int.* **2008**, *2*, 116–123. [CrossRef]
8. Bernardi, P.; Pisa, S. Specific Absorption Rate and Temperature Elevation in a Subject Exposed in the Far-Field of Radio-Frequency Sources Operating in the 10–900-MHz Range. *IEEE Trans. Biomed. Eng.* **2003**, *50*, 295–304. [CrossRef]
9. Shao, Y.L.; Arjun, B.; Leo, H.L.; Chua, K.J. Nano-assisted radiofrequency ablation of clinically extracted irregularly shaped liver tumors. *J. Therm. Biol.* **2017**, *66*, 101–113. [CrossRef]
10. Kennedy, L.C.; Bickford, L.R.; Lewinski, N.A.; Coughlin, A.J.; Hu, Y.; Day, E.S.; West, J.L.; Drezek, R.A. A New Era for Cancer Treatment: Gold-Nanoparticle Mediated Thermal Therapies. *Phototherm. Ther.* **2011**, *7*, 169–183. [CrossRef]
11. Guo, M.; Sun, Y.; Zhang, X. Enhanced radiation therapy of gold nanoparticles in liver cancer. *Appl. Sci.* **2017**, *7*, 232. [CrossRef]
12. Abadeer, N.S.; Murphy, C.J. Recent Progress in Cancer Thermal Therapy Using Gold Nanoparticles. *J. Phys. Chem. C* **2016**, *120*, 4691–4716. [CrossRef]

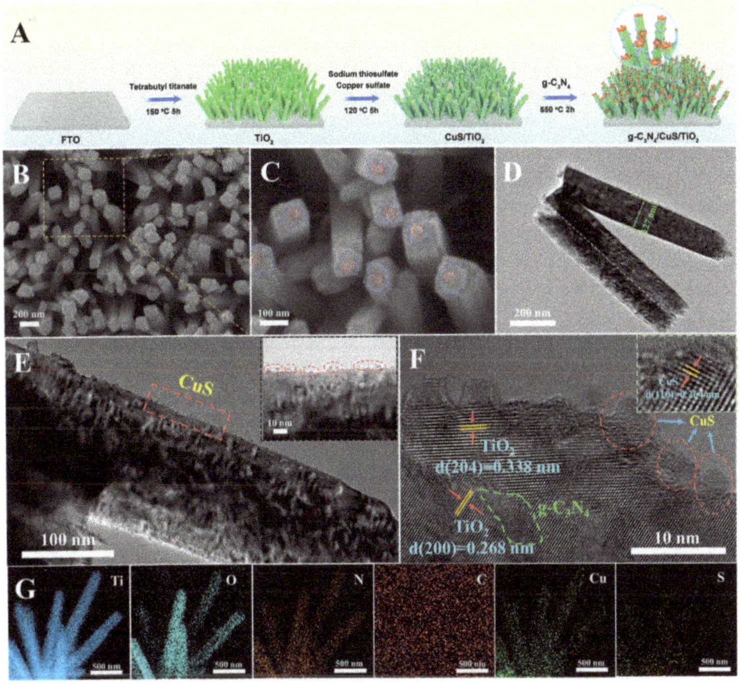

Figure 1. (**A**) Schematic showing the preparation of g-C_3N_4/CuS/TiO_2. (**B**,**C**) Scanning electron microscopy (SEM) images, (**D**,**E**) transmission electron microscopy (TEM) images (inset: corresponding magnified picture of CuS nanoparticles), (**F**) high-resolution transmission electron microscopy (HR-TEM) images (inset: corresponding magnified HR-TEM image of CuS nanoparticles) and (**G**) energy-dispersive X-ray spectroscopy (EDX) of g-C_3N_4/CuS/TiO_2.

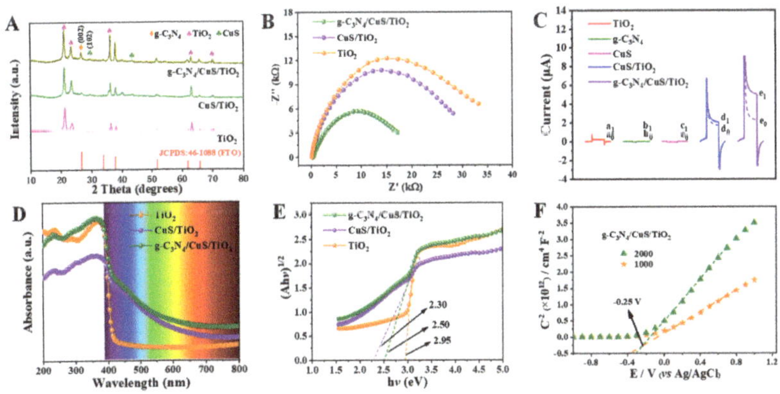

Figure 2. (**A**) X-ray powder diffractometry (XRD) and (**B**) Electrochemical impedance spectroscopy (EIS) plots of TiO_2, CuS/TiO_2, and g-C_3N_4/CuS/TiO_2. (**C**) Photocurrent responses of TiO_2, CuS, g-C_3N_4, CuS/TiO_2, and g-C_3N_4/CuS/TiO_2-modified FTO electrodes in the (a_0–e_0) absence and (a_1–e_1) presence of 123.46 μmol L^{-1} SM. (**D**) UV-Vis absorbance spectra and (**E**) Kubelka–Munk plots of TiO_2, CuS/TiO_2, and g-C_3N_4/CuS/TiO_2. (**F**) Mott–Schottky plots of g-C_3N_4/CuS/TiO_2 at frequencies of 1000 and 2000 Hz.

3.2. Photoelectrochemical Properties of the g-C_3N_4/CuS/TiO_2 PEC Sensing Platform

Next, electrochemical impedance spectroscopy (EIS) was performed to determine the conductivities of the prepared samples. As shown in Figure 2B, g-C_3N_4/CuS/TiO_2 yielded the smallest Nyquist semicircle radius, indicating that the composite had the lowest electron transfer resistance and higher conductivity than the other samples. This result can be attributed to the formation of a large number of heterointerfaces between the ternary semiconductor materials, which is conducive to charge transport.

The photocurrent response provides information about the optoelectronic properties and detection sensitivity of a material. Therefore, we investigated the photocurrents of the TiO_2, g-C_3N_4, CuS, CuS/TiO_2, and g-C_3N_4/CuS/TiO_2-modified FTO electrodes (Figure 2C). Owing to the slow separation of photoinduced electron–hole pairs, the single semiconductor materials (i.e., TiO_2, g-C_3N_4, and CuS) showed relatively low photocurrents both in the presence and absence of antioxidants. In contrast, the CuS/TiO_2 electrode exhibited higher photoelectric activity: its photocurrent increased by 1.49 μA with respect to that of TiO_2 alone, without the addition of the antioxidant. This is because CuS nanoparticles have high absorbance in the visible-light region. However, the photocurrent only increased by 0.32 μA after the addition of 123.46 μmol L^{-1} SM. Therefore, the sensitivity of the CuS/TiO_2 electrode is low and is not sufficient to detect SM. Thus, further modification ofCuS/TiO_2 by loading with g-C_3N_4 nanosheets is required. As expected, the photoelectric response and detection sensitivity of the ternary semiconductor composite (g-C_3N_4/CuS/TiO_2) were significantly improved. Notably, the photocurrent response of the ternary semiconductor composite increased by 0.62 μA in relation to that of the binary semiconductor composite, and the photocurrent was increased by 2.68 μA in the presence of 123.46 μmol L^{-1} SM, suggesting that the ternary heterojunction PEC sensing platform is suitable for evaluating the antioxidant properties of SM. The significant increase in photocurrent can be attributed to two factors: (1) loading with g-C_3N_4 nanosheets increased the number of active sites for the reaction and enhanced light absorption, and (2) the heterojunctions in the ternary semiconductor composites promoted the reaction between antioxidants and holes and greatly improved the electron–hole separation.

To determine the optical properties, UV-Vis absorbance spectroscopy was carried out, as shown in Figure 2D. TiO_2 nanorods can only absorb light between 200 and 400 nm because of their wide band gap. However, the band gap was narrowed and more active sites for the reaction were produced after doping with CuS and g-C_3N_4, which increased the collection and utilization of light. Therefore, the ternary g-C_3N_4/CuS/TiO_2 semiconductor exhibited greater light absorption at 200–800 nm, which is consistent with the EIS results. The ability of a sensor to detect antioxidants is strongly dependent on the band structure of the composites. Therefore, to achieve selective detection, the careful design of the valence band (VB) position of the semiconductor is required. As shown in Figure 2E, the relationship between the Kubelka–Munk function and the photon energy of the samples can be obtained from the UV-Vis absorbance spectra. The VB values ofTiO_2, CuS/TiO_2, and g-C_3N_4/CuS/TiO_2 are 2.95, 2.30, and 2.50 eV, respectively. The Mott–Schottky (M-S) plots ofg-C_3N_4/CuS/TiO_2 semiconductor exhibited positive slopes (Figure 2F), suggesting that the composites are n-type heterostructures, and the flat-band potential (V_{fb}) was −0.25 V (vs. Ag/AgCl). In addition, the conduction band (CB) of the n-type semiconductor is 0.1 V smaller than its V_{fb}; thus, the CB of g-C_3N_4/CuS/TiO_2 is −0.35 V (vs. Ag/AgCl).

3.3. Optimization of Experimental Conditions

The intrinsic properties of photosensitive materials are the main factors affecting the photocurrent signal, but other factors, such as the degree of doping of the semiconductor material, laser wavelength, and applied voltage, cannot be ignored. Optimization experiments were performed inPBS solution containing 123.46 μmol L^{-1} SM. As shown in Figure S3A, we investigated the influence of the amount of deposited CuS on the photocurrent. Crucially, an appropriate amount of CuS nanoparticles will increase the photocurrent, but an excessive amount could result in a drastic decrease in photocurrent. Based on the experiments, a doping level of 5% was selected to prepare the CuS/TiO_2 composite. The mass of$g$-C_3N_4 is another key factor that affects the properties of semiconductors. As shown in Figure S3B, g-C_3N_4 nanosheets provide more active sites for the reaction, and those prepared using a mass of 10 mg generated the highest photocurrent. As shown by the UV-Vis spectra, the g-$C_3N_4/CuS/TiO_2$ exhibited different absorbances in different bands. In particular, as an excitation light source, 630 nm red light has strong penetration. In addition, red light accounts for a large proportion of sunlight, suggesting its suitability for enhancing the sensitivity of the sensor for SM detection (Figure S3C). As shown in Figure S3D, there is a significant gradient between the semiconductor and FTO electrode in the initial state. When a potential is applied, electrons flow rapidly from the conductor band of the semiconductor to the FTO electrode to generate a photocurrent signal. The current at 0 V was 77.3% of that at +0.08 V, indicating that the PEC sensor has adequate sensitivity for SM detection. Therefore, 0 V was chosen as the working voltage for further experiments.

3.4. Antioxidant Assay and Detection Mechanism of the g-$C_3N_4/CuS/TiO_2$ PEC Sensing Platform

To analyze the detection performance of the as-prepared g-$C_3N_4/CuS/TiO_2$ PEC sensor upon 630 nm laser irradiation, the photocurrent signal was measured in PBS solutions (0.1 mol L^{-1}) having different concentrations of SM, PG, and the SM+PG mixture under the optimized assay conditions. As shown in Figure 3A_2, the photoresponse in SM detection exhibited a linear relationship with a detection limit (S/N = 3) of 1.8 μmol L^{-1}, with three linear regression equations obtained for three concentration ranges, as follows: $y = 0.005x + 2.121$, $R^2 = 0.983$ (2.157–100.196 μmol L^{-1}), $y = 0.003x + 2.353$, $R^2 = 0.997$ (100.196–725.196 μmol L^{-1}), and $y = 0.002x + 3.512$, $R^2 = 0.999$ (725.196–1276.667 μmol L^{-1}). Compared with the SM detection performance of previously reported methods (Table S2), the SM detection performance of the g-$C_3N_4/CuS/TiO_2$ PEC sensing platform is excellent. In particular, the proposed sensing platform exhibits a distinctly higher detection sensitivity and a wider linear detection range, which completely satisfy actual sample detection requirements. The photocurrent response of the g-$C_3N_4/CuS/TiO_2$ PEC sensor to PG was lower than that of the SM sensor (Figure 3B_1), and two linear relationships were obtained (Figure 3B_2): $y = 0.003x + 2.832$, $R^2 = 0.991$ (71.787–484.812 μmol L^{-1}) and $y = 0.001x + 3.853$, $R^2 = 0.987$ (484.812–1276.667 μmol L^{-1}). Concerning the antioxidant mixtures of SM+PG, the photoresponse displayed favorable linear relationships in the three concentration ranges (Figure 3C_1), and the corresponding linear equations are $y = 0.079x + 3.606$, $R^2 = 0.983$ (15.990–94.627 μmol L^{-1}); $y = 0.017x + 9.492$, $R^2 = 0.991$ (94.627–735.132 μmol L^{-1}); and $y = 0.005x + 18.292$, $R^2 = 0.989$ (735.132–1276.667 μmol L^{-1}), respectively. These results indicate that the photoelectric response from the two-component mixture of antioxidants was significantly greater than the combined photocurrents of individual components at the same concentration. This indicates a synergistic effect between SM and PG, especially at high concentrations. Based on these findings, we investigated the synergistic effects of SM and VE, BHA, BHT, and TBHQ. As shown in Figure S4, SM was mixed in pairs with these four typical antioxidants, and the obtained trends are similar to that of the two-component mixtures of SM+PG. Presumably, an additive effect occurs at low concentrations, and a synergistic effect occurs at high concentrations. It has been reported that long distances suppress electron transport between antioxidants at low concentrations. Therefore, at relatively high concentrations, frequent electron transfer and the coupling with the oxidant may promote the regeneration of antioxidants, thus resulting in a more pronounced synergistic effect [29–31].

To understand the sensing principle of our sensor, as well as the observed synergistic effects, the electrochemical properties and photoresponses of the antioxidants were investigated in detail. As shown in Table S1, cyclic voltammetry of the antioxidants at a concentration of 484.812 μmol L^{-1} was carried out, yielding the following redox potentials: SM (0.373 V) < VE (0.464 V) < TBHQ (0.531 V) < BHA (0.539 V) < BHT (0.550 V) < PG (0.657 V). SM exhibited the smallest redox potential, suggesting its ability to reduce antioxidants having higher redox potentials [29,32]. As shown in Figure 4 and Table S1, the synergistic effect of binary antioxidant mixtures SM+VE, SM+TBHQ, SM+BHA, SM+BHT, and SM+PG resulted in increases of 1.57-, 1.38-, 2.31-, 2.42-, and 2.67-times, respectively, compared to the sum of the single antioxidant. This result is similar to the trend in the redox potential of the antioxidants. However, the synergistic effect of SM+VE was more obvious than that of SM+TBHQ, which is inconsistent with the increasing trend in redox potential. This is because both SM and VE are natural antioxidants, and their molecular structure and electronic properties (nucleophilic, electrophilic, ionization potential, and phenolic bond dissociation energy) are similar, thereby showing better performance [33]. Furthermore, we investigated the antioxidant effects of mixtures of natural antioxidants (SM+VE), four artificial antioxidants (BHA+BHT+TBHQ+PG), and six natural and artificial antioxidants (SM+VE+BHA+BHT+TBHQ+PG). All the samples showed improvements in the photocurrent responses, SM+VE, BHA+BHT+TBHQ+PG, and SM+VE+BHA+BHT+TBHQ+ PG showed increases of 1.57-, 1.64-, and 1.56-times, respectively, in relation to the sum of photoresponses of all single antioxidants. These results indicate that there are synergistic and antagonistic effects between the antioxidant molecules. In particular, there is a negative correlation between complexity of the antioxidant components and the antioxidant effect; thus, it is crucial to select a suitable formula to maximize the antioxidant performance [34]. These data also indicate that natural antioxidants exhibit good antioxidant properties and could be used to replace synthetic antioxidants as food stabilizers.

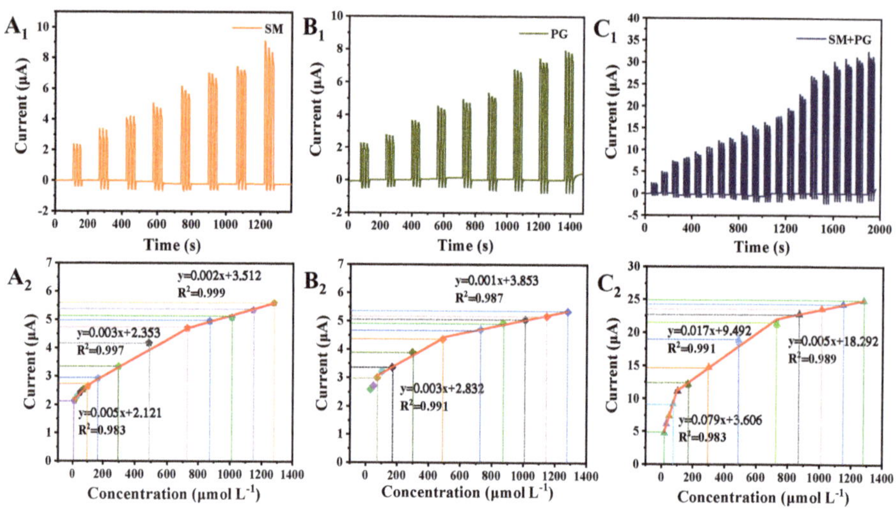

Figure 3. Photocurrent response curves of SM (**A$_1$**), PG (**B$_1$**), and SM+PG in equal proportions (**C$_1$**) generated for the g-C_3N_4/CuS/TiO_2-based PEC sensing platform at different analyte concentrations. Calibration plot of photocurrent versus different concentrations of SM (**A$_2$**), PG (**B$_2$**), and SM+PG (**C$_2$**).

Based on the above results, we conclude that coupling oxidation plays a key role in the synergism of the studied antioxidants: specifically, it reduces the potential difference between pairs of antioxidants and promotes their regeneration, thus improving the performance of binary mixtures of antioxidants. Additionally, the heterojunctions in the g-C_3N_4/CuS/TiO_2 composite promote the reaction between antioxidants and holes, thus improving the separation and transition of electrons and holes (Scheme 1) and amplifying the detectable signal. Therefore, the g-C_3N_4/CuS/TiO_2-modified platform could be applied for the detection and evaluation of the synergistic effect of SM based antioxidants.

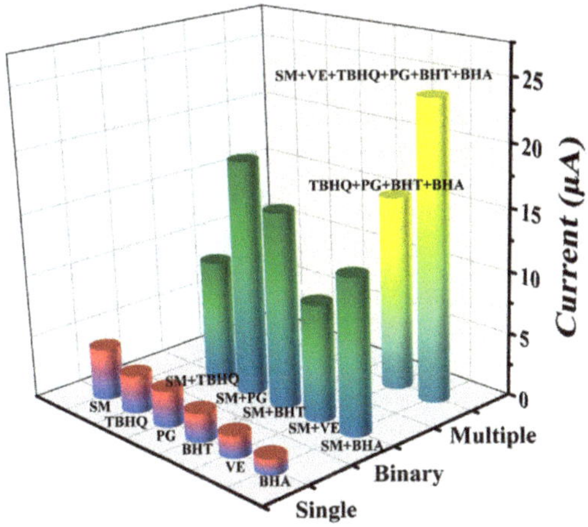

Figure 4. Photocurrents of the g-C_3N_4/CuS/TiO_2 PEC sensor in the presence of 484.812 μmol L^{-1} SM, VE, TBHQ, BHA, BHT, PG, and a mixture of SM and other antioxidants in equal molar ratios.

3.5. Detection Selectivity Stability and Reusability ofg-C_3N_4/CuS/TiO_2 PEC Sensing Platform

Sensor selectivity depends largely on the band structure of the composite. Specifically, an appropriate VB can oxidize target species with a lower redox potential, but does not react with the sugars and amino acids in food. As shown in Figure 5A, we tested the effect of 18 times of fructose, glucose, sucrose, L-malic acid, L-citric acid, ethanol, L-threonine, L-proline, L-lysine, and L-histidine, as well as 700 times of Na^+, K^+, Mg^{2+} and Ca^{2+} on photocurrent based on g-C_3N_4/CuS/TiO_2 PEC sensor with 123.46 μmol L^{-1} SM. No detectable signals (<10%) corresponding to the interfering agents were detected, suggesting that the PEC sensor hashigh SM detection selectivity. Furthermore, we had tested the photocurrent of real samples (soybean oil and peanut oil) in the same conditions of SM, there are insignificant change in the photocurrent response, which further verified the feasibility of the sensor in the analysis of antioxidant capacity of food.

Sensor stability and reusability are also important criterions to evaluate the performance ofPEC sensor, especially for practical, long-term detection in real samples. As shown in Figure 5B, the g-C_3N_4/CuS/TiO_2 PEC sensor was monitored using amperometry with 500 s of "on–off–on" cycling in the presence of 1270.574 μmol L^{-1} SM, which showed no obvious decrease in the photoelectric signal. After 15 days, the photocurrent signal was reduced to 96.8% of the initial state, indicating that the sensor has excellent stability and reusability.

3.6. SM Detection in Soybean and Peanut Oils

Oils and fats naturally become rancid as a result of oxidation during storage or processing. The addition of synthetic or natural antioxidants can slow down lipid peroxidation and speed up the scavenging of free radicals [11,35,36]. In view of the high selectivity of the g-C_3N_4/CuS/TiO_2 PEC sensor platform, the applicability of the sensor for use with real oil samples was investigated using soybean and peanut oils containing various SM concentrations. The corresponding calibration curve is shown in Figure 3A. As listed in Table 1, both soybean and peanut oil samples exhibited high recoveries and low relative standard deviations (RSD; <3%) at different SM concentrations, indicating excellent detection precision and accuracy. Therefore, the natural antioxidant SM showed excellent antioxidant properties and could replace synthetic antioxidants as quality stabilizers for edible oils, which is important for food safety, quality control, and ingredient screening.

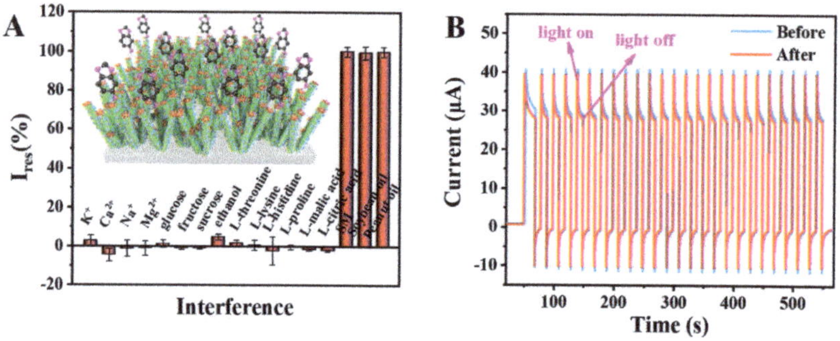

Figure 5. (**A**) Samples and interferences test in the presence of fructose, glucose, sucrose, L-malic acid, L-citric acid, ethanol, L-threonine, L-proline, L-lysine, and L-histidine, as well as Na^+, K^+, Mg^{2+}, and Ca^{2+}, on the photocurrent response of the g-C_3N_4/CuS/TiO_2-based PEC sensor. (**B**) Detection stability and reusability (after 15 days) of the g-C_3N_4/CuS/TiO_2-modified FTO electrode in the presence of 1270.574 μmol L^{-1} SM. The PEC sensor was operated at 0 V under 630 nm light excitation in 0.1 mol L^{-1} PBS (pH = 7.4).

Table 1. SM Concentrations in Soybean and Peanut Oil Samples Detected Using the g-C_3N_4/CuS/TiO_2-based PEC Sensor.

Sample	SM Added (μM)	SM Found (μM)	RSD (%)	Recovery (%)
Soybean oil	4	3.93 ± 0.01	0.31	98.28
	20	20.40 ± 0.20	0.98	102.00
Peanut oil	4	3.97 ± 0.09	2.30	99.16
	20	20.40 ± 0.53	2.59	102.00

4. Conclusions

We developed a novel red-light-mediated signal-on PEC sensor based on a g-C_3N_4/CuS/TiO_2 ternary heterojunction for a highly sensitive analysis of the synergistic antioxidant effects of SM and other antioxidants. The g-C_3N_4/CuS/TiO_2 heterojunction forms a Schottky barrier that facilitates electron transfer. When SM and other antioxidants are present, they scavenge the holes in the VB of g-C_3N_4/CuS/TiO_2, suppressing the recombination of photogenerated charge carriers, thus increasing the PEC photocurrent. The sensor also allowed the rapid evaluation of the synergistic effects of SM and VE, BHA, BHT, TBHQ, and PG. The synergistic antioxidant reaction of multiple antioxidants is mainly driven by coupling oxidation, which enhances the total antioxidant performance by lowering the potential difference between the antioxidants. We found that SM reduces the free radicals of other antioxidants and promotes their regeneration because of its low redox potential. Overall, the g-C_3N_4/CuS/TiO_2 PEC sensor exhibited high sensitivity, good selectivity, and stability, and was successfully applied to the detection of SM in soybean and peanut oils. The reported sensor has applications in nutrition analysis and the development of nutritional additives, which could help improve the quality of life of consumers. In addition, the sensor has potential uses in food fraud analysis, as well as medical research.

Supplementary Materials: The following supporting information can be downloaded at: https://www.mdpi.com/article/10.3390/bios13090859/s1: Reagents; Apparatus; Figure S1: SEM images of TiO_2 and CuS/TiO_2; Figure S2: XP spectra of g-C_3N_4/CuS/TiO_2; Figure S3: Optimization of experimental conditions; Figure S4: Photocurrent response curves; Table S1: Electrochemical characteristics and synergistic effects; Table S2: Comparison of the SM detection performance of various methods. References [10,37–39] are cited in Supplementary Materials.

Author Contributions: Conceptualization, L.H.; methodology, L.H.; validation, Z.L., H.L. and Z.S.; formal analysis, L.H., J.Y. and R.L.; investigation, L.H. and L.N; data curation, J.Y., Z.S., Z.L., R.L. and H.L.; writing—original draft preparation, D.H.; writing—review and editing, L.H.; supervision, L.N. All authors have read and agreed to the published version of the manuscript.

Funding: This research was funded by the following sources: the National Key R&D Program of China, grant number 2022YFD2100304; the National Natural Science Foundation of China, grant numbers 22172040 and 21974031; the University-Industry Collaborative Education Program of the Ministry of Education of China, grant number 220605940231526; the Guangdong Basic and Applied Basic Research Foundation, grant number 2023B1515040004; the Department of Science and Techniques of Guangdong Province, grant numbers 2021A1515010180 and 2019B010933001; the Department of Guangdong Provincial Public Security, grant number GZQC20-PZ11-FD084; the Science and Technology Projects of Guangzhou, grant number 202102010449; the Department of Science & Technology of Guangdong Province, grant number 2022A156; the Guangzhou Municipal Science and Technology Bureau, grant number 202102010449; the Tertiary Education Scientific Research Project of Guangzhou Municipal Education Bureau, grant number 202235344; and the China Postdoctoral Science Foundation, grant number 2022M720867.

Institutional Review Board Statement: Not applicable.

Data Availability Statement: Not applicable.

Conflicts of Interest: The authors declare no conflict of interest.

References

1. Motaghed Mazhabi, R.; Ge, L.; Jiang, H.; Wang, X.M. A facile photoelectrochemical sensor for high sensitive ROS and AA detection based on graphitic carbon nitride nanosheets. *Biosens. Bioelectron.* **2018**, *107*, 54–61. [CrossRef]
2. Xu, C.; Wang, S.H.; Wang, H.; Liu, K.; Zhang, S.Y.; Chen, B.; Liu, H.; Tong, F.; Peng, F.; Tu, Y.F.; et al. Magnesium-Based micromotors as hydrogen generators for precise rheumatoid arthritis therapy. *Nano Lett.* **2021**, *21*, 1982–1991. [CrossRef] [PubMed]
3. Han, X.M.; Liu, L.; Gong, H.Y.; Luo, L.P.; Han, Y.R.; Fan, J.W.; Xu, C.F.; Yue, T.L.; Wang, J.L.; Zhang, W.T. Dextran-stabilized Fe-Mn bimetallic oxidase-like nanozyme for total antioxidant capacity assay of fruit and vegetable food. *Food Chem.* **2022**, *371*, 131115. [CrossRef]
4. Abbas, S.; Sharif, M.K.; Sibt-e-Abbas, M.; Fikre Teferra, T.; Sultan, M.T.; Anwar, M.J. Nutritional and therapeutic potential of sesame seeds. *J. Food Qual.* **2022**, *2022*, 6163753. [CrossRef]
5. Fabre, G.; Bayach, I.; Berka, K.; Paloncýová, M.; Starok, M.; Rossi, C.; Duroux, J.-L.; Otyepka, M.; Trouillas, P. Synergism of antioxidant action of vitamins E, C and quercetin is related to formation of molecular associations in biomembranes. *Chem. Commun.* **2015**, *51*, 7713–7716. [CrossRef]
6. Yuan, T.; Chu, C.; Shi, R.; Cui, T.; Zhang, X.; Zhao, Y.; Shi, X.; Hui, Y.; Pan, J.; Qian, R.; et al. ApoE-Dependent protective effects of sesamol on high-fat diet-induced behavioral disorders: Regulation of the microbiome-gut–brain axis. *J. Agri. Food Chem.* **2019**, *67*, 6190–6201. [CrossRef]
7. Majdalawieh, A.F.; Mansour, Z.R. Sesamol, a major lignan in sesame seeds (sesamum indicum): Anti-cancer properties and mechanisms of action. *Eur. J. Pharmacol.* **2019**, *855*, 75–89. [CrossRef]
8. Liu, Q.; Xie, T.Z.; Xi, Y.J.; Li, L.; Mo, F.F.; Liu, X.B.; Liu, Z.G.; Gao, J.M.; Yuan, T. Sesamol attenuates amyloid peptide accumulation and cognitive deficits in APP/PS1 mice: The mediating role of the gut–brain axis. *J. Agri. Food Chem.* **2021**, *69*, 12717–12729. [CrossRef]
9. Han, L.P.; Li, J.H.; Wang, S.J.; Cheng, W.W.; Ma, L.K.; Liu, G.Q.; Han, D.X.; Niu, L. Sesamol can inhibit the formation of glycidyl ester in deep frying palm oil. *J. Food Process. Pres.* **2022**, *46*, e16236. [CrossRef]
10. Ohno, K.-i.; Sato, K.; Kumano, M.; Watanabe, K.; Fujimura, T. Electrochemical detection of sesamol dimer and its application to measurement of radicals. *Anal. Sci.* **2021**, *37*, 633–635. [CrossRef]
11. Huang, J.; Sun, Q.; Song, G.; Qi, S.; Chen, J.; Zhang, P.; Geng, T.; Lin, Q.; Duan, Y. Antioxidant and anti-isomerization effects of sesamol and resveratrol on high oleic acid peanut oil. *LWT* **2020**, *123*, 109077. [CrossRef]
12. Sallam, K.I.; Abd-Elghany, S.M.; Imre, K.; Morar, A.; Herman, V.; Hussein, M.A.; Mahros, M.A. Ensuring safety and improving keeping quality of meatballs by addition of sesame oil and sesamol as natural antimicrobial and antioxidant agents. *Food Microbiol.* **2021**, *99*, 103834. [CrossRef]
13. El-Bindary, M.A.; Shahat, A.; El-Deen, I.M.; El-Afify, M.A.M.; Hassan, N. Synthesis and application of a novel self-smart sensor based on a modified amino-functionalized Zr-metal–organic framework for rapid and selective detection of some toxic metals in wastewater. *Appl. Organomet. Chem.* **2023**, *37*, e7029. [CrossRef]

14. Huang, L.K.; Liang, Z.S.; Zhang, F.; Luo, H.; Liang, R.L.; Han, F.J.; Wu, Z.F.; Han, D.X.; Shen, J.; Niu, L. Upconversion $NaYF_4$:Yb/Er–TiO_2–Ti_3C_2 heterostructure-based near-infrared light-driven photoelectrochemical biosensor for highly sensitive and selective d-serine detection. *Anal. Chem.* **2022**, *94*, 16246–16253. [CrossRef] [PubMed]
15. Zayats, M.; Kharitonov, A.B.; Pogorelova, S.P.; Lioubashevski, O.; Katz, E.; Willner, I. Probing photoelectrochemical processes in Au−CdS nanoparticle arrays by surface plasmon resonance: application for the detection of acetylcholine esterase inhibitors. *J. Am. Chem. Soc.* **2003**, *125*, 16006–16014. [CrossRef] [PubMed]
16. Zhang, Q.Q.; Yang, H.K.; Du, C.C.; Liu, S.Y.; Zhang, X.H.; Chen, J.H. Bifunctional magnetic Fe_3O_4@Cu_2O@TiO_2 nanosphere-mediated dual-mode assay of PTP1B activity based on photocurrent polarity switching and nanozyme-engineered biocatalytic precipitation strategies. *Anal. Chem.* **2022**, *94*, 13342–13349. [CrossRef] [PubMed]
17. Shu, J.; Tang, D.P. Recent advances in photoelectrochemical sensing: From engineered photoactive materials to sensing devices and detection modes. *Anal. Chem.* **2020**, *92*, 363–377. [CrossRef]
18. Dai, W.X.; Zhang, L.; Zhao, W.W.; Yu, X.D.; Xu, J.J.; Chen, H.Y. Hybrid PbS quantum dot/nanoporous NiO film nanostructure: Preparation, characterization, and application for a self-powered cathodic photoelectrochemical biosensor. *Anal. Chem.* **2017**, *89*, 8070–8078. [CrossRef]
19. Xiao, F.X.; Zeng, Z.; Liu, B. Bridging the gap: Electron relay and plasmonic sensitization of metal nanocrystals for metal clusters. *J. Am. Chem. Soc.* **2015**, *137*, 10735–10744. [CrossRef]
20. Zhao, W.W.; Yu, X.D.; Xu, J.J.; Chen, H.Y. Recent advances in the use of quantum dots for photoelectrochemical bioanalysis. *Nanoscale* **2016**, *8*, 17407–17414. [CrossRef]
21. Chen, C.; Cai, W.M.; Long, M.C.; Zhou, B.X.; Wu, Y.H.; Wu, D.Y.; Feng, Y.J. Synthesis of visible-light responsive graphene oxide/TiO_2 composites with p/n heterojunction. *ACS Nano* **2010**, *4*, 6425–6432. [CrossRef]
22. Qi, H.; Hou, Y.; Wang, W.J.; Deng, W.; Tang, L.; Zhang, C.Y. Single-crystalline nanoflakes assembled CuS microspheres with improved sodium ion storage. *J. Alloys Compd.* **2023**, *942*, 168884. [CrossRef]
23. Zhao, X.; Fan, Y.Y.; Zhang, W.S.; Zhang, X.J.; Han, D.H.; Niu, L.; Ivaska, A. Nanoengineering construction of Cu_2O nanowire arrays encapsulated with g-C_3N_4 as 3D spatial reticulation all-solid-state direct Z-scheme photocatalysts for photocatalytic reduction of carbon dioxide. *ACS Catal.* **2020**, *10*, 6367–6376. [CrossRef]
24. Dekkouche, S.; Morales-Torres, S.; Ribeiro, A.R.; Faria, J.L.; Fontàs, C.; Kebiche-Senhadji, O.; Silva, A.M.T. In situ growth and crystallization of TiO_2 on polymeric membranes for the photocatalytic degradation of diclofenac and 17α-ethinylestradiol. *Chem. Eng. J.* **2022**, *427*, 131476. [CrossRef]
25. Lei, G.C.; Zhao, W.T.; Shen, L.J.; Liang, S.J.; Au, C.; Jiang, L.L. Isolated iron sites embedded in graphitic carbon nitride (g-C_3N_4) for efficient oxidative desulfurization. *Appl. Catal. B-Environ.* **2020**, *267*, 118663. [CrossRef]
26. Xiao, Y.T.; Guo, S.E.; Tian, G.H.; Jiang, B.J.; Ren, Z.Y.; Tian, C.G.; Li, W.; Fu, H.G. Synergetic enhancement of surface reactions and charge separation over holey C_3N_4/TiO_2 2D heterojunctions. *Sci. Bull.* **2021**, *66*, 275–283. [CrossRef] [PubMed]
27. Liu, H.Q.; He, Y.N.; Zhang, H.; Cao, K.Z.; Wang, S.D.; Jiang, Y.; Jing, Q.S.; Jiao, L.F. Lowering the voltage-hysteresis of CuS anode for Li-ion batteries via constructing heterostructure. *Chem. Eng. J.* **2021**, *425*, 130548. [CrossRef]
28. Xiao, L.M.; Liu, T.F.; Zhang, M.; Li, Q.Y.; Yang, J.J. Interfacial construction of Zero-dimensional/one-dimensional g-C_3N_4 nanoparticles/TiO_2 nanotube arrays with Z-scheme heterostructure for improved photoelectrochemical water splitting. *ACS Sustain. Chem. Eng.* **2019**, *7*, 2483–2491. [CrossRef]
29. Liu, D.; Li, Y.; Qian, Y.; Xiao, Y.; Du, S.J.; Qiu, X.Q. Synergistic antioxidant performance of lignin and quercetin mixtures. *ACS Sustain. Chem. Eng.* **2017**, *5*, 8424–8428. [CrossRef]
30. Iglesias, J.; Pazos, M.; Andersen, M.L.; Skibsted, L.H.; Medina, I. Caffeic acid as antioxidant in fish muscle: Mechanism of synergism with endogenous ascorbic acid and α-Tocopherol. *J. Agri. Food Chem.* **2009**, *57*, 675–681. [CrossRef]
31. Han, F.J.; Song, Z.Q.; Xu, J.N.; Dai, M.J.; Luo, S.L.; Han, D.X.; Niu, L.; Wang, Z.X. Oxidized titanium carbide MXene-enabled photoelectrochemical sensor for quantifying synergistic interaction of ascorbic acid based antioxidants system. *Biosens. Bioelectron.* **2021**, *177*, 112978. [CrossRef] [PubMed]
32. Zanfini, A.; Corbini, G.; La Rosa, C.; Dreassi, E. Antioxidant activity of tomato lipophilic extracts and interactions between carotenoids and α-tocopherol in synthetic mixtures. *LWT-Food Sci. Technol.* **2010**, *43*, 67–72. [CrossRef]
33. Palheta, I.C.; Borges, R.S. Sesamol is a related antioxidant to the vitamin E. *Chem. Data Collect.* **2017**, *11–12*, 77–83. [CrossRef]
34. Sahu, S.C. Food additives: A special issue of the journal Food and Chemical Toxicology. *Food Chem. Toxicol.* **2017**, *107*, 529. [CrossRef]
35. El-Bindary, M.A.; El-Desouky, M.G.; El-Bindary, A.A. Metal–organic frameworks encapsulated with an anticancer compound as drug delivery system: Synthesis, characterization, antioxidant, anticancer, antibacterial, and molecular docking investigation. *Appl. Organomet. Chem.* **2022**, *36*, e6660. [CrossRef]
36. Nawade, B.; Mishra, G.P.; Radhakrishnan, T.; Dodia, S.M.; Ahmad, S.; Kumar, A.; Kumar, A.; Kundu, R. High oleic peanut breeding: Achievements, perspectives, and prospects. *Trends Food Sci. Tech.* **2018**, *78*, 107–119. [CrossRef]
37. Morawska, K.; Festinger, N.; Chwatko, G.; Głowacki, R.; Ciesielski, W.; Smarzewska, S. Rapid electroanalytical procedure for sesamol determination in real samples. *Food Chem.* **2020**, *309*, 125789. [CrossRef]

38. Liu, W.; Zhang, K.D.; Qin, Y.Q.; Yu, J.J. A simple and green ultrasonic-assisted liquid–liquid microextraction technique based on deep eutectic solvents for the HPLC analysis of sesamol in sesame oils. *Anal. Method.* **2017**, *9*, 4184. [CrossRef]
39. Liu, H.L.; Wu, D.; Liu, Y.L.; Zhang, H.J.; Ma, T.Z.; Aidaerhan, A.; Wang, J.; Sun, B.G. Application of an optosensing chip based on molecularly imprinted polymer coated quantum dots for the highly selective and sensitive determination of sesamol in sesame oils. *J. Agric. Food Chem.* **2015**, *63*, 2545–2549. [CrossRef]

biosensors

MDPI

Article

Hierarchical Ti-MOF Microflowers for Synchronous Removal and Fluorescent Detection of Aluminum Ions

Jianguo Zhou [1,2,†], Jieyao Song [2,3,†], Guangqiang Ma [4], Yongjian Li [2], Yanan Wei [2], Fei Liu [2] and Hongjian Zhou [1,2,3,*]

1 Institutes of Physical Science and Information Technology, Anhui University, Hefei 230601, China
2 Key Laboratory of Materials Physics, Centre for Environmental and Energy Nanomaterials, Anhui Key Laboratory of Nanomaterials and Nanotechnology, Institute of Solid State Physics, HFIPS, Chinese Academy of Sciences, Hefei 230031, China
3 Department of Materials Science and Engineering, University of Science and Technology of China, Hefei 230026, China
4 School of Mechanical and Electrical Engineering, Xi'an University of Architecture and Technology, Xi'an 710055, China
* Correspondence: hjzhou@issp.ac.cn; Tel.: +86-55165596305; Fax: +86-55165591434
† These authors contributed equally to this work.

Abstract: Bifunctional luminescence metal-organic frameworks with unique nanostructures have drawn ongoing attention for simultaneous determination and elimination of metal ions in the aqueous environment, but still remain a great challenge. In this work, three-dimensional hierarchical titanium metal-organic framework (Ti-MOF) microflowers were developed by a secondary hydrothermal method for not only highly sensitive and selective detection of Al(III), but also simultaneously efficient decontamination. The resulting Ti-MOF microflowers with a diameter of 5–6 μm consisted of nanorods with a diameter of ~200 nm and a length of 1–2 μm, which provide abundant, surface active sites for determination and elimination of Al(III) ions. Because of their substantial specific surface area and superior fluorescence characteristics, Ti-MOF microflowers are used as fluorescence probes for quantitative determination of Al(III) in the aqueous environment. Importantly, the specific FL enhancement by Al(III) via a chelation-enhanced fluorescence mechanism can be utilized for selective and quantitative determination of Al(III). The Al(III) detection has a linear range of 0.4–15 μM and a detection limit as low as 75 nM. By introducing ascorbic acid, interference of Fe(III) can be avoided to achieve selective detection of Al(III) under various co-existing cations. It is noteworthy that the Ti-MOF microflowers exhibit excellent adsorption capacity for Al(III) with a high adsorption capacity of 25.85 mg g^{-1}. The rapid adsorption rate is consistent with a pseudo-second order kinetic model. Ti-MOF is a promising contender as an adsorbent and a fluorescent chemical sensor for simultaneous determination and elimination of Al(III) due to its exceptional water stability, high porosity, and intense luminescence.

Keywords: luminescence; metal-organic frameworks; hierarchical nanostructures; fluorescent sensors; metal ions

1. Introduction

Aluminum is a substance that is commonly present in the environment and in human activities, is thought to hinder the growth of plants, and is linked to diseases such as Alzheimer's and Parkinson's diseases [1]. The World Health Organization (WHO) recommends that the maximum aluminum content in drinking water should be limited to 7.4 μM, and the daily consumption of aluminum for humans should be between 3 and 10 mg [2]. Measurement and purification of Al(III) is essential for environmental monitoring and preserving human health, since aluminum often exists in the body and environment as the Al(III) ion [3]. Consequently, significant efforts are being undertaken to develop novel approaches for the selective detection and removal of Al(III) because of its pervasive contamination and high toxicity. For sensitive detection of Al(III) pollution in the aqueous environment, a variety of atomic absorption or emission spectrometry methods, inductively coupled plasma mass spectrometry (ICP-MS), and fluorescence methods, have been utilized [4,5]. Due to its ease of use, cheap cost, convenience, sensitivity, real-time monitoring, and naked-eye visibility, fluorescence based chemosensors have been identified as promising tools for detecting Al(III). Nevertheless, sensitive and selective detection of Al(III) is still tedious, with several disadvantages due to its lack of spectroscopic features, sluggish coordination, and a great hydration capacity.

Toxic metals must be removed from polluted water. Many methods have been developed recently to remove harmful metals from diverse substances, including ion displacement, membrane filtration, redox co-precipitation, chemical deposition, absorption methods, and solid-phase extraction [6,7]. Among these technologies, adsorption is believed to be the most cost effective method to remove Al(III) from contaminated water because of its simplicity of use and the availability of a variety of adsorbents, such as *Streptomyces rimosus* biomass, Fe_3O_4/TEOS/AMEO/GA, activated carbon, and polyacrylonitrile beads [8]. Unfortunately, most of the previous studies either concentrated on functional adsorption materials to improve selectivity and sensitivity towards metal ion detection or centered on absorbing materials to boost their uptake capacity for metal ions, which seriously hindered their practical applications. Developing a novel bifunctional material that can both identify and eliminate metal ions from environmental samples is of ongoing interest, and would significantly increase the fluorescence signal due to effective adsorption. Qian et al. reported a covalently grafted naphthalimide derivative of 2,6-bis(aminomethyl)pyridine on the surface of silica particles to create a bifunctional fluorescence sensor for the simultaneous detection and separation of trace Hg^{2+} in polluted water samples [9]. Manos et al. developed a microporous metal-organic framework ($H_{16}[Zr_6O_{16}(H_2PATP)_4]Cl_8 \cdot xH_2O$) that has an exceptional capacity to quickly collect and selectively detect hexavalent chromium in aqueous solution [10]. Our group also reported the selective detection and simple removal of arsenate from contaminated water using amino-functionalized iron-based metal organic framework (NH_2-MIL-88(Fe)) nanooctahedra [11]. However, to the best of our knowledge, such bifunctional materials that are capable of detecting and removing Al(III) from environmental samples simultaneously have not been reported to date.

To meet this demand, we attempted development of a bifunctional material with the following characteristics: suitable recognition sites to provide specific receptor-target interactions for luminescence signal responses, strong metal chelating groups with high affinity toward specific target metal ions, ordered and high-density accessible binding sites, exceptional water and chemical stability to facilitate multiple uses, and high efficiency in converting light into electrical energy. To fulfill this goal, luminescence metal-organic framework (LMOF) materials are ideal candidates that offer a special scaffold and tailored functionality with luminescence in integrating two functions of sensing and adsorption into a single material, in contrast to conventional sensing and remediation materials [12–15]. Numerous experiments using MOF-based sensors have recently been published to detect Al(III) [16,17]. For instance, Zhan, et al. developed a water-resistant terbium-MOF sensor for the accurate, precise, and recyclable detection of Al (III) [17]. To ratiometrically detect Al^{3+} ions in an aqueous solution, Zheng et al. created a new zirconium-based MOF composite material called UiO-$(OH)_2$@RhB [18]. Design of an MOF-based sensor for simultaneous detection and removal of Al(III) in an aquatic environment, however, remains a formidable issue.

Adsorption efficiency can be significantly impacted by changing the geometry of adsorbents. This suggests that in order to reduce the diffusion barrier and increase adsorption efficiency, research on the design of adsorbents at the geometrical level should be pursued. The integration of nano-building blocks with the proper order and flexibility to create hierarchical nanostructures has proven to be a successful strategy for improving adsorption performance [19]. Hierarchical nanostructures have been successfully created using a wide range of materials, including metals, metal oxides and sulfides, carbons, and certain organic polymers [20]. Metal-organic hierarchical nanostructures, however, are seldom explored. Even though MOF sensing applications have been of interest, it is often challenging to create a free-standing, three-dimensional (3D) hierarchical micro-/nano-structured MOF that has nanoscale thickness and microsized lateral dimensions. A collection of hierarchical metal-organic framework nanostructures has recently been published by Wang et al. [21,22]. These substances differ from their pure organic or inorganic equivalents in terms of functions, and they may find use in a variety of industries. Based on these advantages, we assumed that the integration of 3D hierarchical micro-/nano-structure and bifunctional LMOF would be highly desired and could improve fluorescence sensing and adsorption performance.

Following these criteria, we developed a novel approach for building three-dimensional (3D) hierarchical titanium-based MOF (Ti-MOF) microflowers by a secondary hydrothermal process for potential use dual-functional fluorescence sensors and adsorbents (Scheme 1). Using a variety of methods, including scanning electron microscopy (SEM), transmission electron microscopy (TEM), X-ray diffractometer (XRD), and fluorescence spectroscopy, the structural features of the obtained product were carefully examined. The sensing characteristics of the developed materials toward Al(III) were investigated, including sensitivity and selectivity in the presence of various interfering ions, to illustrate possible bifunctional applications. Meanwhile, the concomitant adsorption performance of Al(III) onto hierarchical Ti-MOF microflowers was carefully evaluated. Using X-ray photoelectron spectroscopy (XPS) and Fourier transform infrared (FT-IR) spectroscopic studies, the fundamental sensing and adsorption mechanism of hierarchical Ti-MOF microflowers towards Al(III) was investigated.

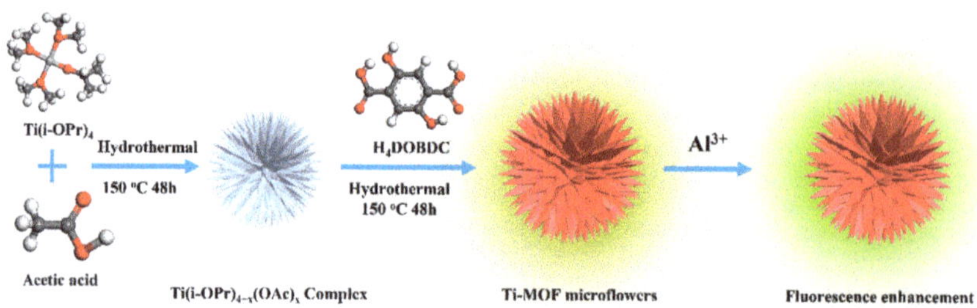

Scheme 1. Schematic illustration for manufacturing hierarchical Ti-MOF microflowers and Al(III) ions luminous sensing.

2. Experimental Section

2.1. Synthesis of Hierarchical Ti-MOF Microflowers

According to a previously published approach, with certain modifications, hierarchical Ti-MOFs microflowers were created by a secondary hydrothermal process [23,24]. First, 178 mg $Ti(i\text{-}OPr)_4$ was dissolved in 10 mL acetic acid with stirring until a milk white emulsion formed. Then, after being transferred, the mixture was heated at 150 °C for 48 h in a 30 mL Teflon-lined stainless-steel autoclave. The autoclave was allowed to cool naturally to ambient temperature before the solution was centrifuged for six minutes at 8000 rpm to remove the supernatant solution, and washed with deionized water three times. Secondly, the achieved white precipitate was re-dispersed in 10 mL of deionized water, and 124 mg H_4DOBDC was injected to the above solution with stirring. A 30 mL Teflon-lined stainless-steel autoclave was used to transfer the mixture, which was heated at 150 °C for 48 h. Finally, the dark red crystals were centrifuged at 8000 rpm for 6 min and rinsed with deionized water and ethanol three times.

To confirm that the synthesis of Ti-MOFs microflowers was achieved by employing the secondary hydrothermal method, direct hydrothermal reaction of $Ti(i\text{-}OPr)_4$ (178 mg) and H_4DOBDC (124 mg) in 10 mL acetic acid was carried out in a 48-h experiment in a 30 mL Teflon-lined stainless-steel autoclave. The produced samples were centrifuged at 8000 rpm for 15 min and rinsed with deionized water and ethanol three times.

2.2. *Fluorescence Determination of Aluminium Ion*

In a typical procedure, the dispersed solutions were made by combining 200 mL of deionized water with 83.5 mg of Ti-MOFs microflowers, followed by 30 min of ultrasonic agitation. Various concentrations of aluminum ion solutions were prepared in deionized water solution. The fluorescence spectrophotometer excitation wavelength was set at 355 nm, and the excitation and emission slit widths were both 2.0. A volume of 3 mL of the sample in a quartz cuvette was used for the fluorescence measurements. For fluorescence detection of Al(III), 60 μL of Al(III) solutions at various concentrations were individually added to 2940 μL stock solutions of Ti-MOFs microflowers, the final Al(III) concentration range being from 0.4 to 100 μM.

To investigate the selectivity of the Ti-MOFs microflowers fluorescent probe towards Al(III) determination, various interference ions, such as Cu(II), Cd(II), Ag(I), Co(II), Ni(II), Ca(II), Mg(II), Fe(II), Fe(III), Hg(II), Mn(II), Zn(II), NH_4^+, were separately added to the 2970 μL stock solutions of Ti-MOFs microflowers containing 30 μL of 1 mM Al(III). The concentration of all interference ions was 100 μM, i.e., 10-fold higher than that of Al(III), and the same detection conditions were employed as in the aforementioned Al(III) sensing tests.

The following figures and tables provide averages and standard deviations from all selectivity and selectivity studies, which were carried out in triplicate.

2.3. *Adsorption of Aluminum Ion*

For the kinetic adsorption experiment, 300 mg of hierarchical Ti-MOFs microflowers were mixed with 300 mL of 50 mg L^{-1} Al(III) solution. The mixture was agitated at 150 rpm at 25 °C, and the pH of the solution was kept constant at 4.7 ± 0.1 by adding tiny amounts of either HCl (0.1 M) or NaOH (0.1 M). At predetermined intervals, 3 mL samples of the suspension were taken. A 0.22 μm membrane filter was used to filter the samples before inductively coupled plasma optical emission spectrometry (ICP-OES) analysis (ICP-6000, Thermal Electron, Waltham, MA, USA). Thus, the ICP-OES could independently acquire the amount of aluminum present in the Al(III) containing solution.

Studies of isothermal adsorption were carried out at pH 4.7 ± 0.1 and 25 °C. To attain adsorption equilibration, 20 mg of the as-prepared hierarchical Ti-MOF microflowers were introduced to 20 mL of Al(III) solution (2–100 mgL^{-1}) and shaken for 96 h. The aforementioned solution was filtered using a 0.22 μm membrane filter after the liquid and solid phases were separated by centrifugation. The concentration of Al(III) was analyzed by ICP-OES after filtration. The Al(III) adsorption amount was calculated by Equation (1):

$$q_e = \frac{(C_0 - C_e) \times V}{m} \tag{1}$$

where q_e is the amount of Al(III) adsorbed on adsorbent at equilibrium time ($mg{\cdot}g^{-1}$), and C_0 and C_e represent the initial and equilibrium Al(III) concentrations ($mg{\cdot}L^{-1}$) in the solution, respectively, V indicates the solution volume (mL) and m is the mass of the adsorbent (mg).

A pseudo-second-order kinetics model was used to analyze the adsorption kinetic data on the assumption that chemisorption was the rate-determining phase. The pseudo-second-order kinetics mode is presented in Equation (2):

$$\frac{t}{Q} = \frac{1}{Q_e}t + \frac{1}{k_2 Q_e^2} \tag{2}$$

where k_2 is the pseudo-second-order rate constant ($g{\cdot}mg^{-1}h^{-1}$), Q_e is the equilibrium adsorption capacity ($mg{\cdot}g^{-1}$), and Q is the amount of Al(III) ions adsorbed on the surface of Ti-MOF at time t (min), respectively. A linear plot of t/Q yields the values of Q_e and k_2.

The capacities of absorbents were determined by equilibrium adsorption isotherms. The experimental data are fitted to Langmuir models.

$$Q_e = \frac{Q_m K_L C_e}{1 + K_L C_e} \tag{3}$$

where C_e is the equilibrium concentration of Al(III) in aqueous solution ($mg{\cdot}L^{-1}$), Q_e represents the amount of Al(III) adsorbed on the surface of Ti-MOF ($mg{\cdot}g^{-1}$), Q_m indicates the maximum amount of Al(III) adsorbed per unit weight of Ti-MOF to form a complete monolayer coverage on the surface, and K_L is the ratio of the rate constants of adsorption and desorption.

3. Results and Discussion

Monodisperse hierarchical Ti-MOF microflowers were produced by a quick, easy, and surfactant-free secondary hydrothermal technique. Through initial hydrothermal synthesis, the precursor of Ti-MOF was fabricated using $Ti(i\text{-}OPr)_4$ as a metal precursor, and acetic acid as the linker and solvent. Acetate groups (OAc^-) interact with titanium alkoxides as complexing nucleophilic ligands. Acetate groups function as bidentate bridging ligands during the stoichiometric process, which also results in an increase in the coordination number of Ti from four to six, and the formation of oligomeric species $Ti(i\text{-}OPr)_{4-x}/(AcOH)_x$. Using SEM and TEM measurements, the shape and structure of the $Ti(i\text{-}OPr)_{4-x}/(AcOH)_x$ complex were identified. As illustrated in Figure 1A, the synthesized $Ti(i\text{-}OPr)_{4-x}/(AcOH)_x$ complex exhibited a great number of nanoscale petals combine to produce a monodispersed flower-like hierarchical structure, which typically had a diameter between 1–3 μm. The hierarchical microflowers are shown in a high-magnification TEM image in Figure 1B. It is evident that these microflowers are made up of different 2D nanosheets (petals), which grew from the flower center in all directions to construct 3D hierarchical nanostructures. The petals of the flower-like $Ti(i\text{-}OPr)_{4-x}/(AcOH)_x$ complex have a slightly curved compact structure and an average length of about 500 nm (Figure 1C). It is important to note that the production of the aforementioned $Ti(i\text{-}OPr)_{4-x}/(AcOH)_x$ complex with flower-like hierarchical structures depends critically on acetic acid. By contrast, by using pure deionized water to replace acetic acid, only highly aggregated, irregular particles were produced. Subsequently, the prepared flower-like $Ti(i\text{-}OPr)_{4-x}/(AcOH)_x$ complex was a precursor in a reaction with H_2DOBDC to produce Ti-MOF through a second hydrothermal synthesis in which AcOH was completely substituted by H_2DOBDC. The Ti-MOF completely inherited the original flower-like morphology from the $Ti(i\text{-}OPr)_{4-x}/(AcOH)_x$ complex. Interestingly, the petals of $Ti(i\text{-}OPr)_{4-x}/(AcOH)_x$ complex had smooth surfaces, while the petal building block of Ti-MOF, owing to the substitution of acetic acid by H_2DOBDC, had a perforated structure made of linked or stacked nanoparticles (10–20 nm in size) with a rough surface, as shown in Figure S1. Overall, the microflower morphologies were conserved during the hydrothermal process even if expanding pore channels affected the internal architecture. However, the direct hydrothermal reaction of $Ti(i\text{-}OPr)_4$ and H_4DOBDC in acetic acid resulted in pseudo-spherical Ti-MOF nanoparticles 200 ± 20 nm in size (Figure S2). The Ti-MOF microflowers precursor was $Ti_2(HDOBDC)_2(H_2DOBDC)$ (NTU-9) [24]. This has two-dimensional (2D) hybrid layers that are aligned perpendicular to the c-axis in hexagonal prism crystals. Each layer has two-dimensional honeycomb-like layers and is built from the bond between Ti^{4+} and DOBDC ligands. The Ti^{4+} is octahedrally coordinated with six oxygen atoms from the hydroxide and carboxylate groups of the ligand DOBDC, with Ti-O bond lengths of 1.858 and 2.037 Å, respectively. Inside the layers, there are 11×11 Å^2 one-dimensional hexagonal channels that all the uncoordinated oxygen atoms point to. Due to the full exposure of active sites and the low mass transfer diffusion barrier, such a hierarchical, porous structure would, in theory, would be very advantageous for sensing and adsorption processes [19].

Figure 1. SEM images of $Ti(i\text{-}OPr)_{4-x}/(AcOH)_x$ complex (**A**), and hierarchical Ti-MOF microflowers (**D**). TEM images of $Ti(i\text{-}OPr)_{4-x}/(AcOH)_x$ complex (**B**,**C**), and hierarchical Ti-MOF microflowers (**E**,**F**).

The crystalline structure and surface analysis of hierarchical Ti-MOF microflowers were investigated by XRD and FT-IR. As illustrated in Figure 2A, the principal diffraction peaks in the experimental XRD pattern of the hierarchical Ti-MOF microflowers demonstrate a clearly defined crystal structure, and are in good agreement with the simulated patterns of the NTU-9 family [24]. The sharp peaks show the successful fabrication of Ti-MOF and indicate the exceptional crystallinity of the framework. Moreover, the FT-IR measurement was employed to confirm the surface functional groups of Ti-MOF. As illustrated in Figure 2B, Ti-MOF microflowers clearly displayed typical peaks at 3149–3435, 1650, 1496, 1429, 1360 and 1209 cm^{-1} due to the vibrational stretching modes of hydroxyl groups, the C=O group, carbon atoms in the phenyl ring, the O−H group, in-plane O–H bending of phenolic hydroxy, and asymmetric stretching of the C−C−O group, in the same peak position as the H_4DOBDC ligand. The FT-IR analysis indicates that hierarchical Ti-MOF microflowers are inherently hydrophilic since the structures contain a large number of oxygen-rich functional groups.

A hierarchical structure with a porous structure is required for an adsorbent to function, and would be beneficial to mass transport. Thus, to characterize the BET surface area and internal pore superstructure of pseudo-spherical Ti-MOF NPs and hierarchical Ti-MOF microflowers, nitrogen adsorption-desorption measurements were performed. As shown in Figure 2C, the isotherm was classified as type IV, with a distinct hysteresis loop at relative pressures ranging from 0.5 to 1.0. The BET-specific surface area of hierarchical Ti-MOF microflowers was estimated to be approximately 264.47 $m^2 \cdot g^{-1}$, which is over 5-fold larger than that of pseudo-spherical Ti-MOF NPs (46.63 $m^2 \cdot g^{-1}$). Moreover, the hysteresis isotherms indicated that Ti-MOF microflowers have a micro/nano-porous structure, which is in line with SEM observations. The hysteresis loop in the Ti-MOF microflowers can be classified as Type H1, indicating a narrow pore size distribution (pore diameter on average: 1.8 nm) according to the Barrett-Joyner-Halenda (BJH) model, as illustrated in the inset of Figure 2C.

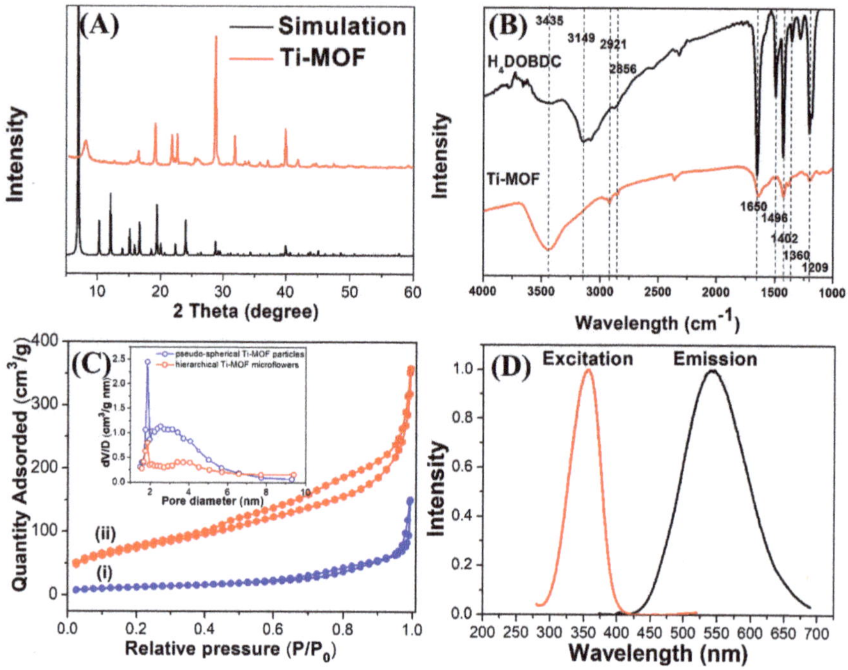

Figure 2. (**A**) Simulated and experimental powder XRD patterns for Ti-MOF microflowers. (**B**) FT-IR spectra of H_4DOBDC and Ti-MOF microflowers. (**C**) N_2 adsorption-desorption isotherms and the pore size distribution (inset image) of (i) pseudo-spherical Ti-MOF particles and (ii) hierarchical Ti-MOF microflowers. (**D**) Fluorescence excitation (red line) and emission (blue line) spectra of Ti-MOF microflowers.

To further investigate the optical properties of Ti-MOF microflowers, fluorescence spectrum studies were carried out. As shown in Figure 2D, the maximal excitation and emission peaks were located at 355 and 538 nm, respectively. A large Stokes shift of 187 nm occurred in the Ti-MOF microflowers to avoid crosstalk between excitation and emission signals. The quantum yield of Ti-MOF microflowers was estimated at 7.49%. Furthermore, due to their insufficient water solubility, most Al(III)-selective fluorescence chemosensors are studied in pure organic (such as DMSO and THF) or organic-water mixed solutions [25]. Therefore, the effects of various conditions on Ti-MOF microflowers should be investigated in order to determine fluorescence stability. As displayed in Figure S3, no distinct photobleaching under UV lamp irradiation over 5 h was observed, demonstrating the remarkable optical stability of Ti-MOF microflowers compared to traditional fluorescent dyes. The fluorescence properties of Ti-MOF microflowers were also investigated at different ionic strengths (modulated by 0–1000 mM $NaNO_3$) and different pH (4.0–9.0) conditions, as illustrated in Figure S4. There was no discernible change in fluorescence intensity in aqueous solutions under different ionic strengths and pH conditions, indicating that Ti-MOF microflowers are extremely stable even under high ionic strength and extreme pH conditions. Moreover, the Ti-MOF microflowers solution remained homogenous for over 1 month at room temperature (without any aggregation or color change). Their excellent water solubility and optical stability makes them potential candidates for a new class of fluorophores for determination of metal ions in aqueous solution without organic co-solvents.

Owing to their excellent fluorescence property and water stability, Ti-MOF microflowers are proposed to be used as fluorescent probes for quantitative determination of Al(III) ions. Because response rate is important in sensing performance, the time-response characteristics of hierarchical Ti-MOF microflowers towards Al(III) were investigated, in which the fluorescence signals of the reaction mixture were collected immediately and consecutively upon the addition of 10 M Al(III) into the Ti-MOF microflowers suspension. Figure S5 shows the difference in FL intensity before and after addition of 10 M Al(III). The fluorescence intensity of Ti-MOF microflowers significantly increased within 2 min after Al(III) addition and remained essentially constant afterward, indicating that the Al(III)-induced fluorescence enhancement reaction is quick. The FL intensity-Al(III) concentration ([Al(III)]) relationship was then established. Figure 3 shows the change of fluorescence intensity of Ti-MOF microflowers with gradually increasing Al(III) concentration from 0 to 100 μM for evaluating their fluorescence sensitivity. Figure 3A shows that the maximum emission peak of Ti-MOF microflowers appears at 538 nm, and is blue-shifted to 510 nm with remarkable fluorescence enhancement with increasing concentrations of Al(III). Figure 3B shows a plot of fluorescence enhancement rate (FER, derived from the peak intensity data in Figure 3A) against Al(III) concentration. There is an excellent linear correlation ($Y = 0.114X - 0.024$, $R^2 = 0.999$) between the FER and [Al(III)] in the low concentration range of 0.4–15 μM. The real limit of detection (LOD) is 0.4 μM, based on Figure 3B. Furthermore, the maximum FER could reach 2.34 as the concentration of Al(III) increased to 100 μM. On the basis of a signal-to-noise ratio of $3\sigma/\kappa$ (κ: slope; σ: standard error), the theoretical LOD was estimated to be 75 nM, which is approximately 100-fold lower than the permitted level (7.4 μM) of aluminum in drinking water by the WHO. Ti-MOF analytical performance is comparable to, or better than, that of previously reported FL sensors for Al(III) detection, as shown in Table S1. The poor coordination ability and strong hydration ability of Al(III) has hampered the development of a suitable fluorescence sensor [26]. A Job's plots experiment was carried out to demonstrate the coordination of Al(III) and Ti-MOF microflowers. The maximum fluorescence intensity as a function of molar fraction of Al(III) was used in the Job's method, and the total concentration of Ti-MOF and Al(III) ion was 20 μM. As shown in Figure S6, 3:2 stoichiometric complexations between Ti-MOF microflowers and Al(III) was confirmed.

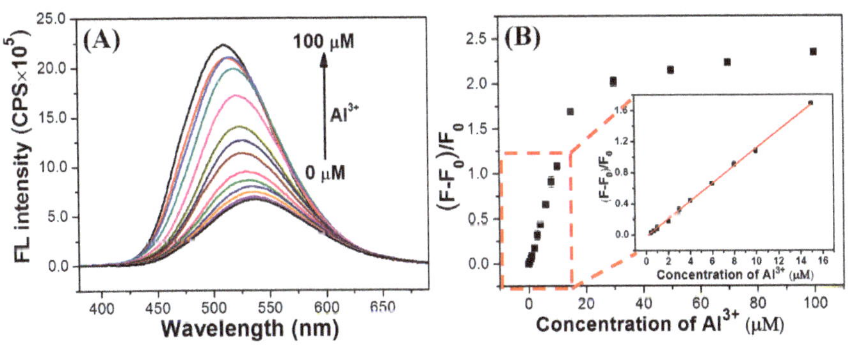

Figure 3. (**A**) Fluorescence response of Ti-MOF microflowers with various concentration of Al(III) (from bottom to top: 0, 0.4, 0.7, 1, 2, 3, 4, 6, 8, 10, 15, 30, 50, 70, 100 μM). (**B**) Fluorescence enhancement rate ($(F-F_0)/F_0$ at 510 nm) of Ti-MOF microflowers to Al(III) at various concentrations (0 to 100 μM). The linear detection range for 0.4–15 μM. Al(III) is shown in the inset image. The error bars represent standard deviations from three independent measurements.

Because of the similar electron configuration in the recognition process, other trivalent ions such as Fe^{3+} and Cr^{3+} frequently interfere with Al(III) detection [27]. Thus, Al(III)-selective fluorescent probes are still in high demand. The Al(III) ion selective detection ability of Ti-MOF microflowers was evaluated towards different interference ions including Cu(II), Cd(II), Cr(III), Co(II), Ni(II), Ca(II), Mg(II), Fe(II), Fe(III), Hg(II), Mn(II), Zn(II) and NH_4^+ under similar conditions. The fluorescence intensity ratios (F/F_0) of Ti-MOF microflowers in the absence and presence of interference metal ions are shown in Figure 4.

As expected, the addition of 10 μM Al(III) resulted in over a two-fold FL enhancement compared with that of the blank. Except for Fe^{3+} ions, there was no significant change in FL intensity based on emission at 510 nm for all interference ions at 10-fold higher concentrations than Al(III). Subsequently, the fluctuation of the F/F_0 was recorded after 10 μM of Al(III) was injected into the above solution to form a competing mixture. The obtained value of the F/F_0 after the solution was added with Al(III), as shown by the green bars in Figure 4, was consistent with the control group, in which Ti-MOF microflowers were simply mixed with Al(III). Furthermore, Fe(III) ions could reduce the fluorescence of Ti-MOF microflowers, while Fe(II) ions resulted in no change in FL intensity of Ti-MOF microflowers. Thus, we employed reductant (e.g., ascorbic acid, AA) to reduce Fe(III) ions to shield the interference of Fe(III) ions towards FL detection of Al(III) ions by Ti-MOF microflowers. As illustrated in Figure S7A, the fluorescence response of Ti-MOF microflowers to Al(III) ions was unaffected by AA. Due to the reduction of Fe(III) ions by AA, the fluoresce intensity of Ti-MOF with Fe(III) ions in the presence of AA showed no significant difference compared to the control group (Figure S7B). All in all, the Al(III) induced fluorescence enhancement of Ti-MOF was not obviously influenced by the interference ions, and further illustrates that Ti-MOF microflowers display an extremely selective response to Al(III) ions.

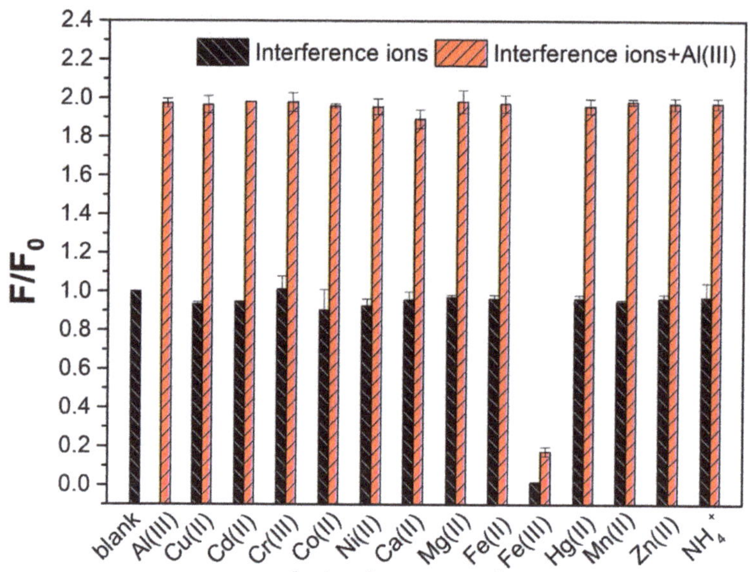

Figure 4. Ion selectivity ability of Ti-MOF microflowers towards various cations. The blank and Al(III) bar represent the Ti-MOF microflowers and the Ti-MOF microflowers with 10 μM Al(III), respectively. The black bars represent the Ti-MOF microflowers with a variety of single interference ions (100 μM). The red bars indicate the Ti-MOF microflowers with mixtures containing 10 μM Al(III) and 100 μM various interference ions. F_0 and F denote FL intensity of the blank in the absence and presence of the target ions, respectively.

3.1. Removal of Al(III) by Ti-MOF Microflowers

In addition to accurate and dependable detection, effective removal of Al(III) from polluted water is critical in wastewater treatment to reduce Al(III) accumulation below levels that pose a serious threat to humans. Time-dependent Al(III) adsorption kinetics experiments were performed to gain insight into the adsorption behavior of Ti-MOF microflowers towards Al(III), with an initial concentration of Al(III) set at 50 mgL^{-1} and a sorbent dosage of 1 g·L^{-1}. To reveal the outstanding characters of the hierarchical structure, the adsorption performance of Ti-MOF microflowers was contrasted with that of pseudo-spherical Ti-MOF nanoparticles synthesized by a direct hydrothermal reaction. As illustrated in Figure 5, it is clear that the concentration of Al(III) decreased rapidly at the beginning (2 h), and then approached a constant value after a long contact time (approximately 60 h) for both types of Ti-MOF. In the case of hierarchical Ti-MOF microflowers, the adsorption of Al(III) was as high as 21.78 mg·g^{-1} within just 12 h. With increasing treatment time, the adsorption of Al(III) on the hierarchical Ti-MOF microflowers was increased to 25.5 mg·g^{-1}, which is higher than that of pseudo-spherical Ti-MOF nanoparticles (20.8 mg·g^{-1}). A pseudo-second-order kinetic model was used to describe the kinetic characteristics of the adsorption sites, which was proportional to the square of the number of unoccupied sites, according to Equation (2), and as shown in Figure 5B. The parameters Q_e and k_2 for Al(III) adsorption were estimated by using linear fitting with high correlation coefficients (R^2 = 0.998), as shown in Table 1. The low value of k_2 indicates that the removal rates of Al(III) by both type of Ti-MOF were fast. Furthermore, $h=k_2Q_e{}^2$ was calculated to quantitatively describe the initial removal rate using Equation (1). The initial removal rate (h) of Al(III) on the hierarchical Ti-MOF microflowers was larger than that of pseudo-spherical Ti-MOF nanoparticles, demonstrating a higher removal rate of Al(III) by hierarchical Ti-MOF microflowers than that of pseudo-spherical Ti-MOF nanoparticles. These results might be attributed to the distinctive hierarchical structure of Ti-MOF microflowers, the thin nanosheets reducing the diffusion resistance of Al(III) ions and allowing for faster kinetics.

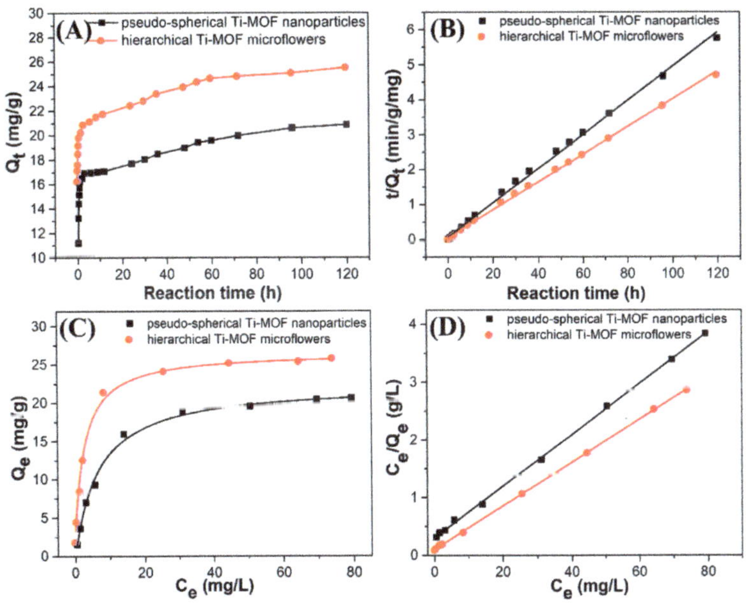

Figure 5. (**A**) Kinetics of Al(III) adsorption and (**B**) the corresponding fitting curves for pseudo-spherical Ti-MOF particles and hierarchical Ti-MOF microflowers via a pseudo-second-order kinetic model (50 mg·L^{-1} of the initial concentration of Al(III), 1 g·L^{-1} of adsorbents with initial solution pH 4.7 ± 0.1 at 25 °C). (**C**) Isothermal adsorption of Al(III), and (**D**) the corresponding plots of C_e/Q_e versus C_e for pseudo-spherical Ti-MOF nanoparticles and hierarchical Ti-MOF microflowers (2–100 mg·L^{-1} of initial Al(III) concentration, 1 g·L^{-1} of adsorbents, 96 h of equilibrium time at 25 °C and initial pH 4.7 ± 0.1).

Table 1. Adsorption parameters of Al(III) on both type of Ti-MOF.

Adsorbents	Kinetic Adsorption (Pseudo-Second-Order Model)			Isotherms (Langmuir Model)	
	K_2 (g·mg^{-1}·min^{-1})	Q_e (mg·g^{-1})	h (mg·g^{-1}·min^{-1})	q_m (mg·g^{-1})	K_L (L·mg^{-1})
pseudo-spherical Ti-MOF NPs	0.0324	20.86	13.67	20.66	0.1468
hierarchical Ti-MOF microflowers	0.0343	25.53	21.92	25.85	0.3831

Isothermal adsorption experiments were carried out based on kinetic adsorption behavior, and a reaction time of 96 h was used for the equilibrium adsorption time. Figure 5C shows the adsorption isotherm of Al(III) on pseudo-spherical Ti-MOF nanoparticles and hierarchical Ti-MOF microflowers by varying initial Al(III) concentrations between 2–100 mg·L^{-1} under initial solution pH of 4.7 ± 0.1. Adsorption of Al(III) increased with increasing concentrations of Al(III) in aqueous solution for both types of Ti-MOF. Table 1 shows the fitted results and calculated parameters for the Langmuir model, which was used to further analyze the Al(III) adsorption isotherms. Obviously, the Al(III) adsorption results for the two adsorbents with a high correlation coefficient, could be well fitted by the Langmuir model (R^2 = 0.999), which applies to monolayer adsorption on a surface with a finite number of identical active sites. Figure 5D shows the corresponding plots, which indicate good linear relationships for both types of Ti-MOF. The substantial maximum Al(III) removal capacities (q_m) were up to 20.66 mg·g^{-1} for pseudo-spherical Ti-MOF nanoparticles and 25.85 mg·g^{-1} for hierarchical Ti-MOF microflowers. Based on the Langmuir model these outperformed the adsorption capacities of other adsorbents listed in Table S2. The superior adsorption capacity of hierarchical Ti-MOF microflowers towards Al(III) can be attributed to their higher specific surface area in comparison to pseudo-spherical Ti-MOF nanoparticles, as well as their unique hierarchically flower-like morphology composed of ultrathin nanosheets that provide abundant adsorption sites for Al(III) capture.

3.2. Possible Mechanism for Selective Determination of Al(III) Ions

According to previous research [28], the effect of metal cations on luminescent MOFs can be attributed to three factors: (1) the collapse of the MOFs framework, (2) cation exchange between the MOFs' central cations and the target cations, and (3) interactions between the metal cations and organic ligands. XRD and SEM measurements were used to confirm whether the crystal structure of the original framework was affected in order to further understand and elucidate the possible sensing mechanism of the phenomenon, in which Al(III)-enhances the FL of Ti-MOF microflowers. The XRD patterns (Figure S8) and SEM images (Figure S9) revealed that the sample obtained by immersing Ti-MOF microflowers in Al(III) ions was nearly identical to the original Ti-MOF, indicating that the Al(III) ions do not cause the Ti-MOF framework to collapse or change.

To verify whether the cation exchange between the central cations of Ti-MOFs and Al(III) occurred, ICP-OES was used to determine the concentrations of Ti(IV) and Al(III) in the solid MOF and supernatant after Ti-MOF treatment with various concentrations of Al(III) ions. As illustrated in Table S3, there were no Ti(IV) in all supernatants after Ti-MOF treatment with different concentration of Al(III) ions, which precluded the possibility of cation exchange between Ti(IV) and Al(III). However, Al(III) was found in all solid Ti-MOF after treatment with different concentrations of Al(III) ions. The summation of Al(III) ions in the supernatant and solid Ti-MOF were equal to the initial concentration of Al(III) ions, which is consistent with the results of the adsorption experiment. This verifies that Ti-MOF microflowers can form a complex with Al(III) ions.

Because of the interaction between metal cations and organic ligands, we propose that the Al(III)-induced fluorescence enhancement of Ti-MOF microflowers can be attributed to the chelation-enhanced fluorescence (CHEF) mechanism [29]. As previously stated, FT-IR spectral analysis shows that the prepared Ti-MOF microflowers have phenolic hydroxyl and carboxyl groups. Metal ions can be chelated by these functional groups to form stable coordination complexes. The fluorescent selectivity of Ti-MOF microflowers towards Al(III) may be attributed to the smaller ionic radius (0.5 Å) of Al(III), which allows for a suitable coordination geometry with Ti-MOF, and a higher charge density, causing strong coordination between Al(III) and Ti-MOF. After adding Al(III), the carbonyl O, and hydroxyl O on DOBDC of Ti-MOF microflowers can coordinate with the Al(III) center atom, which increases the energies of the n–π^* transitions compared to the corresponding π–π^* transitions, and the photoinduced electron transfer process is interrupted, and the FL enhanced [30,31]. As illustrated in Figure 3A, the Al(III) ions induced a blue-shift with remarkable fluorescence enhancement of Ti-MOF microflowers. The wavelength shift indicates that the structure of DOBDC ligands in the surface of Ti-MOF microflowers may be changed in the presence of Al(III), which could result from phenolic hydroxy and carboxyl of DOBDC strongly bound with Al(III). Figure 6A shows the UV-vis absorption spectra of Ti-MOF microflowers in the present of Al(III) ions. Ti-MOF microflowers have four absorption peaks at 248, 290, 350 and 458 nm, while all of four absorption peaks are red-shifted (253, 291.5, 368.5 and 486 nm) after adding Al(III) ions. It is well known that the binding of an auxochrome to a chromophore causes increased absorption and a red shift of the chromophore. This further illustrates that Ti-MOF microflowers can form a complex with Al(III) ions.

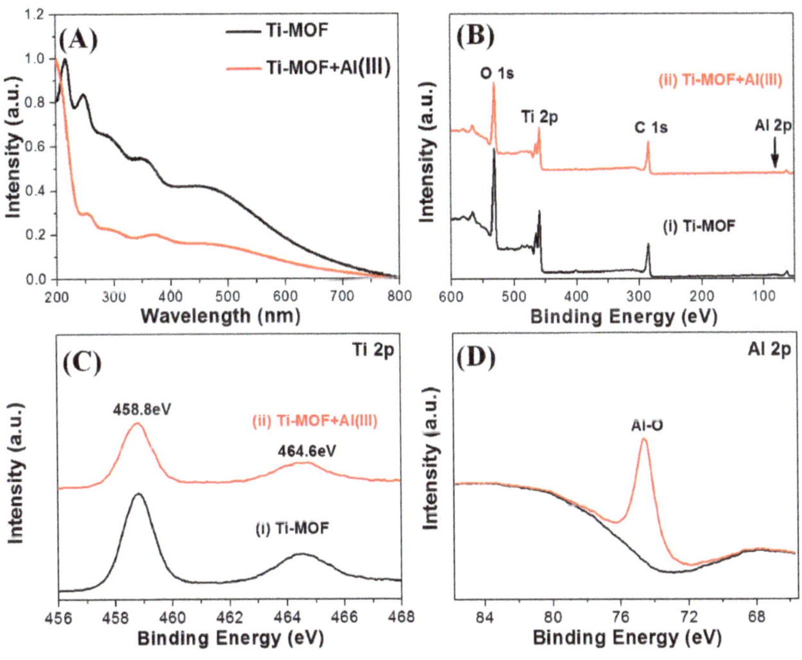

Figure 6. (**A**) UV-vis absorption spectra of Ti-MOF microflowers before and after Al(III) addition. (**B**) XPS spectra of Ti-MOF microflowers before and after Al(III) addition. (**C**) Ti2p spectra of Ti-MOF microflowers before and after adding Al(III). (**D**) Al2p spectra of Ti-MOF microflowers after addition of Al(III).

The coordination interaction between the Al(III) and the carbonyl O, and hydroxyl O on DOBDC of Ti-MOF microflowers was further validated by XPS. Figure 6B–D shows the survey spectra, Ti 2p and Al 2p spectra of Ti-MOF microflowers in the absence and present of Al(III). According to the XPS spectra, the binding energies around 286, 459 and 531 eV corresponded to the C 1s, Ti 2p and O 1s, which are the basic elements of the Ti-MOF microflowers (Figure 6B). After the interaction with Al(III), the emergence of the Al 2p peak at 74.6 eV (Al-O) verified that Al(III) is loaded into the framework of Ti-MOF microflowers (Figure 6C).

4. Conclusions

A secondary hydrothermal method was used to successfully prepare three-dimensional hierarchical titanium-based metal-organic framework (Ti-MOF) microflowers for the simultaneous detection and removal of Al(III) ions in aqueous solution. The as-prepared Ti-MOF microflowers are made up of flowerlike micro/nanostructured particles with sizes ranging from 5 to 6 μm. The particles are made of 200 nm nanorods with a length of 1–2 μm and a porous structure with a pore diameter of 1.8 nm and a specific surface area of 264.47 $m^2 \cdot g^{-1}$, which provides abundant surface active sites for detection and adsorption of Al(III) ions. The resultant Ti-MOF microflowers exhibit excellent stability, high water dispensability, and dramatical fluorescence characteristic, making them useful as fluorescence probes for quantitative detection of Al(III) in aqueous solution. Importantly, specific FL enhancement by Al(III) via the chelation-enhanced fluorescence (CHEF) mechanism can be utilized to selectively and accurately determine Al(III) due to the coordination interaction between the Al(III) and the carbonyl O, and hydroxyl O on DOBDC of Ti-MOF microflowers. Linear concentration ranges of up to 15 μM and a detection limit of 75 nM are easily achievable, the latter being approximately 100-fold lower than the WHO permitted level of aluminum in drinking water (7.4 μM). By introducing ascorbic acid, the interference of Fe(III) can be shielded to achieve selective detection of Al(III) in the presence of various cations. Ti-MOF microflowers have an excellent adsorption capacity for Al(III), with a maximum adsorption capacity of 25.85 $mg \cdot g^{-1}$. The adsorption rate is rapid and corresponds to a pseudo-second-order kinetic model. The Ti-MOF is an excellent candidate as a fluorescent chemical sensor and adsorbent for aqueous contaminants due to its impressive water stability, high porosity, and strong luminescence.

Supplementary Materials: The following supporting information can be downloaded at: https://www.mdpi.com/article/10.3390/bios12110935/s1, Figure S1: HRTEM images of hierarchical Ti-MOF nanoflowers through secondary hydrothermal synthesis; Figure S2: SEM image of pseudo-spherical Ti-MOF by direct hydrothermal reaction; Figure S3: Photostability of Ti-MOF nanoflowers under UV lamp irradiation; Figure S4: Relative fluorescence intensity (F/F_0) of Ti-MOF nanoflowers at different ionic strengths (0–1000 mM $NaNO_3$) and pH values (4.0–9.0); Figure S5: Fluorescence response—time profile of Ti-MOF nanoflowers before and after adding 10 μM Al(III); Figure S6: Job's plot for stoichiometric determination of Ti-MOF and Al(III); Figure S7: Fluorescence spectra of Ti-MOF in the presence of Al^{3+} and AA, and Fluorescence spectra of Ti-MOF in the presence of Fe^{2+}, Fe^{3+} and AA; Figure S8: PXRD pattern of Ti-MOF nanoflowers after treatment with Al^{3+} ions; Figure S9: SEM image of Ti-MOF nanoflowers after treatment with Al^{3+} ions; Table S1: Comparison of different fluorescent material sensors for Al(III) determination [32–44]; Table S2: Comparison of the Al(III) adsorption capacities for different adsorbent materials [35,45–51]; Table S3: ICP-OES analysis of the content of Ti^{4+} and Al^{3+} in the solid MOF and supernatant after Ti-MOF treatment with different concentration of Al^{3+} ions.

Author Contributions: Conceptualization, H.Z.; Data curation, J.Z., J.S., Y.L. and Y.W.; Formal analysis, J.S., G.M., Y.L. and Y.W.; Funding acquisition, H.Z.; Investigation, J.Z. and F.L.; Methodology, J.S., Y.L. and Y.W.; Project administration, H.Z.; Software, G.M.; Supervision, H.Z.; Validation, J.S., G.M. and F.L.; Visualization, G.M. and F.L.; Writing–original draft, J.Z.; Writing–review & editing, H.Z. All authors have read and agreed to the published version of the manuscript.

Funding: This research was funded by [the National Natural Science Foundation of China [Grant No. 51502296 and 51872291], the Key research and development projects of Anhui Province [Grant No. 202104a06020028], the Collaborative Innovation Program of Hefei Science Center, CAS [Grant No. 2020HSC-CIP012], and the presidential Foundation of Hefei Institutes of Physical Science, Chinese Academy of Sciences [No. YZJJZX202019].

Institutional Review Board Statement: Not applicable.

Informed Consent Statement: Not applicable.

Data Availability Statement: Not applicable.

Conflicts of Interest: The authors declare no competing financial interest.

References

1. Cronan, C.S.; Walker, W.J.; Bloom, P.R. Predicting aqueous aluminium concentrations in natural waters. *Nature* **1986**, *324*, 140–143. [CrossRef]
2. World Health Organization. *Guidelines for Drinking-Water Quality*, 1st ed.; World Health Organization: Geneva, Switzerland, 2004.
3. Walton, J.R. Aluminum in hippocampal neurons from humans with Alzheimer's disease. *NeuroToxicology* **2006**, *27*, 385–394. [CrossRef] [PubMed]
4. LLabrecque, C.; Larivière, D. Quantification of rare earth elements using cloud point extraction with diglycolamide and ICP-MS for environmental analysis. *Anal. Methods* **2014**, *6*, 9291–9298. [CrossRef]
5. Fang, B.-Y.; Li, C.; Song, Y.-Y.; Tan, F.; Cao, Y.-C.; Zhao, Y.-D. Nitrogen-doped graphene quantum dot for direct fluorescence detection of Al^{3+} in aqueous media and living cells. *Biosens. Bioelectron.* **2018**, *100*, 41–48. [CrossRef]
6. Sheoran, A.; Sheoran, V. Heavy metal removal mechanism of acid mine drainage in wetlands: A critical review. *Miner. Eng.* **2006**, *19*, 105–116. [CrossRef]
7. Ding, Y.; Zhu, W.; Xu, Y.; Qian, X. A small molecular fluorescent sensor functionalized silica microsphere for detection and removal of mercury, cadmium, and lead ions in aqueous solutions. *Sensors Actuators B Chem.* **2015**, *220*, 762–771. [CrossRef]
8. Septhum, C.; Rattanaphani, S.; Bremner, J. An adsorption study of Al(III) ions onto chitosan. *J. Hazard. Mater.* **2007**, *148*, 185–191. [CrossRef]
9. He, C.; Zhu, W.; Xu, Y.; Chen, T.; Qian, X. Trace mercury (II) detection and separation in serum and water samples using a reusable bifunctional fluorescent sensor. *Anal. Chim. Acta* **2009**, *651*, 227–233. [CrossRef]
10. Rapti, S.; Sarma, D.; Diamantis, S.A.; Skliri, E.; Armatas, G.S.; Tsipis, A.C.; Hassan, Y.S.; Alkordi, M.; Malliakas, C.D.; Kanatzidis, M.G.; et al. All in one porous material: Exceptional sorption and selective sensing of hexavalent chromium by using a Zr^{4+} MOF. *J. Mater. Chem. A* **2017**, *5*, 14707–14719. [CrossRef]
11. Xie, D.; Ma, Y.; Gu, Y.; Zhou, H.; Zhang, H.; Wang, G.; Zhang, Y.; Zhao, H. Bifunctional NH_2-MIL-88(Fe) metal–organic framework nanooctahedra for highly sensitive detection and efficient removal of arsenate in aqueous media. *J. Mater. Chem. A* **2017**, *5*, 23794–23804. [CrossRef]
12. Hu, Z.; Deibert, B.J.; Li, J. Luminescent metal-organic frameworks for chemical sensing and explosive detection. *Chem. Soc. Rev.* **2014**, *43*, 5815–5840. [CrossRef] [PubMed]
13. Shi, M.; Fu, C.; Yu, J.; Yang, Y.; Shi, P. A Novel 2D Metal-Organic Framework Probe: Highly Sensitive and Visual Fluorescent Sensor for Al^{3+}, Cr^{3+}, Fe^{3+} Ions. *N. J. Chem.* **2022**, *46*, 18911–18916. [CrossRef]
14. Zhao, D.; Yu, S.; Jiang, W.-J.; Cai, Z.-H.; Li, D.-L.; Liu, Y.-L.; Chen, Z.-Z. Recent Progress in Metal-Organic Framework Based Fluorescent Sensors for Hazardous Materials Detection. *Molecules* **2022**, *27*, 2226. [CrossRef] [PubMed]
15. Du, T.; Wang, J.; Zhang, L.; Wang, S.; Yang, C.; Xie, L.; Liu, Z.; Ni, Y.; Xie, X.; Sun, J.; et al. Missing-linker engineering of Eu (III)-doped UiO-MOF for enhanced detection of heavy metal ions. *Chem. Eng. J.* **2022**, *431*, 134050. [CrossRef]
16. Wu, L.; Yao, S.; Xu, H.; Zheng, T.; Liu, S.; Chen, J.; Li, N.; Wen, H. Highly selective and turn on fluorescence probe with red shift emission for naked-eye detecting Al^{3+} and Ga^{3+} based on metal-organic framework. *Chin. Chem. Lett.* **2022**, *33*, 541–546. [CrossRef]
17. Zhan, Z.; Jia, Y.; Li, D.; Zhang, X.; Hu, M. A water-stable terbium-MOF sensor for the selective, sensitive, and recyclable detection of Al^{3+} and CO_3^{2-} ions. *Dalton Trans.* **2019**, *48*, 15255–15262. [CrossRef]
18. Zheng, X.; Zhao, Y.; Jia, P.; Wang, Q.; Liu, Y.; Bu, T.; Zhang, M.; Bai, F.; Wang, L. Dual-emission Zr-MOF-based composite material as a fluorescence turn-on sensor for the ultrasensitive detection of Al^{3+}. *Inorg. Chem.* **2020**, *59*, 18205–18213. [CrossRef]
19. Zhang, S.; Li, L.; Zhao, S.; Sun, Z.; Hong, M.; Luo, J. Hierarchical metal-organic framework microflowers for effective CO_2 transformation driven by visible light. *J. Mater. Chem. A* **2015**, *3*, 15764–15768. [CrossRef]
20. Kim, T.K.; Lee, K.J.; Cheon, J.Y.; Lee, J.H.; Joo, S.H.; Moon, H.R. Nanoporous Metal Oxides with Tunable and Nanocrystalline Frameworks via Conversion of Metal–Organic Frameworks. *J. Am. Chem. Soc.* **2013**, *135*, 8940–8946. [CrossRef]
21. Zhang, Z.; Chen, Y.; He, S.; Zhang, J.; Xu, X.; Yang, Y.; Nosheen, F.; Saleem, F.; He, W.; Wang, X. Hierarchical Zn/Ni-MOF-2 Nanosheet-Assembled Hollow Nanocubes for Multicomponent Catalytic Reactions. *Angew. Chem. Int. Ed.* **2014**, *53*, 12517–12521.
22. Zhang, Z.; Chen, Y.; Xu, X.; Zhang, J.; Xiang, G.; He, W.; Wang, X. Well-Defined Metal-Organic Framework Hollow Nanocages. *Angew. Chem. Int. Ed.* **2014**, *53*, 429–433. [CrossRef] [PubMed]
23. Xu, H.; Gao, J.; Qian, X.; Wang, J.; He, H.; Cui, Y.; Yang, Y.; Wang, Z.; Qian, G. Metal-organic framework nanosheets for fast-response and highly sensitive luminescent sensing of Fe^{3+}. *J. Mater. Chem. A* **2016**, *4*, 10900–10905. [CrossRef]
24. Gao, J.; Miao, J.; Li, P.-Z.; Teng, W.Y.; Yang, L.; Zhao, Y.; Liu, B.; Zhang, Q. A p-type Ti(iv)-based metal-organic framework with visible-light photo-response. *Chem. Commun.* **2014**, *50*, 3786–3788. [CrossRef] [PubMed]

25. Liao, Z.-C.; Yang, Z.-Y.; Li, Y.; Wang, B.-D.; Zhou, Q.-X. A simple structure fluorescent chemosensor for high selectivity and sensitivity of aluminum ions. *Dye. Pigment.* **2013**, *97*, 124–128. [CrossRef]
26. Dhara, A.; Jana, A.; Guchhait, N.; Ghosh, P.; Kar, S.K. Rhodamine-based molecular clips for highly selective recognition of Al^{3+} ions: Synthesis, crystal structure and spectroscopic properties. *New J. Chem.* **2014**, *38*, 1627–1634. [CrossRef]
27. Das, S.; Goswami, S.; Aich, K.; Ghoshal, K.; Quah, C.K.; Bhattacharyya, M.; Fun, H.-K. ESIPT and CHEF based highly sensitive and selective ratiometric sensor for Al^{3+} with imaging in human blood cells. *New J. Chem.* **2015**, *39*, 8582–8587. [CrossRef]
28. Hao, J.-N.; Yan, B. Amino-decorated lanthanide(iii) organic extended frameworks for multi-color luminescence and fluorescence sensing. *J. Mater. Chem. C* **2014**, *2*, 6758–6764. [CrossRef]
29. Bhamore, J.R.; Jha, S.; Singhal, R.K.; Park, T.J.; Kailasa, S.K. Facile green synthesis of carbon dots from Pyrus pyrifolia fruit for assaying of Al^{3+} ion via chelation enhanced fluorescence mechanism. *J. Mol. Liq.* **2018**, *264*, 9–16. [CrossRef]
30. Firdaus, F.; Farhi, A.; Faraz, M.; Shakir, M. Benzidine based fluorescent probe for the sensitive detection of heavy metal ions via chelation enhanced fluorescence mechanism—A multiplexed sensing platform. *J. Lumin.* **2018**, *199*, 475–482. [CrossRef]
31. Naksen, P.; Boonruang, S.; Yuenyong, N.; Lee, H.L.; Ramachandran, P.; Anutrasakda, W.; Amatatongchai, M.; Pencharee, S.; Jarujamrus, P. Sensitive detection of trace level Cd (II) triggered by chelation enhanced fluorescence (CHEF) "turn on": Nitrogen-doped graphene quantum dots (N-GQDs) as fluorometric paper-based sensor. *Talanta* **2022**, *242*, 123305. [CrossRef]
32. Zhou, T.-Y.; Lin, L.-P.; Rong, M.-C.; Jiang, Y.-Q.; Chen, X. Silver–gold alloy nanoclusters as a fluorescence-enhanced probe for aluminum ion sensing. *Anal. Chem.* **2013**, *85*, 9839–9844. [CrossRef] [PubMed]
33. Ku, K.-S.; Muthukumar, P.; Angupillai, S.; Son, Y.-A. A new rhodamine 6G based chemosensor for trace level Al3+ and its thin film application in 100% aqueous medium. *Sens. Actuators B Chem.* **2016**, *236*, 184–191. [CrossRef]
34. Wang, H.; He, F.; Yan, R.-J.; Wang, X.-Y.; Zhu, X.; Li, L.-D. Citrate-induced aggregation of conjugated polyelectrolytes for Al3+-ion-sensing assays. *ACS Appl. Mater. Interfaces* **2013**, *5*, 8254–8259. [CrossRef]
35. Xu, L.; Han, S.; Hu, Y.; Dynes, J.; Zhang, L. Rhodamine B-based ordered mesoporous organosilicas for the selective detection and adsorption of Al (III). *N. J. Chem.* **2016**, *40*, 6752–6761. [CrossRef]
36. Xiao, H.; Chen, K.; Jiang, N.; Cui, D.; Yin, G.; Wang, J.; Wang, R. A highly selective turn-on fluorescent probe for Al (III) based on coumarin and its application in vivo. *Analyst* **2014**, *139*, 1980–1986. [CrossRef] [PubMed]
37. Lv, R.; Chen, Z.; Fu, X.; Yang, B.; Li, H.; Su, J.; Gu, w.; Liu, X. A highly selective and fast-response fluorescent probe based on Cd-MOF for the visual detection of Al3+ ion and quantitative detection of Fe3+ ion. *J. Solid State Chem.* **2018**, *259*, 67–72. [CrossRef]
38. Lee, S.A.; You, G.; Choi, Y.; Jo, H.; Kin, A.; Noh, I.; Kin, S.; Kim, Y.; Kim, C. A new multifunctional Schiff base as a fluorescence sensor for Al3+ and a colorimetric sensor for CN- in aqueous media: An application to bioimaging. *Dalton Trans.* **2014**, *43*, 6650–6659. [CrossRef]
39. Kim, Y.; Jang, G.; Lee, T. New fluorescent metal-ion detection using a paper-based sensor strip containing tethered rhodamine carbon nanodots. *ACS Appl. Mater. Interfaces* **2015**, *7*, 15649–15657. [CrossRef]
40. Liang, C.; Bu, W.; Li, C.; Men, G.; Deng, M.; Jiang, Y.; Sun, H.; Jiang, S. A highly selective fluorescent sensor for Al3+ and the use of the resulting complex as a secondary sensor for PPi in aqueous media: Its applicability in live cell imaging. *Dalton Trans.* **2015**, *44*, 11352–11359. [CrossRef]
41. Chen, W.; Meng, X.; Zhuang, G.; Wang, Z.; Kurmoo, M.; Zhao, Q.; Wang, X.; Shan, B.; Tung, C.; Sun, D. A superior fluorescent sensor for Al3+ and UO22+ based on a Co (ii) metal–organic framework with exposed pyrimidyl Lewis base sites. *J. Mater. Chem. A* **2017**, *5*, 13079–13085. [CrossRef]
42. Mi, Y.; Cao, Z.; Chen, Y.; Long, S.; Xie, Q.; Liang, D.; Zhu, W.; Xiang, J. A reversible switch as highly selective sequential chemosensor for Al3+ cation followed by F− anion. *Sens. Actuators B Chem.* **2014**, *192*, 164–172. [CrossRef]
43. Hu, X.; Mao, X.; Zhang, X.; Huang, Y. One-step synthesis of orange fluorescent copper nanoclusters for sensitive and selective sensing of Al3+ ions in food samples. *Sens. Actuators B Chem.* **2017**, *247*, 312–318. [CrossRef]
44. Mu, X.; Qi, L.; Qiao, J.; Ma, H. One-pot synthesis of tyrosine-stabilized fluorescent gold nanoclusters and their application as turn-on sensors for Al3+ ions and turn-off sensors for Fe3+ ions. *Anal. Methods* **2014**, *6*, 6445–6451. [CrossRef]
45. Tassist, A.; Lounici, H.; Abdi, N.; Mameri, N. Equilibrium, kinetic and thermodynamic studies on aluminum biosorption by a mycelial biomass (Streptomyces rimosus). *Journal of Hazardous Materials* **2010**, *183*, 35–43. [CrossRef]
46. Guan, X.; Yan, S.; Xu, Z.; Fan, H. Gallic acid-conjugated iron oxide nanocomposite: An efficient, separable, and reusable adsorbent for remediation of Al(III)-contaminated tannery wastewater. *J. Environ. Chem. Eng.* **2017**, *5*, 479–487. [CrossRef]
47. Lodeiro, P.; Gudina, A.; Herrero, L.; Herrero, R.; Manuel, E. Aluminium removal from wastewater by refused beach cast seaweed. Equilibrium and dynamic studies. *J. Hazard. Mater.* **2010**, *178*, 861–866. [CrossRef]
48. Al-Muhtaseb, S.A.; El-Naas, M.H.; Abdallah, S. Removal of aluminum from aqueous solutions by adsorption on date-pit and BDH activated carbons. *J. Hazard. Mater.* **2008**, *158*, 300–307. [CrossRef]
49. Andaç, M.; Özyapı, E.; Şenel, S.; Say, R.; Denizli, A. Ion-selective imprinted beads for aluminum removal from aqueous solutions. *Ind. Eng. Chem. Res.* **2006**, *45*, 1780–1786. [CrossRef]

50. Islam, A.; Ahmad, H.; Zaidi, N.; Yadav, S. Selective separation of aluminum from biological and environmental samples using glyoxal-bis (2-hydroxyanil) functionalized Amberlite XAD-16 resin: Kinetics and equilibrium studies. *Ind. Eng. Chem. Res.* **2013**, *52*, 5213–5220. [CrossRef]
51. Aly, Z.; Graulet, A.; Scales, N.; Hanley, T. Removal of aluminium from aqueous solutions using PAN-based adsorbents: Characterisation, kinetics, equilibrium and thermodynamic studies. *Environ. Sci. Pollut. Res.* **2014**, *21*, 3972–3986. [CrossRef]

biosensors

MDPI

Article

Terahertz Fingerprint Metasurface Sensor Based on Temperature Variation for Trace Molecules

Weijin Wang , Mingjun Sun , Jie Lin , Ying Xue and Yanpeng Shi *

School of Integrated Circuits, Shandong University, Jinan 250100, China
* Correspondence: ypshi@sdu.edu.cn

Abstract: Terahertz (THz) spectroscopy has demonstrated significant potential for substance detection due to its low destructiveness and due to the abundance of molecular fingerprint absorption signatures that it contains. However, there is limited research on the fingerprint detection of substances at different temperatures. Here, we propose a THz metamaterial slit array sensor that exploits localized surface plasmons to enhance the electric field within the slit. The transmission peak frequency can be modulated via temperature adjustments. This method enables the detection of molecular absorption characteristics at multiple spectral frequency points, thereby achieving a specific and highly sensitive detection of characteristic analyte fingerprint spectra. Additionally, the sensor supports the detection of substances at multiple temperatures and sensitively identifies changes in their absorption properties as a function of temperature. Our research has employed temperature variation to achieve a highly sensitive and specific detection of trace analytes, offering a new solution for THz molecular detection.

Keywords: Terahertz; metasurface; sensing; temperature variation; plasmons

1. Introduction

Terahertz (THz) waves possess ultra-low photon energy, which eliminates the risk of causing damaging ionization in analytes [1]. Many chemical molecules exhibit resonance and rotational frequencies within the THz range, resulting in distinct absorption resonances and intricate characteristic fingerprint spectra. Analyzing the interaction between incident THz waves and molecules provides valuable insights into molecular configuration, conformation, and environmental effects [2], thus enabling specific detection capabilities [3]. THz sensing holds significant promise for the non-destructive, rapid, and precise identification of various chemical molecules, with wide-ranging applications spanning across physics, chemistry, biomedicine, and related fields [4]. Nevertheless, a significant challenge arises from the mismatch between the THz wavelength, typically at the micrometer scale, and the absorption cross-section of molecules, which operates at the nanometer scale [5]. Most molecules exhibit weak interactions with THz waves. Traditional THz spectroscopy methods typically require large analyte quantities to observe their characteristic fingerprint spectra, leading to low sensing sensitivity [6,7]. Enhancing the sensing sensitivity of analytes in practical applications has become imperative.

The application and sensing of high-energy explosive has been one of the international research hotspots. Hexogen (RDX), also known as cyclonite, is a powerful military explosive. It is a white crystalline powder and is widely used in various fields such as military, aerospace, mining, and chemical industries due to its high explosive speed and power [8]. It is almost impossible to detect through X-ray imaging, necessitating the development of new non-contact and non-destructive techniques for effective detection. THz waves, due to their non-destructive nature, offer unique advantages for RDX detection. The broad frequency range and weak absorption of RDX fingerprints limit the practical applications of THz sensing. Enhancing the sensitivity of THz sensing for RDX is the primary focus of current research. HNIW (CL-20), chemically known as 2,4,6,8,10,12-hexanitro-2,4,6,8,10,12-hexaazaisowurtzitane, is a highly energetic and potent non-nuclear elementary explosive. It has wide-ranging applications in military and civilian sectors, garnering significant attention from researchers. CL-20 undergoes an irreversible ε–γ phase transition induced by temperature, a process that is closely linked to its explosive behavior. The precise detection of CL-20 and its phase transition are crucial to ensure its safe utilization [9,10]. Unfortunately, achieving a rapid and highly sensitive sensing of CL-20 and its phase transition process remains challenging. Specifically, detecting fingerprints at different temperatures before and after the ε–γ phase transition of CL-20 presents significant difficulties.

In previous studies, researchers have proposed various artificially designed metasurface structures, including open-ring resonators [11], metal gratings [12], photonic crystal structures [13], waveguide configurations [14], and structures based on bound states in the continuum [15,16], to achieve heightened sensitivity in material detection [17–19]. The majority of these structures are limited to detecting material absorption characteristics at individual frequency points via resonance peaks. They cannot accurately analyze the absorption spectra curve within a specific frequency range to identify particular substances [20]. In recent years, researchers have employed strategies such as structural scanning, angle scanning, stretchable flexible materials, and adjustable graphene to modulate the resonance peak frequency [21]. This advancement allows for the detection of characteristic absorption curves within a defined frequency band, facilitating material analysis and specific substance identification [22,23]. Nonetheless, limitations persist, including narrow sensing frequency bands and the inconvenience of frequency modulation strategies. Current research on the spectral response and sensing capabilities of metasurfaces at varying temperatures remains inadequate.

In this paper, we propose a THz metamaterial nano-slit array sensor based on temperature variation to enhance the detection of molecular fingerprint spectra. The sensor utilizes the temperature-sensitive semiconductor material InSb to form the nano-slit array structure, successfully generating localized surface plasmons (LSPs) [24]. When a surface plasmon is confined to a particle of a size comparable with the wavelength of light, the particle's free electrons participate in the collective oscillation, and it is termed an LSP. LSPs significantly enhance the electric field near the particle's surface, with the maximum enhancement at the surface rapidly diminishing with distance. The particle's optical extinction peaks at the plasmon resonance frequency. This extinction peak depends on the refractive index of the surrounding medium [25]. The sensor achieves localized electric field enhancement based on LSP, significantly improving sensing sensitivity [26,27]. InSb is a frequently utilized temperature-sensitive material. As the temperature increases, both the real and imaginary components of the InSb dielectric constant in the terahertz band significantly and continuously increase. This leads to alterations in the metasurface resonance frequency [28]. When the temperature varies from 275 K to 520 K, the resonance peak frequency gradually shifts from 0.78 THz to 1.6 THz, with a change in amplitude of 0.82 THz. By continuously changing temperatures, the resonance peak can be located at different frequency points. This allows for the sensitive sensing of molecular absorption features at multiple frequencies and the formation of an envelope curve depicting characteristic absorption spectra. Taking RDX detection as an example, this sensor has a detection limit of 1.61 $\mu g/cm^2$. The proposed sensor can specifically identify trace substances. It provides a reliable approach for trace molecule sensing using THz technology [29]. In addition, the sensor can operate at multiple temperatures. By altering the slit length to modify the resonance frequency at a constant temperature, characteristic absorption spectra at multiple temperatures can be detected. By coating the sensor surface with CL-20, distinct characteristic absorption curves of CL-20 were detected at 298 K and 453 K. This enables the observation of changes in CL-20 absorption features before and after the ε–γ phase transition. The sensor can detect changes in molecular fingerprint spectra with temperature. This advances the application of THz sensing in monitoring temperature-dependent absorption characteristics of substances.

2. Structure and Design

Figure 1a illustrates the structural diagram of the THz metamaterial nano-slit array sensor. THz waves, polarized along the x-axis, propagate vertically into the metamaterial. The substrate material utilized is silicon dioxide. The relative permittivity of SiO_2 shows no significant differences at various operating temperatures. We approximate that the relative permittivity of SiO_2 remains constant at 1.73 across all operating temperatures [30–32]. The nano-slit array structure comprises the temperature-sensitive InSb, whose dielectric constant can be approximated by the Drude model as [33].

$$\varepsilon(\omega) = \varepsilon_\infty - \frac{\omega_p^2}{\omega^2 + i\gamma\omega} \quad (1)$$

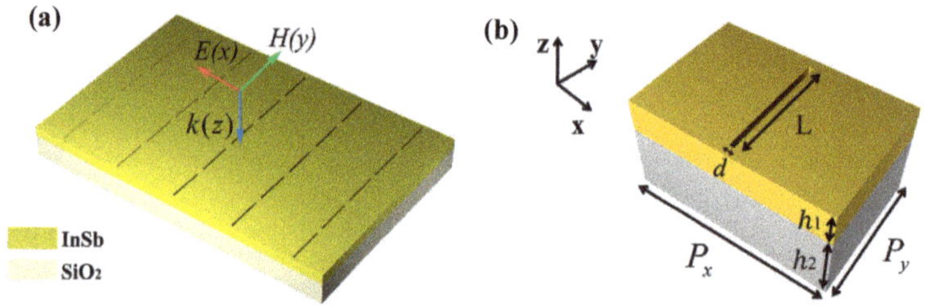

Figure 1. (**a**) Structure diagram of the metasurface sensor. (**b**) Metasurface cell and its structural parameters.

In Equation (1), ε_∞ = 15.68 denotes the dielectric constant at infinite angular frequency, ω represents the angular frequency, $\omega_p = \sqrt{Ne^2/\varepsilon_0 m^*}$ denotes the plasma frequency, $\gamma = 0.1\pi$ THz signifies the damped vibration frequency, N denotes the intrinsic carrier concentration, ε_0 denotes the vacuum dielectric constant, e represents the electron charge, and $m^* = 0.015m_e$ (m_e denotes the free electron mass) represents the effective mass of the free carriers [34].

N can be expressed by the following relation [35]:

$$N = 5.76 \times 10^{20} T^{1.5} \exp(-0.26/2k_B T) \quad (2)$$

In Equation (2), k_B denotes the Boltzmann constant, and T represents the temperature in Kelvin. Hence, the intrinsic carrier concentration N of InSb is influenced by the external temperature, subsequently affecting its dielectric constant. Therefore, InSb serves as a temperature-sensitive semiconductor material, facilitating the convenient modulation of the sensor's resonance peak via changes in the external environment, ultimately leading to the formation of a broadband transmission envelope.

Figure 1b depicts the unit cell of the metamaterial. The geometric parameters of the unit cell are set as follows: period P_x = 120 μm and P_y = 80 μm; slit width d = 3 μm and length L = 68 μm; thickness of InSb and slit h_1= 20 μm; and thickness of the SiO_2 substrate h_2 = 40 μm. To investigate the performance of the sensor, the optical characteristics of the metamaterial are simulated using three-dimensional finite-difference time-domain (FDTD) software. In the simulation, perfect matching layers are applied along the z-direction, while periodic boundary conditions are applied along the x- and y-directions.

3. Result and Discussions

Figure 2a depicts the transmission spectrum of the proposed metasurface sensor when THz waves, polarized along the x-axis, vertically impinge upon the metasurface at an ambient temperature of 310 K. As observed from Figure 2a, a prominent resonance transmission peak is evident at 1 THz. Meanwhile, Figure 2b illustrates the distribution of electric and magnetic fields at the resonance frequency. The electric field (x-z plane) is entirely localized at the slit and is significantly enhanced. The magnetic field is localized at the ends of the slit [36]. Such a field distribution markedly enhances the interaction between THz waves and analytes, thereby enhancing the sensitivity of the sensor [37].

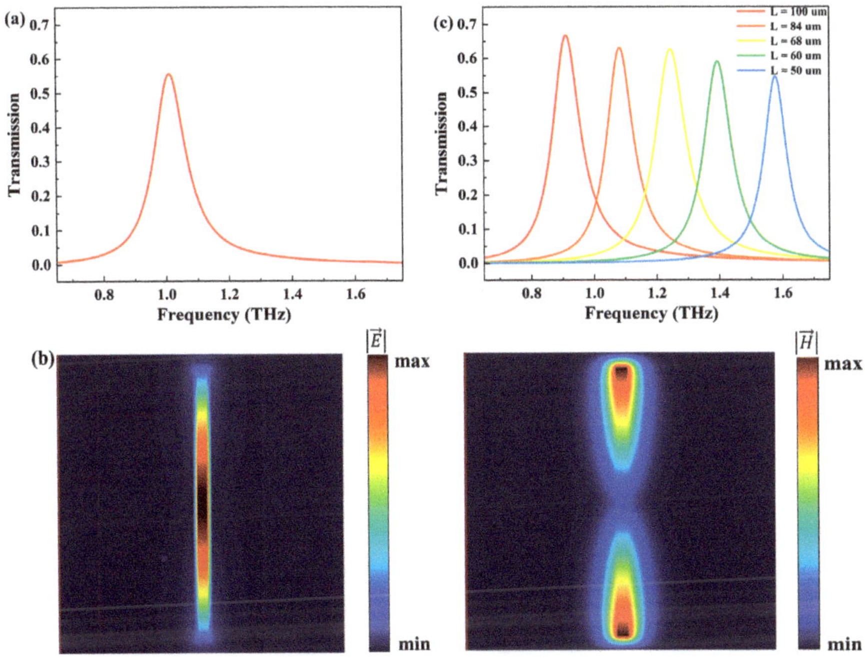

Figure 2. (**a**) Transmission spectrum of the metasurface at T = 310 K. (**b**) Electric and magnetic field distributions at f = 1 THz. (**c**) Transmission spectra for different slit lengths L at T = 360 K.

To comprehensively understand the impact of slit length L on the resonance transmission peak, the transmission spectra of sensors with different L ranging from 50 μm to 100 μm were analyzed. Simultaneously, while adjusting the slit length L, the distance between slits (the period P_y) was adjusted to prevent adjacent slits from contacting each other. Specifically, when L = 50, 60, 68, 84, 100 μm, the corresponding periods P_y were 70, 80, 80, 110, and 120 μm, respectively. As depicted in Figure 2c, with increasing L, the resonance peak frequency gradually decreases, while the transmission peak value slightly increases [38]. The slit length L can influence the resonance frequency and alter the position of the resonance peak. The change in transmission peak value can be explained by the dielectric properties of InSb. As the frequency decreases, the extinction coefficient k of InSb gradually increases. This significantly suppresses THz transmission, thereby reducing the peak transmission.

To characterize the spectral response of the sensor, the relationship between the structural transmission curve and temperature T was investigated. Figure 3a illustrates the variation of the sensor's transmission spectra with temperature T ranging from 275 K to 520 K. As the temperature increases, the resonance angle experiences a blueshift, and the transmission peak value continuously increases. Under slit length L = 68 μm, the resonance peak can shift from 0.78 THz to 1.6 THz, achieving an ultra-wideband shift of 0.82 THz. Figure 3b demonstrates the relationship between the dielectric constant of InSb and temperature, explaining the variation in resonance frequency and transmission peak value [39]. Figure 3c shows the electric field distribution of the metasurface at 1 THz from 300 K to 330 K. The electric field within the slits is significantly enhanced, with the maximum electric field intensity occurring at 310 K. Both increasing and decreasing the temperature lead to a reduction in electric field intensity, corresponding to changes in transmittance at 1 THz for the metasurface. We selected a room temperature range of 260 K to 340 K for molecular fingerprint sensing [40]. Figure 3d presents the transmission spectra (from 260 K to 340 K, with a step size of 2 K) and corresponding metasurface envelope curves. The resonance angles of the transmission spectra at different temperatures were extracted and further fitted into the envelope curves through interpolation (as shown by the red curve in the figure) [41]. With a temperature variation of only 80 K, this envelope curve can cover a broad frequency range from 0.68 THz to 1.16 THz, which is advantageous for the characteristic fingerprint sensing of substances. In this scenario, when the analyte is placed at the slit of the metasurface, an absorption peak matching the fingerprint of the analyte will be observed in the envelope curve [42,43].

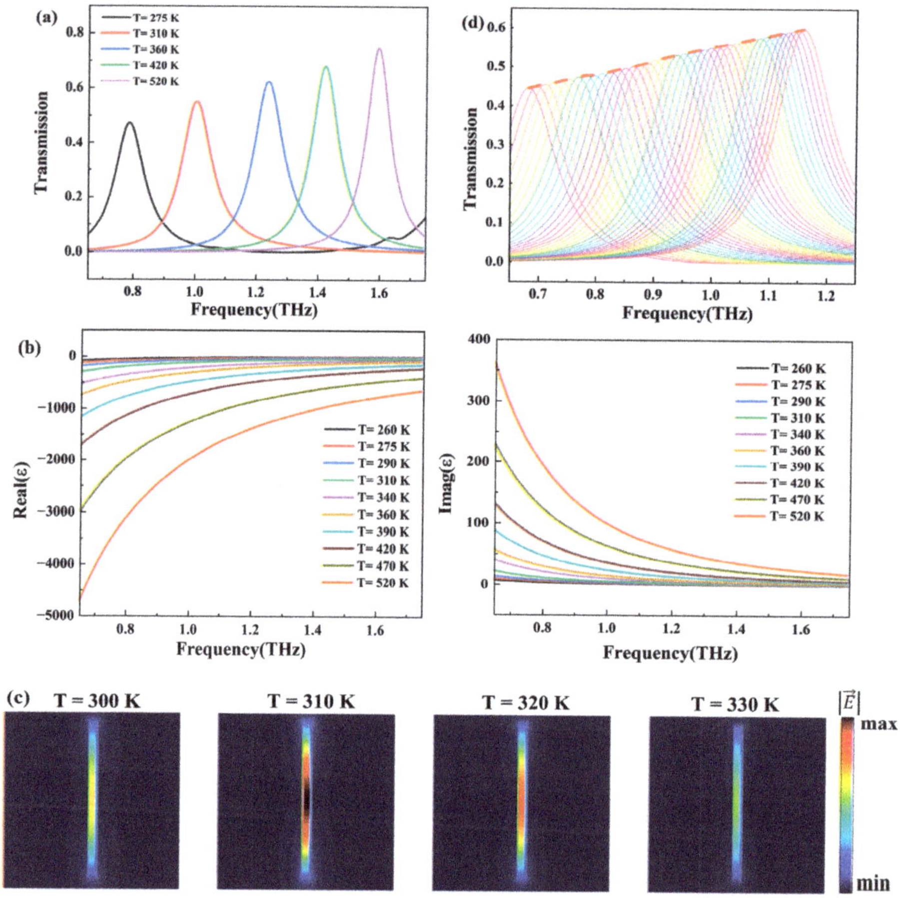

Figure 3. (**a**) Transmission spectra of the metasurface within the ambient temperature range of 275 K to 520 K. (**b**) Dielectric constant of InSb at different ambient temperatures. (**c**) The electric field distribution of the metasurface at 1 THz from 300 K to 330 K. (**d**) Transmission spectra of the metasurface and its envelope curves (the red curve) within the ambient temperature range of 260 K to 340 K (increment of 2 K).

To study its sensing performance, we covered the sensor surface with RDX. Based on the existing THz time-domain spectroscopy analysis results of RDX, we obtained the refractive index n and extinction coefficient k of RDX in the THz range [44], as shown in

Figure 4a. Due to weak intermolecular interactions and lattice vibrations, RDX exhibits a weak absorption peak at 0.88 THz [45]. Additionally, the absorption peak of RDX is relatively broad, necessitating a sufficiently wide sensing frequency bandwidth. Figure 4b illustrates the transmission spectra of the metasurface covered with 8.03 μg/cm^2 RDX as a function of temperature (260–340 K) and the corresponding envelope curves. Because RDX exhibits excellent chemical stability at room temperature and can stably exist for long periods below 400 K, detection at temperatures ranging from 260 K to 340 K is reasonable. With increasing temperature, a significant blueshift in the resonance angle is observed, and a distinct absorption peak appears at 0.88 THz in all spectra. The maximum transmission rate at the same frequency point for different temperatures was extracted, and the results were further fitted into envelope curves through interpolation. It can be observed that the envelope curve exhibits the minimum transmission rate at 0.88 THz and forms an envelope peak at this frequency, corresponding to the fingerprint spectrum of RDX [46]. Let the envelope curve of the sensor without RDX coverage be denoted as T0 and the envelope curve of the sensor covered with 8.03 μg/cm^2 RDX be denoted as Ts. The change in transmission rate was normalized to Tn = (T0 − Ts)/T0. The normalized envelope curve Tn shown in Figure 4c closely follows the trend of the extinction coefficient k of RDX. This indicates that our proposed nano-slit array sensor can accurately amplify and detect the characteristic absorption fingerprint of RDX [47,48], achieving a specific and highly sensitive sensing of RDX.

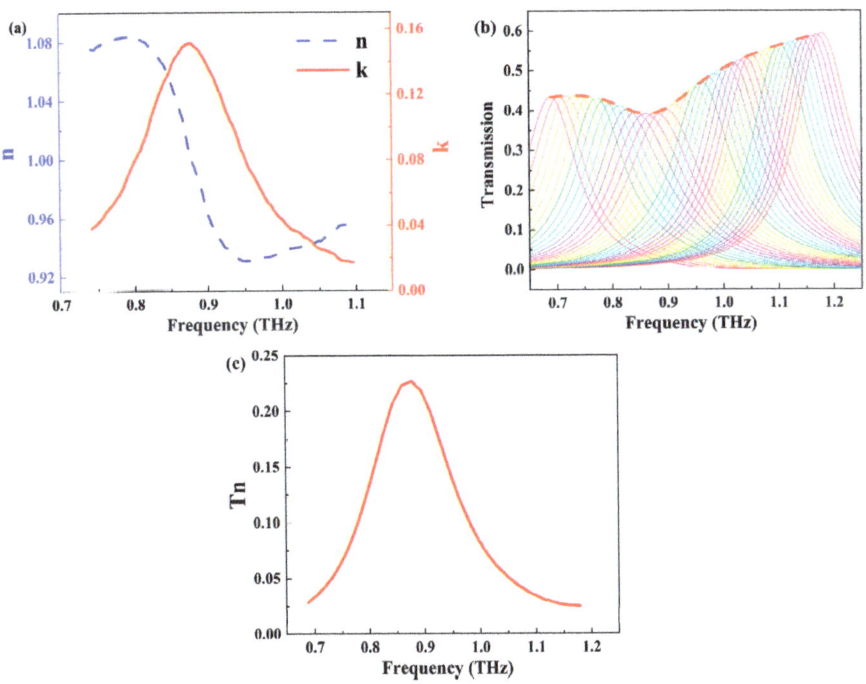

Figure 4. (**a**) Refractive index n and extinction coefficient k of RDX in the THz range. (**b**) Transmission spectra of the metasurface covered with 8.03 μg/cm^2 RDX as a function of ambient temperature (260–340 K) and the corresponding envelope curves (the red curve). (**c**) Envelope curves of the transmission spectra of the metasurface after normalization.

To further investigate the system's capability to detect trace analytes, we covered the metasurface with different concentrations of RDX. Figure 5a illustrates the normalized envelope curves deposited with varying concentrations of RDX on the metasurface. As the RDX concentration increases from 1.61 μg/cm^2 to 40.16 μg/cm^2, the corresponding normalized transmission rate change also gradually increases. Figure 5b demonstrates the relationship between the normalized transmission rate change at 0.88 THz and the RDX concentration in the range of 1.61 μg/cm^2 to 40.16 μg/cm^2. The linear fitting equation of the curve in the figure is approximately y = 0.01519x + 0.08182, where y represents the normalized transmission rate change at 0.88 THz and x denotes the RDX concentration in μg/cm^2. The correlation coefficient of this fitting curve is 0.98, indicating a good linear correlation between the normalized transmission rate change at 0.88 THz and the RDX concentration. In our experiments, the minimum RDX concentration featured a clearly observable envelope absorption peak, and the detection limit of our sensor was 1.61 μg/cm^2. These studies suggest that RDX content can be detected based on the normalized transmission rate change of the envelope peak, enabling a quantitative sensing of RDX with a very low detection limit.

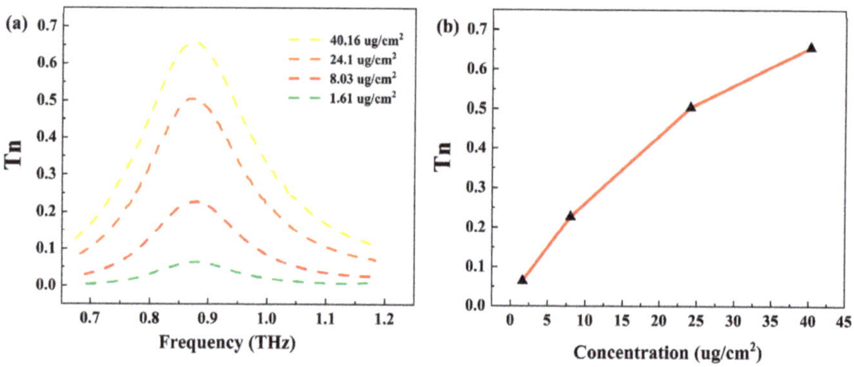

Figure 5. (**a**) Normalized envelope curves deposited with different concentrations of RDX on the metasurface. (**b**) Relationship between the normalized transmission rate change at 0.88 THz and the RDX concentration.

In addition, the sensor can track the characteristic absorption spectra changes of CL-20 before and after the temperature-induced phase transition. Previous studies have demonstrated that CL-20 can undergo an irreversible ε–γ phase transition triggered by temperature variation. At 298 K, the ε-CL-20 phase exhibits two absorption peaks at 1.31 THz and 0.99 THz. As the temperature increases to 453 K, CL-20 transitions into the γ phase. The characteristic absorption peaks at 1.31 THz and 0.99 THz diminish, while a strong absorption occurs at 1.53 THz [9,10]. When the ambient temperature of the metasurface is maintained at 298 K, altering the slit length L results in a redshift in resonance frequency and a slight increase in transmission peak value. Different slit lengths are employed across various regions of the metasurface to encompass the resonance inclination angles of different frequency positions, forming corresponding envelope curves [22]. Figure 6a illustrates the transmission spectra and their envelope curve (depicted in red) for slit lengths ranging from 100 μm to 36 μm. This envelope curve spans the 0.72–1.45 THz range. As the slit length L varies, the period P_y changes with it, ensuring that the distance between the edges of two adjacent slits in the y-direction remains at 40 μm. Figure 6b displays the transmission spectra and their envelope curve (depicted in red) following the introduction of trace CL-20. Notably, distinct absorption peaks emerge at 1.31 THz and 0.99 THz, aligning with the position of the characteristic absorption peak of ε-CL-20. Figure 6c presents the normalized envelope curve, effectively describing the absorption characteristics of ε-CL-20 at 298 K. As the ambient temperature rises to 453 K, Figure 6d showcases the transmission spectra and their envelope curve, with only the transmission spectra from 1 THz to 1.8 THz being displayed, omitting the formant at higher frequency positions. Figure 6e exhibits the transmission spectra and its envelope curve following the introduction of an equal content of CL-20, while Figure 6f represents the normalized envelope curve. At 453 K, the envelope curve exhibits strong absorption at 1.53 THz, while the absorption peak at 1.31 THz disappears. This is consistent with the absorption characteristics of γ-CL-20, effectively describing the absorption features of CL-20 at 453 K. The comparative analysis of the transmission spectra and envelope curves at 298 K and 453 K, covering trace amounts of CL-20, shows a shift in the characteristic absorption peak frequencies from 0.99 THz and 1.31 THz to 1.53 THz. This allows for a clear observation of CL-20's absorption characteristics before and after the phase transition, facilitating the sensing and study of its phase transition mechanism. These findings validate the system's capability to perform highly sensitive sensor detection of material feature absorption fingerprints at multiple temperatures and track changes in material feature absorption fingerprints with temperature.

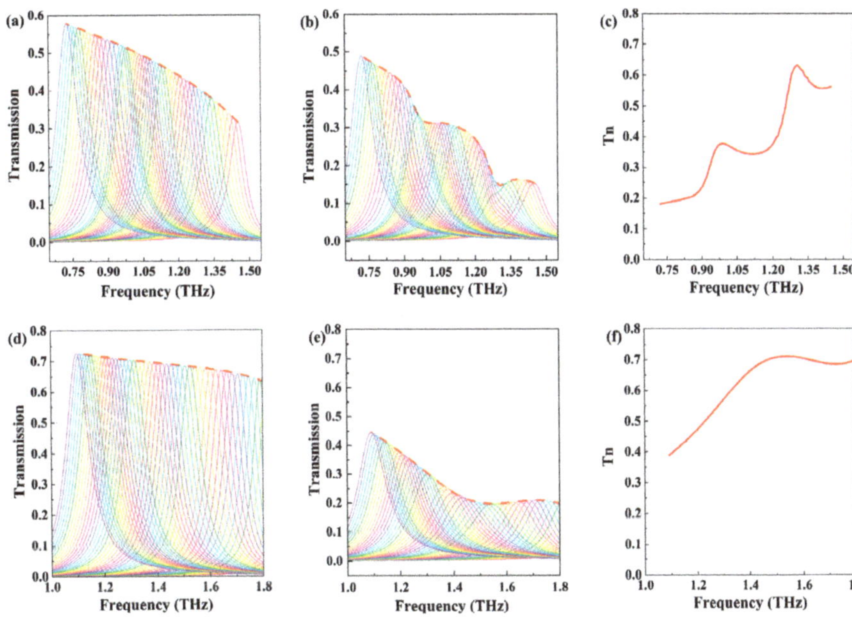

Figure 6. (**a**) Metasurface transmittance spectra and envelope curve of slit length L from 36 μm to 100 μm at 298 K. (**b**) Metasurface transmittance spectra and envelope curves after covering trace CL-20 at 298 K. (**c**) The envelope curve after normalization at 298 K. (**d**) Metasurface transmittance spectra and envelope curve at 453 K. (**e**) Metasurface transmittance spectra and envelope curves after covering trace CL-20 at 453 K. (**f**) The envelope curve after normalization at 453 K.

4. Conclusions

In summary, we propose a nano-slit array sensor based on temperature variation that is capable of a highly sensitive and specific sensing of molecular fingerprints. By integrating structural parameter scanning, this sensor allows for the detection of substance characteristics at various temperatures. The metasurface consists of periodically arranged InSb slits, with SiO_2 as the substrate. When THz waves are vertically incident on the metasurface, LSPs are successfully excited, leading to the localized enhancement of the electric field at the slits. This enhanced electric field significantly amplifies the interaction between the analyte and THz waves, effectively improving sensing sensitivity. With the variation in temperature, the dielectric properties of the temperature-sensitive InSb continuously vary, causing the transmission resonance angle to shift, thereby generating transmission envelope curves covering a wide frequency range. Based on the temperature-scanning strategy, we successfully delineate the characteristic fingerprint spectra of RDX, enabling qualitative and quantitative sensing of RDX with a detection limit of 1.61 μg/cm^2. Simultaneously, employing the structural parameter scanning method and adjusting the slit length L, we successfully depicted the characteristic fingerprint spectra of ε-CL-20 at 298 K and γ-CL-20 at 453 K. This approach achieved a high sensitivity detection of CL-20 absorption features at different temperatures, allowing for the observation of the variation in CL-20 ε–γ phase transition absorption characteristics. Our research indicates that the proposed temperature-scanning metasurface sensor holds significant potential for the wideband fingerprint sensing of trace analytes, thus promoting the development of THz sensing in various fields, including civilian and military security, biomedicine, and beyond.

Author Contributions: Conceptualization, W.W. and Y.S.; methodology, W.W. and Y.S.; software, W.W., M.S. and J.L.; validation, W.W., M.S. and Y.X.; formal analysis, W.W., M.S. and Y.X.; investigation, W.W. and J.L.; resources, Y.S.; data curation, W.W., M.S., Y.X. and Y.S.; writing—original draft preparation, W.W.; writing—review and editing, W.W. and Y.S.; visualization, W.W. and M.S.; supervision, Y.S.; project administration, Y.S.; funding acquisition, Y.S. All authors have read and agreed to the published version of the manuscript.

Funding: This work is supported by the Natural Science Foundation of Shandong Province under Grant ZR2019BF014, the National Natural Science Foundation of China under Grant 61805127, Jinan-SDU Co-development under Project JNSX2023007, and the Fundamental Research Fund of Shandong University under Grant 2018TB002.

Institutional Review Board Statement: Not applicable.

Informed Consent Statement: Not applicable.

Data Availability Statement: The data underlying the results presented in this paper may be obtained from the authors upon reasonable request.

Conflicts of Interest: The authors declare no conflicts of interest.

References

1. Ryu, Y.S.; Lee, D.K.; Kang, J.H.; Lee, S.H.; Yu, E.S.; Seo, M. Ultrasensitive terahertz sensing of gold nanoparticles inside nano slot antennas. *Opt. Express* **2017**, *25*, 30591–30597. [CrossRef] [PubMed]
2. Han, B.; Han, Z.; Qin, J.; Wang, Y.; Zhao, Z. A sensitive and selective terahertz sensor for the fingerprint detection of lactose. *Talanta* **2019**, *192*, 1–5. [CrossRef] [PubMed]
3. Zhu, J.; Wang, Z.; Lin, S.; Jiang, S.; Liu, X.; Guo, S. Low-cost flexible plasmonic nanobump metasurfaces for label-free sensing of serum tumor marker. *Biosens. Bioelectron.* **2020**, *150*, 111905. [CrossRef]
4. Dong, L.; Yang, X.; Zhang, C.; Cerjan, B.; Zhou, L.; Tseng, M.L.; Zhang, Y.; Alabastri, A.; Nordlander, P.; Halas, N.J. Nanogapped Au Antennas for Ultrasensitive Surface-Enhanced Infrared Absorption Spectroscopy. *Nano Lett.* **2017**, *17*, 5768–5774. [CrossRef] [PubMed]
5. Koalla, R.; Mukherjee, S.; Mangali, S.; Vaitheeswaran, G.; Kumar Chaudhary, A. Experimental and theoretical investigation of Low-Frequency vibrational modes of 4-Amino 3,5 Dinitro Pyrazole in terahertz frequency domain. *Spectrochim. Acta Part A Mol. Biomol. Spectrosc.* **2024**, *313*, 124092. [CrossRef]
6. Naftaly, M.; Miles, R.E. Terahertz Time-Domain Spectroscopy for Material Characterization. *Proc. IEEE* **2007**, *95*, 1658–1665. [CrossRef]
7. Koumans, M.; Meulendijks, D.; Middeljans, H.; Peeters, D.; Douma, J.C.; van Mechelen, D. Physics-assisted machine learning for THz time-domain spectroscopy: Sensing leaf wetness. *Sci. Rep.* **2024**, *14*, 7034. [CrossRef]
8. Wu, J.; Wu, J.; Li, J.; Shang, Y.; Chen, L. Molecular Dynamics Simulations of the Thermal Decomposition of RDX/HTPB Explosives. *ACS Omega* **2023**, *8*, 18851–18862. [CrossRef]
9. He, X.; Zhang, Q.; Wang, M.; Liu, Q.; Liu, K.; Li, G.; Kang, Y.; Deng, H.; Shang, L. Terahertz spectral properties of temperature induced phase transition of CL-20. *Infrared Laser Eng.* **2022**, *51*, 287–293. [CrossRef]
10. Duan, Y.; Liu, Q.; Wang, M.; Zhu, Y.; Zhang, Q.; Duan, X.; Deng, H.; Shang, L. Insights into temperature-induced phase transition mechanism of CL-20 using terahertz spectroscopy. *Spectrochim. Acta Part A Mol. Biomol. Spectrosc.* **2024**, *307*, 123640. [CrossRef]
11. Cubukcu, E.; Zhang, S.; Park, Y.-S.; Bartal, G.; Zhang, X. Split ring resonator sensors for infrared detection of single molecular monolayers. *Appl. Phys. Lett.* **2009**, *95*, 043113. [CrossRef]
12. Qin, J.; Cheng, W.; Han, B.; Du, Y.; Han, Z.; Zhao, Z. Ultrasensitive detection of saccharides using terahertz sensor based on metallic nano-slits. *Sci. Rep.* **2020**, *10*, 3712. [CrossRef] [PubMed]

13. Cheng, W.; Han, Z.; Du, Y.; Qin, J. Highly sensitive terahertz fingerprint sensing with high-Q guided resonance in photonic crystal cavity. *Opt. Express* **2019**, *27*, 16071–16079. [CrossRef]
14. Shi, X.; Qin, J.; Han, Z. Enhanced terahertz sensing with a coupled comb-shaped spoof surface plasmon waveguide. *Opt. Express* **2017**, *25*, 278–283. [CrossRef]
15. Cen, W.; Lang, T.; Wang, J.; Xiao, M. High-Q Fano Terahertz resonance based on bound states in the continuum in All-dielectric metasurface. *Appl. Surf. Sci.* **2022**, *575*, 151723. [CrossRef]
16. Wang, Y.; Han, Z.; Du, Y.; Qin, J. Ultrasensitive terahertz sensing with high-Q toroidal dipole resonance governed by bound states in the continuum in all-dielectric metasurface. *Nanophotonics* **2021**, *10*, 1295–1307. [CrossRef]
17. He, Y.; Guo, G.; Feng, T.; Xu, Y.; Miroshnichenko, A.E. Toroidal dipole bound states in the continuum. *Phys. Rev. B* **2018**, *98*, 161112. [CrossRef]
18. Zhao, X.; Chen, C.; Kaj, K.; Hammock, I.; Huang, Y.; Averitt, R.D.; Zhang, X. Terahertz investigation of bound states in the continuum of metallic metasurfaces. *Optica* **2020**, *7*, 1548–1554. [CrossRef]
19. Li, R.; Li, Z.; Jiang, Y. Terahertz biosensor integrated with Au nanoparticles to improve the sensing performance. *Appl. Opt.* **2023**, *62*, 5069–5076. [CrossRef]
20. Han, S.; Rybin, M.V.; Pitchappa, P.; Srivastava, Y.K.; Kivshar, Y.S.; Singh, R. Guided-Mode Resonances in All-Dielectric Terahertz Metasurfaces. *Adv. Opt. Mater.* **2020**, *8*, 1900959. [CrossRef]
21. Xue, Y.; Zhang, S.; Lin, J.; Wang, W.; Chai, Z.; Sun, M.; Shi, Y.; Zhang, Y. Metasurface-based sensor with terahertz molecular fingerprint enhancement in trace additives identification. *J. Phys. D Appl. Phys.* **2024**, *57*, 235104. [CrossRef]
22. Tittl, A.; Leitis, A.; Liu, M.; Yesilkoy, F.; Choi, D.Y.; Neshev, D.N.; Kivshar, Y.S.; Altug, H. Imaging-based molecular barcoding with pixelated dielectric metasurfaces. *Science* **2018**, *360*, 1105–1109. [CrossRef] [PubMed]
23. Zhu, J.; Jiang, S.; Xie, Y.; Li, F.; Du, L.; Meng, K.; Zhu, L.; Zhou, J. Enhancing terahertz molecular fingerprint detection by a dielectric metagrating. *Opt. Lett.* **2020**, *45*, 2335–2338. [CrossRef] [PubMed]
24. Lee, D.K.; Kang, J.H.; Lee, J.S.; Kim, H.S.; Kim, C.; Kim, J.H.; Lee, T.; Son, J.H.; Park, Q.H.; Seo, M. Highly sensitive and selective sugar detection by terahertz nano-antennas. *Sci. Rep.* **2015**, *5*, 15459. [CrossRef] [PubMed]
25. Seo, M.A.; Park, H.R.; Koo, S.M.; Park, D.J.; Kang, J.H.; Suwal, O.K.; Choi, S.S.; Planken, P.C.M.; Park, G.S.; Park, N.K.; et al. Terahertz field enhancement by a metallic nano slit operating beyond the skin-depth limit. *Nat. Photonics* **2009**, *3*, 152–156. [CrossRef]
26. Xie, Y.; Zakharian, A.R.; Moloney, J.V.; Mansuripur, M. Transmission of light through periodic arrays of sub-wavelength slits in metallic hosts. *Opt. Express* **2006**, *14*, 6400–6413. [CrossRef] [PubMed]
27. Ahmadivand, A.; Gerislioglu, B.; Ahuja, R.; Kumar Mishra, Y. Terahertz plasmonics: The rise of toroidal metadevices towards immunobiosensings. *Mater. Today* **2020**, *32*, 108–130. [CrossRef]
28. Lin, S.; Bhattarai, K.; Zhou, J.; Talbayev, D. Thin InSb layers with metallic gratings: A novel platform for spectrally-selective THz plasmonic sensing. *Opt. Express* **2016**, *24*, 19448–19457. [CrossRef] [PubMed]
29. Chen, G.; Liu, B.; Lu, P.; Peng, Y. Rapid Determination of Ochratoxin A in Black Tea Using Terahertz Ultrasensitive Biosensor. *Photonics* **2024**, *11*, 9. [CrossRef]
30. Tan, C.Z.; Arndt, J. Temperature dependence of refractive index of glassy SiO_2 in the infrared wavelength range. *J. Phys. Chem. Solids* **2000**, *61*, 1315–1320. [CrossRef]
31. Davies, C.L.; Patel, J.B.; Xia, C.Q.; Herz, L.M.; Johnston, M.B. Temperature-Dependent Refractive Index of Quartz at Terahertz Frequencies. *J. Infrared Millim. Terahertz Waves* **2018**, *39*, 1236–1248. [CrossRef]
32. Gong, J.; Dai, R.; Wang, Z.; Zhang, C.; Yuan, X.; Zhang, Z. Temperature dependent optical constants for SiO_2 film on Si substrate by ellipsometry. *Mater. Res. Express* **2017**, *4*, 085005. [CrossRef]
33. Howells, S.C.; Schlie, L.A. Transient terahertz reflection spectroscopy of undoped InSb from 0.1 to 1.1 THz. *Appl. Phys. Lett.* **1996**, *69*, 550–552. [CrossRef]
34. Oszwałdowski, M.; Zimpel, M. Temperature dependence of intrinsic carrier concentration and density of states effective mass of heavy holes in InSb. *J. Phys. Chem. Solids* **1988**, *49*, 1179–1185. [CrossRef]
35. Dai, X.; Xiang, Y.; Wen, S.; He, H. Thermally tunable and omnidirectional terahertz photonic bandgap in the one-dimensional photonic crystals containing semiconductor InSb. *J. Appl. Phys.* **2011**, 109. [CrossRef]
36. Seo, M.; Park, H.R. Terahertz Biochemical Molecule-Specific Sensors. *Adv. Opt. Mater.* **2019**, *8*, 1900662. [CrossRef]
37. Novitsky, A.; Zalkovskij, M.; Malureanu, R.; Lavrinenko, A. Microscopic model of the THz field enhancement in a metal nanoslit. *Opt. Commun.* **2011**, *284*, 5495–5500. [CrossRef]
38. Park, H.-R.; Ahn, K.J.; Han, S.; Bahk, Y.-M.; Park, N.; Kim, D.-S. Colossal Absorption of Molecules Inside Single Terahertz Nanoantennas. *Nano Lett.* **2013**, *13*, 1782–1786. [CrossRef] [PubMed]
39. Kang, J.H.; Choe, J.-H.; Kim, D.S.; Park, Q.H. Substrate effect on aperture resonances in a thin metal film. *Opt. Express* **2009**, *17*, 15652–15658. [CrossRef]
40. Bahk, Y.-M.; Choi, J.-W.; Kyoung, J.; Park, H.-R.; Ahn, K.J.; Kim, D.-S. Selective enhanced resonances of two asymmetric terahertz nano resonators. *Opt. Express* **2012**, *20*, 25644–25653. [CrossRef]
41. Zhong, Y.; Du, L.; Liu, Q.; Zhu, L.; Meng, K.; Zou, Y.; Zhang, B. Ultrasensitive specific sensor based on all-dielectric metasurfaces in the terahertz range. *RSC Adv.* **2020**, *10*, 33018–33025. [CrossRef]

42. Xie, Y.; Ma, Y.; Liu, X.; Khan, S.A.; Chen, W.; Zhu, L.; Zhu, J.; Liu, Q.H. Dual-Degree-of-Freedom Multiplexed Metasensor Based on Quasi-BICs for Boosting Broadband Trace Isomer Detection by THz Molecular Fingerprint. *IEEE J. Sel. Top. Quantum Electron.* **2023**, *29*, 1–10. [CrossRef]
43. Chen, H.; Han, J.; Liu, J.; Gao, L.; Ma, S. Identification of chiral lansoprazole drugs using THz fingerprint spectroscopy. *Chem. Pap.* **2023**, *77*, 887–893. [CrossRef]
44. Dorney, T.D.; Baraniuk, R.G.; Mittleman, D.M. Material parameter estimation with terahertz time-domain spectroscopy. *J. Opt. Soc. Am. A* **2001**, *18*, 1562–1571. [CrossRef] [PubMed]
45. Chen, J.; Chen, Y.; Zhao, H.; Bastiaans, G.J.; Zhang, X.C. Absorption coefficients of selected explosives and related compounds in the range of 0.1–2.8 THz. *Opt. Express* **2007**, *15*, 12060–12067. [CrossRef] [PubMed]
46. Li, X.; Nie, L.; Wu, H.; Zhang, L.; Yan, D. Enhancing THz fingerprint detection by the stretchable substrate with a dielectric metagrating. *Appl. Opt.* **2023**, *62*, 9028–9035. [CrossRef] [PubMed]
47. Sun, L.; Xu, L.; Wang, J.; Jiao, Y.; Ma, Z.; Ma, Z.; Chang, C.; Yang, X.; Wang, R. A pixelated frequency-agile metasurface for broadband terahertz molecular fingerprint sensing. *Nanoscale* **2022**, *14*, 9681–9685. [CrossRef]
48. Meng, D.; Liu, J.; Chen, W.; Cheng, Y.-Y.; You, K.-W.; Fan, Z.-C.; Ye, Q.; Huang, P.-H.; Chen, Y.-S. Study on the enhancement mechanism of terahertz molecular fingerprint sensing. *Results Phys.* **2022**, *39*, 105766. [CrossRef]

biosensors

MDPI

Article

Fiber-Optic Distributed Sensing Network for Thermal Mapping of Gold Nanoparticles-Mediated Radiofrequency Ablation

Akbota Sametova [1], Sabit Kurmashev [1], Zhannat Ashikbayeva [1], Aida Amantayeva [1], Wilfried Blanc [2], Timur Sh. Atabaev [3] and Daniele Tosi [1,4,*]

1 School of Engineering and Digital Sciences, Nazarbayev University, Nur-Sultan 010000, Kazakhstan; akbota.sametova@nu.edu.kz (A.S.); sabit.kurmashev@nu.edu.kz (S.K.); zhashikbayeva@nu.edu.kz (Z.A.); aida.amantayeva@alumni.nu.edu.kz (A.A.)
2 Université Côte d'Azur, INPHYNI, CNRS UMR7010, Avenue Joseph Vallot, 06108 Nice, France; wilfried.blanc@inphyni.cnrs.fr
3 Department of Chemistry, Nazarbayev University, 53 Kabanbay Batyr Avenue, Nur-Sultan 010000, Kazakhstan; timur.atabaev@nu.edu.kz
4 National Laboratory Astana, Laboratory of Biosensors and Bioinstruments, Nur-Sultan 010000, Kazakhstan
* Correspondence: daniele.tosi@nu.edu.kz

Abstract: In this work, we report the design of an optical fiber distributed sensing network for the 2-dimensional (2D) in situ thermal mapping of advanced methods for radiofrequency thermal ablation. The sensing system is based on six high-scattering MgO-doped optical fibers, interleaved by a scattering-level spatial multiplexing approach that allows simultaneous detection of each fiber location, in a 40 × 20 mm grid (7.8 mm^2 pixel size). Radiofrequency ablation (RFA) was performed on bovine phantom, using a pristine approach and methods mediated by agarose and gold nanoparticles in order to enhance the ablation properties. The 2D sensors allow the detection of spatiotemporal patterns, evaluating the heating properties and investigating the repeatability. We observe that agarose-based ablation yields the widest ablated area in the best-case scenario, while gold nanoparticles-mediated ablation provides the best trade-off between the ablated area (53.0–65.1 mm^2, 61.5 mm^2 mean value) and repeatability.

Keywords: radiofrequency ablation; fiber optic shape sensors; optical fiber sensor; distributed sensors; gold nanoparticles; biomedical sensors

1. Introduction

The role of minimally invasive cancer thermotherapies is substantially extended in the latest years, as clinical trends show strong progress toward methods with limited invasiveness and consistent repeatability in the replacement of more invasive methods such as surgical resection [1–3]. Methods based on thermal ablation find consistent progress as the technological advances in miniaturization of the devices [4], sensing [5], and real-time imaging [6] allow for achieving substantial clinical improvements while maintaining minimal invasiveness and percutaneous insertion of the surgical device.

Thermal ablation procedures accomplish a successful cancer treatment by transferring energy in the form of electromagnetic waves from a source placed outside of the patient into an applicator, percutaneously inserted into the tumor location [7] or externally placed [8]; the energy dissipated into the applicator is converted into thermal energy, resulting in a localized rise of temperature within the tumoral tissue that spreads from the applicator to the peripheral side. Temperature values over 60 °C result in a nearly instantaneous mortality of cancer cells, while cytotoxic effects are recorded for temperature values over 42 °C [9]; the overall effect is evaluated by the thermal dosimetry [10,11], which integrates the temperature over the exposure time.

Thermal ablation methods differ on the frequency f of the electromagnetic waves, and consequently on the type of applicator involved in the energy delivery [5]. The four main methods that involve thermal heating are based on: (1) radiofrequency ablation (RFA, f~450 kHz), which uses an electrical RF generator and a miniature applicator with an electrode mounted on a single tip [12] or multiple tips [13]; (2) high-intensity focused ultrasound (HIFU, f~1.2 MHz), which uses an ultrasound generator and an external transducer that focuses the incoming waves into the tissue, with a non-contact applicator that ablates the tissue by a combination of thermal effects and cavitation [14]; (3) microwave ablation (MWA, f~2.4 GHz), which uses a microwave generator and an applicator shaped as a transmission line and operating as a near-field antenna [15]; and (4) laser ablation (LA, f~300 THz), which uses a mid-power laser source either externally firing into the tissue surface [9] or coupled into a large-core optical fiber for in situ delivery [16].

Among these procedures, RFA is highly versatile, and achieves growing success rates in the treatment of solid tumors such as hepatocellular carcinoma [2,17], and spinal tumors [18]. An RFA needle can integrate multiple miniature electrodes [19], or a single electrode with a high contact surface for rapid thermotherapies [20]; some applicators incorporate micro-thermocouples for sensing [21]. Outside of cancer care, RFA can be applied in the treatment of cardiac arrhythmia [22], and in interventional pain management [23,24].

The main research trends aimed at the technological improvement of RFA point in two different directions. The first area of interest involves the use of the advances in material sciences to improve the heating efficacy. A critical factor for the cancer treatment via RFA relies on the changes of electrical impedance of the tissues as the temperature approaches the 100 °C value [5]; the vaporization of the inner part of the tissue causes the impedance to abruptly rise, and therefore only a small portion of the electrical power is dissipated onto the tissue. Under this condition, medical generators enter a "safe mode", discontinuing the power supply and interrupting the RFA procedure [25]. In order to improve over this effect, both agarose and chitosan gels [26] have been employed in order to reduce the tissue impedance. In addition, gold and silver nanomaterials [27] have been used in combination with agarose with the purpose of improving the thermal delivery, and ultimately increasing the ablated region.

The second area of interest is the use of advanced biosensors for the detection of physical parameters in situ prior to the ablation (for diagnostic purposes) and during the thermal treatment. Optical fiber sensors serve this purpose much better than electrical or mechanical sensors [28], as they can incorporate biological [29] and biophysical sensing [30], as well as owning key properties in terms of biocompatibility, miniaturization of the footprint, disposability, response time, and spatial distribution.

Several optical fiber biosensors have been reported for this purpose. Evers et al. [31] reported fiber-optic biosensors for liver tissue identification, discriminating the healthy and tumoral tissues with a spectroscopic probe. Loyez et al. [29] reported a plasmonic tilted fiber Bragg grating with the capability of in situ detection of cytokeratin biomarkers for the identification of tumoral cells for following treatments. Tosi et al. [32] reported a fiber-optic dual pressure and temperature sensor for measuring the pressure in the proximity of the RFA tip during an ex vivo procedure. Several authors also reported the use of fiber Bragg gratings (FBGs) [12,33] and distributed sensors [34,35], for the temperature detection during hyperthermia, exploiting the spatial resolution of optical fiber sensors.

From an application standpoint, and from the point of view of controlling the thermal ablation process in real-time [36], temperature sensing plays the largest role in measuring the effectiveness of the procedure since thermal damage is largely depending upon the instantaneous temperature and also the variability of the electrical and thermal properties of the tissues prevents from obtaining ablation patterns with high repeatability in the experimental conditions. Densely arrayed FBG sensors [37], chirped FBGs [30], long period gratings [38,39], and distributed sensors, particularly when arranged into a multi-fiber grid-shaped arrangement [34], have the possibility of sub-centimeter spatially resolved thermal sensing. However, while distributed sensors interrogated via optical backscatter reflectometry [33] can use inexpensive single-mode fibers or high scattering fibers as sensors without any additional manufacturing, FBG sensors are more expensive as they require inscription of multiple devices into the fiber, making the system less compatible with a disposable use.

Prior works were performed on RFA using ferromagnetic nanoparticles [26], and silver nanoparticles with a green-oriented synthesis method [40], showing an increase of the treated region when nanoparticles are inserted in situ in a solution of agarose or chitosan gel. However, gold nanoparticles (AuNP) represent the most popular method to extend the performance of thermotherapies, as AuNP have been demonstrated in radiofrequency [41] and laser [42] ablation. AuNP have been investigated as a method for improving the thermal treatments, since they combine their properties of ease of synthesis and biocompatibility [43] with their advantageous electrothermal effects: lowering the impedance of the tissue at the electrode contact point, hence extending the duration of RFA [44], and improving the heating process at the peripheral side of the tumor [45,46], hence targeting a wider region. In addition, AuNP improves the drug delivery systems within cancer tissues, as shown in previous works [27,47,48].

In this work, we consolidate the design of a distributed fiber-optic sensing network for real-time, mini-invasive, and spatially resolved thermal detection for AuNP-mediated RFA; results are presented by comparing the thermal response of pristine ablation with AuNP-mediated ablation with two different density levels (1 and 4 mg/mL), and agarose-mediated RFA. The biosensing system is based on a network of six optical fibers having high scattering, arranged in a spatial division multiplexing setup. The positioning of the fibers in the tissue allows sensing over a 40 × 20 mm grid, with 7.8 mm^2 pixel size. RFA experiments have been performed ex vivo on bovine phantom, in pristine mode, mediated by agarose, and finally mediated by Au nanoparticles in different densities. The resulting 2-dimensional (2D) thermal maps allow for the recording of the different heating patterns and spatiotemporal trends, and investigate the efficacy of each ablation process.

2. Materials and Methods

2.1. Experimental Setup

The experimental setup of the thermal ablation procedure shown in Figure 1 comprises the following parts: (a) optical backscatter reflectometer (OBR, Luna 4600, Roanoke, VA, USA) with a computer used to collect and process the data; (b) distributed MgO nanoparticle-doped optical fibers spliced to single-mode fibers; (c) commercially obtained bovine liver; (d) RF applicator, with cylindrical shape (160 mm length, 3 mm diameter) and an active brass electrode on the conical tip of ~10 mm length; and (e) RF/MWA Hybrid Generator (LEANFA S.r.l., Ruvo di Puglia, Italy) that launches a 450-kHz radiofrequency signal to the applicator. The OBR is used in distributed sensing mode, measuring the frequency shift of each fiber signature within the network with millimeter-level spatial resolution [49]. Thermal ablation experiments employed six separate high-scattering optical fibers with different lengths plugged with the OBR Luna 4600 to measure temperature change during radiofrequency ablation.

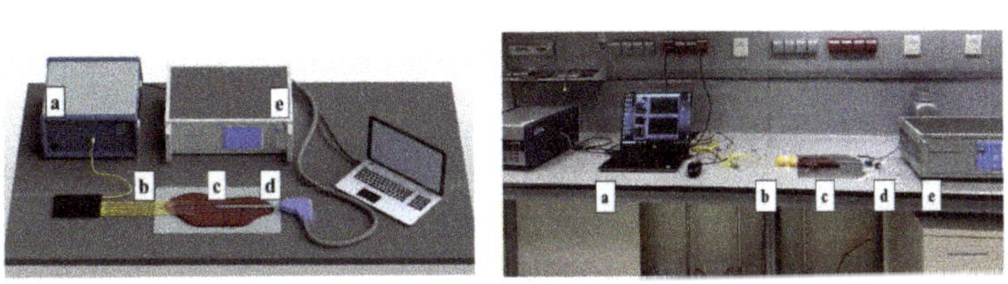

Figure 1. Schematic (**left**) and photographic view (**right**) of the radiofrequency ablation setup of the parenchymal tissue. The setup consists of: (a) OBR Luna 4600 with the computer used for data acquisition, (b) optical fibers, (c) bovine liver, (d) the RFA applicator, and (e) a hybrid RF/MWA generator used in RF mode.

2.2. Fiber Calibration

The presented work applied six Mg-silicate NPs fibers tagged M01 containing erbium in the core. The fabrication of M01 fiber includes solvents such as erbium (III) chloride hexahydrate and magnesium chloride in volumes of 10—4 mol/L and 0.1 mol/L, respectively [50]. Six MgO nanoparticle-doped optical fibers with the core diameter of 10 μm and the cladding diameter of 125 μm, matching the size and compound of single-mode glass fibers, have been used for the detection of the temperature change. Figure 2 shows the photographs of the used fibers.

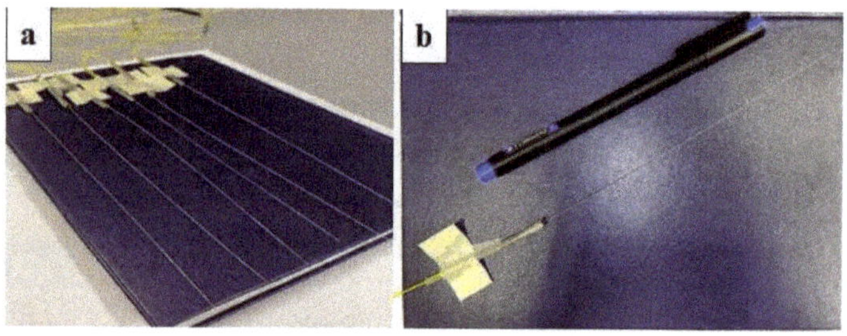

Figure 2. Photographic view of the MgO-doped fibers employed for temperature sensing. (**a**) Photos of the fibers displayed on a grid; (**b**) view of a single nanoparticle-doped optical fiber.

We report in Figure 3 the thermal calibration of the fiber, obtained by exposing the fiber in a water bath heated by a thermal plate (IKA magnetic stirrer hot-plate, IKA-Werke Gmbh, Staufen, Germany), at temperatures ranging from 21 to 57 °C, and recording the temperature with a commercial fiber Bragg grating (Technica Optical Components LLC, Beijing, China) interrogated by an FBG (Fiber Bragg Grating) interrogator (si255 model, Micron Optics/Luna, Roanoke, VA, United States). The thermal coefficient is estimated as 9.18 pm/°C, which is very similar to FBG sensors working at 1550 nm (~10 pm/°C [32]).

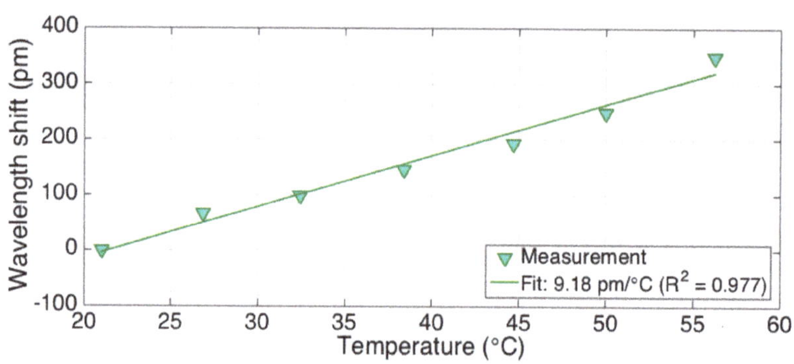

Figure 3. Calibration of the temperature coefficient of the optical fibers used for sensing.

2.3. Gold Nanoparticle Preparation

The presented study uses AuNPs of 15–20 nm size and in different densities introduced directly into the phantom. The synthesis of AuNPs was conducted using the citrate reduction method as presented by Turkevic et al. [51]. The 20 nm AuNPs were achieved by adding 2.0 mL of 34 mM trisodium citrate solution into a boiling solution, containing 0.5 mL of 1% hydrogen tetrachloroaurate (III) trihydrate and 50 mL of deionized water. The color of the solvent changed from bright yellow to dark violet in a few minutes and turned to ruby red after 15 min of stirring. The obtained solution was cooled at room temperature for the next 20 min and then cleaned by deionized water using a centrifuge operating at

15,000 rpm in 2-mL tubes (Figure 4). The size and shape of the synthesized GNPs were characterized using transmission electron microscopy (TEM) (Figure 5).

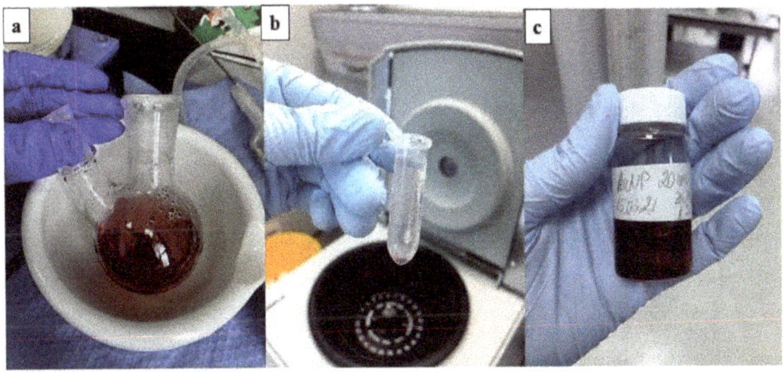

Figure 4. Gold nanoparticle cooling process (**a**) and after centrifuge (**b**,**c**).

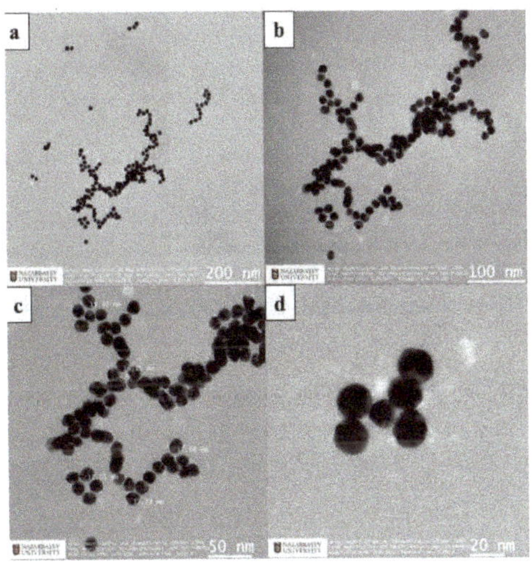

Figure 5. TEM images of AuNPs in different zooms: 200 nm (**a**), 100 nm (**b**), 50 nm (**c**), and 20 nm (**d**). The shape and size of AuNPs equivalent to 15–20 nm spheres in different zooms.

Agarose gel was prepared by mixing 1 g of agarose powder with Tris Acetate EDTA (TAE buffer) in a 50 mL volume and heating in the microwave until it dissolved completely. In order to obtain different densities of nanoparticles starting from 1 mg/mL to 4 mg/mL, AuNPs were dissolved in 0.2% agarose solution at a 1:1 to 1:4 ratio, respectively.

2.4. RF Ablation Experiments

Gold nanoparticles with different densities were prepared before thermal ablation experiments. The calibration of MgO-doped optical fibers was conducted before the experiments. Gold nanoparticles at different densities such as 1 mg/mL and 4 mg/mL were introduced ex vivo in the parenchymal tissue surface and around the applicator. The hybrid generator and reflectometer turned on at the same time: during the first phase, the tissue was ablated; then, the generator was automatically turned off by reaching the safe mode impedance value at 800 Ω, while the measurement was continued for the next 50 s. The applicator and fibers were positioned on opposite sides to each other on the *y*-axis; the distance between fibers was 4 mm on the *x*-axis and the tip of the applicator was placed between the third and fourth fibers (Figure 6). Thermal ablation experiments for each condition were repeated four times: pristine, pure agarose, gold nanoparticles with the density starting at 1 mg/mL, and gold nanoparticles with a density of 4 mg/mL to avoid contamination of fibers.

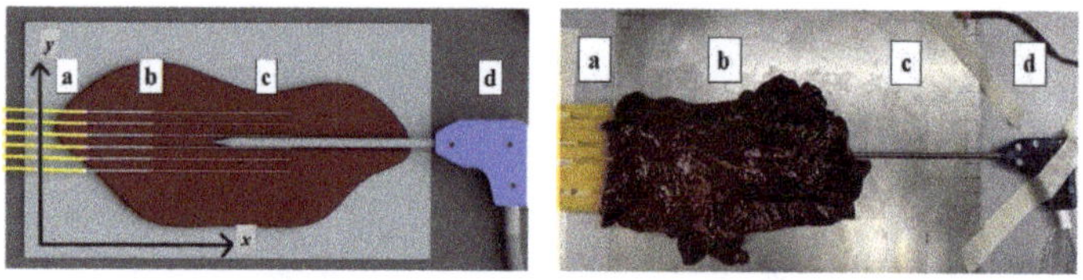

Figure 6. The scheme and photographic view fibers (a), bovine liver (b), negatively charged plate (c), and the applicator (d) in the vicinity.

In the experiments, the applicator and fibers have been positioned manually by the operator in order to maintain the relative distance between the electrode tip and each fiber of the grid; agarose and agarose/AuNP were introduced after the fiber insertion, through a syringe. In order to preserve the fidelity of fiber insertion through the tissue, the fibers were inserted through the phantom, from one side; Figure 6a shows the cross-section of the tissue, while in Figure 6b, the photograph shows that in the actual experiments, the liver tissue was positioned around the fibers. Post-ablation photographs are displayed upon cutting the tissue in the cross-section.

MgO-doped optical fibers were spliced to single-mode fibers using a standard telecom splicer, resulting in a sensing system that detects the temperature change during radiofrequency ablation with 1-s speed. Figure 7 shows the resulting amplitude trace recorded on the OBR when all six fibers are connected, forming a scattering-level multiplexing network that allows simultaneous temperature recording on each location. The scattering gain of each nanoparticle-doped fiber is about 40 dB. The received data were first processed using a threshold algorithm which can identify sudden dips in signal amplitude and, thus, exact locations of all six fibers' ends were discovered. Then, the peak temperature reading, which is in correspondence of the vertical coordinate of the applicator tip, was determined for each fiber and 20 data points around it were used for interpolation (using spline functions). Finally, interpolated domains were converted to thermal maps to accurately monitor the temperature change area.

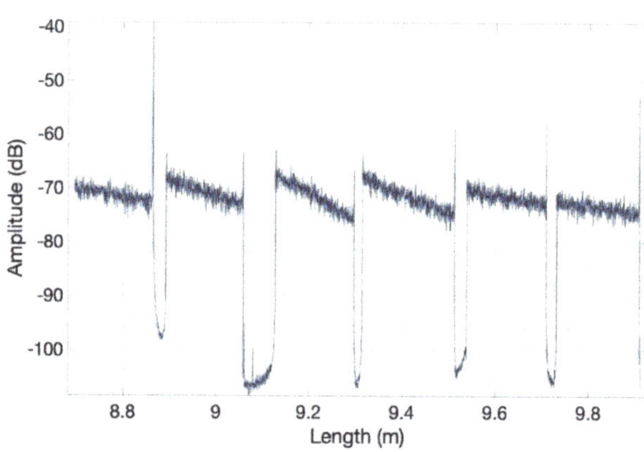

Figure 7. Backscattering trace, recording the Rayleigh backscattering intensity at every point along the fiber sensing network. The six regions with high intensity correspond to each nanoparticle-doped fiber span, while the low-intensity regions correspond to single-mode fibers used to distribute the signal to each sensing fiber.

3. Experimental Results

3.1. Thermal Maps

We report in Figure 8 an example of a thermal map, recorded with the previously described setup for RFA ablation carried out with AuNP with a density of 4 mg/mL. Data are recorded on a xy grid with size 40 mm × 20 mm; the pixel of the grid is 2.5 mm (OBR spatial resolution after processing) × 5.0 mm (distance between each MgO-doped fiber), for a total of 102 sensing points (one sensor per each 7.8 mm^2). Data reported in the figure show the isothermal curves, spacing the temperature data by 10 °C.

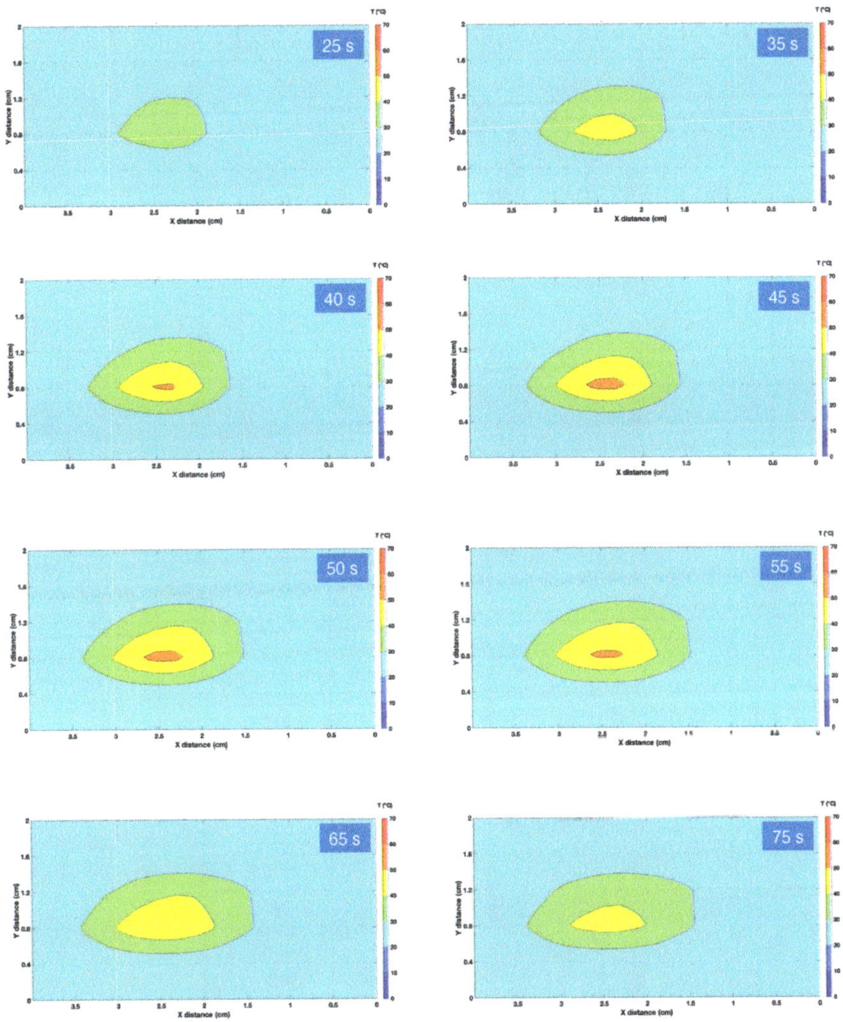

Figure 8. Thermal map recorded for an RFA experiment, using AuNP with a density of 4 mg/mL. The chart reports the data recorded on the xy plane (x = direction parallel to the RFA applicator and to the sensing fibers). The colorimetric map reported the isothermal curves, with 10 °C separation between each layer. Eight different maps are reported, during the heating phase (with times 25, 35, 40, 45, and 50 s from the RF generator power on), and during the cooling phase when the RF power is discontinued (55, 65, and 75 s elapsed time). The horizontal lines on each chart report the position of the six fibers.

During the first part of the ablation, we observe a progressive heating as the temperature rises, peaking in correspondence to the active electrode. After 40 s, the inner temperature overcomes the 60 °C threshold, reaching the maximum extension after 50 s. At this point, the RF generator gets discontinued, causing the temperature to drop.

In Figure 9, we compare the thermal map observed at the peak temperature, in correspondence to the RF power discontinued from the generator. Thermal maps are showed in comparison with the pictures of the ablated tissue, for each different experiment: pristine ablation, agarose-mediated ablation, and AuNP-mediated ablation with densities of 1 mg/mL and 4 mg/mL. We observe that the experiments return a similar result in terms of the shape of the ablated region; AuNP-mediated RFA, however, results in a slightly elongated pattern of the isothermal curves.

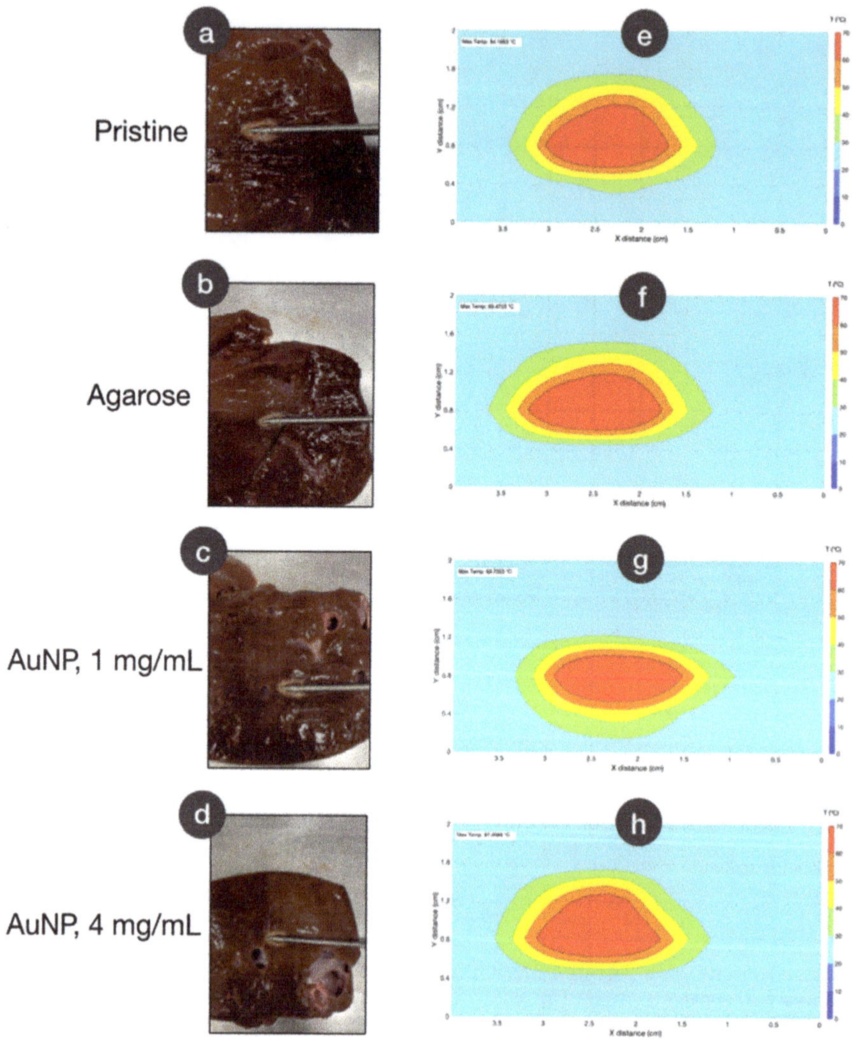

Figure 9. Thermal maps observed for different conditions, compared to the photograph of the ablated tissue. (**a–d**) Photographs of the tissue after RFA procedure; (**e–h**) thermal maps observed at the peak temperature condition. Experiments have been performed in pristine condition (**a**,**e**), with agarose gel inserted in the tissue (**b**,**f**), and with AuNP with densities of 1 mg/mL (**c**,**g**) and 4 mg/mL (**d**,**h**).

The experiments carried out during RFA are subjected to a high variability of results [27]. While the distribution of the nanoparticles in the tissue appears to play a role [40], the major source of variability is due to the different properties of the tissues: this is both attributable to the differences in electrical and thermal properties of the parenchyma, which also tend to increase in tumoral tissues [31], and also to the presence of blood vessels that prevent an efficient heat distribution [52]. In order to account for the repeatability of the ablation process, we performed four experiments for each ablation type, all using a similar phantom. The results are shown in Figure 10, where the thermal maps at the temperature peak are reported. We observe a difference in the shape and extension of the thermal zones: in particular, the AuNP-mediated isothermal curves appear to have a rounder contour, more similar to an elliptical shape; conversely, pristine ablation returns isothermal shapes characterized by more irregularities.

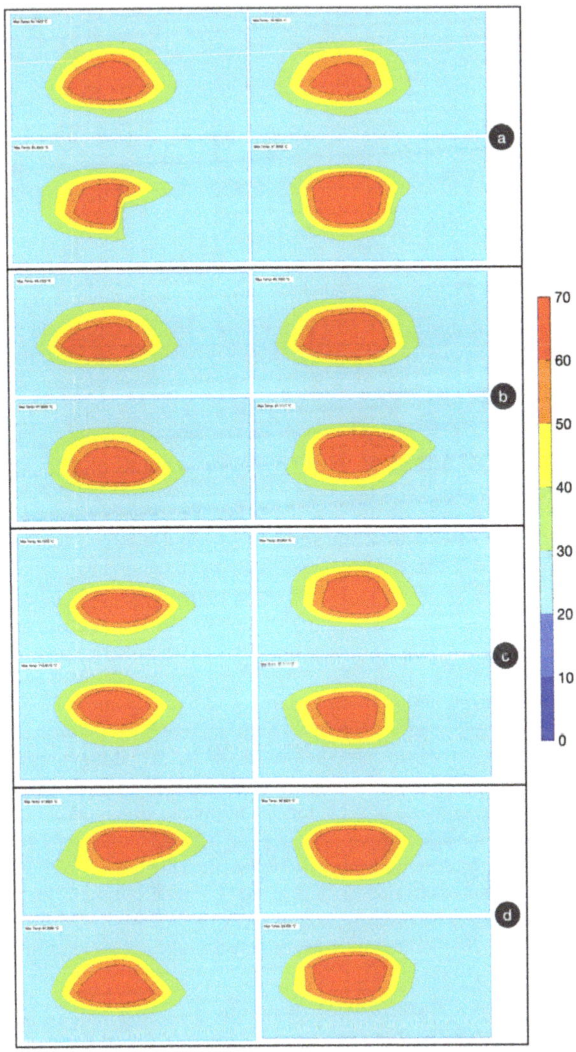

Figure 10. Evaluation of all thermal maps for four experiments, under each condition: (**a**) pristine; (**b**) agarose; (**c**) AuNP, 1 mg/mL; (**d**) AuNP, 4 mg/mL. The chart reports isothermal curves, according to the color bar on the right. Horizontal axis: x direction (4 cm range); vertical axis: y direction (2 cm range).

3.2. Maximum Temperature

The maximum temperature is an important indicator of the heating process occurring during RFA. According to Rhim et al. [53], temperature values exceeding 120 °C should be avoided, while the 80–100 °C peak temperature ensures the most efficient ablation process. Figure 11 shows the peak temperature recorded for all experiments, reporting both the average and the standard deviation. Pristine ablation shows an average temperature of 89.0 °C (the lowest value); whole agarose-mediated RFA shows 95.3 °C average (the highest value); and AuNP-mediated RFA returns similar temperature values (92.2 and 92.3 °C, respectively). AuNP-mediated RFA with 4 mg/mL shows the best repeatability of peak temperature (standard deviation = 4.0 °C), while AuNP-mediated ablation shows the worst repeatability (standard deviation = 13.7 °C).

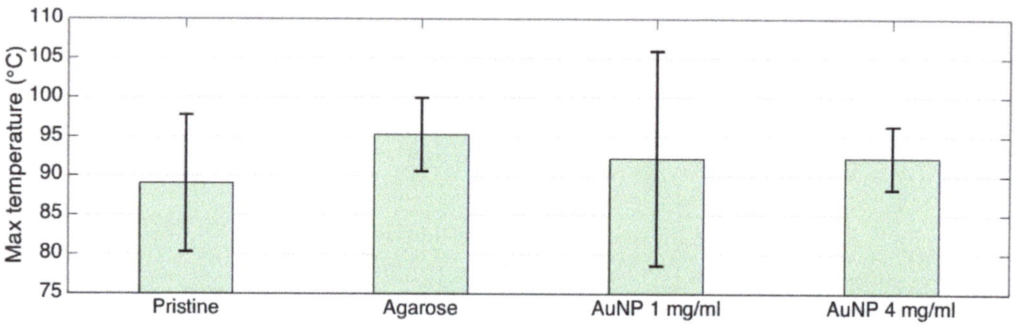

Figure 11. Peak temperature recorded during the RFA experiments under each condition; bars = average of four experiments; error bars = ±standard deviation.

The temporal evolution of the peak temperature shows insights on the heating process and the duration of the ablation procedure [54]. In Figure 12, the temporal trend of the peak temperature is reported for each experimental condition, for the first 50 s of the ablation process. We observe that the duration of the ablation process is similar for all conditions, about 37 s with a larger variability for the AuNP-mediated condition, 1 mg/mL and pristine ablation, and a smaller extension of the standard deviation region for the AuNP-mediated ablation, 4 mg/mL and agarose-mediated. Looking at the average trend, the agarose-mediated ablation reaches the 60 °C threshold in the fastest time (15 s), about 2 s faster than the pristine ablation.

3.3. Evaluation of Thermal Damage Regions

Thermal dosimetry estimates the amount of thermal damage induced by the ablation process [10]. The mortality rate of cancer cells is nearly instantaneous for temperature values higher than 60 °C, while no damage is induced below 42–44 °C depending on the type of tumor. For intermediate values, the mortality rate is a function of the temperature and the exposure time, and a common estimate is 52 °C for one minute of exposure. However, since RFA is a faster process with a duration of ~35 s, we can highlight three regions: (1) maximum thermal damage (temperature >60 °C), where the mortality rate is almost ideal; (2) cytotoxic region (temperature between 42 °C and 60 °C), where temperature induces a partial damage of the tissue; and (3) safe region (temperature below 42 °C), unaffected by RFA. The possibility of drawing isothermal curves at the moment of maximum heating determines the width of each region.

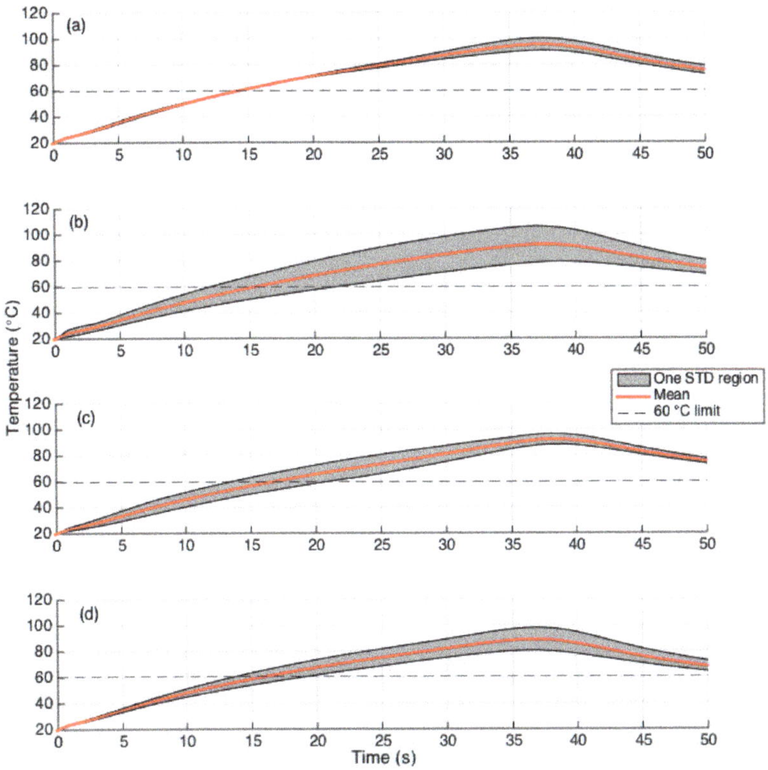

Figure 12. Temporal evolution of the maximum temperature over time, for each RFA ablation. The chart reports the results of four experiments, displaying the mean value (red curve) and the range containing ± one standard deviation (grey interval). RFA conditions: (**a**) agarose; (**b**) AuNP, 1 mg/mL; (**c**) AuNP, 4 mg/mL; (**d**) pristine.

Figure 13 shows the quantification of the thermal damage condition. Pristine ablation shows an ablated surface of 55.0 mm^2 (mean value), and has the widest extension (33.9–66.4 mm^2). Agarose-mediated RFA shows the highest extension of the ablated region (67.3 mm^2 mean value, 74.7 mm^2 maximum). On the other side, the AuNP-mediated ablation with 4 mg/mL density shows the best trade-off between amount of ablated tissue (61.5 mm^2 mean value) and repeatability (53.0–65.1 mm^2 minimum−maximum values). The percentual indicators displayed on the right chart show the quantification of the increase or decrease of ablated tissue with respect to pristine ablation (mean value): it shows that pristine ablation has the worst amount of ablated tissue and the widest repeatability range, while agarose-mediated RFA can ablate up to 19.8% more of tissue.

In Figure 14, the cytotoxicity regions are shown, reporting the areas of the tissue exposed to temperatures >42 °C. We observe a similar trend with respect to the thermal damage results, with agarose performing the widest region of damage (mean value 132.3 mm^2) but with the widest range, and AuNP-mediated nanoparticles with the higher density to represent the best trade-off between amount of thermal damage (mean value 127.9 mm^2, +8.5% with respect to pristine conditions) and repeatability (116.7–141.1 mm^2 minimum−maximum values).

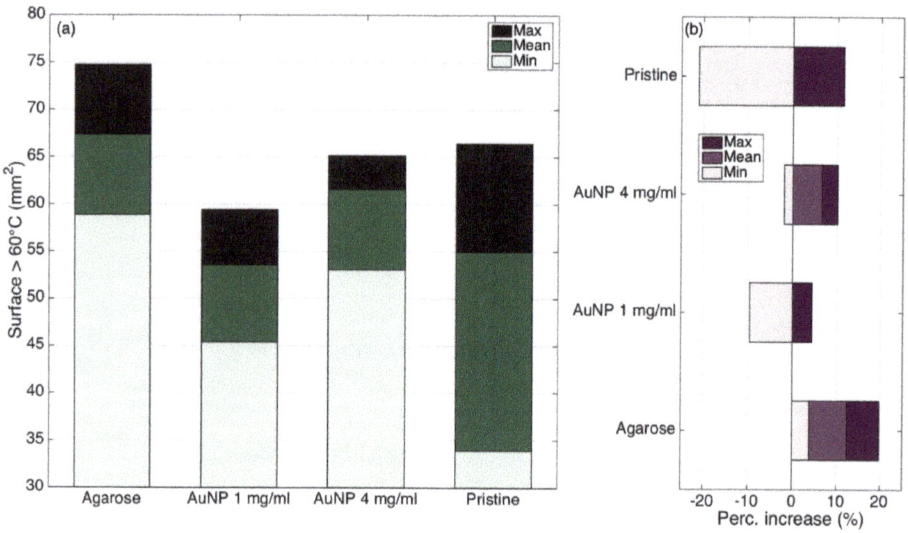

Figure 13. Quantification of the thermal damage region (temperature >60 °C) for each experimental condition. (**a**) Evaluation of the surface exposed to temperature exceeding 60 °C at the maximum ablation temperature; bar charts show the maximum (dark), minimum (bright), and mean (intermediate) values of the areas, over four experiments for each RFA condition. (**b**) Percentual increase or decrease for the ablated surface, with respect to the reference condition (pristine RFA, mean surface).

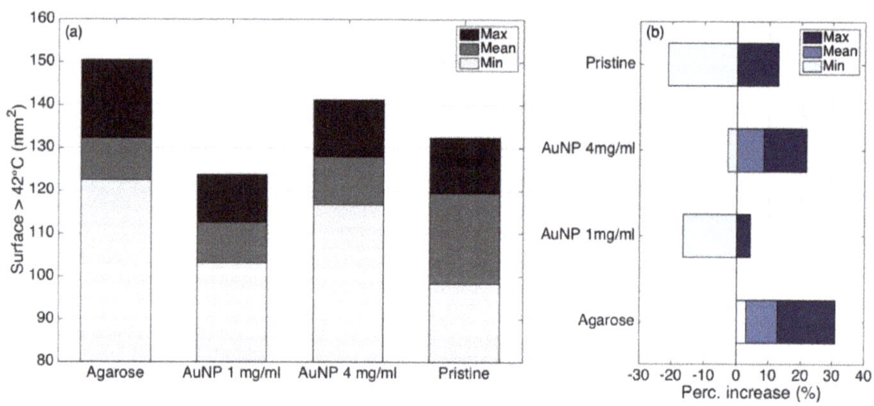

Figure 14. Quantification of the cytotoxic region (temperatures >42 °C) for each experimental condition. (**a**) Evaluation of the surface exposed to temperatures exceeding 42 °C at the maximum ablation temperature; bar charts show the maximum (dark), minimum (bright), and mean (intermediate) values of the areas, over four experiments for each RFA condition. (**b**) Percentual increase or decrease for the cytotoxic surface, with respect to the reference condition (pristine RFA, mean surface).

4. Discussion

The display of results shows the effectiveness of the ablation method, and the importance of the in-situ sensing device for the real-time detection of temperature patterns. A first point of discussion involves the efficacy of temperature sensing in the context of thermal ablation. As highlighted in Figure 10, the lack of repeatability of RFA requires a precise thermal control in situ: distributed sensing in this case is a very effective technology, as it ensures not only rapid sensing with accurate detection (about 1 pm uncertainty, corresponding to ~0.1 °C), but the entire sensing network is designed for the possible in vivo use, since the fibers have miniature form factor, biocompatibility, and the sensing mechanism is robust with respect to the catheterization options [28] and possible strains occurring during the insertion [55]. The proposed sensing network has the potential of being much more cost-effective than FBG arrays, in terms of disposable use. It is in fact essential to ensure that the cost of the whole sensing device that complements the RFA device is a fraction of the applicator itself, in order to ensure an affordable treatment; while FBG arrays have a cost of a few hundreds of dollars per array, the MgO-doped fiber can be fabricated with the same technology in terms of doping, preforming, and drawing of SMF fibers that cost a few dollar cents per meter (about $0.08/m in current markets). Additionally, MgO-doped fibers can be spooled and spliced to SMF fibers using standard splicers, without the need to develop new splicing methods. In addition, FBG sensors encode the sensing in a specific location, and therefore they are both limited by the grating length and misalignment of each grating position [12]. Distributed sensing, on the other hand, can improve the spatial resolution down to 10 μm (theoretical value of the OBR [56]), and since the entire fiber acts as a sensor, it is possible to align thermal maps using signal processing methods or other artefacts [57]. On the other hand, distributed sensing provides a much more detailed detection with respect to thermocouples, even in miniaturized formats [58], or fluorescence-based detectors [59], as these methods are intended for single-point sensing while the proposed method can resolve several tens of sensors per each cm^2.

Regarding the thermal ablation process, the analysis of cytotoxicity and thermal damage levels shows that the RFA process improves when using agarose and Au nanoparticles with the appropriate density. While agarose-mediated RFA achieves the best results in terms of heat delivery, guaranteeing the largest thermal ablation results, the use of AuNP shows the best compromise between the width of the ablated area and the need for a repeatable process. As validated by experiments, by using 4 mg/mL density, the best results can be achieved. The possibility of extending the ablated region is an important asset in advancing RFA to treat larger types of cancer, particularly for hepatic tumors [60].

5. Conclusions

In conclusion, we reported the use of a distributed fiber-optic sensing network optimized for the detection of thermal patterns in RFA procedure, mediated by the use of agarose and gold nanoparticles for the improvement of the heat delivery. The fiber-optic sensing network is formed by a set of six MgO-NP fibers, deployed over a grid of 40 × 20 mm, forming 102 sensing points spanning the spatial resolution of the interrogator on the horizontal axis and the spacing between each fiber on the vertical axis.

Several experiments of RFA were performed, in pristine condition as well as mediated by agarose and AuNP (1 and 4 mg/mL density). Thermal maps and isothermal representation allow precise identification of the ablated region, determining the area exposed to the ideal thermal damage (60 °C and above) and cytotoxicity levels for partial mortality (between 42 and 60 °C). Experimental results show that agarose-mediated thermal ablation yields the widest treated area, but with low repeatability; on the other hand, AuNP-mediated ablation using 4 mg/mL density provides the best trade-off between efficacy of ablation and repeatability.

Future work will further consolidate the RFA applicator, the on-board sensors for real-time detection, and the nanoparticle delivery into a single device, scaling up the possibility of using this arrangement in clinical settlements.

Author Contributions: Conceptualization, A.S., S.K., Z.A. and D.T.; methodology, all authors; investigation, A.S., Z.A. and S.K.; data curation, S.K., A.A. and D.T.; software, S.K. and A.A.; resources, W.B., T.S.A. and D.T.; visualization, S.K.; supervision, T.S.A. and D.T.; writing—original draft, D.T., A.S. and S.K.; writing—review and editing, all authors; funding acquisition, D.T. All authors have read and agreed to the published version of the manuscript.

Funding: The research was funded through Nazarbayev University, under grants SMARTER (code: 091019CRP2117) and EPICGuide (code: 240919FD3908), and grant Nanoslim ANR-17-CE08-0002-05 project granted by The French National Research Agency.

Institutional Review Board Statement: Not applicable.

Informed Consent Statement: Not applicable.

Data Availability Statement: Data presented in this work are not publicly available at this time, but can be obtained upon reasonable request from the authors.

Conflicts of Interest: The authors declare no conflict of interest.

References

1. Li, L.; Zhang, J.; Liu, X.; Li, X.; Jiao, B.; Kang, T. Clinical outcomes of radiofrequency ablation and surgical resection for small hepatocellular carcinoma: A meta-analysis. *J. Gastroenterol. Hepatol.* **2012**, *27*, 51–58. [CrossRef] [PubMed]
2. Cho, Y.K.; Rhim, H.; Noh, S. Radiofrequency Ablation versus Surgical Resection as Primary Treatment of Hepatocellular Carcinoma Meeting the Milan Criteria: A Systematic Review. *J. Gastroenterol. Hepatol.* **2011**, *26*, 1354–1360. [CrossRef] [PubMed]
3. HirokiNishikawa, H.; Inuzuka, T.; Takeda, H.; Nakajima, J.; Matsuda, F.; Sakamoto, A.; Henmi, S.; Hatamaru, K.; Ishikawa, T.; Saito, S.; et al. Comparison of percutaneous radiofrequency thermal ablation and surgical resection for small hepatocellular carcinoma. *BMC Gastroenterol.* **2011**, *11*, 143. [CrossRef]
4. Solmi, L.; Nigro, G.; Roda, E. Therapeutic effectiveness of echo-guided percutaneous radiofrequency ablation therapy with a LeVeen needle electrode in hepatocellular carcinoma. *World J. Gastroenterol.* **2006**, *12*, 1098–1104. [CrossRef] [PubMed]
5. Schena, E.; Tosi, D.; Saccomandi, P.; Lewis, E.; Kim, T. Fiber Optic Sensors for Temperature Monitoring during Thermal Treatments: An Overview. *Sensors* **2016**, *16*, 1144. [CrossRef]
6. Clasen, S.; Pereira, P.L. Magnetic resonance guidance for radiofrequency ablation of liver tumors. *J. Magn. Reson. Imaging* **2008**, *27*, 421–433. [CrossRef]
7. Buscarini, L.; Buscarini, E.; Di Stasi, M.; Vallisa, D.; Quaretti, P.; Rocca, A. Percutaneous radiofrequency ablation of small hepatocellular carcinoma: Long-term results. *Eur. Radiol.* **2001**, *11*, 914–921. [CrossRef]
8. MacDonell, J.; Patel, N.; Rubino, S.; Ghoshal, G.; Fischer, G.; Burdette, E.C.; Hwang, R.; Pilitsis, J.G. Magnetic resonance–guided interstitial high-intensity focused ultrasound for brain tumor ablation. *Neurosurg. Focus* **2018**, *44*, E11. [CrossRef]
9. Ashikbayeva, Z.; Aitkulov, A.; Jelbuldina, M.; Issatayeva, A.; Beisenova, A.; Molardi, C.; Saccomandi, P.; Blanc, W.; Inglezakis, V.; Tosi, D. Distributed 2D temperature sensing during nanoparticles assisted laser ablation by means of high-scattering fiber sensors. *Sci. Rep.* **2020**, *10*, 12593. [CrossRef]
10. Sapareto, S.A.; Dewey, W.C. Original Contribution Thermal Dose Determination in Cancer Therapy. *Int. J. Radiat. Oncol. Biol. Phys.* **1984**, *10*, 787–800. [CrossRef]
11. Sapareto, S.A. Thermal isoeffect dose: Addressing the problem of thermotolerance. *Int. J. Hyperth.* **1987**, *3*, 297–305. [CrossRef] [PubMed]
12. Tosi, D.; Macchi, E.; Braschi, G.; Gallati, M.; Cigada, A.; Poeggel, S.; Leen, G.; Lewis, E. Monitoring of radiofrequency thermal ablation in liver tissue through fibre Bragg grating sensors array. *Electron. Lett.* **2014**, *50*, 981–983. [CrossRef]
13. Varadarajulu, S.; Jhala, N.C.; Drelichman, E.R. EUS-guided radiofrequency ablation with a prototype electrode array system in an animal model (with video). *Gastrointest. Endosc.* **2009**, *70*, 372–376. [CrossRef] [PubMed]
14. Ter Haar, G. HIFU Tissue Ablation: Concept and Devices. In *Advances in Experimental Medicine and Biology*; Springer: New York, NY, USA, 2016; Volume 880, pp. 3–20. [CrossRef]
15. Jelbuldina, M.; Korobeinyk, A.; Korganbayev, S.; Tosi, D.; Dukenbayev, K.; Inglezakis, V.J. Real-Time Temperature Monitoring in Liver During Magnetite Nanoparticle-Enhanced Microwave Ablation with Fiber Bragg Grating Sensors: Ex Vivo Analysis. *IEEE Sens. J.* **2018**, *18*, 8005–8011. [CrossRef]
16. Gassino, R.; Liu, Y.; Konstantaki, M.; Vallan, A.; Pissadakis, S.; Perrone, G. A Fiber Optic Probe for Tumor Laser Ablation with Integrated Temperature Measurement Capability. *J. Light. Technol.* **2017**, *35*, 3447–3454. [CrossRef]
17. Nault, J.-C.; Sutter, O.; Nahon, P.; Ganne-Carrié, N.; Seror, O. Percutaneous treatment of hepatocellular carcinoma: State of the art and innovations. *J. Hepatol.* **2018**, *68*, 783–797. [CrossRef]
18. Mayer, T.; Cazzato, R.; De Marini, P.; Auloge, P.; Dalili, D.; Koch, G.; Garnon, J.; Gangi, A. Spinal metastases treated with bipolar radiofrequency ablation with increased (>70 °C) target temperature: Pain management and local tumor control. *Diagn. Interv. Imaging* **2021**, *102*, 27–34. [CrossRef]
19. Wood, B.; Locklin, J.K.; Viswanathan, A.; Kruecker, J.; Haemmerich, D.; Cebral, J.; Sofer, A.; Cheng, R.; McCreedy, E.; Cleary, K.; et al. Technologies for Guidance of Radiofrequency Ablation in the Multimodality Interventional Suite of the Future. *J. Vasc. Interv. Radiol.* **2007**, *18*, 9–24. [CrossRef]
20. Hong, K.; Georgiades, C. Radiofrequency Ablation: Mechanism of Action and Devices. *J. Vasc. Interv. Radiol.* **2010**, *21* (Suppl. S8), S179–S186. [CrossRef]

21. Wielandts, J.-Y.; Almorad, A.; Hilfiker, G.; Gillis, K.; El Haddad, M.; Vijgen, J.; Berte, B.; Waroux, J.-B.L.P.D.; Tavernier, R.; Duytschaever, M.; et al. Biosense Webster's QDOT Micro™ radiofrequency ablation catheter. *Futur. Cardiol.* **2021**, *17*, 817–825. [CrossRef]
22. Red, F.; Orady, M. Drug Therapy R Adio-F Requency a Blation as T Reatment for C Ardiac A Rrhythmias. *N. Engl. J. Med.* **1999**, *340*, 534–544.
23. Bhandari, A.; Ellias, M. Loin pain hematuria syndrome: Pain control with RFA to the splanchanic plexus. *Pain Clin.* **2000**, *12*, 323–327. [CrossRef]
24. Orgera, G.; Krokidis, M.; Matteoli, M.; Varano, G.M.; La Verde, G.; David, V.; Rossi, M. Percutaneous Vertebroplasty for Pain Management in Patients with Multiple Myeloma: Is Radiofrequency Ablation Necessary? *Cardiovasc. Interv. Radiol.* **2014**, *37*, 203–210. [CrossRef] [PubMed]
25. Thiagalingam, A.; D'Avila, A.; McPherson, C.; Malchano, Z.; Ruskin, J.; Reddy, V. Impedance and Temperature Monitoring Improve the Safety of Closed-Loop Irrigated-Tip Radiofrequency Ablation. *J. Cardiovasc. Electrophysiol.* **2007**, *18*, 318–325. [CrossRef]
26. Ashikbayeva, Z.; Aitkulov, A.; Wolf, A.; Dostovalov, A.; Amantayeva, A.; Kurbanova, A.; Inglezakis, V.; Tosi, D. Investigation of Thermal Effects of Radiofrequency Ablation Mediated with Iron Oxide Nanoparticles Dispersed in Agarose and Chitosan Solvents. *Appl. Sci.* **2021**, *11*, 2437. [CrossRef]
27. Ashikbayeva, Z.; Tosi, D.; Balmassov, D.; Schena, E.; Saccomandi, P.; Inglezakis, V. Application of Nanoparticles and Nanomaterials in Thermal Ablation Therapy of Cancer. *Nanomaterials* **2019**, *9*, 1195. [CrossRef]
28. Tosi, D.; Schena, E.; Molardi, C.; Korganbayev, S. Fiber optic sensors for sub-centimeter spatially resolved measurements: Review and biomedical applications. *Opt. Fiber Technol.* **2018**, *43*, 6–19. [CrossRef]
29. Loyez, M.; Larrieu, J.-C.; Chevineau, S.; Remmelink, M.; Leduc, D.; Bondue, B.; Lambert, P.; Devière, J.; Wattiez, R.; Caucheteur, C. In situ cancer diagnosis through online plasmonics. *Biosens. Bioelectron.* **2019**, *131*, 104–112. [CrossRef]
30. Korganbayev, S.; Orazayev, Y.; Sovetov, S.; Bazyl, A.; Schena, E.; Massaroni, C.; Gassino, R.; Vallan, A.; Perrone, G.; Saccomandi, P.; et al. Detection of thermal gradients through fiber-optic Chirped Fiber Bragg Grating (CFBG): Medical thermal ablation scenario. *Opt. Fiber Technol.* **2018**, *41*, 48–55. [CrossRef]
31. Evers, D.; Nachabé, R.; Hompes, D.; van Coevorden, F.; Lucassen, G.; Hendriks, B.; van Velthuysen, M.-L.; Wesseling, J.; Ruers, T. Optical sensing for tumor detection in the liver. *Eur. J. Surg. Oncol. (EJSO)* **2013**, *39*, 68–75. [CrossRef]
32. Tosi, D.; Macchi, E.G.; Braschi, G.; Cigada, A.; Gallati, M.; Rossi, S.; Poeggel, S.; Leen, G.; Lewis, E. Fiber-optic combined FPI/FBG sensors for monitoring of radiofrequency thermal ablation of liver tumors: Ex vivo experiments. *Appl. Opt.* **2014**, *53*, 2136. [CrossRef] [PubMed]
33. Tosi, D.; Macchi, E.G.; Gallati, M.; Braschi, G.; Cigada, A.; Rossi, S.; Leen, G.; Lewis, E. Fiber-optic chirped FBG for distributed thermal monitoring of ex-vivo radiofrequency ablation of liver. *Biomed. Opt. Express* **2014**, *5*, 1799. [CrossRef] [PubMed]
34. Beisenova, A.; Issatayeva, A.; Sovetov, S.; Korganbayev, S.; Jelbuldina, M.; Ashikbayeva, Z.; Blanc, W.; Schena, E.; Sales, S.; Molardi, C.; et al. Multi-fiber distributed thermal profiling of minimally invasive thermal ablation with scattering-level multiplexing in MgO-doped fibers. *Biomed. Opt. Express* **2019**, *10*, 1282. [CrossRef] [PubMed]
35. Beccaria, A.; Bellone, A.; Mirigaldi, A.; Serafini, V.; Olivero, M.; Vallan, A.; Perrone, G. Temperature monitoring of tumor hyperthermal treatments with optical fibers: Comparison of distributed and quasi-distributed techniques. *Opt. Fiber Technol.* **2020**, *60*, 102340. [CrossRef]
36. Korganbayev, S.; Orrico, A.; Bianchi, L.; De Landro, M.; Wolf, A.; Dostovalov, A.; Saccomandi, P. Closed-Loop Temperature Control Based on Fiber Bragg Grating Sensors for Laser Ablation of Hepatic Tissue. *Sensors* **2020**, *20*, 6496. [CrossRef]
37. Morra, F.; De Landro, M.; Korganbayev, S.; Wolf, A.; Dostovalov, A.; Cigada, A.; Saccomandi, P. Spatially resolved thermometry during laser ablation in tissues: Distributed and quasi-distributed fiber optic-based sensing. *Opt. Fiber Technol.* **2020**, *58*, 102295. [CrossRef]
38. Esposito, F.; Srivastava, A.; Iadicicco, A.; Campopiano, S. Multi-parameter sensor based on single Long Period Grating in Panda fiber for the simultaneous measurement of SRI, temperature and strain. *Opt. Laser Technol.* **2019**, *113*, 198–203. [CrossRef]
39. Esposito, F.; Srivastava, A.; Sansone, L.; Giordano, M.; Campopiano, S.; Iadicicco, A. Label-Free Biosensors Based on Long Period Fiber Gratings: A Review. *IEEE Sens. J.* **2020**, *21*, 12692–12705. [CrossRef]
40. Ashikbayeva, Z.; Aitkulov, A.; Atabaev, T.S.; Blanc, W.; Inglezakis, V.J.; Tosi, D. Green-Synthesized Silver Nanoparticle–Assisted Radiofrequency Ablation for Improved Thermal Treatment Distribution. *Nanomaterials* **2022**, *12*, 426. [CrossRef]
41. Gannon, C.J.; Patra, C.R.; Bhattacharya, R.; Mukherjee, P.; Curley, S.A. Intracellular gold nanoparticles enhance non-invasive radiofrequency thermal destruction of human gastrointestinal cancer cells. *J. Nanobiotechnol.* **2008**, *6*, 2. [CrossRef]
42. Mooney, R.; Schena, E.; Saccomandi, P.; Zhumkhawala, A.; Aboody, K.; Berlin, J.M. Gold nanorod-mediated near-infrared laser ablation: In vivo experiments on mice and theoretical analysis at different settings. *Int. J. Hyperth.* **2017**, *33*, 150–159. [CrossRef] [PubMed]
43. Tshikhudo, T.; Wang, Z.; Brust, M. Biocompatible gold nanoparticles. *Mater. Sci. Technol.* **2004**, *20*, 980–984. [CrossRef]
44. Kruse, D.E.; Stephens, D.N.; Lindfors, H.A.; Ingham, E.S.; Paoli, E.E.; Ferrara, K.W. A Radio-Frequency Coupling Network for Heating of Citrate-Coated Gold Nanoparticles for Cancer Therapy: Design and Analysis. *IEEE Trans. Biomed. Eng.* **2011**, *58*, 2002–2012. [CrossRef] [PubMed]

45. Letfullin, R.R.; Iversen, C.B.; George, T.F. Modeling nanophotothermal therapy: Kinetics of thermal ablation of healthy and cancerous cell organelles and gold nanoparticles. *Nanomed. Nanotechnol. Biol. Med.* **2011**, *7*, 137–145. [CrossRef]
46. Pedro, R.N.; Thekke-Adiyat, T.; Goel, R.; Shenoi, M.; Slaton, J.; Schmechel, S.; Bischof, J.; Anderson, J.K. Use of Tumor Necrosis Factor–alpha-coated Gold Nanoparticles to Enhance Radiofrequency Ablation in a Translational Model of Renal Tumors. *Urology* **2010**, *76*, 494–498. [CrossRef]
47. Lacroce, E.; Saccomandi, P.; Rossi, F. Can gold nanoparticles improve delivery performance of polymeric drug-delivery systems? *Ther. Deliv.* **2021**, *12*, 489–492. [CrossRef]
48. Cardinal, J.; Klune, J.R.; Chory, E.; Jeyabalan, G.; Kanzius, J.S.; Nalesnik, M.; Geller, D.A. Non-Invasive Radiofrequency Ablation of Cancer Targe. *Surgery* **2008**, *144*, 125–132. [CrossRef]
49. Molardi, C.; Korganbayev, S.; Blanc, W.; Tosi, D. Characterization of a nanoparticles-doped optical fiber by the use of optical backscatter reflectometry. In Proceedings of the Advanced Sensor Systems and Applications VIII, Beijing, China, 11–13 October 2018; Volume 10821, p. 76. [CrossRef]
50. Molardi, C.; Blanc, W.; Tosi, D. (INVITED) Fiber loop resonator sensor achieved by high-scattering MgO nanoparticle-doped fibers. *Opt. Mater. X* **2020**, *7*, 100057. [CrossRef]
51. Turkevich, J. Colloidal Gold: Part II Colour, Coagulation, Adhesion, Alloying and Catalytic Properties. *Gold Bull.* **1985**, *18*, 125–131. [CrossRef]
52. Huang, H.-W. Influence of blood vessel on the thermal lesion formation during radiofrequency ablation for liver tumors. *Med. Phys.* **2013**, *40*, 073303. [CrossRef]
53. Rhim, H.; Goldberg, S.N.; Dodd, G.D.; Solbiati, L.; Lim, H.K.; Tonolini, M.; Cho, O.K. Essential Techniques for Successful Radio-frequency Thermal Ablation of Malignant Hepatic Tumors. *RadioGraphics* **2001**, *21*, S17–S35. [CrossRef] [PubMed]
54. Macchi, E.G.; Tosi, D.; Braschi, G.; Gallati, M.; Cigada, A.; Busca, G.; Lewis, E. Optical fiber sensors-based temperature distribution measurement inex vivoradiofrequency ablation with submillimeter resolution. *J. Biomed. Opt.* **2014**, *19*, 117004. [CrossRef] [PubMed]
55. Froggatt, M.; Moore, J. High-spatial-resolution distributed strain measurement in optical fiber with Rayleigh scatter. *Appl. Opt.* **1998**, *37*, 1735–1740. [CrossRef] [PubMed]
56. Issatayeva, A.; Amantayeva, A.; Blanc, W.; Molardi, C.; Tosi, D. Temperature compensation of the fiber-optic based system for the shape reconstruction of a minimally invasive surgical needle. *Sens. Actuators A Phys.* **2021**, *329*, 112795. [CrossRef]
57. Bao, X.; Chen, L. Recent Progress in Distributed Fiber Optic Sensors. *Sensors* **2012**, *12*, 8601–8639. [CrossRef]
58. Saccomandi, P.; Schena, E.; Silvestri, S. Techniques for temperature monitoring during laser-induced thermotherapy: An overview. *Int. J. Hyperth.* **2013**, *29*, 609–619. [CrossRef]
59. Zhang, Z.; Grattan, K.T.V.; Palmer, A.W. Fiber-optic high-temperature sensor based on the fluorescence lifetime of alexandrite. *Rev. Sci. Instrum.* **1992**, *63*, 3869–3873. [CrossRef]
60. Solbiati, L.; Livraghi, T.; Goldberg, S.N.; Ierace, T.; Meloni, F.; Dellanoce, M.; Cova, L.; Halpern, E.F.; Gazelle, G.S. Percutaneous Radio-frequency Ablation of Hepatic Metastases from Colorectal Cancer: Long-term Results in 117 Patients. *Radiology* **2001**, *221*, 159–166. [CrossRef]

biosensors

MDPI

Review

Cost-Effective Fiber Optic Solutions for Biosensing

Cátia Leitão [1,*], Sónia O. Pereira [1], Carlos Marques [1], Nunzio Cennamo [2], Luigi Zeni [2], Madina Shaimerdenova [3], Takhmina Ayupova [3] and Daniele Tosi [3,4,*]

1 i3N, Department of Physics, University of Aveiro, 3810-193 Aveiro, Portugal; sonia.pereira@ua.pt (S.O.P.); carlos.marques@ua.pt (C.M.)
2 Department of Engineering, University of Campania Luigi Vanvitelli, Via Roma 29, 81031 Aversa, Italy; nunzio.cennamo@unicampania.it (N.C.); luigi.zeni@unicampania.it (L.Z.)
3 School of Engineering and Digital Sciences, Nazarbayev University, Nur-Sultan 010000, Kazakhstan; madina.shaimerdenova@nu.edu.kz (M.S.); takhmina.ayupova@nu.edu.kz (T.A.)
4 Laboratory of Biosensors and Bioinstruments, National Laboratory Astana, Nur-Sultan 010000, Kazakhstan
* Correspondence: catia.leitao@ua.pt (C.L.); daniele.tosi@nu.edu.kz (D.T.)

Abstract: In the last years, optical fiber sensors have proven to be a reliable and versatile biosensing tool. Optical fiber biosensors (OFBs) are analytical devices that use optical fibers as transducers, with the advantages of being easily coated and biofunctionalized, allowing the monitorization of all functionalization and detection in real-time, as well as being small in size and geometrically flexible, thus allowing device miniaturization and portability for point-of-care (POC) testing. Knowing the potential of such biosensing tools, this paper reviews the reported OFBs which are, at the moment, the most cost-effective. Different fiber configurations are highlighted, namely, end-face reflected, unclad, D- and U-shaped, tips, ball resonators, tapered, light-diffusing, and specialty fibers. Packaging techniques to enhance OFBs' application in the medical field, namely for implementing in subcutaneous, percutaneous, and endoscopic operations as well as in wearable structures, are presented and discussed. Interrogation approaches of OFBs using smartphones' hardware are a great way to obtain cost-effective sensing approaches. In this review paper, different architectures of such interrogation methods and their respective applications are presented. Finally, the application of OFBs in monitoring three crucial fields of human life and wellbeing are reported: detection of cancer biomarkers, detection of cardiovascular biomarkers, and environmental monitoring.

Keywords: POC monitoring; smartphone optical biosensors; optical interrogation methods; cancer biomarkers; cardiovascular biomarkers; environmental monitoring

1. Introduction

Biosensors are devices that quantify (or semi-quantify) a biological or chemical analyte by generating a measurable signal proportional to its concentration [1]. They are usually applied in different fields, from biomedical to environmental, allowing monitoring of specific disease biomarkers in body fluids (blood, urine, saliva, and sweat) [2] and detection of micro-organisms [3] and pollutants [4] in the environment, among other things. In the case of optical biosensors, the signal generated can be coded in wavelength, phase, and signal intensity, among other optical features [5]. Optical absorbance, fluorescence, bioluminescence, interferometry, ellipsometry, reflectometric interference spectroscopy, and surface-enhanced Raman scattering are also examples of highly applied optical techniques to biosensing [6].

In relation optical fiber biosensors, usually, the biosensing process relies on the interaction of the evanescent wave with the fiber's surroundings. Therefore, in order to turn the fiber sensitive to the environmental refractive index (RI) changes, fibers can have gratings inscribed or can be altered in different geometries. Moreover, noble metal coatings with thicknesses around 30–70 nm, enabling the surface plasmon resonance (SPR) phenomena, or nanoparticles, and thus generating a localized SPR (LSPR), can be applied. The use of optical fibers brings important advantages, namely: small dimensions—allowing them to be used in biomedical applications in invasive and non-invasively approaches; dielectric nature—making them electrically safe; and low signal attenuation [7]. The main disadvantages to optical fiber biosensors are the fragility and high-cost interrogation approaches [8]. However, as research on optical fiber sensors (OFS) has been evolving, a large number of studies reporting robust and cost-effective optical fiber biosensing (OFB) approaches has been published in the past few years. The sensing principles that will be described are mainly based on interrogation in the visible spectrum due to the well-known cost-effectiveness of such an approach. However, it should not be forgotten that, in the last few years, efforts that have been made in improving interrogation costs of infrared and gratings spectral analysis. Currently, this is a hot topic in this research field with excellent outputs [9–11].

This review focuses on the use of OFB approaches that are cost-effective. As can be verified in Figure 1, different fiber geometries and structures will be addressed, namely, U-bent, D-shaped, unclad fibers, tapered fibers, ball resonators, end-face reflection, and light-diffusing fibers (LDFs). Different OFB packaging methods for robustness improvement will be also presented, as well as examples of sensors coupled with smartphones. Cost-effective biofunctionalization methods, mainly based on aptamers and molecularly imprinted polymers (MIPs), will also be discussed. Different applications on cancer and cardiovascular biomarkers' detection and environmental monitoring will be presented, and finally, a critical analysis of the state-of-the-art and prospects will be made.

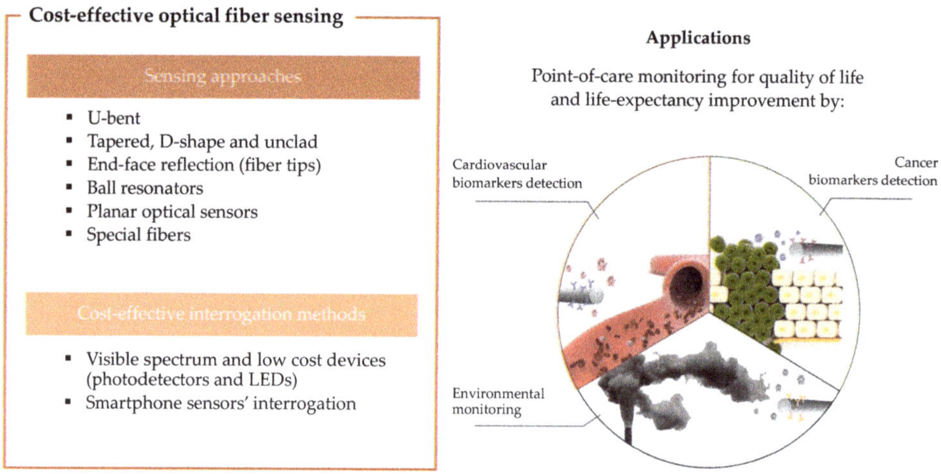

Figure 1. Schematic illustration of the main contents of this review.

2. Cost-Effective Optical Fiber Configurations for Biosensing

Starting from the optical fibers that are usually applied in cost-effective biosensing approaches, different kinds of fibers can be used, from regular silica single-mode fibers (SMFs) [12] to multimode fibers (MMFs) such as plastic-clad silica (PCS) [13] and plastic optical fibers (POFs) [14]. The different kinds of optical fibers, conventional silica-based, microstructured, or specialty fibers in terms of materials and designs offer several useful advantages in developing sensors/biosensors. For instance, when telecommunication wavelength range and setups are used, the silica-based fibers are more appropriate, such as to exploit the monomodal characteristic for several sensing approaches. In contrast, POFs are especially advantageous due to their excellent flexibility, lower attenuation in the visible range, easy handling, multimodal characteristic, and the fact that plastic can withstand smaller bend radii than glass [15].

As regards the optical signal trajectory, as shown in Figure 2, the different sensing configurations, which will be described as follows, can work on transmission (a) or in reflection (b). Additionally, OFBs can be used as intrinsic and extrinsic OFS. In particular, when the optical fiber interacts with the analyzed medium directly, it is defined as an intrinsic OFS, whereas, when it is used as an optical waveguide allowing the launch of the light to the sensing region and its collection, it is defined as an extrinsic OFS.

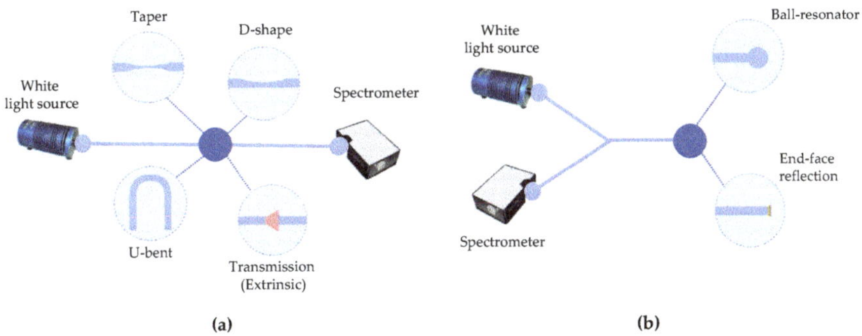

Figure 2. Schematics of OFB working in (**a**) transmission and (**b**) reflection.

The interrogation methods in the scope of this review are mainly based on the analysis of the visible spectrum, with data codification in wavelength, especially when using SPR and LSPR, and in intensity variations. Although both SMF and MMF can be suitable for wavelength interrogation, as concerns intensity-based approaches, OFBs find great advantage in the use of fibers, having a large core, which implies a high numerical aperture and strongly multimodal operation. The possibility of applying MMFs enables the use of inexpensive light-emitting diodes (LEDs) [16] and simple photodetectors such as charge-coupled devices (CCDs) or photodetectors (PDs) [17], as light can be easily coupled into the fibers using basic optics and cheap interconnections. In contrast, SMFs require hardware with higher complexity and costs.

While the use of a large numerical aperture and strongly multicore fibers is typically a disadvantage in the design of physical sensors, several biosensing structures exploit the favorable properties of MMFs as part of the detection systems. From a light propagation standpoint, in a multimode fiber, modes with a lower index are strongly guided, while higher-order modes are weakly guided and become more sensitive to geometrical modifications of the fiber, such as thinning the fiber core [17], removing the cladding in a D-shaped form [18] or completely, through an etching [19,20] or mechanical [21] process, or bending the fiber in curved shapes [22]. The two most notable approaches for low-cost biosensors based on geometrically modified MMFs are based on U-bent fibers [22,23] and tapered fibers [24].

2.1. U-Bent Biosensors

The U-bent fiber is a transmission device operating in the visible spectral range that can be made of a SMF [25] or MMF [15], in which the fiber is curved with a specific bend radius, as highlighted in Figure 3a. During the bending process to obtain the U-shape, the fiber experiences mechanical stress, which causes a variation in its optical properties due to the photoelastic effect. The induced stress will yield a non-uniform RI of the core, with a higher RI at the inner bend and lower at the outer bend of the core. As the bend radius decreases, the RI profile changes more significantly due to the increased stress. Lower order modes are well confined within the fiber and propagate through with minimal losses; higher-order modes, due to the increased angle of incidence at the interface between core and cladding, do not satisfy the total internal condition anymore and are strongly attenuated. Consequently, the intensity of the field increases at the outer region of the core and, at the core-cladding interface, the Fresnel transmission coefficient approximates 1, leading to greater radiation loss. As regards the inner interface of the bent region, more light satisfies the total internal reflection principle, with less light lost in that region. In U-shape fibers, there is a critical radius of curvature, below which the fiber no longer transmits light. The critical radius of curvature decreases for a large difference of RI or large NA values. Larger NA fibers can bend with a smaller bend radius without significant loss of light given the same wavelength of light [26].

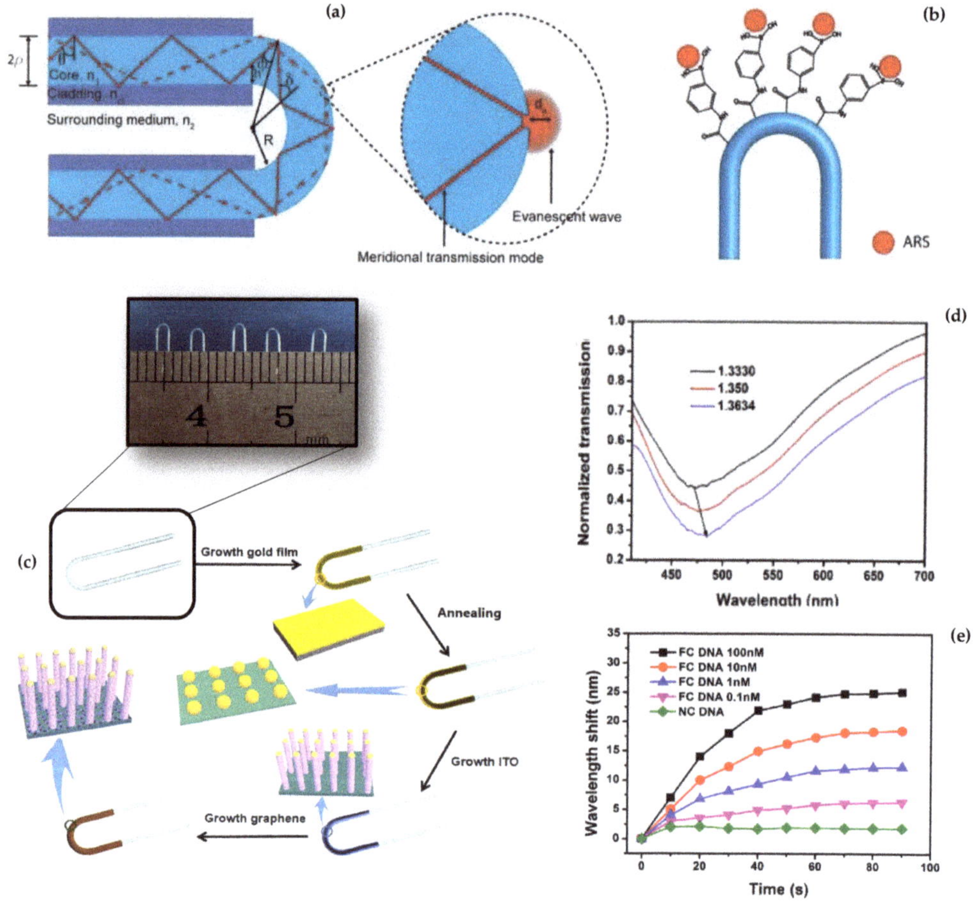

Figure 3. Schematic and implementation of U-shape fiber biosensors: schematic of a U-bent fiber section with light propagation within the curved portion of the fiber (**a**) with inset showing the surface-penetrating meridional rays providing RI sensitivity, and functionalization of the biosensor for glucose detection (**b**) (ARS—Alizarin Red S) (images reproduced from [27]); (**c**) U-bent LSPR sensor proposed by Yang et al. with the respective production steps, namely, U-shape production and photographs, and formation of ITO coating for plasmonic-based sensing, with results of the change in transmission spectrum for three different values of RI (**d**) and the response of the sensor, measuring the wavelength shift over time, for different DNA concentrations (**e**) (images adapted from [28]).

In order to transform this structure into an efficient RI sensor capable of sensing the surrounding environment, the design [22] makes the use of fiber with an exposed core (hence, with no cladding or depleted cladding) in U-shape. This sensing structure is shown in Figure 3a and was optimized for biosensing according to Azkune et al. [27]. The bent portion of the fiber has no cladding, depleted through wet-etching, dry etching, or by simply fabricating a cladding-less fiber. POF fibers in this regard have been substantially used in U-bent fiber sensors [27,29–31] as they are more stretchable, having a lower Young's modulus, and millions of modes are propagated, hence, guaranteeing a measurable change of transmission when the RI in the surrounding medium changes.

In addition, the U-bent fiber structure shows the best versatility when packaging the device, as the fiber can be tightly folded within a narrow space. Chen et al. [32], among others, reported a U-bent fiber sensor folded in a 1.1 mm tube structure, which is compatible with most medical needles, to fabricate a device with a high level of miniaturization.

The possible functionalization of a U-bent fiber section, as reported in Figure 3b [27], relies on immobilization of bioreceptors on the bent portion of the fiber, in correspondence with the region with the maximum sensitivity. The authors propose a glucose detection system based on the release of ARS (Alizarin Red S) occurring when glucose is bound through competitive binding to phenylboronic acid.

U-bent fibers are often used in conjunction with SPR effects to improve detection. While a U-bent fiber provides a significant transmission change, the wavelength shift observed due to the plasmonic resonance can be detected in a more robust way. This principle was implemented by Yang et al. [28], who reported a DNA biosensor by a thin-film U-bent plasmonic graphene/ITO (indium tin oxide) biosensor. The main achievements reported by Yang et al. are also shown in Figure 3. The Figure 3c shows the graphene/ITO layer deposition around the core-exposed section of a POF, folded around a 1.3 mm bending diameter in the inner region. The transmission spectrum through the bent section of the fiber shows both an intensity change and a wavelength shift due to the combined effect of the fiber losses of meridional rays and localized SPR at the outer interface (Figure 3d. By measuring the wavelength shift, Yang et al. estimated the sensor response to different full complementary DNA and non-complementary DNA at 0.1–100 nM concentrations, as displayed in Figure 3e.

Several biosensors have been reported as U-bent fiber probes based on MMFs and inexpensive spectrometers and hardware based on these principles. George et al. [29] reported an immunobiosensor for the detection of chikungunya (an infectious disease) through the non-structural protein 3 (CHIKV-nsP3) biomarker, achieving a limit-of-detection (LOD) of 0.52 ng/mL. Bandaru et al. [33] reported the production of a plasmonic U-bent fiber biosensor capable of detecting human immunoglobulin G with ultra-low LOD ($\sim$7 aM). Manoharan et al. [34] described a similar architecture for detecting bacterial endotoxins such as lipopolysaccharides with a LOD of 0.4 ng/mL.

2.2. Tapered Fiber Biosensors

Tapered fibers have been used to implement fiber-optic interferometers [35] or multi-parametric sensors [36] in SMFs. In MMFs, tapers are used to induce losses in the fiber without the need for bending, which allows for easier packaging of the sensor in medical devices [37]. The fabrication of fiber tapers is implemented by modern CO_2 or fiber laser splicers [38] as a rapid process; it is, therefore, compatible with large volume manufacturing. Figure 4 shows a tapered silica SMF with LSPR for biosensing, developed by C. Huang et al. [39]. As can be seen, in a tapered fiber, the diameter of the fiber is progressively thinned until it reaches a minimum diameter in the so-called waist region, that corresponds to the minimum thickness (region A), with an intermediate region with variable diameter (region B). Unlike the etching process, tapering maintains the proportion between core and cladding diameter. When exposed to a thickness change, the lower radius of the core filters out the higher-order modes as the V number of the fiber decreases, reaching its minimum in the waist region [37]. For the case of an LSPR-based taper, it can be seen in Figure 4d,e that its response changes in wavelength and relative intensity/transmittance in the visible region.

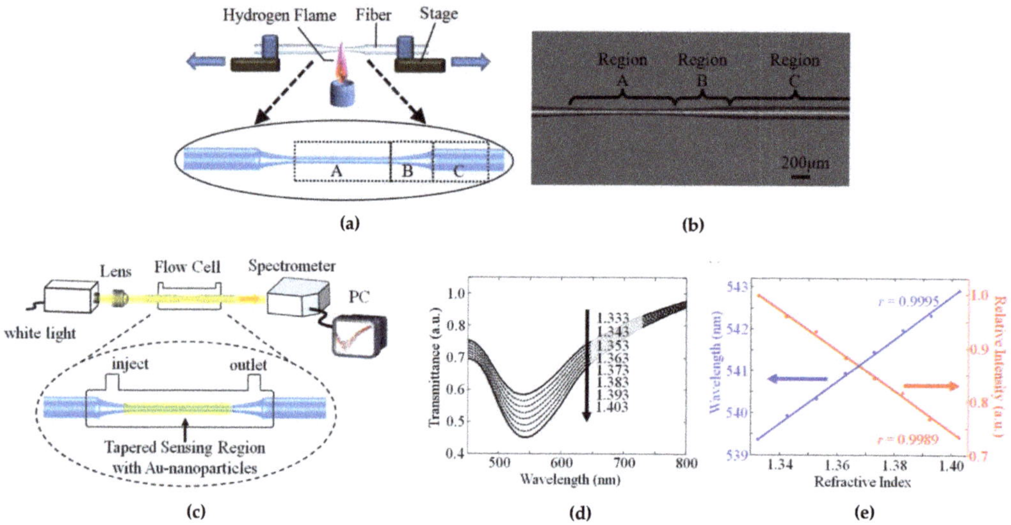

Figure 4. Schematic and implementation of tapered OFB: (**a**) taper production highlighting the respective fiber taper regions and (**b**) optical microscopy image of the tapered fiber showing the three regions; (**c**) taper coated with Au NPs and respective interrogation scheme, and (**d**) results showing the variation in wavelength and transmittance with RI variation with respective linear fittings (**e**) (images reproduced from [39]).

POF are also good candidates for low-cost tapered fiber structures, as the fiber has very thin cladding and, therefore, allows the core to be exposed to the outer environment. Dash et al. [38] reported a fiber taper in a POF having an initial diameter of 250 μm tapered to 175 μm and coated with a layer of gold nanoparticles (Au NPs) over a 4 mm length. By measuring the output power after the tapered section, a drop is observed when the outer RI increases; a power change of about 5% has been observed when measuring up to 0.2 mM of bovine serum albumin (BSA) concentration. A similar system was reported by Rahman et al. [40] for water salinity detection.

2.3. D-Shaped and Unclad Biosensors

As U-bent and tapered OF sensing structures, D-shaping or uncladding an optical fiber aims to increase the interaction of the evanescent field with the surrounding medium. The D-shaped OFS, or side-polished, can be made in SMF and MMF, usually by polishing. As can be seen in Figure 5, the polishing is made until the core surface is exposed [41], but, in other cases, the core can be completely removed in that region [42]. The distance from the polished surface to the fiber surface is usually referred to as the residual thickness and is of utmost importance in the strength of the interaction of the evanescent wave fields with the external medium. For this specific sensor, developed by Zheng et al. [42] and represented in Figure 5a, the optimal RI sensitivity was achieved by a residual thickness of 34.09 μm, attaining a value of 10,243 nm/RIU (RI units). In this kind of structure, the surface roughness also seems to have a crucial impact on the sensor's response [43]. A great feature that D-shaped OFS have is the flat surface that can be achieved, facilitating deposition of different materials [44] and, also, biosensing testing with only a sample drop [45].

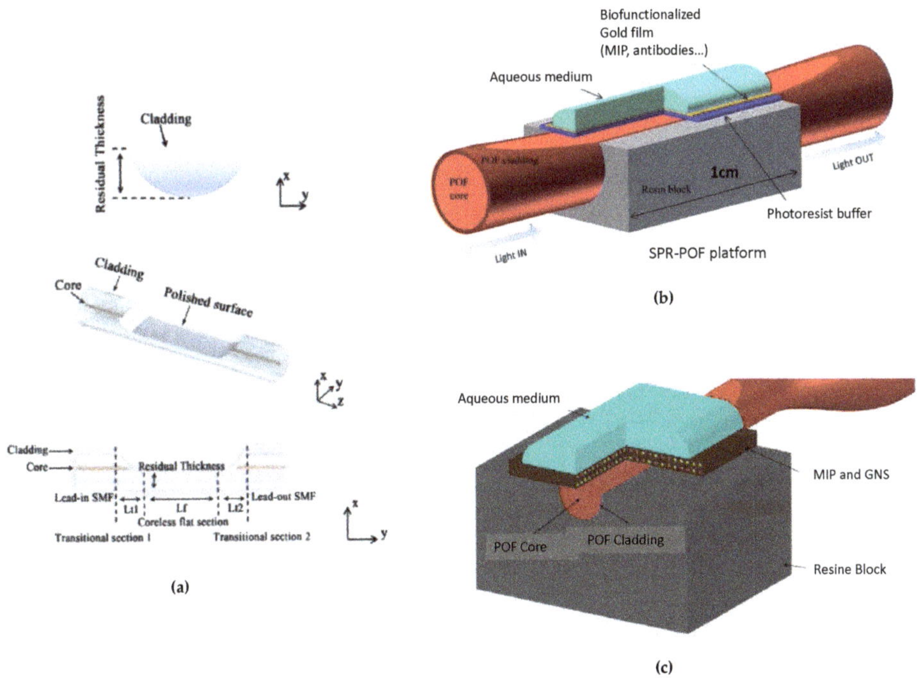

Figure 5. Schematics of D-shaped OFBs: (**a**) core-less D-shape in different perspectives (images adapted from [42]); (**b**) SPR-POF D-shaped biosensing platform (image adapted from [46]. Copyright ©2021, with permission from Elsevier); (**c**) LSPR D-shaped tapered-POF biosensor based on gold nanostars (GNS) and MIPs (image reprinted from [47]. Copyright ©2015, with permission from Elsevier).

Cennamo et al. [14,18] have presented SPR sensors produced by side removing the cladding and part of the core of the POF, spinning a thin layer on the exposed core with a RI major of the core, and finally sputtering a gold nano-film. The buffer layer under the gold nano-film improves the performance of the SPR sensor and the gold film's adhesion on the platform. Figure 5b outlines these SPR–POF platforms, in which the gold layer can be functionalized with different bioreceptors, namely MIPs [48], aptamers [18], and/or antibodies [49]. Moreover, D-shaped tapered-POFs have also been used to accomplish different kinds of plasmonic bio/chemical sensor configurations, as depicted in Figure 5c, for the detection of trinitrotoluene (TNT) [47]. On the other hand, unclad OFBs have the advantage of a higher interaction of the evanescent wave with the surrounding environment. The removal of the cladding can be achieved by polishing using sandpapers in rotative movements [21] or, in the case of silica fibers, by chemical etching [19].

2.4. End-Face Reflection/Optical Fiber Tips

Optical fiber tips are sensors that work on reflection, usually by having the fiber end coated with a reflective/plasmonic material such as silver or gold. In its simpler form, fiber tips are flat at the end and are usually created in MMFs, for instance, PCSs [50], which are interrogated using bifurcated optical fibers connected to the sensor by means of a bare fiber terminator, such as the BFT1 from Thorlabs. To produce flat tips, the fibers are usually cleaved by precision cutters, since a flat surface is crucial for assuring the collected signal spectra reflectivity [51]. Nevertheless, a wide range of tip forms and functionalities can be achieved employing different fabrication technologies, such as mechanical processing, chemical etching, laser processing, self-assembly, and chemical or physical vapor deposition processing, as well as material transfer methods [52].

Fiber tips-based biosensors are expected to be an important tool in POC monitoring due to their high RI sensitivity, compactness, and ease of operation. For example, for flat fiber tips coated with gold by sputtering, it is possible to achieve RI sensitivities between 700 and 2000 nm/RIU depending on the gold film thickness [50]. An alternative approach was reported by Lee et al. [53], who developed a tip aptasensor based on LSPR created with gold nanorods (GNR). The sensor structure and response in the detection of ochratoxin A (OTA) can be seen in Figure 6.

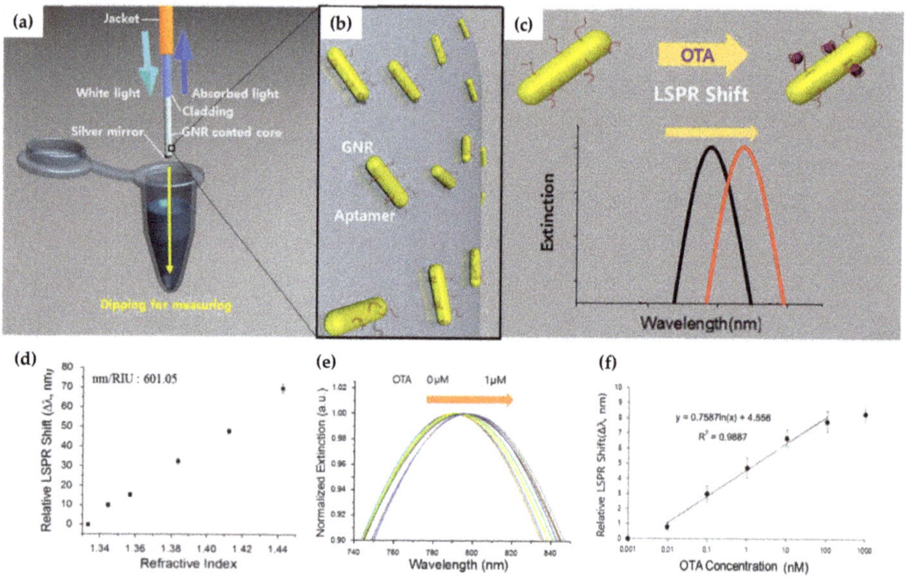

Figure 6. Schematic of a tip OFB: (**a**) components and measurement procedure; (**b**) magnification of the sensor surface biofunctionalized with aptamer-modified GNR; (**c**) schematization of LSPR wavelength shift when the analyte, in this case, OTA, is recognized by the specific aptamer; (**d**) GNR-coated tip OFS response to RI; (**e**) LSPR wavelength shift with different concentrations of OTA, and (**f**) experimental response of the biosensor to different concentrations of OTA with respective linear fitting (images adapted with permission from ref. [53]. Copyright ©2018, with permission from Elsevier).

2.5. Ball Resonators

A ball resonator is a device where a circularly shaped microsphere is fabricated on the tip of an optical fiber of approximately a hundred micrometers in diameter. The RI inside the ball resonator is higher than the environment around the spherical structure. In this case, the light that goes through optical fiber and reaches the edges of the microsphere is reflected inside the ball by total internal reflection and, therefore, cannot escape to the surrounding environment [54]. Nevertheless, due to the fact that the surface of the ball resonator has a curved interface, its total internal reflection is never complete; therefore, bending losses could occur, which can lead to the light leakage out of the ball resonator. This leads to an enhanced interaction of samples outside the ball resonator with light thanks to the confinement and rebounding of light inside the circular cavity [55,56]. These evanescent waves are able to interact with both the surface of the microsphere and the environment that surrounds it. Each waveguide of optical field circulation allows it to contact the surrounding environment, which subsequently increases the sensitivity as the amount of circulating rounds rises. The photon lifetime in the microspheric cavity is able to measure it, which is called the quality (Q) factor. The higher the lifetime of the photon, the lower the limit of detection. As previously stated, evanescent waves enable some of the light inside the ball resonator to escape into the outer environment. This allows a microsphere to acquire an increased sensitivity, which makes it useful for a host of applications. Interaction of the surface of the ball resonator with an analyte outside it results in changes in the effective RI. The experience of the analytes of various refractive indexes results in spectral changes such as wavelength and frequency shift. These favorable properties make ball resonators promising as biosensors for the detection of various analytes.

The advantage of microspheres is that they can be easily fabricated in the laboratory from various materials, organic and inorganic; they are simple in handling and have a high-quality factor [56]. The ball resonator is a device that requires a quick manufacturing process. A CO_2 laser splicer allows it to be manufactured in a controllable way in a single step by fusing two single-mode fibers and then tearing them off to make a spherical tip (Figure 7) [57]. Fabrication of ball resonators is advantageous over the fabrication of other optical devices thanks to its duration, which takes from ~60 s up to a couple of minutes depending on the speed of the motors within the machine. Moreover, the fabricated device demonstrates robustness, stability, accuracy in measurements, and higher tensile strength compared to etched fibers, which easily break [57–59].

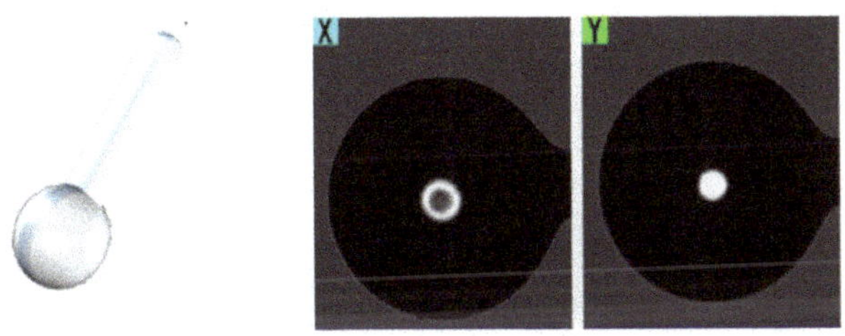

Figure 7. Image of the ball resonator: (**a**) 3D picture of optical ball resonator; (**b**) photographs taken by X and Y cameras of CO_2 laser splicer. Image reproduced from [57].

In comparison to previously studied optical fiber-based biosensors such as etched Fiber Bragg grating (eFBG), etched tilted Fiber Bragg grating (eTFBG), and etched MgO nanoparticles (NPs) doped fiber, ball resonators show a rather pronounced cost-effectiveness [58–60]. The use of a CO_2 laser splicer enables the achievement of ball resonators with an accurate diameter and low alignment error in a highly repeatable and fast manner within a few seconds. However, there are some limitations in the use of optical ball resonators. Low reflectivity is one of them, requiring the use of an optical backscatter reflectometer which has excellent sensitivity. Moreover, the weak interference phenomenon occurring in the micro-sphere between the multiple reflective paths results in poor visibility of the spectral fringes.

2.6. *Planar Optical Sensor Chips Monitored via POFs*

POFs can be used to implement simple and highly sensitive optical fiber sensors by exploiting their advantages, such as the excellent flexibility, the easy manipulation and modification, the large diameter, the fact that plastic can withstand smaller bend radii than glass, and the great numerical aperture (NA). In particular, this last characteristic is very useful in comprising extrinsic optical fiber sensor configurations, for instance, for light injection in plasmonic thin bacterial cellulose slab waveguides [61] and InkJet-printed optical waveguides [62].

Figure 8 shows several plasmonic schemes based on extrinsic POF sensors. A sensing approach based on a nano-plasmonic sensor chip monitored via a custom 3D-printed holder combined with POFs is reported in Figure 8a. The nano-plasmonic chip is based on a gold nanograting (GNG) fabricated on a poly(methyl methacrylate) (PMMA) substrate by electron beam lithography (EBL) [63]. As shown, the light source is connected through a POF optical coupler (50:50) to two POF patches, one illuminating the sensor with nanograting and the other illuminating the reference sensor (a PMMA chip with the same gold film but without the nanograting). At the output of the holder, there are two POF patches used to collect the transmitted light and send it into two similar spectrometers. In this configuration, the PMMA substrate is considered a transparent substrate.

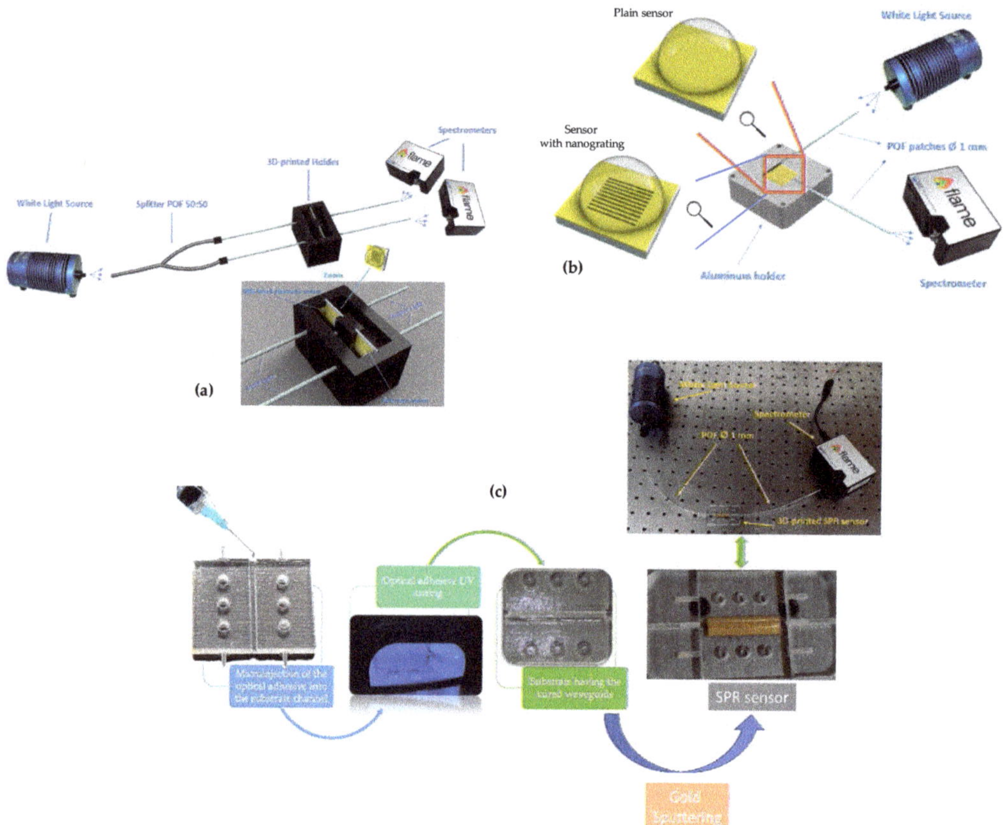

Figure 8. Plasmonic sensor configurations based on extrinsic POF sensors: (**a**) PMMA-based nanoplasmonic sensor chip monitored by exploiting the transparency of the substrate (image adapted from [63]); (**b**) PMMA-based plasmonic chips monitored by a custom setup produced to excite the plasmonic phenomena via the multimode slab waveguide (image adapted from [64]); (**c**) 3D printed SPR sensor chip with the respective setup (image adapted from [62]).

Alternatively, the PMMA substrate can be regarded as a multimode slab waveguide, as shown in Figure 8b, similar to other plasmonic sensor chips [65,66]. In particular, a gold nano-film can be deposited on the PMMA chip, obtaining an SPR sensor chip [67,68] or producing a gold nanograting on the PMMA chip surface via EBL, generating a nano-plasmonic sensor chip [69]. By using the sensor setup illustrated in Figure 8b, when a gold nano-film is present, the sensor's performance is similar to that obtained by an SPR D-shaped POF sensor [64], whereas, when a nano-plasmonic chip is used, the orientation of the nanostripes forming the grating pattern, with respect to the direction of the input light (longitudinal or orthogonal), influences the biosensing performances [69].

Figure 8c shows another low-cost 3D printed SPR sensor chip monitored via POFs. This SPR sensor chip has been designed as a disassembled component composed of four different parts by using Autodesk® Fusion 360, and then the STL files were generated [62]. The sensor's production was performed by a 3D printing method. The material used was a liquid photopolymer ink. Once the SPR sensor parts' construction was completed, the waveguide core of the 3D printed optical device was fabricated. Thus, a UV photopolymer adhesive was microinjected into the sensor channel and cured for 10 min by means of a lamp bulb with UVA emission at 365 nm. Finally, on the cured core, a gold nanofilm was deposited by a sputtering. The thickness of the sputtered gold was about 60 nm.

2.7. Special Fibers

LDF-Based SPR Sensors

Recently, an attractive kind of special optical fibers, named Light-Diffusing Fibers (LDF), has been used to generate highly sensitive SPR sensors. In LDF, the light is not confined to the core; on the contrary, it is diffused on the external medium all along their length. Thus, Cennamo et al. have used this aspect to excite the plasmonic phenomenon on metal nano-films deposited on the LDF [70–74]. In particular, this type of SPR sensor can be produced using silica [70,71] or polymer [72] by coating the fiber with metal nano-films, such as gold [70,72] or silver [71], by sputtering. This aspect allows an easier fabrication procedure since only a metal deposition step is required to build the SPR sensor. Moreover, SPR sensors based on silica LDF have already demonstrated high performances in bio/chemical sensing applications, as reported in [73], when the SPR–LDF probe is combined with a bioreceptor (e.g., antibody). For instance, Figure 9a shows an outline of a SPR sensor system based on a silica LDF covered by gold nano-film connected to a white light source and a spectrometer [70]. SPR spectra obtained studying the sensor RI sensitivity, using different concentrations of water–glycerin solutions, are represented in Figure 9b. The experimental results reveal a high sensitivity of the SPR wavelength to the outer medium's refractive index (bulk sensitivity), with values ranging from 1500 to about 4000 nm/RIU in the analyzed range. Moreover, it was seen in [71] that these bulk sensitivity values can be improved by using silver nano-films instead of the gold.

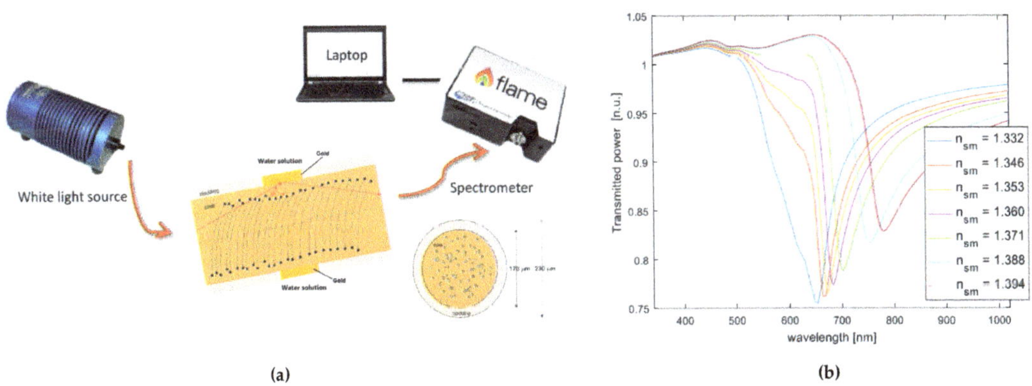

Figure 9. Silica LDF-based SPR sensor system: (**a**) data acquisition and working scheme; (**b**) SPR spectra obtained at different refractive indices from 1.332 to 1.394. Image reproduced from [70].

Finally, concerning these types of silica SPR–LDF sensors, Cennamo et al. have demonstrated that the tapering process produces a significant worsening of the bulk sensitivity and a slight decrease in the full width at half maximum (FWHM) of the SPR spectra [74]. Therefore, in this SPR–LDF sensor, a possible alternative consists in using the tapered LDF as a modal filter (after the sensitive region) by determining a trade-off between the loss in the sensitivity and the FWHM decrease [74], in a similar way to other configurations, based on different modal filters [75].

Furthermore, low-cost polymeric LDF have been also explored to produce SPR-based sensors [72]. In particular, by sputtering a gold nano-film on the PMMA-based LDF, the obtained bulk sensitivity ranges from 1000 to almost 3000 nm/RIU in the refractive index range from 1.332 to 1.392. In the same work, a novel kind of modal filter has been shown, accomplished by covering the PMMA-based LDF with an aqueous solution to improve the signal-to-noise ratio (SNR) of the SPR sensor, thanks to the filtering of the higher modes, without losing its sensitivity.

3. Packaging of Optical Fiber Biosensors

In order to move from the optical biosensing system to practical devices, miniaturization, portability, and easy handling are required. Therefore, strategies for the encapsulation and packaging of the optical components should be considered in order to meet the demands of the final device, namely by preserving the sensing properties. Fiber-optic biosensors are designed to perform real-time detection in medical applications; the possibility to design packaging structures and embed the sensing elements within medical devices allows performing subcutaneous, percutaneous, or endoscopic operations [76]. In addition, wearable technologies designed with optical fiber biosensors embed the fibers within fabric, forming permanent sensing structures [77]. Figure 10 displays some of the options for packaging, targeting different in situ sensing applications.

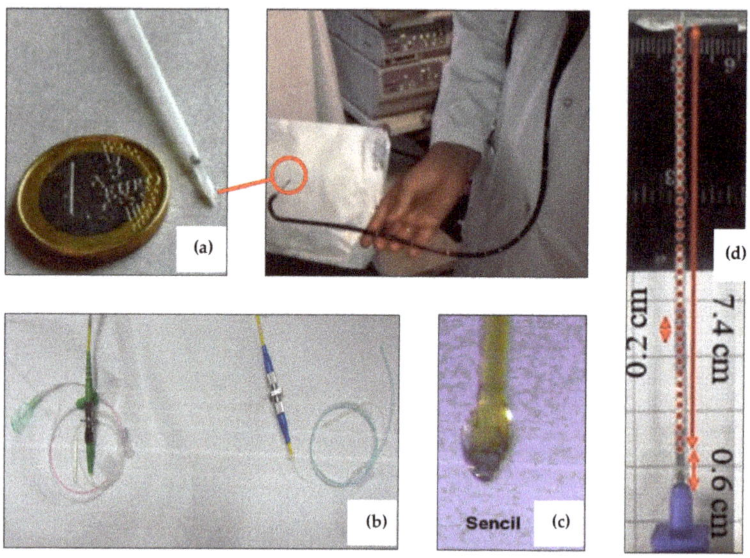

Figure 10. Example of proposed packaging for fiber optic biosensing applications: (**a**) Endoscopic packaging for in situ detection of cancer biomarkers (image reproduced from [78]); (**b**) Urologic packaging, with pressure sensors embodied in a rectal catheter (left fiber) and an abdominal catheter (right fiber) (image reproduced from [79]); (**c**) SencilTM (sensory cilia) packaging for subcutaneous fluorescence sensing (image reproduced from [80]); (**d**) Epidural needle for percutaneous insertion embodying a network of fiber sensors (image reproduced from [81]).

Endoscopic packaging has been proposed by Guo et al. [78] (Figure 10a) and Loyez et al. [82] for in situ detection of cytokeratin-7 cancer biomarkers. The proposed design integrates the plasmonic biosensor used for detection into a commercially available endoscope that maintains its original functions and allows the precise positioning of the device into the target zone. The work reported in [82] integrates the sensor in a 1.2/1.6 mm (inner/outer diameter) hollow endoscope made with a thermoplastic that maintains a semi-rigid profile. The catheter is terminated by a conical section that allows penetration into the tissue. The active part of the sensor is exposed to the tissue through a fence that allows direct contact between the sensor and its coating and the detection area. The in situ detection reported in this work allows for obtaining discrimination between healthy and tumoral cells.

An example of packaging suited for the urologic diagnostic is shown in Figure 10b [79]; other notable packaging options have been reported by Poeggel et al. [83] and Guo et al. [84]. Urologic diagnostic requires the measurement of biological [84] or biophysical [83] parameters within the bladder and requires additional elements to compensate for external factors, usually requiring a rectal catheter. Urologic devices are flexible silicon rubber hollow catheters that are positioned by the surgeon within the bladder through the urinary tract. The sensors can detect the surrounding environment through a set of side holes in the proximity of the tip.

A form of less invasive packaging, suitable for subcutaneous sensing (underneath the skin), has been reported by Liao et al. in [80] (as shown in Figure 10c) and [85]. The packaging form proposed in these works is labelled as Sencil^{TM}, or sensory cilia. The sensor is based on quantum-dot fluorescence made with multimode fiber. The low-cost packaging is based on immersing the sensing tip in a hydrogel: by using a polyethylene glycol acrylate precursor, the fiber forms a thicker tip that can sustain the penetration underneath the skin, reaching a subcutaneous position where the fluorescence-based glucose measurement can be performed.

Percutaneous packaging formats are in use in biomedical applications that require deep-seated diagnostic or therapeutic devices [76]; in this scenario, a rigid needle is used to pierce through the skin for several centimeters, reaching the target area by a straight penetration. Epidural sensors, such as the implementation shown in Figure 10d, proposed by Issatayeva et al. [81], and reported in [86], make use of a Tuohy needle bent around the tip and with fibers mounted on the outer surface.

The packages proposed in [76–86] integrate the sensing tasks with additional purposes, such as delivering a therapy [76], tracing the needle position through X-ray imaging [77], or performing differential diagnostic [83], to form multifunctional, theranostic medical devices.

The integration of sensors in wearable technologies [77], on the other hand, poses the challenge of integrating one or multiple sensors on a device that is worn for a long term. Wearable biosensors based on optical fibers follow two main approaches for fiber packaging [87]. A first method relies on the integration of the fibers into an external element, to be worn separately from the apparel: Issatayeva et al. [81] integrated an array of sensors in two wearable bands, designed to transfer the strain due to respiratory movements to the sensors.

However, the main trend in wearable devices is to integrate the fibers directly on clothing, where they can continuously monitor vital parameters during sports activities or specific actions. Lo Presti et al. [88] proposed a sensory system based on several gratings integrated into a t-shirt for the measurement of physical parameters during daily activities. Li et al. [89] performed a study of the integration of sensors into textiles by sewing an optical fiber through a piece of fabric and allowing a highly efficient thermal transfer of the body temperature to the sensor. In addition, Esmaeilzadeh et al. [90] reported a biosensor based on the surface plasmon resonance, having the sensing element integrated within a textile for contact sensing.

4. Smartphone Fiber Optic Sensors

Smartphones are ubiquitous pocket devices that offer unprecedented diagnostic opportunities for the real-time detection of vital parameters [91]. Modern off-the-shelf phones have several instruments that can be used for sensing purposes [92], and mobile apps can be developed to use each component of the phone to create sensing systems. In addition, apps can use features such as geolocalization, internet of things, or access to external machine learning methods [93] to enhance the potential for diagnostic and biomedical devices.

The potential of smartphones has been exploited within a number of optical sensors [94–98]; the main purpose of the research carried out by several groups in the last few years is driven by replicating diagnostic methods based on fluorescence spectroscopy [94], fluorescence resonance energy transfer [95], colorimetry [96], or spectrophotometry [97], among others, but by replacing the hardware with the existing components on the smartphone, with the minimal addition of external hardware [98].

Overall, optical biosensors based on smartphone platforms are an excellent diagnostic tool and use mounts that can be used as phone accessories in order to integrate the pre-existing devices of the smartphones (such as the flashlight LED or the camera) with additional components such as lenses, spatial filters, diffraction gratings, or additional laser sources [99,100].

Within optical sensors for smartphones, a new generation of optical fiber sensors for smartphones has been developed and reported within the latest years, and it is consistently being applied in the detection of biological analytes, biohazards, and biophysical parameters. While smartphone optical sensors can analyze previously collected samples or can inspect samples positioned in the proximity of the phone's camera, optical fiber sensors for smartphones introduce the remote sensing possibility. Thanks to the low losses of optical fibers and the availability of multimode fibers that operate in the visible wavelength range, it is possible to design systems that implement the remote sensing feature that is typical of an optical fiber sensor, but where the hardware is partially or entirely replaced by the smartphone itself.

Figure 11 illustrates the main implementations and applications of smartphone optical fiber sensors.

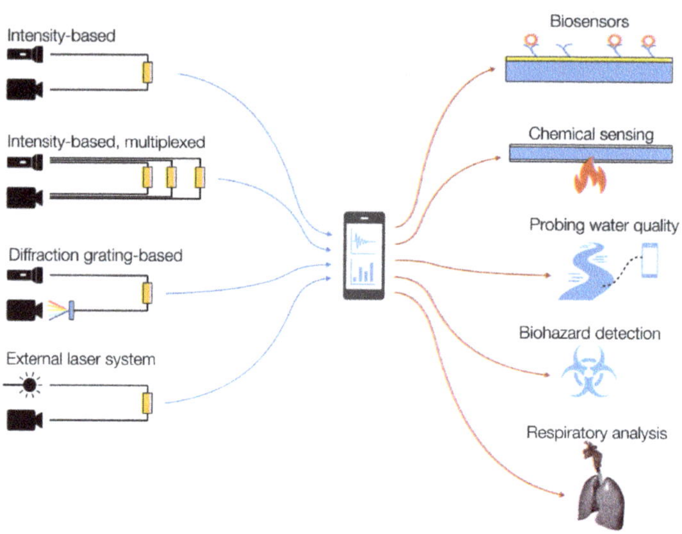

Figure 11. Illustration of the main architectures (**left**) and applications (**right**) of optical fiber biosensors based on smartphones.

4.1. *Intensity-Based All-Fiber Smartphone Sensors*

Intensity-based sensors represent the simplest implementation, as they can be defined as all-fiber sensors; in these architectures, reported in [101–103], the sensing element modulates the light transmitted through the fiber, and both the light source (flashlight LED off) and the light detector (camera) belong to the phone. The only external hardware is a 3D-printed connector that holds the fibers in place and performs light coupling from the flashlight to the input fiber and from the output fiber to the camera.

All systems based on intensity-varying sensors operate with POF fibers, with a core/cladding diameter of 0.98/1 mm and numerical aperture (NA) of 0.47; this wide acceptance angle allows easy coupling of the light in and out of the fibers, delivering a sufficient amount of light to the detector.

The first system was reported by Sultangazin et al. [101] for biochemical detection of hydrogen sulfide (H_2S), which is a potentially dangerous biohazard for oil and gas and mining workers; the system uses the flashlight LED from the phone (a phosphor white-light LED illuminating the whole visible wavelength range) and the camera as a pixel-based intensity detector, which converts the output beam from the fiber into greyscale intensity level, and integrates this on the whole set of available pixels. The working principle is based on the change of attenuation in an Ag-coated POF when exposed to hydrogen sulfide; the formation of silver sulfide (Ag_2S) decreases the wall reflectivity of the outer fiber, increasing evanescent losses and, therefore, reducing the transmission. By integrating, via an app developed in Java, the power incident on the camera, Sultangazin et al. reported an intensity drop of 0.54%/min to 4.9%/min when the sensor is exposed to hydrogen sulfide.

A similar concept was reported by Aitkulov and Tosi [102], extending the sensing mechanism to biophysical sensing by measuring the breathing rate pattern. The sensing element is constituted by a pair of POF fibers, cleaved and spaced by a set distance; when fibers are aligned, they exhibit the maximum transmission, while during breathing movements, the angle between the transmitting and receiving fiber increases the coupling losses, decreasing the detected power. The sensing element is then embodied in an elastic band and worn on the chest for a wearable application. By means of a Fourier analysis of the obtained intensity time signals, the respiratory rate is determined.

The first attempt at a multiplexed fiber-optic sensing system for smartphones was reported in [103]; by merely changing the connector type from the prior system [102], the authors demonstrated a three-fold increase in sensing points. The principle of operation relies on splitting the LED light into multiple fibers and collecting each output into a different portion of the camera. This setup, labelled camera-division multiplexing, allows for isolating each light spot collected on the camera by a tessellation, with minimal crosstalk.

All-smartphone-POF systems [101–103] make full use of the phone's optical circuitry and application programming interfaces for each component. By setting the exposure time, controlled by the ISO of the camera, and disabling the automated routines for the flashlight, it is possible to detect the intensity through photo or video acquisition [102,103] or through a separate mobile app [101]. In this case, no further components are needed to implement the sensing mechanism, as the software can set ISO and calculate the intensity at saturation point, adapting to the power level at the detector. The only external device serves the purpose of mating the phone to fiber and is a phone accessory that can be 3D-printed to be customized to each smartphone model.

4.2. *Diffraction-Grating Assisted Smartphone Optical Fiber Sensors*

The main alternative to power detection via smartphone is to use the camera of the phone as a spectrometer and detect the wavelength shift of spectrally sensitive elements such as surface plasmon resonance (SPR) or fiber Bragg gratings (FBGs). In order to convert the camera to a spectrometer, a diffractive element, such as a grating, or a low-cost element, for instance, a piece of DVD disc, can be used. Thanks to this arrangement, it is then possible to measure the wavelength shift of transmission spectra, making a sensing system that is insensitive to power level and, therefore, potentially more robust to the fiber coupling.

Bremer et al. [104] presented the first fiber-optic RI sensor adapted for smartphones by means of the SPR effect. The system proposed in this work converts the phone camera to a spectrometer by means of a diffractive element, a holographic grating with 1200 lines/mm mounted at a 45° angle with respect to the camera lens. Then, by detecting the output spectrum sampled by each array of pixels, the spectrum of the SPR sensor was detected, obtaining a sensitivity of 1678 pixel/RIU after optimizing the resolution of the phone's camera.

The system proposed by Bremer and Roth was adapted in [105] to the detection of SPR in a planar waveguide configuration. Here, the fiber sensor is replaced by a tapered waveguide geometry with an initial thickness of 1200 μm, reduced to 200 μm in the proximity of the sensor. Two SPR sensors have been designed in the waveguide system, enabling multiplexed detection; the waveguide is coated with a thin film of gold in order to immobilize aptamers, which are used as specific bioreceptors for the detection of 25-hydroxyvitamin D. The waveguide system demonstrates the possibility to detect this vitamin D with a device mounted as an external accessory for the phone.

A different approach was proposed by Markvart et al. [11,106], who showed the possibility to interrogate traditional wavelength-selective fiber-optic sensors, such as interferometers [106] or gratings [11], using an architecture based on the smartphone. The first system, shown in Figure 12, is designed for the interrogation of Fabry–Perot (FP) interferometers. The system uses a diffraction element in front of the camera, collecting the light at a 60° angle. The spectrometer is used as a colorimeter through a signal processing method that converts the output image into the corresponding transmission spectrum by means of signal processing.

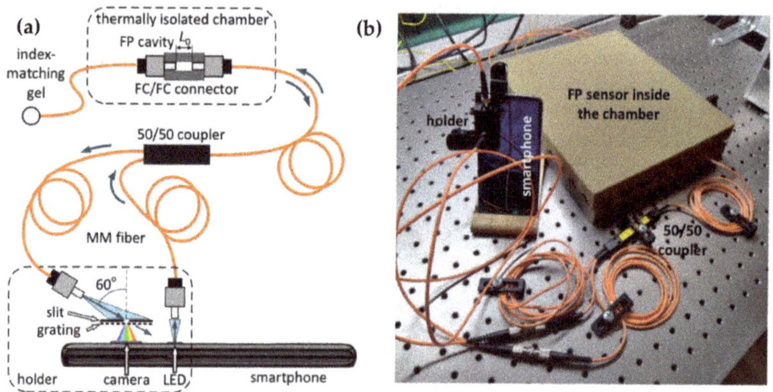

Figure 12. Schematic representation (**a**) and photograph (**b**) of the smartphone optical fiber sensing system for Fabry–Perot sensor interrogation (image reproduced from [106]).

A similar system was reported in [11] for the interrogation of chirped fiber Bragg grating sensors, inscribed on a multimode fiber for strain and temperature sensing. The chirped grating, with bandwidth ~ 8 nm at the red wavelength, exhibits a wavelength shift that can be detected by estimating the pixel-based spectrum via signal processing. Furthermore, Pan et al. [107] reported a temperature sensor based on a multimode fiber, detected in reflection through a smartphone system similar to the previous ones.

4.3. Smartphone Optical Fiber Sensors with External Sources

While the previous architectures makes use of both the light source and detector of the phone, other systems proposed in [107–111] maintain a fiber-optic probe but use a different light source. In these cases, the system is still fiber-coupled and compatible with a smartphone, but significant external hardware is required.

Kamizi et al. [107] developed a fiber-optic sensing network that monitors walking patterns. To do so, a series of single-mode fibers have been placed underneath insoles, with sensors acting as macro-bending units that increase the propagation losses when pressed during the walking motion. The system relies on external hardware that incorporates the LED light source and all the coupling mechanisms, while the output beam is detected through the camera of the phone.

Liu et al. [109] reported a fluorescence fiber-optic sensing system based on a smartphone assisted by an external laser source. The system uses a red laser in order to excite fluorescence in a multimode glass fiber functionalized with quantum dots. The reflected light is collected by the phone's camera by means of an external module customized for the smartphone's size and camera position. An Android app is used to calibrate the system and collect the data. Here, the proposed application is the detection of mercury cations (Hg^{2+}) in the 1–1000 nM range. A similar system was described in [110] for pH detection and in [111] for dual-channel fluorescence monitoring.

The different main implementation strategies of smartphone-based optical fiber sensors are summarized in Table 1.

Table 1. Review of the fiber optic smartphone sensors and their applications.

Ref.	Application	Smartphone	Operative System	Optical Fiber	Internal Devices	External Devices
[101]	Remote detection of hydrogen sulfide	Samsung Galaxy S6 Edge	Android 7.1.1	POF, Ø0.98/1 mm, NA 0.47	Flashlight, camera (set ISO)	3D-printed connector
[102]	Breathing pattern detection	Redmi Note 4	Android 7	POF, Ø0.98/1 mm, NA 0.47	Flashlight, camera (set ISO)	3D-printed connector
[103]	Multiplexed Breathing pattern detection	Redmi Note 4	Android 7	3xPOF, Ø0.98/1 mm, NA 0.47	Flashlight, camera (set ISO)	3D-printed connector
[104]	SPR sensor for refractive index	Huawei Ascend Y300	Android 4.1.1	Thorlabs BFL48, Ø400 μm, NA = 0.48	Flashlight, camera	Diffraction grating, couplers
[105]	Vitamin D Detection	Apple iPhone 6s	iOS	Tapered waveguide, 200–1200 μm	Flashlight, camera	Diffraction grating, external mount
[106]	Interrogation of Fabry–Perot sensor	Huawei P20 Pro	Android	Graded-index MMF, core Ø62.5 μm	Flashlight, camera	Slit, grating, 2× FC/PC connectors
[11]	Interrogation of chirped fiber Bragg grating	Huawei P20 P50	Android	Graded-index MMF, core Ø62.5 μm	Flashlight, camera	Slit, grating, 2× FC/PC connectors
[107]	Temperature Sensing			Multimode glass fiber, Ø300 μm	Flashlight, camera	Diffraction grating, pedestals, couplers
[108]	Identification of walking pattern	Motorola, Moto G3 Turbo	Android	Glass single-mode fiber	Camera	External phone mount, LED
[109]	Detection of mercuric cations	Nubia Z17 Mini	Android	Multimode glass fiber, Ø105/125 μm	Camera	Red laser, connecting module
[110]	On-site pH detection	Nubia Z17 Mini	Android	Multimode glass fiber, Ø105/125 μm	Camera	Red laser, connecting module
[111]	Dual-channel fluorescence detection	Nubia Z17 Mini	Android	Multimode glass fiber, Ø105/125 μm	Camera	Red laser, connecting module

5. Biosensing Applications

As introduced before, due to their unique characteristics, OFBs have enormous potential to be used as devices for POC monitoring of biochemical components in diverse fields. This review is focused on three areas that are of utmost importance for the health and wellbeing of the population in general: cancer and cardiovascular biomarkers detection and environmental monitoring.

Real-time, label-free, and high specificity towards a biomarker of interest are a few of the fundamental features of the modern biosensor [6]. Among other essential traits are the miniaturized size of the device and high sensitivity [6]. In the sensing technology, optical fibers demonstrate adaptability for the recognition of numerous analytes and allow the combination of diverse detection techniques. They can monitor refractive index change in the media from basic buffers and water to a complex serum, urine, and blood when detecting proteins, small biomolecules, cells, and other biomarkers [112]. Thanks to strong evanescent fields, optical fiber biosensors modified with biorecognition elements are a promising tool for sensing target molecules [113]. With an increasing demand for enhanced selectivity, specificity, and sensitivity, it is necessary to continue improving optical systems for biosensing applications.

5.1. *Cancer Biomarkers Detection*

Detection of cancer biomarkers plays a significant role in the prognosis and treatment of various cancers [114]. At the early stage of illness, the concentration of biomarkers is extremely low, and this requires highly sensitive diagnostics methods with very low LODs. Although there is ongoing progress in the standard detection and treatment methods, the mortality due to advancement and metastasis of cancer is still high [115].

Conventional detection procedures for identifying cancer biomarkers require patients to provide a biopsy of the tumor. Subsequently, the tested sample needs to undergo either fluorescence in situ hybridization (FISH) or immunohistochemistry (IHC) [116,117]. Both testing procedures involve the use of fluorescent tags to label chromosomes or DNA sequences [118] in the former and proteins [119] in the latter case. Since the above testing methods are expensive, take long processing times, and, above all, involve invasive procedures [120], alternative measures need to be called for. The research in the field of cancer detection using optical fiber sensors has made considerable progress where non-invasive, real-time, label-free methodologies are proposed [50,121–124]. Different optical fiber based systems have been designed and developed to specifically detect cancer biomarkers, aiming towards a fast diagnostic method for POC testing, as will be presented in the following sections.

5.1.1. Human Epidermal Growth Factor Receptor 2 (HER2)

HER2 (185 kDa) is part of the human epidermal growth factor receptor family, which participates in cell growth and differentiation. However, its overexpression plays an active role in the inception and proliferation of aggressive forms of breast cancer [125]. Patients who undergo costly trastuzumab treatment have a high chance of cardiac toxicity. Thus, taking into account the risk associated with the above therapy, detection of the biomarker is crucial for the prediction and therapy of tumors [126].

In 2017, Sun et al. [124] reported the detection of HER2 on a taper interferometer embedded in FBG. The device was sensitive to RI changes, and wavelength shift was recorded while detecting HER2. The lowest concentration of HER2 detected was at 2 ng/mL. Moreover, the FBG part of the compact device was suggested to be used as an independent thermometer during the sensing process.

An optical fiber-based surface plasmon resonance (OF-SPR) optrode for detection of HER2 was proposed by Loyez et al. [50]. In the experiment, a multimode optical fiber with a diameter of 400 μm was sputtered with gold (thickness 45 nm), followed by silanization and attachment of anti-HER2 aptamers. HER2 biomarkers were detected in phosphate buffered saline (PBS) solution presenting a LOD of 0.6 μg/mL. CK17 protein was tested as a negative control, and no significant shift was observed. Moreover, a sandwich assay experiment using anti-HER2 antibody further improved the LOD to 9.3 ng/mL (~86 pM).

5.1.2. Cluster of Differentiation 44 (CD44)

CD44 is a transmembrane protein with a molecular weight of 85–200 kDa present in human cells, embryonic stem cells, cancer cells, and cancer stem cells [127]. Its overexpression promotes the migration, proliferation, metastasis, and reoccurrence of cancer [128]. Numerous variant forms of CD44 (CD44v) were confirmed to be related with several malignant tumors and metastasis [129]. Additionally, there is a correlation between the role of CD44 in circulating stem cells and the impact it has on the development of cancer and therapeutic outcome [130].

In a recent work from Bekmurzayeva et al. [131], a spherical fiber optic tip fabricated using a CO_2 laser splicer was investigated. The fiber surface was pre-treated with (3-aminopropyl)trimethoxysilane (APTMS), following several steps to functionalize it with anti-CD44 antibodies, as shown in Figure 13, aiming to detect various levels of CD44. The determined LOD was of 17 pM. In order to demonstrate the specificity of the studies, thrombin and interleukin 4 were used as controls, reporting a minor change in signal.

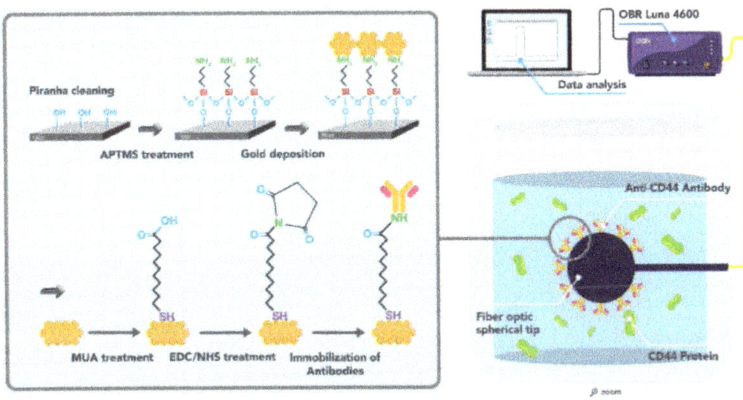

Figure 13. Schematic of functionalization of spherical tip-biosensor used for detection of CD44 biomarker (image reproduced from [131]).

5.1.3. Thyroglobulin (Tg)

Differentiated thyroid cancer (DTC) is one of the most common types of endocrine cancer [132]. While the progression rate and death incidence for DTC are indolent, the relapse cases after primary therapy with the need for thyroidectomy are high, which are related to the increased instances of long-term morbidity [133–135]. Hence, a lifetime check-up is advised for timely diagnosis to avoid further progression and subsequent death [136]. Thyroglobulin (Tg) is a dimeric precursor glycoprotein weighing 660 kDa, made by the thyroid follicular cells, and is necessary for thyroid hormone synthesis (T_3 and T_4). Its levels can be measured in the serum of a healthy person up to 40 ng/mL. The elevated amount of Tg, which is a highly specific biomarker to DTC, is an indication of the latter [137].

In work performed by Kim et al. [138], a fiber-optic LSPR sensor combined with a microfluidic channel was proposed to detect Tg. One of the preparation steps for optical fibers involved treating them with Au NPs (average diameter of 50 nm). The device was exposed to a range of concentrations of Tg, giving the LOD of 93.11 fg/mL. Comparatively, the response time was 10 min, whereas conventional techniques such as chemiluminescence assay (CLIA), immunoenzymometric assay (IEMA), and immunoradiometric assay (IRMA) take a day or more [138].

5.1.4. Cytokeratin 7 (CK7)

There is a discrepancy between the biomarkers expressed by tumors at the early and metastatic stages. Cytokeratin 7 (CK7), which is usually expressed by epithelial cells coating the cavities of internal organs, can assist in the diagnosis of carcinomas. Cancer cells of the lung, breast, thyroid and salivary glands, and female reproductive organs will express CK7. Besides finding the biomarker in primary lung cancer, fragments of cytokeratin can be released into the bloodstream by malignant cells [139].

Ribaut et al. [123], detected CK7 protein (78 kDa) and CK7 (2.6 kDa) peptide using an SPR optical biosensor based on TFBG. In this work, TFBGs with tilt angles between 7° and 9° were coated with gold (thickness of 50 nm) using a high vacuum sputter followed by surface pre-treatment and immobilization of anti-CK7 antibodies. Detection of both CK7 protein and CK7 peptide attained a LOD of 1 pM in PBS and 0.4 nM in serum, respectively.

5.1.5. Cytokeratin 17 (CK17)

Nearly a third of cases of cancer mortality worldwide are caused by late diagnosis and aggressive growth of lung cancer [140]. The average span of life after confirmation of disease and its subsequent treatment with chemotherapy is less than a year [141]. To enable early-stage diagnosis and to enhance recovery and survival rate, the identification of biomarkers is highly in demand, since various types of lung cancer require various treatment procedures. Generally, an overexpression of cytokeratins is associated with carcinogenesis [142,143]. Among them, Cytokeratin 17 (CK17) promotes cancer progression and epithelial proliferation [144].

Ribaut et al. [121], in 2017, reported a gold-coated (50 nm) TFBG immunosensor with 7° for detection of the CK17 biomarker. The sensor surface was functionalized with a self-assembled monolayer of S_2-PEG_6-COOH followed by immobilization of the anti-CK17 antibody. Furthermore, the sensor was packaged into a hollow cylindrical needle, and a window was cut with a laser to expose the fiber to the surrounding environment. The packaged biosensor was used to detect biomarkers in PBS and CK17-encapsulated gel matrix, giving a LOD of 1 pM and 0.1 ng/mL, respectively. In addition, the recognition of analyte was conducted ex vivo in healthy and tumoral tissue samples. Compared to healthy tissue, the sensor's response to protein showed a linear trend in a biopsy sample of CK17+ positive patients.

Similarly, Loyez et al. [122] detected CK17 comparing four biofunctionalized TFBG based sensors (without gold—using silanization, gold-coated via sputtering, gold-coated with electroless deposition (ELP), and hybrid gold-coated—a mix of sputtering and ELP) that underwent various pre-treatment procedures. Sensing of analyte showed a linear trend in all four TFBGs. Fiber prepared using silanization showed the lowest sensitivity comparing to the others. However, it was stated that it could be used without taking into account polarization effects. The LOD for the gold-sputtered TFBG was close to 1 pg/mL, and it was the most sensitive device. The sensor was further inserted in the tumorous part of the lung biopsy, showing a positive trend for CK17. Fibers were gold-coated using ELP and hybrid methods and were both sensitive to the analyte, with the former showing sensitivity down to 1ng/mL and the latter being as sensitive as the gold-sputtered TFBG.

5.2. Cardiovascular Biomarkers Detection

Cardiovascular diseases (CVDs) are the prime cause of death in the world. It is estimated that, in 2019, more than 19 million people died from CVDs, corresponding to 32% of all global deaths [145]. Most CVDs are related to lifestyle and modifiable factors [146]. Therefore, prevention is the greatest weapon for reducing CVDs and improving wellbeing. Factors such as stress, sedentary lifestyle, obesity, metabolic diseases, and hypertension are preponderant in the development of such epidemy [147]. Hence, their monitoring and control play a fundamental role in the decreasing of CVDs. Population-level interventions always have to consider the cost-effectiveness of the applied measures and monitoring techniques [147]. In a disease with such a large prevalence, this aspect is of utmost importance.

In CVD management, each person is different, and individual assessment should be personalized according to age, co-morbidities, and lifestyle. Moreover, monitoring should be cost-effective and, when possible, at point-of-care [148]. Recently, different cost-effective optical fiber sensors were developed to monitor different cardiovascular biomarkers, namely, blood pressure [149], cholesterol levels [150] and diabetes mellitus (glucose levels) [151,152], heart acute myocardial infarction (AMI) by Troponins quantification [151,153], heart and kidney failure [154], and stress [21].

5.2.1. Detection of Cholesterol and Glucose

Dyslipidemia, which is the presence of abnormal lipidic levels in the blood, and glucose metabolism disorders, such as diabetes, are independent risk factors of CVDs [147]. Therefore, monitoring of such analytes in POC locations is of outmost importance in CVD prevention. Kumar et al. [150] developed a label-free reflectance-based cholesterol biosensor consisting of SMF with a hollow core fiber working in reflection, coated with Au NPs and with diameters of around 11 nm, for sensitivity enhancement using LSPR phenomenon. The biorecognition molecule was cholesterol oxidase enzyme. With this approach, a LOD of 25.5 nM and a sensitivity of 16.15 nm/μM were attained. In 2019, an optical tapered fiber structure coated with Au NPs, working on transmission, was developed [155]. The taper, created on SMF fiber, achieved a minimum diameter of 40 μm, a total length of 4 mm, and a 5 mm transition stretching region on each side of the taper. The biorecognition molecule was glucose oxidase (GOx) enzyme. This sensing approach was studied in a detection range up to 10 mM, achieving a LOD of 322 μM and a sensitivity of 0.93 nm/mM. In 2020, Zheng et al. [156] described a highly-sensitive glucose sensor, based on back-reflection configuration, using plastic cladding fiber of 600 μm. The fiber tip was coated with gold by sputtering and was then biofunctionalized with GOx enzyme. A measurement range of 0–0.5 mg/mL and a resolution of 0.0004 mg/mL were achieved.

5.2.2. Detection of Acute Myocardial Infarction Biomarkers

Acute myocardial infarction (AMI) refers to the death of myocardial cells due to ischemia or the imbalance between the blood supply and demand within the coronary arteries as a result of an acute thrombotic process [157]. As a life-threatening situation, the fast diagnosis of AMI is vital for the early initiation of appropriate therapeutic measures. The main tools to diagnose AMI are electrocardiography (ECG) and cardiac troponins detection [158]. Cardiac troponins are regulatory proteins specific to the myocardium that are released into the circulation when myocytes are being damaged [157]. There are three kinds of troponins (Tn): I, T, and C, with troponin T and I being cardiac-specific, while troponin C is expressed by both cardiac and skeletal muscle.

An ultrasensitive label-free optical microfiber coupler biosensor based on interference turning point effect was developed to detect cardiac troponin I. The interrogation setup consisted of a halogen light source, with a microscope objective to focus the light into the fiber. The sensor consisted of a microfiber coupler fixed in a fluid cell integrated into a polydimethylsiloxane (PDMS) chamber for sample solutions delivery. A visible spectrometer was used as a signal analyzer. With this sensing architecture, an RI sensitivity of 91,777.9 nm/RIU was achieved, and after biofunctionalization with anti-cTnI antibodies and testing in cTnI solutions, it reached a LOD of 2 fg/mL [151]. Krupin and Berini [152] developed a long-range surface plasmon–polariton (LRSPP) waveguide biosensor to detect troponin I. The sensors consisted of gold stripes of about 35 nm thickness, embedded in a low-index optical-grade fluoropolymer (CYTOP^{TM}) with fluidic channels etched to the surface of the gold strips. LRSPPs were excited by butt-coupling a polarization-maintaining single-mode fiber (PM-SMF) to the input facet. The optical interrogation included an LED peaking at 1310 nm connected to the PM–SMF. The transmitted light through the LRSPP passes to a 25× objective lens and a 50:50 beam splitter. One part of the split beam is sent to an infrared camera for visual monitoring and alignment, and the other part is sent to a photodetector for evaluation of power variations in time during the detection procedure. The gold waveguide was biofunctionalized with anti-cTnI antibody via Protein G and passivated with BSA. The sensor was tested in a direct and sandwich detection mode, reaching a LOD of 430 pg/mL and 28 pg/mL, respectively.

5.2.3. Heart and Kidney Failure

Heart failure is a cardiovascular disease characterized by a disorder of the heart, that can be structural and/or functional, which can cause increased intracardiac pressure and/or inadequate cardiac output [147]. In heart failure patients, kidney function has to be closely examined, since renal dysfunction due to diuretics, particularly in patients with heart failure, is a common cause of hospitalization [159]. Botewad et al. [154] developed a biosensor for urea detection, a biomarker of kidney function [160]. This study used as transducer a 450 μm diameter core plastic-clad fiber (PCF) uncladded throughout a 2 cm portion. The uncladded region of the fiber was modified with a composite of polyaniline (PANI) with ZnO and biofunctionalized with urease enzyme, which catalyzes the hydrolysis of urea. This intrinsic sensor worked in transmission using a halogen lamp and a spectrophotometer as light source and detector, respectively. When testing in urea solutions, this biosensor reached a LOD of 10 nM.

In 2021, Li et al. [161] reported a biosensing approach comprised by SMFs, a multicore fiber (MCF), and MMF fibers, in the structure SMF–MCF–MMF–SMF, for the detection of creatinine in the human body. This structure, produced by a fusion splicer machine, was etched to a 90 μm diameter and coated with graphene oxide (GO), Au NPs, and molybdenum disulfide NPs (MoS_2-NPs) and then biofunctionalized with creatinase enzyme. When tested in creatinine solutions, the sensor presented a LOD of 128.4 μM in a linear range of 0–2000 μM.

In heart failure, higher enzymatic activity with maintained oxygen consumption contributes to the physiopathology of myocardial insufficiency and appears to be an indicator of oxidative stress [162]. In 2021, Ortega-Gomez et al. [51] developed a plasmonic tip biosensor for the detection of reduced cytochrome c, which is a multifunctional enzyme with a crucial role in electron transfer in the mitochondrial transport chain. This biosensor consisted of a PCF MMF uncladded tip, that was coated with Au NP biofunctionalized with cytochrome c as the biorecognition molecule, achieving a LOD of 60 nM.

5.2.4. Stress

Stress conditions (depending on the degree, duration, and individual response) often lead to maladaptive physiological responses and are associated with CVDs in their acute and chronic forms [163]. Stress also increases the probability of developing CVD risk factors such as hypertension, diabetes, and obesity [164]. Prolonged high levels of cortisol, known as the stress hormone, are related to physical and psychological disorders [165]. Therefore, the development of an easy and low-cost POC monitoring method to detect this stress biomarker is a current hot topic. In 2021, C. Leitão et al. [21] reported a POF uncladded approach for the detection of cortisol. The biosensor used SPR as a sensitivity enhancer by coating the unclad part of the fiber with a gold/palladium (AuPd) alloy by the sputtering technique. The surface was biofunctionalized with anti-cortisol antibodies using cysteamine as an intermediated linker. The final surface was passivated with BSA. When tested in cortisol solutions of concentrations between 0.005 and 10 ng/mL, the proposed sensor had a total of 15 nm wavelength shift. The attained sensitivity and LOD were 3.56 ± 0.20 nm/(log(ng/mL)) and 1 pg/mL, respectively. In this research, control tests were also performed in a sensor functionalized with antibodies for human chorionic gonadotropin (anti-hCG antibodies) in which the variance of the resonance wavelength was only 1 nm, much lower than the sensor modified with anti-cortisol antibodies.

5.3. Environmental Monitoring

Today, more than ever, there is a growing concern about emerging contaminants and pollutants which severely affect the environment and human health. These contaminants and pollutants, which include insecticides, gases, stimulants, antibiotics, anti-inflammatory drugs, and pesticides, among others, are biologically dynamic and highly resistant. As a result, they continue in the environment, instigating harmful effects to non-target organisms and humans, making it imperative to find approaches capable of correctly detecting these substances in order to protect the environment and our lives.

In agricultural and fisheries engineering, a variety of environmental monitoring sensors (quality of air, soil, and water) allow us to understand the environmental impact of agricultural activities and recognize whether the farm conditions are suitable to cultivate. Many studies have focused on the monitoring of farm soil and aquaculture water, since soil fertility is the key to farmers looking for improvement of crop yields and agricultural productivity, and well-controlled fish water tanks are key to the fishery sector. Further, oil/gas industry pieces of equipment (pumps, pipes, and joints, among others) are required to be monitored. In case any of these become damaged due to natural disasters or human intervention, not only are their properties lost but also the surrounding environment can become polluted. The equipment is submerged at the sea bottom or into a downhole, which are high-pressure and environments with corrosive saltwater. In the case of delivery pipes, these are spread over several kilometers and buried underground. This means that oil and gas industries need non-invasive and low-cost sensing solutions to monitor many critical parameters such as gas detection and/or leakage and oil quality, among other things.

5.3.1. Phenolic Compounds

Phenolic compounds are mainly produced by burning wood and coal apart from their existence as sewage and industrial by-products [166]. With water being the most valuable natural resource for mankind, its pollution by organic and inorganic compounds in the current industrial age has been a serious concern and has placed it as one of the world's leading health risks [167]. Phenol wastewater is one of the industrial sewages which causes harm worldwide, with it also being one of the worst sources of environmental water pollution [168]. Drinking water with extremely high concentrations of phenolic compounds can cause muscle convulsions, difficulty in walking, and even death [169]. These phenolic compounds release toxic gases during combustion that are potentially harmful to human health if absorbed. *p*-Cresol can enter human bodies in different ways as it is used as a flavoring agent in foods and in some traditional medicines and is always present in tea, oil, and tap water [170]. Recently, an optical fiber-based lossy mode resonance sensor built by using NPs of ZnO/MoS_2 and the MIP technique was used to detect *p*-cresol [171]. To achieve the specificity of the sensor, tyrosinase enzyme is commonly used for *p*-cresol detection [172]. Tyrosinase (polyphenol oxidase) is an enzyme that catalyzes phenol *o*-hydroxylation yielding *o*-diphenol (monophenolase activity) that subsequently oxidized to *o*-quinone [173]. Very recently, Wang et al. [174] reported that a localized plasmon-based sensor was developed for *p*-cresol detection, consisting of a nonadiabatic 40 μm of tapered optical fiber experimentally fabricated and computationally analyzed using the beam propagation method. For performance optimization of the sensor, two probes were proposed, where probe 1 was immobilized with Au NPs and probe 2 was immobilized with the Au NPs along with ZnO NPs. To increase the specificity of the sensor, the probes were functionalized with tyrosinase enzyme. Different solutions of *p*-cresol in the concentration range of 0–1000 μM were prepared in an artificial urine solution for sensing purposes. Different analytes were prepared for selectivity measurement. The linearity range, sensitivity, and LOD of the probe using ZnO NPs were 0–1000 μM, 5.6 nm/mM (accuracy 0.981), and 57.43 μM, respectively, making the overall performance of such a probe much better, due to the inclusion of ZnO NPs, which increases the biocompatibility of the sensor probe.

5.3.2. Phthalate Esters

Phthalate esters, also known as PAEs, are a type of synthetic chemical substance that has been allocated to the list of priority pollutants because of their endocrine-disrupting and toxic effects on the human body. In 2019, Cennamo et al. [175] reported a D-shaped plasmonic optical fiber biosensor to detect the presence of naphthalene in sea water. The D-shaped configuration was obtained by removing the cladding of the POF by a polishing process. In this study, an antibody specific to the naphthalene molecule was designed and produced and, for that, a retro-synthetic chemical strategy was applied to modify the NAPHTA structure in a derivative structure to obtain the anti-NAPHTA antibodies. This modified NAPHTA structure was coupled to a protein carrier and was used for immunization. The capability of the antibody to bind to naphthalene was assessed by enzyme-linked immunosorbent assay (ELISA) tests. Through *N*-ethyl-*N'*-(3-dimethylaminopropyl)carbodiimide (EDC)/*N*-hydroxysuccinimide (NHS) chemistry, the gold surface was derivatized and functionalized with the produced antibody. Subsequently, tests using real matrices of sea water were performed using the produced biosensor. Through the obtained results, it was possible to observe that the POF biosensor was able to sense the presence of naphthalene in a sea water solution with a LOD of 0.76 ng/mL (0.76 ppb), which is lower than the limit value of naphthalene (0.13 μg/mL). In a 2020 study, Lamarca et al. [176] prepared a label-free U-shaped immunosensor for the detection of ciprofloxacin (CIP) in wastewater samples, as CIP is a broadly utilized antibiotic to treat infections and is a common contaminant of wastewater treatment plants. The glass optical fiber surface was functionalized with PANI, followed by immobilization of anti-CIP antibody. CIP could be detected with a LOD of 3.30×10^{-3} ng/L, in a linear range between 0.01 ng/L and 10,000 ng/L, and with a quantification limit of 0.01 ng/L. In addition, the immunosensor offered a high average recovery of 91%. Benzo(a)pyrene (B(a)P) is one of the most toxic polycyclic aromatic hydrocarbons and a carcinogen, making monitoring its concentration levels essential for human health and environmental contamination avoidance. To monitor B(a)P contamination levels in the water, Gao et al. [177] proposed an in-line fiber optofluidic immunosensor using a hollow-core fiber with its surface immobilized with antibodies. As such, the immunoreaction between the antibody and the B(a)P molecule induced a significant change in the RI inside the in-line optofluidic channel. The attained results presented a LOD of 1.65 pM and sensitivities of up to 23 pm/pM.

5.3.3. Gases and Volatile Compounds

There are myriad gases and volatile materials that researchers and engineers are interested in studying for a number of reasons (ranging from safety issues to general analytical analysis) in the oil/gas industry, from which three will be focused on: formaldehyde (CH_2O), hydrogen sulfide (H_2S), and carbon dioxide (CO_2), where the identification of specific molecules in diverse media and under extreme conditions has been carried out.

González-Vila et al. [178] developed an MIP coating synthesized around a metal-coated optical fiber sensor by an electropolymerization process, working as an electrode, where the oxidation takes place at the surface of the metal. The electrodeposition was figured by the TFBG–SPR sensor and, as a result of the MIP coating, the TFBG-based sensor acquired sensitivity in gaseous atmospheres. A sensitivity of 2.10 pm/ppm when detecting tiny formaldehyde concentrations in the gaseous state was achieved. In addition, the sensor exhibits a selective behavior to this molecule such that the presence of other volatile compounds did not produce a substantial change in the sensor's response.

H_2S is usually found in nature by the decomposition of organic materials, especially in oil and gas production chains. It can also be found in mineral environments such as coal and salt deposits, as well as other mineral extractions containing sulfur. Physical, chemical, and biological agents are the main agents responsible for the formation of this compound. Various systems have been developed using optical fiber sensors due to their large advantages. Sultangazin et al. [101] proposed a low-cost H_2S sensor based on plastic optical fiber functionalized with silver deposition on the fiber's outer surface. The sensor is integrated with a smartphone used as an interrogator unit, and the response time is just over a few minutes. Ke et al. [179] presented an optical fiber evanescent-wave sensor. The sensing probe is fabricated by etching a standard single-mode fiber, where the transmitted optical power centered at 1631.9 nm is monitored by a power meter. As the H_2S gas increases, more energy of the evanescent wave is absorbed, leading to a reduction in the optical power. Prado et al. [180] very recently presented a new study that considers the advantages of optical fiber sensors in conjunction with the colorimetric detection abilities of surface plasmon resonance in Au NPs, reporting the development of a H_2S detector based on optical fiber coated with Au NPs. The proposed configuration and method used make it possible to detect the presence of H_2S in gaseous systems while operating at room temperature and with important advantages regarding easy production and short response time. The developed detector has the ability to sense H_2S levels in the range of 0.4 to 2.0 ppm at room temperature.

CO_2 is another critical parameter in the oil/gas industry. There are many materials to use with an optical fiber that have a specific reaction with CO_2. A range of different materials has been reported in the literature over recent years, such as xerogels doped with 1-hydroxy-3,6,8-pyrenetrisulfonic acid trisodium salt (HPTS) [181], also known as pyranine. This CO_2 sensor is based on pyranine, a pH-sensitive fluorescent indicator dye. In the presence of CO_2, the dye has an ion transfer that alters the absorption features of pyranine and xerogel at 396 nm and 460 nm, effectively reducing absorption at these wavelengths, causing reactions at room temperature, and having rational response times. Another approach uses oxidation or reduction reactions in which an electron transfer process occurs between a gas and a material. An example of this is the hybrid nickel oxide/reduced graphene oxide (NiO/rGO), which is a structured coating material. The reactionary wavelengths are 670 nm and 771 nm, and it is the combination of the nanostructured material and its chemical composition in the presence of CO_2 that triggers the reaction, altering the distribution of the radicals within the materials, changing the electron density and, hence, the permittivity. The result is a small but significant change in the emission wavelengths [182]. A redox reaction using single-wall carbon nanotubes in localized surface plasmon structures has been used for CO_2 detection [183], where the chemical selectivity is conferred in terms of the activation energy, allowing for room temperature operation.

5.3.4. Aquaculture Monitoring

In agricultural and fisheries engineering, many studies have been explored in terms of soil pollution and critical parameters in water fish tanks. When stress is persistent and uncontrollable, it is considered pathological, which can trigger depression and cardiovascular diseases [184], for example, and therefore the development of technology capable of monitoring stress is essential. Stress involves a large number of neuronal circuits and, once it is promoted, leads to the release of glucocorticoids, in particular cortisol [184]. The substantial variation in this hormone occurs due to exposure to psychological, environmental, or emotional stress [185]. As a result, cortisol is one of the most important stress biomarkers.

One area in which stress also presents significant influence is aquaculture; especially, the impact of such a stress hormone is a challenge to be overcome in recirculating aquaculture systems. When small variations in the water composition or quality occur in these systems, stress induction can arise as well as reduced food intake, which leads to reduced fish growth and, consequently, leads to possible mortality when acute or chronic stress is high. For these reasons and others, it is then essential to monitor cortisol in water [186]. In 2020, Sharma et al. [187] simulated an SPR fiber optic immunosensor for cortisol detection at the wavelength of 830 nm. The sensor consisted of an Ag layer with 2D materials, conventional (graphene, tungsten disulfide (WS_2), and MoS_2) and transition metal carbides (MXenes: Ti_3C_2, $Ti_3C_2O_2$, $Ti_3C_2F_2$, and $Ti_3C_2(OH)_2$), considered one at a time. The sensor that showed a superior balanced set of performance parameters under both modes was the $Ti_3C_2O_2$-based probe. Through simulation, this probe was able to achieve a LOD of 15.7 fg/mL. The cortisol sensor developed by Leitão et al. [21], described in the last section, also can be applied to monitor cortisol levels in aquaculture water.

One of the key limiting factors in aquaculture is the presence of ammonia; therefore, its early detection in small concentrations prevents fish mortality and improves the production quality [188]. Yi Zhu et al. [189] reported a combination of optical fibers and tapered optical fibers (SMF–MMF–taper–MMF–taper–SMF) as miniature interferometry-based optical fiber ammonia gas sensors. A range from 0 to 5460 μg/L of ammonia in the gas chamber was performed. The sensing material involved ZnO nanoflowers deposited on the sensing area (middle MMF and tapers), by a drop of a ZnO solution and 6 h drying in a vacuum oven at 60 °C. Ammonia sensitivity performance is compared between ZnO nanoflowers and ZnO nanospheres of around the same size (1 μm), showing sensitivities of 5.75 pm/(μg/L) versus ~2.25 pm/(μg/L), respectively. Shrivastay et al. [190] presented a contemporary approach to design and developed a hypersensitive ammonia gas sensor producing a Mach–Zehnder interferometer (MZI) by an SMF–PCF–SMF fiber substrate to perform the interference by immobilizing PANI@SnO_2 nanocomposite to achieve sensing. In this case, excitation of core and cladding modes of PCF is achieved using collapse region that is formed at the junction of SMF and PCF specialty fiber, achieving very fast response and recovery times of 7 and 2 s, respectively, which can detect as low as 8.09 ppt (47.59 fM). The reusable probe showed the potential for rapid detection of ultra-trace ammonia with high selectivity and reproducible features. In 2020, Leal-Junior et al. [191] presented a low-cost fiber-optic probe for the early detection of ammonia. The sensor was based on the chemical interaction between the Oxazine 170 perchlorate layer, deposited in an uncladded polymer optical fiber, and the ammonia dissolved in water. In addition, a thin metallic layer (composed of gold and palladium) was deposited in the fiber end facet and acted as a reflector for the optical signals, enabling the use of the proposed sensor in reflection mode. Different configurations of the sensor were tested, where the effects of the PDMS protective layer, thermal treatments, and the use of reflection or transmission modes were compared in the assessment of ammonia concentrations in the range of 100 ppb to 900 ppb. Results showed better performance (as a function of the sensor sensitivity and linearity) of the sensor with the annealing thermal treatment and without the PDMS layer. Then, the proposed fiber-optic probe was applied to the ammonia detection in high-salinity water, and ammonia concentrations as low as 100 ppb were detected.

Another problematic factor is the usage of myriad pesticides and insecticides in agriculture that are dangerously applied to the soil, such as organophosphorus pesticides that are used worldwide for agricultural purposes. Miliutina et al. [192] reported a functional plasmonic sensor aimed at the monitoring of pesticide spreading and determination of their concentration. It utilized the functionalization of a plasmon-supported fiber surface decorated with a metal–organic framework compound with the formula $Zn_4O(BDC)_3$, also known as the MOF-5 layer. The MOF-5 layer provides the extraction of pesticides from the surrounding medium, which causes the shift in the plasmon resonance absorption band position. The created system demonstrated high selectivity and sensitivity towards organophosphorus pesticides. Particularly, the sensors were successfully applied for the detection of fenitrothion and paraoxon. The lower detectable concentration was found to be as small as 1 pM, which makes the proposed sensors comparable to common analytical approaches to pesticides detection. The proposed cost-effective technique allows for simple and straightforward pesticide detection, even in the complex samples, without any sample's pretreatment step, and it can be easily and scalable transferred to outdoor conditions. In 2019, Kant et al. [193] presented an SPR based fiber-optic sensor for the pesticide fenitrothion utilizing Ta_2O_5 nanostructures sequestered onto a rGO matrix. A thin layer of silver was deposited on the unclad core of silica optical fiber for SPR generation and was followed by the deposition of a sensing surface comprising a layer of tantalum oxide NPs sequestered in a nano-scaled matrix of rGO. The sensing mechanism is based on the interaction of fenitrothion with the silver film, which leads to a change in the RI. Characterized by a wavelength interrogation scheme, the fiber-optic sensor exhibited a redshift equaling 56 nm corresponding to fenitrothion concentration in the range from 0.25 to 4 μM, including the blank solution. The spectral sensitivity is 24 nm/μM, the limit of detection is 38 nM, and the response time is as short as 23 s. The sensor is selective, repeatable, and works at ambient temperature.

6. Conclusions

This paper presents a review of optical fiber-based cost-effective biosensing platforms, covering different geometries, interrogation techniques, encapsulation methods, and, finally, presenting applications in three crucial fields of sensing for wellbeing: cancer and cardiovascular biomarkers detection and environmental monitoring.

The development and application of different types of optical fiber biosensing has increasing over the years at an impressive rate, with the demystification of some associated prejudices. In the past, optical fiber biosensing was labeled as high cost, and for this reason was only considered for highly specific applications. However, as fiber optic biosensor technologies have continued to show excellent performance, the instrumentation associated with their interrogation has dropped in cost, and essentially new methods of designing interrogating sensors have been developed. Fiber-optic sensors working in the visible range are naturally low-cost. Furthermore, the possibility to use the instrumentation of a smartphone to perform the readout of these types of sensors, in addition to considerably lowering the system cost, can be the basis for more democratic and accessible self-monitoring.

Fragility was also one of the points associated with fiber optic sensors. However, various encapsulation techniques are currently used to make these sensors highly robust while maintaining or even amplifying their sensitivity. Currently, we are able to obtain sensors with very high performance, which are highly robust and with practical and low-cost reading techniques.

The potential of all these cost-effective optical fiber sensing techniques reviewed in this article is supported by the numerous applications and excellent results shown in the three monitoring areas addressed. Nonetheless, as in any other biosensing technique, several aspects still need to be improved and worked on, namely, in the enhancement of their repeatability and reproducibility and by developing techniques to enable their reuse. In the future, it is expected that further research will be devoted to make progress on these topics.

Author Contributions: Conceptualization and methodology, C.L., N.C. and D.T.; writing—original draft preparation, C.L., C.M., N.C., M.S., T.A. and D.T.; writing—review and editing, C.L., S.O.P., C.M., N.C., L.Z., M.S., T.A. and D.T.; supervision, C.L. and D.T. All authors have read and agreed to the published version of the manuscript.

Funding: The authors acknowledge financial support of the project i3N, UIDB/50025/2020, UIDP/50025/2020, and LA/P/0037/2020, and DigiAqua project (PTDC/EEI-EEE/0415/2021) financed by national funds through the FCT/MEC. C. Marques and C. Leitão thank FCT through the CEEIND/00034/2018 (iFish project) and CEEIND/00154/2020 research contracts, respectively. S. O. Pereira thanks i3N for the BPD grant BPD/UI96/5808/2017. The work was also funded by Nazarbayev University, under SMARTER (code: 091019CRP2117) and EPICGuide (code: 240919FD3908) grants.

Institutional Review Board Statement: Not applicable.

Informed Consent Statement: Not applicable.

Conflicts of Interest: The authors declare no conflict of interest.

Abbreviations

The following abbreviations are used in this manuscript:

AMI	Acute myocardial infarction
APTMS	(3-Aminopropyl)trimethoxysilane
ARS	Alizarin Red S
B[a]P	Benzo(a)pyrene
BSA	Bovine serum albumin
CCDs	Charge-coupled devices
CD44	Cluster of differentiation 44
CIP	Ciprofloxacin
CK7	Cytokeratin 7

CK17	Cytokeratin 17
CLIA	Chemiluminescence assay
CVDs	Cardiovascular diseases
DNA	Deoxyribonucleic acid
DTC	Differentiated thyroid cancer
EBL	Electron beam lithography
ECG	Electrocardiography
EDC	*N*-ethyl-*N*′-(3-dimethylaminopropyl)carbodiimide
eFBG	Etched Fiber Bragg grating
ELISA	Enzyme-linked immunosorbent assay
ELP	Electroless deposition
eTFBG	Etched tilted Fiber Bragg grating
FWHM	Full width at half maximum
GNG	Gold nanograting
GNS	Gold nanostars
GNR	Gold nanorods
GO	Graphene oxide
GOx	Glucose oxidase
HER2	Human epidermal growth factor receptor 2
HPTS1	Hydroxy-3,6,8-pyrene trisulfonic acid trisodium salt
HTR	Human transferrin
IEMA	Immunoenzymometric assay
IRMA	Immunoradiometric assay
ITO	Indium tin oxide
LDF	Light diffusing fibers
LEDs	Light-emitting diodes
LOD	Limit-of-detection
LRSPP	Long-range surface plasmon–polariton
LSPR	Localized surface plasmon resonance
MIPs	Molecularly imprinted polymers
MMFs	Multimode fibers
MZI	Mach-Zehnder interferometer
NA	Numerical aperture
NHS	*N*-hydroxysuccinimide
NPs	Nanoparticles
OFB	Optical fiber biosensor
OFS	Optical fiber sensor
OF-SPR	Optical fiber-based surface plasmon resonance
OTA	Ochratoxin A
PANI	Polyaniline
PBS	Phosphate buffered saline
PCS	Plastic-clad silica
PDs	Photodetectors
PDMS	Polydimethylsiloxane
PMMA	Poly(methyl methacrylate)
PM-SMF	Polarization-maintaining single-mode fiber
POC	Point-of-care
POFs	Plastic optical fibers
RI	Refractive index
RIU	Refractive index units
rGO	Reduced graphene oxide
SMFs	Single-mode fibers
SNR	Signal-to-noise ratio
SPR	Surface plasmon resonance
Tn	Troponin
Tg	Thyroglobulin
TNT	Trinitrotoluene
VEGF	Vascular Endothelial Growth Factor

References

1. Bhalla, N.; Jolly, P.; Formisano, N.; Estrela, P. Introduction to Biosensors. *Essays Biochem.* **2016**, *60*, 1–8. [CrossRef] [PubMed]
2. Senf, B.; Yeo, W.-H.; Kim, J.-H. Recent Advances in Portable Biosensors for Biomarker Detection in Body Fluids. *Biosensors* **2020**, *10*, 127. [CrossRef] [PubMed]
3. Kotsiri, Z.; Vidic, J.; Vantarakis, A. Applications of Biosensors for Bacteria and Virus Detection in Food and Water–A Systematic Review. *J. Environ. Sci.* **2022**, *111*, 367–379. [CrossRef] [PubMed]
4. Ray, S.; Panjikar, S.; Anand, R. Design of Protein-Based Biosensors for Selective Detection of Benzene Groups of Pollutants. *ACS Sens.* **2018**, *3*, 1632–1638. [CrossRef] [PubMed]
5. Chen, C.; Wang, J. Optical Biosensors: An Exhaustive and Comprehensive Review. *Analyst* **2020**, *145*, 1605–1628. [CrossRef]
6. Damborský, P.; Švitel, J.; Katrlík, J. Optical Biosensors. *Essays Biochem.* **2016**, *60*, 91–100. [CrossRef]
7. Mowbray, S.E.; Amiri, A.M. A Brief Overview of Medical Fiber Optic Biosensors and Techniques in the Modification for Enhanced Sensing Ability. *Diagnostics* **2019**, *9*, 23. [CrossRef]
8. Sabri, N.; Aljunid, S.A.; Salim, M.S.; Fouad, S. Fiber Optic Sensors: Short Review and Applications. In *Springer Series in Materials Science*; Springer: Singapore, 2015; Volume 204, pp. 299–311.
9. Díaz, C.; Leitão, C.; Marques, C.; Domingues, M.; Alberto, N.; Pontes, M.; Frizera, A.; Ribeiro, M.; André, P.; Antunes, P. Low-Cost Interrogation Technique for Dynamic Measurements with FBG-Based Devices. *Sensors* **2017**, *17*, 2414. [CrossRef]
10. Dos Santos, P.S.S.; Jorge, P.A.S.; de Almeida, J.; Coelho, L. Low-Cost Interrogation System for Long-Period Fiber Gratings Applied to Remote Sensing. *Sensors* **2019**, *19*, 1500. [CrossRef]
11. Markvart, A.A.; Liokumovich, L.B.; Medvedev, I.O.; Ushakov, N.A. Smartphone-Based Interrogation of a Chirped FBG Strain Sensor Inscribed in a Multimode Fiber. *J. Light. Technol.* **2021**, *39*, 282–289. [CrossRef]
12. Tien, C.; Lin, H.; Su, S. High Sensitivity Refractive Index Sensor by D-Shaped Fibers and Titanium Dioxide Nanofilm. *Adv. Condens. Matter Phys.* **2018**, *2018*, 1–6. [CrossRef]
13. Wei, W.; Nong, J.; Zhu, Y.; Zhang, G.; Wang, N.; Luo, S.; Chen, N.; Lan, G.; Chuang, C.-J.; Huang, Y. Graphene/Au-Enhanced Plastic Clad Silica Fiber Optic Surface Plasmon Resonance Sensor. *Plasmonics* **2018**, *13*, 483–491. [CrossRef]
14. Cennamo, N.; Chiavaioli, F.; Trono, C.; Tombelli, S.; Giannetti, A.; Baldini, F.; Zeni, L. A Complete Optical Sensor System Based on a POF-SPR Platform and a Thermo-Stabilized Flow Cell for Biochemical Applications. *Sensors* **2016**, *16*, 196. [CrossRef]
15. Lomer, M.; Arrue, J.; Jauregui, C.; Aiestaran, P.; Zubia, J.; López-Higuera, J.M. Lateral Polishing of Bends in Plastic Optical Fibres Applied to a Multipoint Liquid-Level Measurement Sensor. *Sensors Actuators A Phys.* **2007**, *137*, 68–73. [CrossRef]
16. Perrone, G.; Vallan, A. A Low-Cost Optical Sensor for Noncontact Vibration Measurements. *IEEE Trans. Instrum. Meas.* **2009**, *58*, 1650–1656. [CrossRef]
17. Arcas, A.; Dutra, F.; Allil, R.; Werneck, M. Surface Plasmon Resonance and Bending Loss-Based U-Shaped Plastic Optical Fiber Biosensors. *Sensors* **2018**, *18*, 648. [CrossRef]
18. Cennamo, N.; Pasquardini, L.; Arcadio, F.; Vanzetti, L.E.; Bossi, A.M.; Zeni, L. D-Shaped Plastic Optical Fibre Aptasensor for Fast Thrombin Detection in Nanomolar Range. *Sci. Rep.* **2019**, *9*, 18740. [CrossRef]
19. Sypabekova, M.; Aitkulov, A.; Blanc, W.; Tosi, D. Reflector-Less Nanoparticles Doped Optical Fiber Biosensor for the Detection of Proteins: Case Thrombin. *Biosens. Bioelectron.* **2020**, *165*, 112365. [CrossRef]
20. Taffoni, F.; Formica, D.; Saccomandi, P.; Pino, G.; Schena, E. Optical Fiber-Based MR-Compatible Sensors for Medical Applications: An Overview. *Sensors* **2013**, *13*, 14105–14120. [CrossRef]
21. Leitão, C.; Leal-Junior, A.; Almeida, A.R.; Pereira, S.O.; Costa, F.M.; Pinto, J.L.; Marques, C. Cortisol AuPd Plasmonic Unclad POF Biosensor. *Biotechnol. Rep.* **2021**, *29*, e00587. [CrossRef]
22. Gowri, A.; Sai, V.V.R. Development of LSPR Based U-bent Plastic Optical Fiber Sensors. *Sensors Actuators B Chem.* **2016**, *230*, 536–543. [CrossRef]
23. Sai, V.V.R.; Kundu, T.; Mukherji, S. Novel U-bent Fiber Optic Probe for Localized Surface Plasmon Resonance Based Biosensor. *Biosens. Bioelectron.* **2009**, *24*, 2804–2809. [CrossRef]
24. Corres, J.M.; Bravo, J.; Matias, I.R.; Arregui, F.J. Tapered Optical Fiber Biosensor for the Detection of Anti-Gliadin Antibodies. In Proceedings of the 2007 IEEE Sensors, Atlanta, GA, USA, 28–31 October 2007; 2007; pp. 608–611.
25. Wen, H.-Y.; Huang, C.-W.; Li, Y.-L.; Chen, J.-L.; Yeh, Y.-T.; Chiang, C.-C. A Lamping U-Shaped Fiber Biosensor Detector for MicroRNA. *Sensors* **2020**, *20*, 1509. [CrossRef]
26. Tan, A.J.Y.; Ng, S.M.; Stoddart, P.R.; Chua, H.S. Theoretical Model and Design Considerations of U-Shaped Fiber Optic Sensors: A Review. *IEEE Sens. J.* **2020**, *20*, 14578–14589. [CrossRef]
27. Azkune, M.; Ruiz-Rubio, L.; Aldabaldetreku, G.; Arrospide, E.; Pérez-Álvarez, L.; Bikandi, I.; Zubia, J.; Vilas-Vilela, J. U-Shaped and Surface Functionalized Polymer Optical Fiber Probe for Glucose Detection. *Sensors* **2017**, *18*, 34. [CrossRef] [PubMed]
28. Yang, W.; Yu, J.; Xi, X.; Sun, Y.; Shen, Y.; Yue, W.; Zhang, C.; Jiang, S. Preparation of Graphene/ITO Nanorod Metamaterial/U-Bent-Annealing Fiber Sensor and DNA Biomolecule Detection. *Nanomaterials* **2019**, *9*, 1154. [CrossRef] [PubMed]
29. George, A.; Amrutha, M.S.; Srivastava, P.; Sunil, S.; Sai, V.V.R., Srinivasan, R. Development of a U-bent Plastic Optical Fiber Biosensor with Plasmonic Labels for the Detection of Chikungunya Non-Structural Protein 3. *Analyst* **2021**, *146*, 244–252. [CrossRef]
30. Gowri, A.; Rajamani, A.S.; Ramakrishna, B.; Sai, V.V.R. U-bent Plastic Optical Fiber Probes as Refractive Index Based Fat Sensor for Milk Quality Monitoring. *Opt. Fiber Technol.* **2019**, *47*, 15–20. [CrossRef]

31. Divagar, M.; Gowri, A.; John, S.; Sai, V.V.R. Graphene Oxide Coated U-bent Plastic Optical Fiber Based Chemical Sensor for Organic Solvents. *Sensors Actuators B Chem.* **2018**, *262*, 1006–1012. [CrossRef]
32. Chen, K.-C.; Li, Y.-L.; Wu, C.-W.; Chiang, C.-C. Glucose Sensor Using U-Shaped Optical Fiber Probe with Gold Nanoparticles and Glucose Oxidase. *Sensors* **2018**, *18*, 1217. [CrossRef]
33. Bandaru, R.; Divagar, M.; Khanna, S.; Danny, C.G.; Gupta, S.; Janakiraman, V.; Sai, V.V.R. U-bent Fiber Optic Plasmonic Biosensor Platform for Ultrasensitive Analyte Detection. *Sensors Actuators B Chem.* **2020**, *321*, 128463. [CrossRef]
34. Manoharan, H.; Kalita, P.; Gupta, S.; Sai, V.V.R. Plasmonic Biosensors for Bacterial Endotoxin Detection on Biomimetic C-18 Supported Fiber Optic Probes. *Biosens. Bioelectron.* **2019**, *129*, 79–86. [CrossRef]
35. Lu, P.; Men, L.; Sooley, K.; Chen, Q. Tapered Fiber Mach–Zehnder Interferometer for Simultaneous Measurement of Refractive Index and Temperature. *Appl. Phys. Lett.* **2009**, *94*, 131110. [CrossRef]
36. Ayupova, T.; Shaimerdenova, M.; Tosi, D. Shallow-Tapered Chirped Fiber Bragg Grating Sensors for Dual Refractive Index and Temperature Sensing. *Sensors* **2021**, *21*, 3635. [CrossRef]
37. Latifi, H.; Zibaii, M.I.; Hosseini, S.M.; Jorge, P. Nonadiabatic Tapered Optical Fiber for Biosensor Applications. *Photonic Sens.* **2012**, *2*, 340–356. [CrossRef]
38. Dash, S.P.; Patnaik, S.K.; Tripathy, S.K. Investigation of a Low Cost Tapered Plastic Fiber Optic Biosensor Based on Manipulation of Colloidal Gold Nanoparticles. *Opt. Commun.* **2019**, *437*, 388–391. [CrossRef]
39. Lin, H.-Y.; Huang, C.-H.; Cheng, G.-L.; Chen, N.-K.; Chui, H.-C. Tapered Optical Fiber Sensor Based on Localized Surface Plasmon Resonance. *Opt. Express* **2012**, *20*, 21693. [CrossRef]
40. Rahman, H.A.; Harun, S.W.; Yasin, M.; Phang, S.W.; Damanhuri, S.S.A.; Arof, H.; Ahmad, H. Tapered Plastic Multimode Fiber Sensor for Salinity Detection. *Sens. Actuators A Phys.* **2011**, *171*, 219–222. [CrossRef]
41. Pesavento; Profumo; Merli; Cucca; Zeni; Cennamo An Optical Fiber Chemical Sensor for the Detection of Copper(II) in Drinking Water. *Sensors* **2019**, *19*, 5246. [CrossRef]
42. Zheng, H.; Huang, B.; Li, Y.; Zhang, R.; Gu, X.; Li, Z.; Lin, H.; Zhu, W.; Tang, J.; Guan, H.; et al. Residual Thickness Enhanced Core-Removed D-Shaped Single-Mode Fiber and Its Application for VOC Evaporation Monitoring. *Opt. Express* **2020**, *28*, 15641. [CrossRef]
43. Sequeira, F.; Cennamo, N.; Rudnitskaya, A.; Nogueira, R.; Zeni, L.; Bilro, L. D-Shaped POF Sensors for Refractive Index Sensing—The Importance of Surface Roughness. *Sensors* **2019**, *19*, 2476. [CrossRef] [PubMed]
44. Ying, Y.; Si, G.; Luan, F.; Xu, K.; Qi, Y.; Li, H. Recent Research Progress of Optical Fiber Sensors Based on D-Shaped Structure. *Opt. Laser Technol.* **2017**, *90*, 149–157. [CrossRef]
45. Cennamo, N.; D'Agostino, G.; Pasquardini, L.; Arcadio, F.; Perri, C.; Coppola, N.; Angelillo, I.F.; Altucci, L.; Di Marzo, F.; Parisio, E.M.; et al. (INVITED)Quantitative Detection of SARS-CoV-2 Virions in Aqueous Mediums by IoT Optical Fiber Sensors. *Results Opt.* **2021**, *5*, 100177. [CrossRef]
46. Cennamo, N.; Pesavento, M.; Zeni, L. A Review on Simple and Highly Sensitive Plastic Optical Fiber Probes for Bio-Chemical Sensing. *Sens. Actuators B Chem.* **2021**, *331*, 129393. [CrossRef]
47. Cennamo, N.; Donà, A.; Pallavicini, P.; D'Agostino, G.; Dacarro, G.; Zeni, L.; Pesavento, M. Sensitive Detection of 2,4,6-Trinitrotoluene by Tridimensional Monitoring of Molecularly Imprinted Polymer with Optical Fiber and Five-Branched Gold Nanostars. *Sens. Actuators B Chem.* **2015**, *208*, 291–298. [CrossRef]
48. Cennamo, N.; D'Agostino, G.; Perri, C.; Arcadio, F.; Chiaretti, G.; Parisio, E.M.; Camarlinghi, G.; Vettori, C.; Di Marzo, F.; Cennamo, R.; et al. Proof of Concept for a Quick and Highly Sensitive On-Site Detection of SARS-CoV-2 by Plasmonic Optical Fibers and Molecularly Imprinted Polymers. *Sensors* **2021**, *21*, 1681. [CrossRef]
49. Pasquardini, L.; Cennamo, N.; Malleo, G.; Vanzetti, L.; Zeni, L.; Bonamini, D.; Salvia, R.; Bassi, C.; Bossi, A.M. A Surface Plasmon Resonance Plastic Optical Fiber Biosensor for the Detection of Pancreatic Amylase in Surgically-Placed Drain Effluent. *Sensors* **2021**, *21*, 3443. [CrossRef]
50. Loyez, M.; Lobry, M.; Hassan, E.M.; DeRosa, M.C.; Caucheteur, C.; Wattiez, R. HER2 Breast Cancer Biomarker Detection Using a Sandwich Optical Fiber Assay. *Talanta* **2021**, *221*, 121452. [CrossRef]
51. Ortega-Gomez, A.; Barroso, J.; Calatayud-Sánchez, A.; Zubia, J.; Benito-Lopez, F.; Basabe-Desmonts, L.; Villatoro, J. Cytochrome c Detection by Plasmonic Nanospectroscopy on Optical Fiber Facets. *Sensors Actuators B Chem.* **2021**, *330*, 129358. [CrossRef]
52. Xiong, Y.; Xu, F. Multifunctional Integration on Optical Fiber Tips: Challenges and Opportunities. *Adv. Photonics* **2020**, *2*, 064001. [CrossRef]
53. Lee, B.; Park, J.-H.; Byun, J.-Y.; Kim, J.H.; Kim, M.-G. An Optical Fiber-Based LSPR Aptasensor for Simple and Rapid in-Situ Detection of Ochratoxin A. *Biosens. Bioelectron.* **2018**, *102*, 504–509. [CrossRef]
54. Righini, G.; Soria, S. Biosensing by WGM Microspherical Resonators. *Sensors* **2016**, *16*, 905. [CrossRef]
55. Xu, X.; Chen, W.; Zhao, G.; Li, Y.; Lu, C.; Yang, L. Wireless Whispering-Gallery-Mode Sensor for Thermal Sensing and Aerial Mapping. *Light Sci. Appl.* **2018**, *7*, 62. [CrossRef]
56. Chiasera, A.; Dumeige, Y.; Féron, P.; Ferrari, M.; Jestin, Y.; Nunzi Conti, G.; Pelli, S.; Soria, S.; Righini, G.C. Spherical Whispering-Gallery-Mode Microresonators. *Laser Photon. Rev.* **2010**, *4*, 457–482. [CrossRef]
57. Shaimerdenova, M.; Ayupova, T.; Sypabekova, M.; Tosi, D. Fiber Optic Refractive Index Sensors Based on a Ball Resonator and Optical Backscatter Interrogation. *Sensors* **2020**, *20*, 6199. [CrossRef]

58. Sypabekova, M.; Korganbayev, S.; González-Vila, Á.; Caucheteur, C.; Shaimerdenova, M.; Ayupova, T.; Bekmurzayeva, A.; Vangelista, L.; Tosi, D. Functionalized Etched Tilted Fiber Bragg Grating Aptasensor for Label-Free Protein Detection. *Biosens. Bioelectron.* **2019**, *146*, 111765. [CrossRef]
59. Bekmurzayeva, A.; Dukenbayev, K.; Shaimerdenova, M.; Bekniyazov, I.; Ayupova, T.; Sypabekova, M.; Molardi, C.; Tosi, D. Etched Fiber Bragg Grating Biosensor Functionalized with Aptamers for Detection of Thrombin. *Sensors* **2018**, *18*, 4298. [CrossRef]
60. Aitkulov, A.; Sypabekova, M.; Molardi, C.; Blanc, W.; Tosi, D. Fabrication and Performance Evaluation of Reflectorless Refractive Index Fiber Optic Sensors Using Etched Enhanced Backscattering Fibers. *Measurement* **2021**, *172*, 108874. [CrossRef]
61. Cennamo, N.; Trigona, C.; Graziani, S.; Zeni, L.; Arcadio, F.; Xiaoyan, L.; Di Pasquale, G.; Pollicino, A. Green LSPR Sensors Based on Thin Bacterial Cellulose Waveguides for Disposable Biosensor Implementation. *IEEE Trans. Instrum. Meas.* **2021**, *70*, 1–8. [CrossRef]
62. Cennamo, N.; Saitta, L.; Tosto, C.; Arcadio, F.; Zeni, L.; Fragalá, M.E.; Cicala, G. Microstructured Surface Plasmon Resonance Sensor Based on Inkjet 3D Printing Using Photocurable Resins with Tailored Refractive Index. *Polymers* **2021**, *13*, 2518. [CrossRef]
63. Arcadio, F.; Zeni, L.; Montemurro, D.; Eramo, C.; Di Ronza, S.; Perri, C.; D'Agostino, G.; Chiaretti, G.; Porto, G.; Cennamo, N. Biochemical Sensing Exploiting Plasmonic Sensors Based on Gold Nanogratings and Polymer Optical Fibers. *Photonics Res.* **2021**, *9*, 1397. [CrossRef]
64. Zeni, L.; Pesavento, M.; Marchetti, S.; Cennamo, N. [INVITED] Slab Plasmonic Platforms Combined with Plastic Optical Fibers and Molecularly Imprinted Polymers for Chemical Sensing. *Opt. Laser Technol.* **2018**, *107*, 484–490. [CrossRef]
65. Cennamo, N.; Pasquardini, L.; Arcadio, F.; Lunelli, L.; Vanzetti, L.; Carafa, V.; Altucci, L.; Zeni, L. SARS-CoV-2 Spike Protein Detection through a Plasmonic D-Shaped Plastic Optical Fiber Aptasensor. *Talanta* **2021**, *233*, 122532. [CrossRef]
66. Cennamo, N.; Varriale, A.; Pennacchio, A.; Staiano, M.; Massarotti, D.; Zeni, L.; D'Auria, S. An Innovative Plastic Optical Fiber-Based Biosensor for New Bio/Applications. The Case of Celiac Disease. *Sens. Actuators B Chem.* **2013**, *176*, 1008–1014. [CrossRef]
67. Cennamo, N.; Mattiello, F.; Zeni, L. Slab Waveguide and Optical Fibers for Novel Plasmonic Sensor Configurations. *Sensors* **2017**, *17*, 1488. [CrossRef]
68. Arcadio, F.; Zeni, L.; Perri, C.; D'Agostino, G.; Chiaretti, G.; Porto, G.; Minardo, A.; Cennamo, N. Bovine Serum Albumin Protein Detection by a Removable SPR Chip Combined with a Specific MIP Receptor. *Chemosensors* **2021**, *9*, 218. [CrossRef]
69. Arcadio, F.; Zeni, L.; Minardo, A.; Eramo, C.; Di Ronza, S.; Perri, C.; D'Agostino, G.; Chiaretti, G.; Porto, G.; Cennamo, N. A Nanoplasmonic-Based Biosensing Approach for Wide-Range and Highly Sensitive Detection of Chemicals. *Nanomaterials* **2021**, *11*, 1961. [CrossRef]
70. Cennamo, N.; Zeni, L.; Catalano, E.; Arcadio, F.; Minardo, A. Refractive Index Sensing through Surface Plasmon Resonance in Light-Diffusing Fibers. *Appl. Sci.* **2018**, *8*, 1172. [CrossRef]
71. Cennamo, N.; Zeni, L.; Arcadio, F.; Catalano, E.; Minardo, A. A Novel Approach to Realizing Low-Cost Plasmonic Optical Fiber Sensors: Light-Diffusing Fibers Covered by Thin Metal Films. *Fibers* **2019**, *7*, 34. [CrossRef]
72. Cennamo, N.; Arcadio, F.; Prete, D.D.; Buonanno, G.; Minardo, A.; Pirozzi, S.; Zeni, L. A Simple and Efficient Plasmonic Sensor in Light Diffusive Polymer Fibers. *IEEE Sens. J.* **2021**, *21*, 16054–16060. [CrossRef]
73. Cennamo, N.; Trono, C.; Giannetti, A.; Baldini, F.; Minardo, A.; Zeni, L.; Tombelli, S. Biosensors Exploiting Unconventional Platforms: The Case of Plasmonic Light-Diffusing Fibers. *Sens. Actuators B Chem.* **2021**, *337*, 129771. [CrossRef]
74. Cennamo, N.; Arcadio, F.; Zeni, L.; Catalano, E.; Del Prete, D.; Buonanno, G.; Minardo, A. The Role of Tapered Light-Diffusing Fibers in Plasmonic Sensor Configurations. *Sensors* **2021**, *21*, 6333. [CrossRef] [PubMed]
75. Cennamo, N.; Coelho, L.; Santos, D.F.; Baptista, J.M.; Guerreiro, A.; Jorge, P.A.S.; Zeni, L. Modal Filtering for Optimized Surface Plasmon Resonance Sensing in Multimode Plastic Optical Fibers. *IEEE Sens. J.* **2015**, *15*, 6306–6312. [CrossRef]
76. Tosi, D.; Schena, E.; Molardi, C.; Korganbayev, S. Fiber Optic Sensors for Sub-Centimeter Spatially Resolved Measurements: Review and Biomedical Applications. *Opt. Fiber Technol.* **2018**, *43*, 6–19. [CrossRef]
77. Issatayeva, A.; Beisenova, A.; Tosi, D.; Molardi, C. Fiber-Optic Based Smart Textiles for Real-Time Monitoring of Breathing Rate. *Sensors* **2020**, *20*, 3408. [CrossRef] [PubMed]
78. Guo, T.; González-Vila, Á.; Loyez, M.; Caucheteur, C. Plasmonic Optical Fiber-Grating Immunosensing: A Review. *Sensors* **2017**, *17*, 2732. [CrossRef] [PubMed]
79. Poeggel, S.; Tosi, D.; Duraibabu, D.; Leen, G.; McGrath, D.; Lewis, E. Optical Fibre Pressure Sensors in Medical Applications. *Sensors* **2015**, *15*, 17115–17148. [CrossRef] [PubMed]
80. Liao, K.-C.; Chang, S.-C.; Chiu, C.-Y.; Chou, Y.-H. Acute Response in Vivo of a Fiber-Optic Sensor for Continuous Glucose Monitoring from Canine Studies on Point Accuracy. *Sensors* **2010**, *10*, 7789–7802. [CrossRef]
81. Issatayeva, A.; Amantayeva, A.; Blanc, W.; Tosi, D.; Molardi, C. Design and Analysis of a Fiber-Optic Sensing System for Shape Reconstruction of a Minimally Invasive Surgical Needle. *Sci. Rep.* **2021**, *11*, 8609. [CrossRef]
82. Loyez, M.; Larrieu, J.-C.; Chevineau, S.; Remmelink, M.; Leduc, D.; Bondue, B.; Lambert, P.; Devière, J.; Wattiez, R.; Caucheteur, C. In Situ Cancer Diagnosis through Online Plasmonics. *Biosens. Bioelectron.* **2019**, *131*, 104–112. [CrossRef]
83. Poeggel, S.; Duraibabu, D.; Tosi, D.; Leen, G.; Lewis, E.; McGrath, D.; Fusco, F.; Sannino, S.; Lupoli, L.; Ippolito, J.; et al. Differential in Vivo Urodynamic Measurement in a Single Thin Catheter Based on Two Optical Fiber Pressure Sensors. *J. Biomed. Opt.* **2015**, *20*, 037005. [CrossRef]

84. Guo, T.; Liu, F.; Liang, X.; Qiu, X.; Huang, Y.; Xie, C.; Xu, P.; Mao, W.; Guan, B.-O.; Albert, J. Highly Sensitive Detection of Urinary Protein Variations Using Tilted Fiber Grating Sensors with Plasmonic Nanocoatings. *Biosens. Bioelectron.* **2016**, *78*, 221–228. [CrossRef]
85. Liao, K.-C.; Hogen-Esch, T.; Richmond, F.J.; Marcu, L.; Clifton, W.; Loeb, G.E. Percutaneous Fiber-Optic Sensor for Chronic Glucose Monitoring in Vivo. *Biosens. Bioelectron.* **2008**, *23*, 1458–1465. [CrossRef]
86. Beisenova, A.; Issatayeva, A.; Iordachita, I.; Blanc, W.; Molardi, C.; Tosi, D. Distributed Fiber Optics 3D Shape Sensing by Means of High Scattering NP-Doped Fibers Simultaneous Spatial Multiplexing. *Opt. Express* **2019**, *27*, 22074. [CrossRef]
87. Rantala, J.; Hännikäinen, J.; Vanhala, J. Fiber Optic Sensors for Wearable Applications. *Pers. Ubiquitous Comput.* **2011**, *15*, 85–96. [CrossRef]
88. Lo Presti, D.; Massaroni, C.; D'Abbraccio, J.; Massari, L.; Caponero, M.; Longo, U.G.; Formica, D.; Oddo, C.M.; Schena, E. Wearable System Based on Flexible FBG for Respiratory and Cardiac Monitoring. *IEEE Sens. J.* **2019**, *19*, 7391–7398. [CrossRef]
89. Li, H.; Yang, H.; Li, E.; Liu, Z.; Wei, K. Wearable Sensors in Intelligent Clothing for Measuring Human Body Temperature Based on Optical Fiber Bragg Grating. *Opt. Express* **2012**, *20*, 11740. [CrossRef]
90. Esmaeilzadeh, H.; Rivard, M.; Arzi, E.; Légaré, F.; Hassani, A. Smart Textile Plasmonic Fiber Dew Sensors. *Opt. Express* **2015**, *23*, 14981. [CrossRef]
91. Merazzo, K.J.; Totoricaguena-Gorriño, J.; Fernández-Martín, E.; del Campo, F.J.; Baldrich, E. Smartphone-Enabled Personalized Diagnostics: Current Status and Future Prospects. *Diagnostics* **2021**, *11*, 1067. [CrossRef]
92. de Haan, K.; Ceylan Koydemir, H.; Rivenson, Y.; Tseng, D.; Van Dyne, E.; Bakic, L.; Karinca, D.; Liang, K.; Ilango, M.; Gumustekin, E.; et al. Automated Screening of Sickle Cells Using a Smartphone-Based Microscope and Deep Learning. *NPJ Digit. Med.* **2020**, *3*, 76. [CrossRef]
93. Hernández-Neuta, I.; Neumann, F.; Brightmeyer, J.; Ba Tis, T.; Madaboosi, N.; Wei, Q.; Ozcan, A.; Nilsson, M. Smartphone-based Clinical Diagnostics: Towards Democratization of Evidence-based Health Care. *J. Intern. Med.* **2019**, *285*, 19–39. [CrossRef] [PubMed]
94. McCracken, K.E.; Tat, T.; Paz, V.; Yoon, J.-Y. Smartphone-Based Fluorescence Detection of Bisphenol A from Water Samples. *RSC Adv.* **2017**, *7*, 9237–9243. [CrossRef]
95. Severi, C.; Melnychuk, N.; Klymchenko, A.S. Smartphone-Assisted Detection of Nucleic Acids by Light-Harvesting FRET-Based Nanoprobe. *Biosens. Bioelectron.* **2020**, *168*, 112515. [CrossRef] [PubMed]
96. Khachornsakkul, K.; Dungchai, W. Development of an Ultrasound-Enhanced Smartphone Colorimetric Biosensor for Ultrasensitive Hydrogen Peroxide Detection and Its Applications. *RSC Adv.* **2020**, *10*, 24463–24471. [CrossRef]
97. Morosanova, M.A.; Bashkatova, A.S.; Morosanova, E.I. Spectrophotometric and Smartphone-Assisted Determination of Phenolic Compounds Using Crude Eggplant Extract. *Molecules* **2019**, *24*, 4407. [CrossRef]
98. Geng, Z.; Zhang, X.; Fan, Z.; Lv, X.; Su, Y.; Chen, H. Recent Progress in Optical Biosensors Based on Smartphone Platforms. *Sensors* **2017**, *17*, 2449. [CrossRef]
99. Hussain, I.; Das, M.; Ahamad, K.U.; Nath, P. Water Salinity Detection Using a Smartphone. *Sens. Actuators B Chem.* **2017**, *239*, 1042–1050. [CrossRef]
100. Saini, S.S.; Sridhar, A.; Ahluwalia, K. Smartphone Optical Sensors. *Opt. Photonics News* **2019**, *30*, 34–41. [CrossRef]
101. Sultangazin, A.; Kusmangaliyev, J.; Aitkulov, A.; Akilbekova, D.; Olivero, M.; Tosi, D. Design of a Smartphone Plastic Optical Fiber Chemical Sensor for Hydrogen Sulfide Detection. *IEEE Sens. J.* **2017**, *17*, 6935–6940. [CrossRef]
102. Aitkulov, A.; Tosi, D. Optical Fiber Sensor Based on Plastic Optical Fiber and Smartphone for Measurement of the Breathing Rate. *IEEE Sens. J.* **2019**, *19*, 3282–3287. [CrossRef]
103. Aitkulov, A.; Tosi, D. Design of an All-POF-Fiber Smartphone Multichannel Breathing Sensor with Camera-Division Multiplexing. *IEEE Sensors Lett.* **2019**, *3*, 1–4. [CrossRef]
104. Bremer, K.; Roth, B. Fibre Optic Surface Plasmon Resonance Sensor System Designed for Smartphones. *Opt. Express* **2015**, *23*, 17179. [CrossRef]
105. Walter, J.-G.; Alwis, L.S.M.; Roth, B.; Bremer, K. All-Optical Planar Polymer Waveguide-Based Biosensor Chip Designed for Smartphone-Assisted Detection of Vitamin D. *Sensors* **2020**, *20*, 6771. [CrossRef]
106. Markvart, A.; Liokumovich, L.; Medvedev, I.; Ushakov, N. Continuous Hue-Based Self-Calibration of a Smartphone Spectrometer Applied to Optical Fiber Fabry-Perot Sensor Interrogation. *Sensors* **2020**, *20*, 6304. [CrossRef]
107. Pan, T.; Cao, W.; Wang, M. TiO2 Thin Film Temperature Sensor Monitored by Smartphone. *Opt. Fiber Technol.* **2018**, *45*, 359–362. [CrossRef]
108. Kamizi, M.A.; Negri, L.H.; Fabris, J.L.; Muller, M. A Smartphone Based Fiber Sensor for Recognizing Walking Patterns. *IEEE Sens. J.* **2019**, *19*, 9782–9789. [CrossRef]
109. Liu, T.; Wang, W.; Jian, D.; Li, J.; Ding, H.; Yi, D.; Liu, F.; Wang, S. Quantitative Remote and On-Site Hg2+ Detection Using the Handheld Smartphone Based Optical Fiber Fluorescence Sensor (SOFFS). *Sens. Actuators B Chem.* **2019**, *301*, 127168. [CrossRef]
110. Liu, T.; Wang, W.; Ding, H.; Yi, D. Smartphone-Based Hand-Held Optical Fiber Fluorescence Sensor for On-Site PH Detection. *IEEE Sens. J.* **2019**, *19*, 9441–9446. [CrossRef]
111. Liu, T.; Wang, W.; Ding, H.; Liu, Z.; Zhang, S.; Yi, D. Development of a Handheld Dual-Channel Optical Fiber Fluorescence Sensor Based on a Smartphone. *Appl. Opt.* **2020**, *59*, 601. [CrossRef]

112. Peltomaa, R.; Glahn-Martínez, B.; Benito-Peña, E.; Moreno-Bondi, M. Optical Biosensors for Label-Free Detection of Small Molecules. *Sensors* **2018**, *18*, 4126. [CrossRef]
113. Yu, B.; Huang, Y.; Zhou, J.; Guo, T.; Guan, B.-O. Understanding the PH-Dependent Interaction between Graphene Oxide and Single-Stranded DNA through a Fiber-Optic Interferometer. *Phys. Chem. Chem. Phys.* **2016**, *18*, 32266–32271. [CrossRef]
114. Aebersold, R.; Anderson, L.; Caprioli, R.; Druker, B.; Hartwell, L.; Smith, R. Perspective: A Program to Improve Protein Biomarker Discovery for Cancer. *J. Proteome Res.* **2005**, *4*, 1104–1109. [CrossRef]
115. Chen, X.-H.; Huang, S.; Kerr, D. Biomarkers in Clinical Medicine. *IARC Sci. Publ.* **2011**, *163*, 303–322.
116. Mass, R.D.; Press, M.F.; Anderson, S.; Cobleigh, M.A.; Vogel, C.L.; Dybdal, N.; Leiberman, G.; Slamon, D.J. Evaluation of Clinical Outcomes According to HER2 Detection by Fluorescence In Situ Hybridization in Women with Metastatic Breast Cancer Treated with Trastuzumab. *Clin. Breast Cancer* **2005**, *6*, 240–246. [CrossRef]
117. Martinez, P.; Hernández-Losa, J.; Cedrés, S.; Castellví, J.; Martinez-Marti, A.; Tallada, N.; Murtra-Garrell, N.; Navarro-Mendivill, A.; Rodriguez-Freixinos, V.; Canela, M.; et al. Fluorescence In Situ Hybridization and Immunohistochemistry as Diagnostic Methods for ALK Positive Non-Small Cell Lung Cancer Patients. *PLoS ONE* **2013**, *8*, e52261. [CrossRef]
118. Sieben, V.J.; Debes Marun, C.S.; Pilarski, P.M.; Kaigala, G.V.; Pilarski, L.M.; Backhouse, C.J. FISH and Chips: Chromosomal Analysis on Microfluidic Platforms. *IET Nanobiotechnol.* **2007**, *1*, 27. [CrossRef]
119. Ramos-Vara, J.A. Technical Aspects of Immunohistochemistry. *Vet. Pathol.* **2005**, *42*, 405–426. [CrossRef]
120. Mukundan, H.; Kubicek, J.Z.; Holt, A.; Shively, J.E.; Martinez, J.S.; Grace, K.; Grace, W.K.; Swanson, B.I. Planar Optical Waveguide-Based Biosensor for the Quantitative Detection of Tumor Markers. *Sens. Actuators B Chem.* **2009**, *138*, 453–460. [CrossRef]
121. Ribaut, C.; Loyez, M.; Larrieu, J.-C.; Chevineau, S.; Lambert, P.; Remmelink, M.; Wattiez, R.; Caucheteur, C. Cancer Biomarker Sensing Using Packaged Plasmonic Optical Fiber Gratings: Towards in Vivo Diagnosis. *Biosens. Bioelectron.* **2017**, *92*, 449–456. [CrossRef]
122. Loyez, M.; Lobry, M.; Wattiez, R.; Caucheteur, C. Optical Fiber Gratings Immunoassays. *Sensors* **2019**, *19*, 2595. [CrossRef]
123. Ribaut, C.; Voisin, V.; Malachovská, V.; Dubois, V.; Mégret, P.; Wattiez, R.; Caucheteur, C. Small Biomolecule Immunosensing with Plasmonic Optical Fiber Grating Sensor. *Biosens. Bioelectron.* **2016**, *77*, 315–322. [CrossRef] [PubMed]
124. Sun, D.; Ran, Y.; Wang, G. Label-Free Detection of Cancer Biomarkers Using an In-Line Taper Fiber-Optic Interferometer and a Fiber Bragg Grating. *Sensors* **2017**, *17*, 2559. [CrossRef] [PubMed]
125. Ross, J.S.; Fletcher, J.A.; Linette, G.P.; Stec, J.; Clark, E.; Ayers, M.; Symmans, W.F.; Pusztai, L.; Bloom, K.J. The HER-2/ Neu Gene and Protein in Breast Cancer 2003: Biomarker and Target of Therapy. *Oncologist* **2003**, *8*, 307–325. [CrossRef]
126. Doi, T.; Shitara, K.; Naito, Y.; Shimomura, A.; Fujiwara, Y.; Yonemori, K.; Shimizu, C.; Shimoi, T.; Kuboki, Y.; Matsubara, N.; et al. Safety, Pharmacokinetics, and Antitumour Activity of Trastuzumab Deruxtecan (DS-8201), a HER2-Targeting Antibody–Drug Conjugate, in Patients with Advanced Breast and Gastric or Gastro-Oesophageal Tumours: A Phase 1 Dose-Escalation Study. *Lancet Oncol.* **2017**, *18*, 1512–1522. [CrossRef]
127. Basakran, N.S. CD44 as a Potential Diagnostic Tumor Marker. *Saudi Med. J.* **2015**, *36*, 273–279. [CrossRef] [PubMed]
128. Ristamäki, R.; Joensuu, H.; Jalkanen, S. Serum CD44 in Non-Hodgkin's Lymphoma. *Leuk. Lymphoma* **1999**, *33*, 433–440. [CrossRef] [PubMed]
129. Wang, S.J.; Wong, G.; de Heer, A.-M.; Xia, W.; Bourguignon, L.Y.W. CD44 Variant Isoforms in Head and Neck Squamous Cell Carcinoma Progression. *Laryngoscope* **2009**, *119*, 1518–1530. [CrossRef]
130. Yan, Y.; Zuo, X.; Wei, D. Concise Review: Emerging Role of CD44 in Cancer Stem Cells: A Promising Biomarker and Therapeutic Target. *Stem Cells Transl. Med.* **2015**, *4*, 1033–1043. [CrossRef]
131. Bekmurzayeva, A.; Ashikbayeva, Z.; Myrkhiyeva, Z.; Nugmanova, A.; Shaimerdenova, M.; Ayupova, T.; Tosi, D. Label-Free Fiber-Optic Spherical Tip Biosensor to Enable Picomolar-Level Detection of CD44 Protein. *Sci. Rep.* **2021**, *11*, 19583. [CrossRef]
132. Davies, L.; Welch, H.G. Current Thyroid Cancer Trends in the United States. *JAMA Otolaryngol. Neck Surg.* **2014**, *140*, 317. [CrossRef]
133. Benbassat, C.A.; Mechlis-Frish, S.; Guttmann, H.; Glaser, B.; Krausz, Y. Current Concepts in the Follow-up of Patients with Differentiated Thyroid Cancer. *Isr. Med. Assoc. J.* **2007**, *9*, 540–545.
134. McLeod, D.S.A. Current Concepts and Future Directions in Differentiated Thyroid Cancer. *Clin. Biochem. Rev.* **2010**, *31*, 9–19.
135. Eustatia-Rutten, C.F.A.; Smit, J.W.A.; Romijn, J.A.; van der Kleij-Corssmit, E.P.M.; Pereira, A.M.; Stokkel, M.P.; Kievit, J. Diagnostic Value of Serum Thyroglobulin Measurements in the Follow-up of Differentiated Thyroid Carcinoma, a Structured Meta-Analysis. *Clin. Endocrinol.* **2004**, *61*, 61–74. [CrossRef]
136. Pellegriti, G.; Frasca, F.; Regalbuto, C.; Squatrito, S.; Vigneri, R. Worldwide Increasing Incidence of Thyroid Cancer: Update on Epidemiology and Risk Factors. *J. Cancer Epidemiol.* **2013**, *2013*, 965212. [CrossRef]
137. Giovanella, L.; Clark, P.M.; Chiovato, L.; Duntas, L.; Elisei, R.; Feldt-Rasmussen, U.; Leenhardt, L.; Luster, M.; Schalin-Jäntti, C.; Schott, M.; et al. DIAGNOSIS OF ENDOCRINE DISEASE: Thyroglobulin Measurement Using Highly Sensitive Assays in Patients with Differentiated Thyroid Cancer: A Clinical Position Paper. *Eur. J. Endocrinol.* **2014**, *171*, R33–R46. [CrossRef]
138. Kim, H.-M.; Jeong, D.H.; Lee, H.-Y.; Park, J.-H.; Lee, S.-K. Design and Validation of Fiber Optic Localized Surface Plasmon Resonance Sensor for Thyroglobulin Immunoassay with High Sensitivity and Rapid Detection. *Sci. Rep.* **2021**, *11*, 15985. [CrossRef]

139. Man, Y.; Cao, J.; Jin, S.; Xu, G.; Pan, B.; Shang, L.; Che, D.; Yu, Q.; Yu, Y. Newly Identified Biomarkers for Detecting Circulating Tumor Cells in Lung Adenocarcinoma. *Tohoku J. Exp. Med.* **2014**, *234*, 29–40. [CrossRef]
140. Sharma, S.V.; Bell, D.W.; Settleman, J.; Haber, D.A. Epidermal Growth Factor Receptor Mutations in Lung Cancer. *Nat. Rev. Cancer* **2007**, *7*, 169–181. [CrossRef]
141. Schiller, J.H.; Harrington, D.; Belani, C.P.; Langer, C.; Sandler, A.; Krook, J.; Zhu, J.; Johnson, D.H. Comparison of Four Chemotherapy Regimens for Advanced Non–Small-Cell Lung Cancer. *N. Engl. J. Med.* **2002**, *346*, 92–98. [CrossRef]
142. Omary, M.B.; Ku, N.O.N. Intermediate Filament Proteins of the Liver: Emerging Disease Association and Functions. *Hepatology* **1997**, *25*, 1043–1048. [CrossRef]
143. Chu, P.G.; Weiss, L.M. Keratin Expression in Human Tissues and Neoplasms. *Histopathology* **2002**, *40*, 403–439. [CrossRef] [PubMed]
144. DePianto, D.; Kerns, M.L.; Dlugosz, A.A.; Coulombe, P.A. Keratin 17 Promotes Epithelial Proliferation and Tumor Growth by Polarizing the Immune Response in Skin. *Nat. Genet.* **2010**, *42*, 910–914. [CrossRef] [PubMed]
145. WHO. *World Health Statistics 2019: Monitoring Health for the SDGs, Sustainable Development Goals*; WHO: Geneva, Switzerland, 2019.
146. Piepoli, M.F.; Hoes, A.W.; Agewall, S.; Albus, C.; Brotons, C.; Catapano, A.L.; Cooney, M.-T.; Corrà, U.; Cosyns, B.; Deaton, C.; et al. 2016 European Guidelines on Cardiovascular Disease Prevention in Clinical Practice. *Eur. Heart J.* **2016**, *37*, 2315–2381. [CrossRef] [PubMed]
147. Visseren, F.L.J.; Mach, F.; Smulders, Y.M.; Carballo, D.; Koskinas, K.C.; Bäck, M.; Benetos, A.; Biffi, A.; Boavida, J.-M.; Capodanno, D.; et al. 2021 ESC Guidelines on Cardiovascular Disease Prevention in Clinical Practice. *Eur. Heart J.* **2021**, *42*, 3227–3337. [CrossRef] [PubMed]
148. Ouyang, M.; Tu, D.; Tong, L.; Sarwar, M.; Bhimaraj, A.; Li, C.; Coté, G.L.; Di Carlo, D. A Review of Biosensor Technologies for Blood Biomarkers toward Monitoring Cardiovascular Diseases at the Point-of-Care. *Biosens. Bioelectron.* **2021**, *171*, 112621. [CrossRef]
149. Leitão, C.; Ribau, V.; Afreixo, V.; Antunes, P.; André, P.; Pinto, J.L.; Boutouyrie, P.; Laurent, S.; Bastos, J.M. Clinical Evaluation of an Optical Fiber-Based Probe for the Assessment of Central Arterial Pulse Waves. *Hypertens. Res.* **2018**, *41*, 904–912. [CrossRef]
150. Kumar, S.; Singh, R.; Kaushik, B.K.; Chen, N.; Yang, Q.S.; Zhang, X. LSPR-Based Cholesterol Biosensor Using Hollow Core Fiber Structure. *IEEE Sens. J.* **2019**, *19*, 7399–7406. [CrossRef]
151. Zhou, W.; Li, K.; Wei, Y.; Hao, P.; Chi, M.; Liu, Y.; Wu, Y. Ultrasensitive Label-Free Optical Microfiber Coupler Biosensor for Detection of Cardiac Troponin I Based on Interference Turning Point Effect. *Biosens. Bioelectron.* **2018**, *106*, 99–104. [CrossRef]
152. Krupin, O.; Berini, P. Long-Range Surface Plasmon-Polariton Waveguide Biosensors for Human Cardiac Troponin I Detection. *Sensors* **2019**, *19*, 631. [CrossRef]
153. Liu, T.; Liang, L.-L.; Xiao, P.; Sun, L.-P.; Huang, Y.-Y.; Ran, Y.; Jin, L.; Guan, B.-O. A Label-Free Cardiac Biomarker Immunosensor Based on Phase-Shifted Microfiber Bragg Grating. *Biosens. Bioelectron.* **2018**, *100*, 155–160. [CrossRef]
154. Botewad, S.N.; Pahurkar, V.G.; Muley, G.G.; Gaikwad, D.K.; Bodkhe, G.A.; Shirsat, M.D.; Pawar, P.P. PANI-ZnO Cladding-Modified Optical Fiber Biosensor for Urea Sensing Based on Evanescent Wave Absorption. *Front. Mater.* **2020**, *7*, 184. [CrossRef]
155. Yang, Q.; Zhang, X.; Kumar, S.; Singh, R.; Zhang, B.; Bai, C.; Pu, X. Development of Glucose Sensor Using Gold Nanoparticles and Glucose-Oxidase Functionalized Tapered Fiber Structure. *Plasmonics* **2020**, *15*, 841–848. [CrossRef]
156. Zheng, W.; Han, B.; Siyu, E.; Sun, Y.; Li, X.; Cai, Y.; Zhang, Y. Highly-Sensitive and Reflective Glucose Sensor Based on Optical Fiber Surface Plasmon Resonance. *Microchem. J.* **2020**, *157*, 105010. [CrossRef]
157. Jeremias, A. The Utility of Troponin Measurement to Detect Myocardial Infarction: Review of the Current Findings. *Vasc. Health Risk Manag.* **2010**, *6*, 691. [CrossRef]
158. Boeddinghaus, J.; Reichlin, T.; Nestelberger, T.; Twerenbold, R.; Meili, Y.; Wildi, K.; Hillinger, P.; Giménez, M.R.; Cupa, J.; Schumacher, L.; et al. Early Diagnosis of Acute Myocardial Infarction in Patients with Mild Elevations of Cardiac Troponin. *Clin. Res. Cardiol.* **2017**, *106*, 457–467. [CrossRef]
159. Al-Naher, A.; Wright, D.; Devonald, M.A.J.; Pirmohamed, M. Renal Function Monitoring in Heart Failure-What Is the Optimal Frequency? A Narrative Review. *Br. J. Clin. Pharmacol.* **2018**, *84*, 5–17. [CrossRef]
160. Ferguson, M.A.; Waikar, S.S. Established and Emerging Markers of Kidney Function. *Clin. Chem.* **2012**, *58*, 680–689. [CrossRef]
161. Li, M.; Singh, R.; Marques, C.; Zhang, B.; Kumar, S. 2D Material Assisted SMF-MCF-MMF-SMF Based LSPR Sensor for Creatinine Detection. *Opt. Express* **2021**, *29*, 38150. [CrossRef]
162. Vogt, S.; Ruppert, V.; Pankuweit, S.; Paletta, J.P.J.; Rhiel, A.; Weber, P.; Irqsusi, M.; Cybulski, P.; Ramzan, R. Myocardial Insufficiency Is Related to Reduced Subunit 4 Content of Cytochrome c Oxidase. *J. Cardiothorac. Surg.* **2018**, *13*, 95. [CrossRef]
163. Dar, T.; Radfar, A.; Abohashem, S.; Pitman, R.K.; Tawakol, A.; Osborne, M.T. Psychosocial Stress and Cardiovascular Disease. *Curr. Treat. Options Cardiovasc. Med.* **2019**, *21*, 23. [CrossRef]
164. Osborne, M.T.; Shin, L.M.; Mehta, N.N.; Pitman, R.K.; Fayad, Z.A.; Tawakol, A. Disentangling the Links between Psychosocial Stress and Cardiovascular Disease. *Circ. Cardiovasc. Imaging* **2020**, *13*, 10931. [CrossRef]
165. Pulopulos, M.M.; Vanderhasselt, M.-A.; De Raedt, R. Association between Changes in Heart Rate Variability during the Anticipation of a Stressful Situation and the Stress-Induced Cortisol Response. *Psychoneuroendocrinology* **2018**, *94*, 63–71. [CrossRef]
166. Usha, S.P.; Gupta, B.D. Fiber Optic SPR Based P-Cresol Sensor Using Ag/ZnO Nanoparticle-Chitosan/Tyrosinanse. In Proceedings of the JSAP-OSA Joint Symposia 2015; Optical Society of America: Nagoya, Japan, 2015; p. 13p_2C_12.

167. Gupta, V. Removal of Lindane and Malathion from Wastewater Using Bagasse Fly Ash—a Sugar Industry Waste. *Water Res.* **2002**, *36*, 2483–2490. [CrossRef]
168. Gupta, V.K.; Tyagi, I.; Agarwal, S.; Singh, R.; Chaudhary, M.; Harit, A.; Kushwaha, S. Column Operation Studies for the Removal of Dyes and Phenols Using a Low Cost Adsorbent. *Glob. J. Environ. Sci. Manag.* **2016**, *2*, 1–10. [CrossRef]
169. Michałowicz, J.; Duda, W. Phenols—Sources and Toxicity. *Polish J. Environ. Stud.* **2007**, *16*, 347–362.
170. King, R.A.; May, B.L.; Davies, D.A.; Bird, A.R. Measurement of Phenol and P-Cresol in Urine and Feces Using Vacuum Microdistillation and High-Performance Liquid Chromatography. *Anal. Biochem.* **2009**, *384*, 27–33. [CrossRef]
171. Usha, S.P.; Gupta, B.D. Urinary P-Cresol Diagnosis Using Nanocomposite of ZnO/MoS2 and Molecular Imprinted Polymer on Optical Fiber Based Lossy Mode Resonance Sensor. *Biosens. Bioelectron.* **2018**, *101*, 135–145. [CrossRef]
172. Li, Y.-F.; Liu, Z.-M.; Liu, Y.-L.; Yang, Y.-H.; Shen, G.-L.; Yu, R.-Q. A Mediator-Free Phenol Biosensor Based on Immobilizing Tyrosinase to ZnO Nanoparticles. *Anal. Biochem.* **2006**, *349*, 33–40. [CrossRef] [PubMed]
173. Adamski, J.; Nowak, P.; Kochana, J. Simple Sensor for the Determination of Phenol and Its Derivatives in Water Based on Enzyme Tyrosinase. *Electrochim. Acta* **2010**, *55*, 2363–2367. [CrossRef]
174. Wang, Y.; Zhu, G.; Li, M.; Singh, R.; Marques, C.; Min, R.; Kaushik, B.K.; Zhang, B.; Jha, R.; Kumar, S. Water Pollutants P-Cresol Detection Based on Au-ZnO Nanoparticles Modified Tapered Optical Fiber. *IEEE Trans. Nanobioscience* **2021**, *20*, 377–384. [CrossRef] [PubMed]
175. Cennamo, N.; Zeni, L.; Ricca, E.; Isticato, R.; Marzullo, V.M.; Capo, A.; Staiano, M.; D'Auria, S.; Varriale, A. Detection of Naphthalene in Sea-Water by a Label-Free Plasmonic Optical Fiber Biosensor. *Talanta* **2019**, *194*, 289–297. [CrossRef]
176. Lamarca, R.S.; Franco, D.F.; Nalin, M.; de Lima Gomes, P.C.F.; Messaddeq, Y. Label-Free Ultrasensitive and Environment-Friendly Immunosensor Based on a Silica Optical Fiber for the Determination of Ciprofloxacin in Wastewater Samples. *Anal. Chem.* **2020**, *92*, 14415–14422. [CrossRef]
177. Gao, R.; Lu, D.-F.; Zhang, M.-Y.; Qi, Z.-M. Optofluidic Immunosensor Based on Resonant Wavelength Shift of a Hollow Core Fiber for Ultratrace Detection of Carcinogenic Benzo[a]Pyrene. *ACS Photonics* **2018**, *5*, 1273–1280. [CrossRef]
178. González-Vila, Á.; Debliquy, M.; Lahem, D.; Zhang, C.; Mégret, P.; Caucheteur, C. Molecularly Imprinted Electropolymerization on a Metal-Coated Optical Fiber for Gas Sensing Applications. *Sensors Actuators B Chem.* **2017**, *244*, 1145–1151. [CrossRef]
179. Ke, Z.-J.; Tang, D.-L.; Lai, X.; Dai, Z.-Y.; Zhang, Q. Optical Fiber Evanescent-Wave Sensing Technology of Hydrogen Sulfide Gas Concentration in Oil and Gas Fields. *Optik* **2018**, *157*, 1094–1100. [CrossRef]
180. Prado, A.R.; Díaz, C.A.R.; Lyra Nunes, L.G.; Oliveira, J.P.; Guimarães, M.C.C.; Leal-Junior, A.; Ribeiro, M.R.N.; Pontes, M.J. Surface Plasmon Resonance-Based Optical Fiber Sensors for H2S In Situ Detection. *Plasmonics* **2021**, *16*, 787–797. [CrossRef]
181. Chu, C.-S.; Lo, Y.-L. Fiber-Optic Carbon Dioxide Sensor Based on Fluorinated Xerogels Doped with HPTS. *Sens. Actuators B Chem.* **2008**, *129*, 120–125. [CrossRef]
182. Shanavas, S.; Ahamad, T.; Alshehri, S.M.; Acevedo, R.; Anbarasan, P.M. A Facile Microwave Route for Fabrication of NiO/RGO Hybrid Sensor with Efficient CO2 and Acetone Gas Sensing Performance Using Clad Modified Fiber Optic Method. *Optik* **2021**, *226*, 165970. [CrossRef]
183. Allsop, T.; Arif, R.; Neal, R.; Kalli, K.; Kundrát, V.; Rozhin, A.; Culverhouse, P.; Webb, D.J. Photonic Gas Sensors Exploiting Directly the Optical Properties of Hybrid Carbon Nanotube Localized Surface Plasmon Structures. *Light Sci. Appl.* **2016**, *5*, e16036. [CrossRef]
184. Pinto, V.; Sousa, P.; Catarino, S.O.; Correia-Neves, M.; Minas, G. Microfluidic Immunosensor for Rapid and Highly-Sensitive Salivary Cortisol Quantification. *Biosens. Bioelectron.* **2017**, *90*, 308–313. [CrossRef]
185. Holsboer, F.; Ising, M. Stress Hormone Regulation: Biological Role and Translation into Therapy. *Annu. Rev. Psychol.* **2010**, *61*, 81–109. [CrossRef]
186. Usha, S.P.; Shrivastav, A.M.; Gupta, B.D. A Contemporary Approach for Design and Characterization of Fiber-Optic-Cortisol Sensor Tailoring LMR and ZnO/PPY Molecularly Imprinted Film. *Biosens. Bioelectron.* **2017**, *87*, 178–186. [CrossRef]
187. Sharma, A.K.; Kaur, B.; Marques, C. Simulation and Analysis of 2D Material/Metal Carbide Based Fiber Optic SPR Probe for Ultrasensitive Cortisol Detection. *Optik* **2020**, *218*, 164891. [CrossRef]
188. Randall, D.; Tsui, T.K. Ammonia Toxicity in Fish. *Mar. Pollut. Bull.* **2002**, *45*, 17–23. [CrossRef]
189. Zhu, Y.; Fu, H.; Ding, J.; Li, H.; Zhang, M.; Zhang, J.; Liu, Y. Fabrication of Three-Dimensional Zinc Oxide Nanoflowers for High-Sensitivity Fiber-Optic Ammonia Gas Sensors. *Appl. Opt.* **2018**, *57*, 7924. [CrossRef] [PubMed]
190. Shrivastav, A.M.; Sharma, G.; Rathore, A.S.; Jha, R. Hypersensitive and Selective Interferometric Nose for Ultratrace Ammonia Detection with Fast Response Utilizing PANI@SnO 2 Nanocomposite. *ACS Photonics* **2018**, *5*, 4402–4412. [CrossRef]
191. Leal-Junior, A.G.; Frizera, A.; Marques, C. Low-Cost Fiberoptic Probe for Ammonia Early Detection in Fish Farms. *Remote Sens.* **2020**, *12*, 1439. [CrossRef]
192. Miliutina, E.; Guselnikova, O.; Burtsev, V.; Elashnikov, R.; Postnikov, P.; Svorcik, V.; Lyutakov, O. Plasmon-Active Optical Fiber Functionalized by Metal Organic Framework for Pesticide Detection. *Talanta* **2020**, *208*, 120480. [CrossRef]
193. Kant, R. Surface Plasmon Resonance Based Fiber–Optic Nanosensor for the Pesticide Fenitrothion Utilizing Ta2O5 Nanostructures Sequestered onto a Reduced Graphene Oxide Matrix. *Microchim. Acta* **2020**, *187*, 8. [CrossRef]

MDPI AG
Grosspeteranlage 5
4052 Basel
Switzerland
Tel.: +41 61 683 77 34

Biosensors Editorial Office
E-mail: biosensors@mdpi.com
www.mdpi.com/journal/biosensors

www.ingramcontent.com/pod-product-compliance
Lightning Source LLC
Chambersburg PA
CBHW061608210926
51850CB00009B/2250

Mathematical Physics II

Mathematical Physics II

Editor

Enrico De Micheli

MDPI • Basel • Beijing • Wuhan • Barcelona • Belgrade • Manchester • Tokyo • Cluj • Tianjin

Editor
Enrico De Micheli
Consiglio Nazionale delle Ricerche
Italy

Editorial Office
MDPI
St. Alban-Anlage 66
4052 Basel, Switzerland

This is a reprint of articles from the Special Issue published online in the open access journal *Mathematics* (ISSN 2227-7390) (available at: https://www.mdpi.com/journal/mathematics/special_issues/Mathematical_Physic_II).

For citation purposes, cite each article independently as indicated on the article page online and as indicated below:

LastName, A.A.; LastName, B.B.; LastName, C.C. Article Title. *Journal Name* **Year**, *Article Number*, Page Range.

ISBN 978-3-03943-495-4 (Hbk)
ISBN 978-3-03943-496-1 (PDF)

Contents

About the Editor

Enrico De Micheli is currently Senior Researcher at the National Research Council (CNR) of Italy and teaches Quantum Information Theory and Computation at the Information Technology Institute (ISICT) of the University of Genova (Italy). He received his degree in Theoretical Physics from the University of Genova in 1986. He worked as research fellow at the Institute for Scientific Research (IRST) of Trento. He has been visiting scientist at the University of British Columbia in Vancouver (Canada). He is reviewer of the AMS and Editor/Referee of several mathematical journals and has published some 60 papers in peer-reviewed journals. His research interests include Special Functions of Mathematical Physics, Potential Theory, Quantum Mechanics, Thermal Quantum Field Theory, Approximation Theory, Inverse Problems, and Computational Physics.

Preface to "Mathematical Physics II"

The mysterious charm of Mathematical Physics is beautifully represented in the celebrated 1960 paper of Eugene Wigner "The Unreasonable Effectiveness of Mathematics in the Natural Sciences." It is indeed hard for us to understand the astonishing appropriateness of the language of mathematics for the formulation of the laws of physics and its capability to do predictions, appropriateness that emerged immediately at the beginning of the scientific thought and was splendidly depicted by Galileo: "The grand book, the Universe, is written in the language of Mathematics." Paraphrasing the words of Bertrand Russell, in this marriage the supreme beauty, cold and austere, of Mathematics complements the supreme beauty, warm and engaging, of Physics. This book, which consists of nine articles, gives a flavor of this beauty and covers various topics related to physics and engineering. A brief outline of these topics is given hereafter.

The study of free probability in certain probability spaces induced by functions on p-adic number fields is here a very interesting example of the application of p-adic mathematical methods for modeling physical phenomena. Within the quantum information processing framework is presented the analysis of a non-commutative measure of quantum discord in the two-qubit case. The Riemann–Hilbert problem plays relevant but different roles in two papers. In one paper, the Riemann–Hilbert problem, formulated with respect to the spectral parameter, and the prolongation structure theory are used to analyze the modified nonlinear Schrödinger equation. In the other case, the Riemann–Hilbert structure emerges from the holomorphic extension of certain Legendre expansions, leading thus to an explicit connection between spherical Laplace transform and non-Euclidean Fourier transform. Remaining in the spectral analysis field, the study of a deformed wave equation, with the Laplacian being replaced by a differential-difference Dunkl operator, shows its relation with a generalized Fourier transform and the non-existence of the related Huygens principle. The Gibbs phenomenon is then the subject of a paper where the primary tool of analysis is the representation of suitable functions in terms of dual tight framelets. Finally, numerical analysis and optimization methods are the main mathematical devices used to study equations that are relevant in the study of material properties, such as thermomechanical performances, flaw dynamics, and bearing capacity of structures.

In conclusion, as Editor of this Special Issue, I wish to thank the authors of the articles for their valuable contributions, the referees for their precious reviews, and Ms. Julie Shi and Ms. Grace Wang of MDPI for their kind assistance.

Enrico De Micheli
Editor

mathematics

Article

Primes in Intervals and Semicircular Elements Induced by p-Adic Number Fields $\mathbb{Q}_p$ over Primes p

Ilwoo Cho [1,*] **and Palle Jorgensen** [2]

[1] Department of Mathematics & Statistics, Saint Ambrose University, 421 Ambrose Hall, 518 W. Locust St., Davenport, IA 52803, USA

[2] Department of Mathematics, University of Iowa, 14 McLean Hall, Iowa City, IA 52242, USA; palle-jorgensen@uiowa.edu

* Correspondence: choilwoo@sau.edu

Received: 11 December 2018; Accepted: 15 February 2019; Published: 19 February 2019

Abstract: In this paper, we study free probability on (weighted-)semicircular elements in a certain Banach $*$-probability space $(\mathfrak{LS}, \tau^0)$ induced by measurable functions on p-adic number fields $\mathbb{Q}_p$ over primes p. In particular, we are interested in the cases where such free-probabilistic information is affected by primes in given closed intervals of the set $\mathbb{R}$ of real numbers by defining suitable "truncated" linear functionals on $\mathfrak{LS}$.

Keywords: free probability; primes; p-adic number fields; Banach $*$-probability spaces; weighted-semicircular elements; semicircular elements; truncated linear functionals

MSC: 05E15; 11G15; 11R47; 11R56; 46L10; 46L54; 47L30; 47L55

---

1. Introduction

In [1,2], we constructed and studied *weighted-semicircular elements* and *semicircular elements* induced by *p-adic number fields* $\mathbb{Q}_p$, for all $p \in \mathcal{P}$, where $\mathcal{P}$ is the set of all *primes* in the set $\mathbb{N}$ of all *natural numbers*. In this paper, we consider certain "truncated" free-probabilistic information of the weighted-semicircular laws and the semicircular law of [1]. In particular, we are interested in free distributions of certain free reduced words in our (weighted-)semicircular elements under conditions dictated by the primes p in a "suitable" *closed interval* $[t_1, t_2]$ of the set $\mathbb{R}$ of *real numbers*. Our results illustrate how the original (weighted-)semicircular law(s) of [1] is (resp., are) distorted by truncations on $\mathcal{P}$.

1.1. Preview and Motivation

Relations between *primes* and *operators* have been widely studied not only in mathematical fields (e.g., [3–6]), but also in other scientific fields (e.g., [7]). For instance, we studied how primes act on certain *von Neumann algebras* generated by p -adic and Adelic *measure spaces* in [8,9]. Meanwhile, in [10], primes are regarded as *linear functionals* acting on *arithmetic functions*, understood as *Krein-space operators* under the representation of [11]. Furthermore, in [12,13], free-probabilistic structures on *Hecke algebras* $\mathcal{H}\left(GL_2(\mathbb{Q}_p)\right)$ are studied for $p \in \mathcal{P}$. These series of works are motivated by number-theoretic results (e.g., [4,5,7]).

In [2], we constructed weighted-semicircular elements $\{Q_{p,j}\}_{j\in\mathbb{Z}}$ and corresponding semicircular elements $\{\Theta_{p,j}\}_{j\in\mathbb{Z}}$ in a certain Banach $*$-algebra $\mathfrak{LS}_p$ induced from the $*$-*algebra* $\mathcal{M}_p$ consisting of *measurable functions* on a p-adic number field $\mathbb{Q}_p$, for $p \in \mathcal{P}$. In [1], the *free product* Banach $*$-probability space $(\mathfrak{LS}, \tau^0)$ of the measure spaces $\{\mathfrak{LS}_p(j)\}_{p\in\mathcal{P},j\in\mathbb{Z}}$ of [2] were constructed over both primes and integers, and weighted-semicircular elements $\{Q_{p,j}\}_{p\in\mathcal{P},\, j\in\mathbb{Z}}$ and semicircular elements $\{\Theta_{p,j}\}_{p\in\mathcal{P},\, j\in\mathbb{Z}}$ were studied in $\mathfrak{LS}$, as *free generators*.

In this paper, we are interested in the cases where the free product linear functional τ^0 of [1] on the Banach $*$-algebra $\mathfrak{LS}$ is truncated in $\mathcal{P}$. The distorted free-distributional data from such truncations are considered. The main results characterize how the original free distributions on $(\mathfrak{LS}, \tau^0)$ are affected by the given truncations on $\mathcal{P}$.

1.2. Overview

We briefly introduce the backgrounds of our works in Section 2. In the short Sections 3–8, we construct the Banach $*$-probability space $(\mathfrak{LS}, \tau^0)$ and study *weighted-semicircular elements* $Q_{p,j}$ and corresponding *semicircular elements* $\Theta_{p,j}$ in $(\mathfrak{LS}, \tau^0)$, for all $p \in \mathcal{P}$, $j \in \mathbb{Z}$.

In Section 9, we define a free-probabilistic sub-structure $\mathbb{LS} = (\mathbb{LS}, \tau^0)$ of the Banach $*$-probability space $(\mathfrak{LS}, \tau^0)$, having possible non-zero free distributions, and study free-probabilistic properties of $\mathbb{LS}$. Then, *truncated linear functionals of* τ^0 on $\mathbb{LS}$ and truncated free-probabilistic information on $\mathbb{LS}$ are studied. The main results illustrate how our truncations distort the original free distributions on $\mathbb{LS}$ (and hence, on $\mathfrak{LS}$).

In Section 10, we study *free sums* X of $\mathbb{LS}$ having their free distribution, the (weighted-)semicircular law(s), under truncation. Note that, in general, if free sums X have more than one summand as operators, then X cannot be (weighted-)semicircular in $\mathbb{LS}$. However, certain truncations make them be.

In Section 11, we investigate a type of truncation (compared with those of Sections 9 and 10). In particular, certain truncations inducing so-called *prime-neighborhoods* are considered. The *unions* of such prime-neighborhoods provide corresponding distorted free probability on $\mathbb{LS}$ (different from that of Sections 9 and 10).

2. Preliminaries

In this section, we briefly introduce the backgrounds of our proceeding works.

2.1. Free Probability

Readers can review *free probability theory* from [14,15] (and the cited papers therein). *Free probability* is understood as the noncommutative operator-algebraic version of classical *measure theory* and *statistics*. The classical *independence* is replaced by the *freeness*, by replacing *measures* on sets with *linear functionals* on noncommutative ($*$-)algebras. It has various applications not only in pure mathematics (e.g., [16–20]), but also in related topics (e.g., see [2,8–11]). Here, we will use the *combinatorial free probability theory* of *Speicher* (e.g., see [14]).

In the text, without introducing detailed definitions and combinatorial backgrounds, *free moments* and *free cumulants* of operators will be computed. Furthermore, the *free product of $*$-probability spaces in the sense of* [14,15] is considered without detailed introduction.

Note now that one of our main objects, the $*$-algebra $\mathcal{M}_p$ of Section 3, are commutative, and hence, (traditional, or usual "noncommutative") free probability theory is not needed for studying *functional analysis* or *operator algebra theory* on $\mathcal{M}_p$, because the freeness on this commutative structure is trivial. However, we are not interested in the free-probability-depending operator-algebraic structures of commutative algebras, but in statistical data of certain elements to establish (weighted-)semicircular elements. Such data are well explained by the free-probability-theoretic terminology and language. Therefore, as in [2], we use "free-probabilistic models" on $\mathcal{M}_p$ to construct and study our (weighted-)semicircularity by using concepts, tools, and techniques from free probability theory "non-traditionally." Note also that, in Section 8, we construct "traditional" free-probabilistic structures, as in [1], from our "non-traditional" free-probabilistic structures of Sections 3–7 (like the *free group factors*; see, e.g., [15,19]).

2.2. *Analysis of* $\mathbb{Q}_p$

For more about p-adic and Adelic analysis, see [7]. Let $p \in \mathcal{P}$, and let $\mathbb{Q}_p$ be the *p-adic number field.* Under the p-adic addition and the p-adic multiplication of [7], the set $\mathbb{Q}_p$ forms a *field* algebraically. It is equipped with the *non-Archimedean norm* $|.|_p$, which is the inherited *p-norm* on the set $\mathbb{Q}$ of all *rational numbers* defined by:

$$|x|_p = \left|p^k \tfrac{a}{b}\right|_p = \tfrac{1}{p^k},$$

whenever $x = p^k \frac{a}{b}$ in $\mathbb{Q}$, where $k, a \in \mathbb{Z}$, and $b \in \mathbb{Z} \setminus \{0\}$. For instance,

$$|\tfrac{8}{3}|_2 = \left|2^3 \times \tfrac{1}{3}\right|_2 = \tfrac{1}{2^3} = \tfrac{1}{8},$$

and:

$$|\tfrac{8}{3}|_3 = |3^{-1} \times 8|_3 = \tfrac{1}{3^{-1}} = 3,$$

and:

$$|\tfrac{8}{3}|_q = \tfrac{1}{q^0} = 1, \text{ whenever } q \in \mathcal{P} \setminus \{2, 3\}.$$

The *p-adic number field* $\mathbb{Q}_p$ is the maximal p-norm closure in $\mathbb{Q}$. Therefore, under norm topology, it forms a *Banach space* (e.g., [7]).

Let us understand the *Banach field* $\mathbb{Q}_p$ as a *measure space,*

$$\mathbb{Q}_p = (\mathbb{Q}_p, \sigma(\mathbb{Q}_p), \mu_p),$$

where $\sigma(\mathbb{Q}_p)$ is the *σ-algebra of* $\mathbb{Q}_p$ *consisting of all μ_p-measurable subsets,* where μ_p is a left-and-right additive invariant *Haar measure* on $\mathbb{Q}_p$ satisfying:

$$\mu_p(\mathbb{Z}_p) = 1,$$

where $\mathbb{Z}_p$ is the *unit disk of* $\mathbb{Q}_p$, consisting of all *p-adic integers* x satisfying $|x|_p \leq 1$. Moreover, if we define:

$$U_k = p^k \mathbb{Z}_p = \{p^k x \in r\mathbb{Q}_p : x \in \mathbb{Z}_p\}, \tag{1}$$

for all $k \in \mathbb{Z}$ (with $U_0 = \mathbb{Z}_p$), then these μ_p-measurable subsets U_k's of (1) satisfy:

$$\mathbb{Q}_p = \underset{k \in \mathbb{Z}}{\cup} U_k,$$

and:

$$\mu_p (U_k) = \frac{1}{p^k} = \mu_p (x + U_k), \forall x \in \mathbb{Q}_p, \tag{2}$$

and:

$$\cdots \subset U_2 \subset U_1 \subset U_0 = \mathbb{Z}_p \subset U_{-1} \subset U_{-2} \subset \cdots.$$

In fact, the family $\{U_k\}_{k \in \mathbb{Z}}$ forms a *basis* of the Banach topology for $\mathbb{Q}_p$ (e.g., [7]).

Define now subsets ∂_k of $\mathbb{Q}_p$ by:

$$\partial_k = U_k \setminus U_{k+1}, for\ all k \in \mathbb{Z}. \tag{3}$$

We call such μ_p-measurable subsets ∂_k the kth boundaries of U_k in $\mathbb{Q}_p$, for all $k \in \mathbb{Z}$. By (2) and (3), one obtains that:

$$\mathbb{Q}_p = \underset{k \in \mathbb{Z}}{\sqcup} \partial_k,$$

and:

$$\mu_p (\partial_k) = \mu_p (U_k) - \mu_p (U_{k+1}) = \frac{1}{p^k} - \frac{1}{p^{k+1}}, \tag{4}$$

and:

$$\partial_{k_1} \cap \partial_{k_2} = \begin{cases} \partial_{k_1} & \text{if } k_1 = k_2 \\ \varnothing & \text{otherwise,} \end{cases},$$

for all $k, k_1, k_2 \in \mathbb{Z}$, where $\sqcup$ is the *disjoint union* and $\varnothing$ is the *empty set*.

Now, let $\mathcal{M}_p$ be the algebra,

$$\mathcal{M}_p = \mathbb{C}\left[\{\chi_S : S \in \sigma(\mathbb{Q}_p)\}\right], \tag{5}$$

where χ_S are the usual *characteristic functions* of $S \in \sigma\left(\mathbb{Q}_p\right)$.

Then the algebra $\mathcal{M}_p$ of (5) forms a well-defined $*$-*algebra over* $\mathbb{C}$, with its *adjoint*,

$$\left(\sum_{S \in \sigma(G_p)} t_S \chi_S\right)^* \overset{def}{=} \sum_{S \in \sigma(G_p)} \overline{t_S}\, \chi_S,$$

where $t_S \in \mathbb{C}$, having their *conjugates* $\overline{t_S}$ in $\mathbb{C}$.

Let $\sum_{S \in \sigma(G_p)} t_S \chi_S \in \mathcal{M}_p$. Then, one can define the *p-adic integral* by:

$$\int_{\mathbb{Q}_p} \left(\sum_{S \in \sigma(\mathbb{Q}_p)} t_S \chi_S\right) d\mu_p = \sum_{S \in \sigma(\mathbb{Q}_p)} t_S \mu_p(S). \tag{6}$$

Note that, by (4), if $S \in \sigma(\mathbb{Q}_p)$, then there exists a subset Λ_S of $\mathbb{Z}$, such that:

$$\Lambda_S = \{j \in \mathbb{Z} : S \cap \partial_j \neq \varnothing\}, \tag{7}$$

satisfying:

$$\int_{\mathbb{Q}_p} \chi_S \, d\mu_p = \int_{\mathbb{Q}_p} \sum_{j \in \Lambda_S} \chi_{S \cap \partial_j} \, d\mu_p$$

$$= \sum_{j \in \Lambda_S} \mu_p\left(S \cap \partial_j\right)$$

by (6)

$$\leq \sum_{j \in \Lambda_S} \mu_p\left(\partial_j\right) = \sum_{j \in \Lambda_S} \left(\frac{1}{p^j} - \frac{1}{p^{j+1}}\right), \tag{8}$$

by (4), for all $S \in \sigma(\mathbb{Q}_p)$, where Λ_S is in the sense of (7).

Proposition 1. *Let $S \in \sigma(\mathbb{Q}_p)$, and let $\chi_S \in \mathcal{M}_p$. Then, there exist $r_j \in \mathbb{R}$, such that:*

$$0 \leq r_j \leq 1 in \mathbb{R}, for all j \in \Lambda_S, \tag{9}$$

and:

$$\int_{\mathbb{Q}_p} \chi_S \, d\mu_p = \sum_{j \in \Lambda_S} r_j \left(\tfrac{1}{p^j} - \tfrac{1}{p^{j+1}}\right).$$

Proof. The existence of $r_j = \frac{\mu_p(S \cap \partial_j)}{\mu_p(\partial_j)}$, for all $j \in \mathbb{Z}$, is guaranteed by (7) and (8). The p-adic integral in (9) is obtained by (8). □

3. Free-Probabilistic Model on $\mathcal{M}_p$

Throughout this section, fix a prime $p \in \mathcal{P}$, and let $\mathbb{Q}_p$ be the corresponding p-adic number field and $\mathcal{M}_p$ be the $*$-algebra (5) consisting of μ_p-measurable functions on $\mathbb{Q}_p$. Here, we establish a suitable (non-traditional) free-probabilistic model on $\mathcal{M}_p$ implying p-adic analytic data.

Let U_k be the basis elements (1) of the topology for $\mathbb{Q}_p$ with their boundaries ∂_k of (3), i.e.,

$$U_k = p^k \mathbb{Z}_p, forallk \in \mathbb{Z}, \tag{10}$$

and:

$$\partial_k = U_k \setminus U_{k+1}, \text{ for all } k \in \mathbb{Z}.$$

Define a linear functional $\varphi_p : \mathcal{M}_p \to \mathbb{C}$ by the p-adic integration (6),

$$\varphi_p(f) = \int_{\mathbb{Q}_p} f d\mu_p, forallf \in \mathcal{M}_p. \tag{11}$$

Then, by (9) and (11), one obtains:

$$\varphi_p\left(\chi_{U_j}\right) = \frac{1}{p^j} \text{ and } \varphi_p\left(\chi_{\partial_j}\right) = \frac{1}{p^j} - \frac{1}{p^{j+1}},$$

for all $j \in \mathbb{Z}$.

Definition 1. *We call the pair* $(\mathcal{M}_p, \varphi_p)$ *the p-adic (non-traditional) free probability space for* $p \in \mathcal{P}$*, where* φ_p *is the linear functional* (11) *on* $\mathcal{M}_p$.

Remark 1. *As we discussed in Section 2.1, we study the measure-theoretic structure* $(\mathcal{M}_p, \varphi_p)$ *as a free-probabilistic model on* $\mathcal{M}_p$ *for our purposes. Therefore, without loss of generality, we regard* $(\mathcal{M}_p, \varphi_p)$ *as a non-traditional free-probabilistic structure. In this sense, we call* $(\mathcal{M}_p, \varphi_p)$ *the p-adic free probability space for p. The readers can understand* $(\mathcal{M}_p, \varphi_p)$ *as the pair of a commutative* $*$*-algebra* $\mathcal{M}_p$ *and a linear functional* φ_p*, having as its name the p-adic free probability space.*

Let ∂_k be the kth boundary $U_k \setminus U_{k+1}$ of U_k in $\mathbb{Q}_p$, for all $k \in \mathbb{Z}$. Then, for $k_1, k_2 \in \mathbb{Z}$, one obtains that:

$$\chi_{\partial_{k_1}} \chi_{\partial_{k_2}} = \chi_{\partial_{k_1} \cap \partial_{k_2}} = \delta_{k_1,k_2} \chi_{\partial_{k_1}},$$

by (4), and hence,

$$\begin{aligned} \varphi_p\left(\chi_{\partial_{k_1}} \chi_{\partial_{k_2}}\right) &= \delta_{k_1,k_2} \varphi_p\left(\chi_{\partial_{k_1}}\right) \\ &= \delta_{k_1,k_2}\left(\frac{1}{p^{k_1}} - \frac{1}{p^{k_1+1}}\right), \end{aligned} \tag{12}$$

where δ is the *Kronecker delta*.

Proposition 2. *Let* $(j_1, \ldots, j_N) \in \mathbb{Z}^N$*, for* $N \in \mathbb{N}$*. Then:*

$$\prod_{l=1}^{N} \chi_{\partial_{j_l}} = \delta_{(j_1,\ldots,j_N)} \chi_{\partial_{j_1}} \text{ in } \mathcal{M}_p,$$

and hence,

$$\varphi_p\left(\prod_{l=1}^{N} \chi_{\partial_{j_l}}\right) = \delta_{(j_1,\ldots,j_N)}\left(\frac{1}{p^{j_1}} - \frac{1}{p^{j_1+1}}\right), \tag{13}$$

where:

$$\delta_{(j_1,\ldots,j_N)} = \left(\prod_{l=1}^{N-1} \delta_{j_l,j_{l+1}}\right)\left(\delta_{j_N,j_1}\right).$$

Proof. The proof of (13) is done by induction on (12). □

Thus, one can get that, for any $S \in \sigma(\mathbb{Q}_p)$,

$$\varphi_p(\chi_S) = \varphi_p\left(\sum_{j \in \Lambda_S} \chi_{S \cap \partial_j}\right) \tag{14}$$

where Λ_S is in the sense of (7).

$$
\begin{aligned}
&= \sum_{j \in \Lambda_S} \varphi_p\left(\chi_{S \cap \partial_j}\right) = \sum_{j \in \Lambda_S} \mu_p\left(S \cap \partial_j\right) \\
&= \sum_{j \in \Lambda_S} r_j \left(\tfrac{1}{p^j} - \tfrac{1}{p^{j+1}}\right),
\end{aligned} \tag{15}
$$

by (13), where $0 \le r_j \le 1$ are in the sense of (9) for all $j \in \Lambda_S$.

Furthermore, if $S_1, S_2 \in \sigma\left(\mathbb{Q}_p\right)$, then:

$$
\begin{aligned}
\chi_{S_1}\chi_{S_2} &= \left(\sum_{k \in \Lambda_{S_1}} \chi_{S_1 \cap \partial_k}\right)\left(\sum_{j \in \Lambda_{S_2}} \chi_{S_2 \cap \partial_j}\right) \\
&= \sum_{(k,j) \in \Lambda_{S_1} \times \Lambda_{S_2}} \delta_{k,j} \chi_{(S_1 \cap S_2) \cap \partial_j} \\
&= \sum_{j \in \Lambda_{S_1,S_2}} \chi_{(S_1 \cap S_2) \cap \partial_j},
\end{aligned} \tag{16}
$$

where:

$$\Lambda_{S_1,S_2} = \Lambda_{S_1} \cap \Lambda_{S_2}.$$

Proposition 3. *Let $S_l \in \sigma(\mathbb{Q}_p)$, and let $\chi_{S_l} \in \left(\mathcal{M}_p, \varphi_p\right)$, for $l = 1, \ldots, N$, for $N \in \mathbb{N}$. Let:*

$$\Lambda_{S_1,\ldots,S_N} = \bigcap_{l=1}^{N} \Lambda_{S_l} \text{ in } \mathbb{Z},$$

where Λ_{S_l} are in the sense of (7), for $l = 1, \ldots, N$. Then, there exist $r_j \in \mathbb{R}$, such that:

$$0 \le r_j \le 1 \text{ in } \mathbb{R}, \text{ for } j \in \Lambda_{S_1,\ldots,S_N},$$

and:

$$\varphi_p\left(\prod_{l=1}^{N} \chi_{S_l}\right) = \sum_{j \in \Lambda_{S_1,\ldots,S_N}} r_j \left(\frac{1}{p^j} - \frac{1}{p^{j+1}}\right). \tag{17}$$

Proof. The proof of (17) is done by induction on (16) with the help of (15). □

4. Representations of $\left(\mathcal{M}_p, \varphi_p\right)$

Fix a prime p in $\mathcal{P}$, and let $\left(\mathcal{M}_p, \varphi_p\right)$ be the p-adic free probability space. By understanding $\mathbb{Q}_p$ as a measure space, construct the L^2-*space* H_p,

$$H_p \overset{def}{=} L^2\left(\mathbb{Q}_p, \sigma(\mathbb{Q}_p), \mu_p\right) = L^2\left(\mathbb{Q}_p\right), \tag{18}$$

over $\mathbb{C}$. Then, this L^2-space H_p of (18) is a well-defined *Hilbert space* equipped with its *inner product* $<,>_2$,

$$\langle h_1, h_2 \rangle_2 \overset{def}{=} \int_{\mathbb{Q}_p} h_1 h_2^* d\mu_p, \tag{19}$$

for all $h_1, h_2 \in H_p$.

Definition 2. *We call the Hilbert space H_p of (18), the p-adic Hilbert space.*

By the definition (18) of the p-adic Hilbert space H_p, our $*$-algebra $\mathcal{M}_p$ acts on H_p, via an *algebra-action* α^p,

$$\alpha^p(f)(h) = fh, forallh \in H_p, \tag{20}$$

for all $f \in \mathcal{M}_p$.

Notation: Denote $\alpha^p(f)$ of (20) by α_f^p, for all $f \in \mathcal{M}_p$. Furthermore, for convenience, denote $\alpha_{\chi_S}^p$ simply by α_S^p, for all $S \in \sigma(\mathbb{Q}_p)$. □

By (20), the linear morphism α^p is indeed a well-determined $*$-algebra-action of $\mathcal{M}_p$ acting on H_p (equivalently, every α_f^p is a $*$-homomorphism from $\mathcal{M}_p$ into the operator algebra $B(H_p)$ of all bounded operators on H_p, for all $f \in \mathcal{M}_p$), since:

$$\begin{aligned}\alpha_{f_1 f_2}^p(h) &= f_1 f_2 h = f_1(f_2 h) \\ &= f_1\left(\alpha_{f_2}^p(h)\right) = \alpha_{f_1}^p \alpha_{f_2}^p(h),\end{aligned}$$

for all $h \in H_p$, implying that:

$$\alpha_{f_1 f_2}^p = \alpha_{f_1}^p \alpha_{f_2}^p, \tag{21}$$

for all $f_1, f_2 \in \mathcal{M}_p$; and:

$$\begin{aligned}\left\langle \alpha_f^p(h_1), h_2 \right\rangle_2 &= \langle f h_1, h_2 \rangle_2 = \int_{\mathbb{Q}_p} f h_1 h_2^* d\mu_p \\ &= \int_{\mathbb{Q}_p} h_1 f h_2^* d\mu_p = \int_{\mathbb{Q}_p} h_1 (h_2 f^*)^* d\mu_p \\ &= \int_{\mathbb{Q}_p} h_1 (f^* h_2)^* d\mu_p = \left\langle h_1, \alpha_{f^*}^p(h_2) \right\rangle_2,\end{aligned}$$

for all $h_1, h_2 \in H_p$, for all $f \in \mathcal{M}_p$, implying that:

$$\left(\alpha_f^p\right)^* = \alpha_{f^*}, forall f \in \mathcal{M}_p, \tag{22}$$

where $<,>_2$ is the inner product (19) on H_p.

Proposition 4. *The linear morphism α^p of (20) is a well-defined $*$-algebra-action of $\mathcal{M}_p$ acting on H_p. Equivalently, the pair (H_p, α^p) is a Hilbert-space representation of $\mathcal{M}_p$.*

Proof. The proof is done by (21) and (22). □

Definition 3. *The Hilbert-space representation (H_p, α^p) is said to be the p-adic representation of $\mathcal{M}_p$.*

Depending on the p-adic representation (H_p, α^p) of $\mathcal{M}_p$, one can construct the C^**-subalgebra* M_p of the operator algebra $B(H_p)$.

Definition 4. *Define the C^*-subalgebra M_p of the operator algebra $B(H_p)$ by:*

$$M_p \stackrel{def}{=} \overline{\alpha^p(\mathcal{M}_p)} = \overline{\mathbb{C}\left[\alpha_f^p : f \in \mathcal{M}_p\right]}, \tag{23}$$

where $\overline{X}$ mean the operator-norm closures of subsets X of $B(H_p)$. Then, this C^*-algebra M_p is called the p-adic C^*-algebra of the p-adic free probability space $(\mathcal{M}_p, \varphi_p)$.

5. Free-Probabilistic Models on M_p

Throughout this section, let us fix a prime $p \in \mathcal{P}$, and let $(\mathcal{M}_p, \varphi_p)$ be the corresponding p-adic free probability space. Let (H_p, α^p) be the p-adic representation of $\mathcal{M}_p$, and let M_p be the p-adic C^*-algebra (23) of $(\mathcal{M}_p, \varphi_p)$.

We here construct suitable free-probabilistic models on M_p. In particular, we are interested in a system $\{\varphi_j^p\}_{j \in \mathbb{Z}}$ of *linear functionals on* M_p, determined by the jth boundaries $\{\partial_j\}_{j \in \mathbb{Z}}$ of $\mathbb{Q}_p$.

Define a linear functional $\varphi_j^p : M_p \to \mathbb{C}$ by a linear morphism,

$$\varphi_j^p(a) \stackrel{def}{=} \left\langle a\left(\chi_{\partial_j}\right), \chi_{\partial_j} \right\rangle_2, \tag{24}$$

for all $a \in M_p$, for all $j \in \mathbb{Z}$, where $<,>_2$ is the inner product (19) on the p-adic Hilbert space H_p of (18).

Remark that if $a \in M_p$, then:

$$a = \sum_{S \in \sigma(\mathbb{Q}_p)} t_S \, \alpha_S^p, \text{ in } M_p$$

(with $t_S \in \mathbb{C}$), where $\sum$ is a finite or infinite (i.e., limit of finite) sum(s) under the C^*-topology for M_p. Thus, the linear functionals φ_j^p of (24) are well defined on M_p, for all $j \in \mathbb{Z}$, i.e., for any fixed $j \in \mathbb{Z}$, one has that:

$$\begin{aligned} \left|\varphi_j^p(a)\right| &= \left| \sum_{S \in \sigma(\mathbb{Q}_p)} t_S \left\langle \chi_{S \cap \partial_j}, \chi_{\partial_j} \right\rangle_2 \right| \\ &= \left| \sum_{S \in \sigma(\mathbb{Q}_p)} t_S \mu_p\left(\chi_{S \cap \partial_j}\right) \right| \\ &\leq \mu_p\left(\partial_j\right) \left| \sum_{S \in \sigma(\mathbb{Q}_p)} t_S \right| \leq \left(\frac{1}{p^j} - \frac{1}{p^{j+1}}\right) \|a\|, \end{aligned} \tag{25}$$

where:

$$\|a\| = \sup\left\{ \|a(h)\|_2 : h \in H_p \text{ with } \|h\|_2 = 1 \right\}$$

is the C^*-norm on M_p (inherited by the operator norm on the operator algebra $B(H_p)$), and $\|.\|_2$ is the Hilbert-space norm,

$$\|f\|_2 = \sqrt{\langle f, f \rangle_2}, \forall f \in H_p,$$

induced by the inner product $<,>_2$ of (19). Therefore, for any fixed integer $j \in \mathbb{Z}$, the corresponding linear functional φ_j^p of (24) is bounded on M_p.

Definition 5. *Let $j \in \mathbb{Z}$, and let φ_j^p be the linear functional (24) on the p-adic C^*-algebra M_p. Then, the pair $\left(M_p, \varphi_j^p\right)$ is said to be the jth p-adic (non-traditional) C^*-probability space.*

Remark 2. *As in Section 4, the readers can understand the pairs $\left(M_p, \varphi_j^p\right)$ simply as structures consisting of a commutative C^*-algebra M_p and linear functionals φ_j^p on M_p, whose names are j^{th} p-adic C^*-probability spaces for all $j \in \mathbb{Z}$, for $p \in \mathcal{P}$.*

Fix $j \in \mathbb{Z}$, and take the corresponding j^{th} p-adic C^*-probability space $\left(M_p, \varphi_j^p\right)$. For $S \in \sigma(\mathbb{Q}_p)$ and a generating operator α_S^p of M_p, one has that:

$$\begin{aligned} \varphi_j^p(\alpha_S^p) &= \left\langle \alpha_S^p(\chi_{\partial_j}), \chi_{\partial_j} \right\rangle_2 = \left\langle \chi_{S \cap \partial_j}, \chi_{\partial_j} \right\rangle_2 \\ &= \int_{\mathbb{Q}_p} \chi_{S \cap \partial_j} \chi_{\partial_j}^* d\mu_p = \int_{\mathbb{Q}_p} \chi_{S \cap \partial_j} \chi_{\partial_j} d\mu_p \end{aligned} \tag{26}$$

by (19)

$$\begin{aligned} &= \int_{\mathbb{Q}_p} \chi_{S \cap \partial_j} d\mu_p = \mu_p\left(S \cap \partial_j\right) \\ &= r_S \left(\frac{1}{p^j} - \frac{1}{p^{j+1}}\right), \end{aligned} \tag{27}$$

for some $0 \leq r_S \leq 1$ in $\mathbb{R}$, for $S \in \sigma(\mathbb{Q}_p)$.

Proposition 5. *Let $S \in \sigma(\mathbb{Q}_p)$ and $\alpha_S^p = \alpha_{\chi_S}^p \in \left(M_p, \varphi_j^p\right)$, for a fixed $j \in \mathbb{Z}$. Then, there exists $r_S \in \mathbb{R}$, such that:*

$$0 \leq r_S \leq 1 \text{ in } \mathbb{R},$$

and:

$$\varphi_j^p\left(\left(\alpha_S^p\right)^n\right) = r_S\left(\frac{1}{p^j} - \frac{1}{p^{j+1}}\right), \text{for all } n \in \mathbb{N}. \tag{28}$$

Proof. Remark that the generating operator α_S^p is a projection in M_p, in the sense that:

$$\left(\alpha_S^p\right)^* = \alpha_S^p = \left(\alpha_S^p\right)^2, \text{ in } M_p,$$

so,

$$\left(\alpha_S^p\right)^n = \alpha_S^p, \text{ for all } n \in \mathbb{N}.$$

Thus, for any $n \in \mathbb{N}$, we have:

$$\varphi_j^p\left(\left(\alpha_S^p\right)^n\right) = \varphi_j^p(\alpha_S^p) = r_S\left(\tfrac{1}{p^j} - \tfrac{1}{p^{j+1}}\right),$$

for some $0 \leq r_S \leq 1$ in $\mathbb{R}$, by (27). □

As a corollary of (28), one obtains the following corollary.

Corollary 1. *Let ∂_k be the kth boundaries (10) of $\mathbb{Q}_p$, for all $k \in \mathbb{Z}$. Then:*

$$\varphi_j^p\left(\left(\alpha_{\partial_k}^p\right)^n\right) = \delta_{j,k}\left(\frac{1}{p^j} - \frac{1}{p^{j+1}}\right) \tag{29}$$

for all $n \in \mathbb{N}$, for all $j \in \mathbb{Z}$.

Proof. The formula (29) is shown by (28). □

6. Semigroup C^*-Subalgebras $\mathfrak{S}_p$ of M_p

Let M_p be the p-adic C^*-algebra (23) for an arbitrarily-fixed $p \in \mathcal{P}$. Take operators:

$$P_{p,j} = \alpha_{\partial_j}^p \in M_p, \tag{30}$$

where ∂_j are the jth boundaries (10) of $\mathbb{Q}_p$, for the fixed prime p, for all $j \in \mathbb{Z}$.

Then, these operators $P_{p,j}$ of (30) are *projections* on the p-adic Hilbert space H_p in M_p, i.e.,

$$P_{p,j}^* = P_{p,j} = P_{p,j}^2,$$

for all $j \in \mathbb{Z}$. We now restrict our interest to these projections $P_{p,j}$ of (30).

Definition 6. *Fix $p \in \mathcal{P}$. Let $\mathfrak{S}_p$ be the C^*-subalgebra:*

$$\mathfrak{S}_p = C^*\left(\{P_{p,j}\}_{j\in\mathbb{Z}}\right) = \overline{\mathbb{C}\left[\{P_{p,j}\}_{j\in\mathbb{Z}}\right]} \text{ of } M_p, \tag{31}$$

where $P_{p,j}$ are projections (30), for all $j \in \mathbb{Z}$. We call this C^*-subalgebra $\mathfrak{S}_p$ the p-adic boundary (C^*-)subalgebra of M_p.

The p-adic boundary subalgebra $\mathfrak{S}_p$ of the p-adic C^*-algebra M_p satisfies the following structure theorem.

Proposition 6. *Let $\mathfrak{S}_p$ be the p-adic boundary subalgebra (31) of the p-adic C^*-algebra M_p. Then:*

$$\mathfrak{S}_p \overset{*\text{-iso}}{=} \underset{j\in\mathbb{Z}}{\oplus}\left(\mathbb{C}\cdot P_{p,j}\right) \overset{*\text{-iso}}{=} \mathbb{C}^{\oplus|\mathbb{Z}|}, \tag{32}$$

in M_p.

Proof. The proof of (32) is done by the mutual orthogonality of the projections $\{P_{p,j}\}_{j\in\mathbb{Z}}$ in M_p. Indeed, one has:

$$P_{p,j_1}P_{p,j_2} = \alpha^p_{\partial_{j_1}}\alpha^p_{\partial_{j_2}} = \alpha^p_{\partial_{j_1}\cap\partial_{j_2}} = \delta_{j_1,j_2}P_{p,j_1},$$

in $\mathfrak{S}_p$, for all $j_1, j_2 \in \mathbb{Z}$. □

Define now linear functionals φ^p_j (for a fixed prime p) by:

$$\varphi^{(p)}_j = \varphi^p_j \mid_{\mathfrak{S}_p} on\mathfrak{S}_p, \tag{33}$$

where φ^p_j in the right-hand side of (33) are the linear functionals (24) on M_p, for all $j \in \mathbb{Z}$.

7. Weighted-Semicircular Elements

Let M_p be the p-adic C^*-algebra, and let $\mathfrak{S}_p$ be the p-adic boundary subalgebra (31) of M_p, satisfying the structure theorem (32). Fix $p \in \mathcal{P}$. Recall that the generating projections $P_{p,j}$ of $\mathfrak{S}_p$ satisfy:

$$\varphi^{(p)}_j\left(P_{p,j}\right) = \frac{1}{p^j} - \frac{1}{p^{j+1}}, \forall j \in \mathbb{Z}, \tag{34}$$

by (33) (also see (28) and (29)).

Now, let ϕ be the *Euler totient function*, an *arithmetic function:*

$$\phi : \mathbb{N} \to \mathbb{C},$$

defined by:

$$\phi(n) = \left|\{k \in \mathbb{N} : k \leq n,\ \gcd(n,k) = 1\}\right|, \tag{35}$$

for all $n \in \mathbb{N}$, where $|X|$ mean the cardinalities of sets X and gcd is the *greatest common divisor*.

It is well known that:

$$\phi(n) = n\left(\prod_{q\in\mathcal{P},\, q|n}\left(1 - \tfrac{1}{q}\right)\right),$$

for all $n \in \mathbb{N}$, where "$q \mid n$" means "q divides n." For instance,

$$\phi(p) = p - 1 = p\left(1 - \frac{1}{p}\right), \forall p \in \mathcal{P}. \tag{36}$$

Thus:

$$\begin{aligned}\varphi^{(p)}_j\left(P_{p,j}\right) &= \left(\tfrac{1}{p^j} - \tfrac{1}{p^{j+1}}\right) = \tfrac{1}{p^j}\left(1 - \tfrac{1}{p}\right) \\ &= \tfrac{p}{p^{j+1}}\left(1 - \tfrac{1}{p}\right) = \tfrac{\phi(p)}{p^{j+1}},\end{aligned}$$

by (34), (35), and (36), for all $P_{p,j} \in \mathfrak{S}_p$. More generally,

$$\varphi^{(p)}_j\left(P_{p,k}\right) = \delta_{j,k}\left(\frac{\phi(p)}{p^{j+1}}\right), \forall p \in \mathcal{P}, k \in \mathbb{Z}. \tag{37}$$

Now, for a fixed prime p, define new linear functionals τ^p_j on $\mathfrak{S}_p$, by linear morphisms satisfying that:

$$\tau^p_j = \frac{1}{\phi(p)}\varphi^{(p)}_j, on\mathfrak{S}_p, \tag{38}$$

for all $j \in \mathbb{Z}$, where φ^p_j are in the sense of (33).

Then, one obtains new (non-traditional) C^*-probabilistic structures,

$$\{\mathfrak{S}_p(j) = \left(\mathfrak{S}_p,\ \tau_j^p\right) : p \in \mathcal{P}, j \in \mathbb{Z}\}, \tag{39}$$

where τ_j^p are in the sense of (38).

Proposition 7. *Let $\mathfrak{S}_p(j) = (\mathfrak{S}_p,\ \tau_j^p)$ be in the sense of (39), and let $P_{p,k}$ be generating operators of $\mathfrak{S}_p(j)$, for $p \in \mathcal{P}, j \in \mathbb{Z}$. Then:*

$$\tau_j^p\left(P_{p,k}^n\right) = \frac{\delta_{j,k}}{p^{j+1}}, forall n \in \mathbb{N}. \tag{40}$$

Proof. The formula (40) is proven by (37) and (38). Indeed, since $P_{p,k}$ are projections in $\mathfrak{S}_p(j)$,

$$\tau_j^p\left(P_{p,k}^n\right) = \tau_j^p\left(P_{p,k}\right) = \delta_{j,k}\left(\tfrac{1}{p^{j+1}}\right),$$

for all $n \in \mathbb{N}$, for all $p \in \mathcal{P}$, and $j, k \in \mathbb{Z}$. □

7.1. Semicircular and Weighted-Semicircular Elements

Let $(A,\ \varphi)$ be an arbitrary (traditional or non-traditional) *topological $*$-probability space* (C^*-probability space, or W^*-probability space, or Banach $*$-probability space, etc.), consisting of a (noncommutative, resp., commutative) topological $*$-algebra A (C^*-algebra, resp., W^*-algebra, resp., Banach $*$-algebra, etc.), and a (bounded or unbounded) linear functional φ on A.

Definition 7. *Let a be a self-adjoint element in $(A,\ \varphi)$. It is said to be even in $(A,\ \varphi)$, if all odd free moments of a vanish, i.e.,*

$$\varphi\left(a^{2n-1}\right) = 0, forall n \in \mathbb{N}. \tag{41}$$

Let a be a "self-adjoint," and "even" element of $(A,\ \varphi)$ satisfying (41). Then, it is said to be semicircular in $(A,\ \varphi)$, if:

$$\varphi(a^{2n}) = c_n, forall n \in \mathbb{N}, \tag{42}$$

where c_k are the kth Catalan number,

$$c_k = \tfrac{1}{k+1}\begin{pmatrix} 2k \\ k \end{pmatrix} = \tfrac{1}{k+1}\tfrac{(2k)!}{(k!)^2} = \tfrac{(2k)!}{k!(k+1)!},$$

for all $k \in \mathbb{N}_0 = \mathbb{N} \cup \{0\}$.

It is well known that, if $k_n(...)$ is the *free cumulant on A in terms of a linear functional φ* (in the sense of [14]), then a self-adjoint element a is *semicircular* in $(A,\ \varphi)$, if and only if:

$$k_n\left(\underbrace{a,\ a,\ \ldots\ldots,\ a}_{n\text{-times}}\right) = \begin{cases} 1 & \text{if } n = 2 \\ 0 & \text{otherwise,} \end{cases} \tag{43}$$

for all $n \in \mathbb{N}$ (e.g., see [14]). The above equivalent free-distributional data (43) of the semicircularity (42) are obtained by the *Möbius inversion of* [14].

Motivated by (43), one can define the *weighted-semicircularity*.

Definition 8. *Let $a \in (A,\ \varphi)$ be a self-adjoint element. It is said to be weighted-semicircular in $(A,\ \varphi)$ with its weight t_0 (in short, t_0-semicircular), if there exists $t_0 \in \mathbb{C}^\times = \mathbb{C} \setminus \{0\}$, such that:*

$$k_n\left(\underbrace{a,\ a,\ \ldots.,\ a}_{n\text{-times}}\right) = \begin{cases} t_0 & \text{if } n = 2 \\ 0 & \text{otherwise,} \end{cases} \tag{44}$$

for all $n \in \mathbb{N}$, where $k_n(...)$ is the free cumulant on A in terms of φ.

By the definition (44) and by the Möbius inversion of [14], one obtains the following free-moment characterization (45) of the weighted-semicircularity (44): A self-adjoint element a is t_0-*semicircular* in (A, φ), if and only if there exists $t_0 \in \mathbb{C}^{\times}$, such that:

$$\varphi(a^n) = \omega_n \, t_0^{\frac{n}{2}} c_{\frac{n}{2}},$$

where:

$$\omega_n = \begin{cases} 1 & \text{if } n \text{ is even} \\ 0 & \text{if } n \text{ is odd,} \end{cases} \tag{45}$$

for all $n \in \mathbb{N}$, where c_m are the mth Catalan numbers for all $m \in \mathbb{N}_0$.

Thus, from below, we use the weighted-semicircularity (44) and its characterization (45) alternatively.

7.2. Tensor Product Banach $*$-Algebra $\mathfrak{L}\mathfrak{S}_p$

Let $\mathfrak{S}_p(k) = \left(\mathfrak{S}_p, \tau_k^p\right)$ be a (non-traditional) C^*-probability space (39), for $p \in \mathcal{P}$, $k \in \mathbb{Z}$. Define *bounded linear transformations* $\mathbf{c}_p$ and $\mathbf{a}_p$ "acting on the p-adic boundary subalgebra $\mathfrak{S}_p$ of M_p," by linear morphisms satisfying,

$$\mathbf{c}_p\left(P_{p,j}\right) = P_{p,j+1},$$

and:

$$\mathbf{a}_p\left(P_{p,j}\right) = P_{p,j-1}, \tag{46}$$

on $\mathfrak{S}_p$, for all $j \in \mathbb{Z}$.

By (46), these linear transformations $\mathbf{c}_p$ and $\mathbf{a}_p$ are bounded under the operator-norm induced by the C^*-norm on $\mathfrak{S}_p$. Therefore, the linear transformations $\mathbf{c}_p$ and $\mathbf{a}_p$ are regarded as Banach-space operators "acting on $\mathfrak{S}_p$," by regarding $\mathfrak{S}_p$ as a Banach space (under its C^*-norm). i.e., $\mathbf{c}_p$ and $\mathbf{a}_p$ are elements of the *operator space* $B\left(\mathfrak{S}_p\right)$ consisting of all bounded operators on the Banach space $\mathfrak{S}_p$.

Definition 9. *The Banach-space operators $\mathbf{c}_p$ and $\mathbf{a}_p$ of (46) are called the p-creation, respectively, the p-annihilation on $\mathfrak{S}_p$, for $p \in \mathcal{P}$. Define a new Banach-space operator $l_p \in B\left(\mathfrak{S}_p\right)$ by:*

$$l_p = \mathbf{c}_p + \mathbf{a}_p on \mathfrak{S}_p. \tag{47}$$

We call it the p-radial operator on $\mathfrak{S}_p$.

Let l_p be the p-radial operator $\mathbf{c}_p + \mathbf{a}_p$ of (47) on $\mathfrak{S}_p$. Construct a *closed subspace* $\mathfrak{L}_p$ of $B(\mathfrak{S}_p)$ by:

$$\mathfrak{L}_p = \overline{\mathbb{C}[l_p]} \subset B(\mathfrak{S}_p), \tag{48}$$

where $\overline{Y}$ means the operator-norm-topology closure of every subset Y of the operator space $B(\mathfrak{S}_p)$.

By the definition (48), $\mathfrak{L}_p$ is not only a closed subspace, but also a well-defined Banach algebra embedded in the vector space $B(\mathfrak{S}_p)$. On this Banach algebra $\mathfrak{L}_p$, define the adjoint $(*)$ by:

$$\sum_{k=0}^{\infty} s_k l_p^k \in \mathfrak{L}_p \longmapsto \sum_{k=0}^{\infty} \overline{s_k} l_p^k \in \mathfrak{L}_p, \tag{49}$$

where $s_k \in \mathbb{C}$ with their conjugates $\overline{s_k} \in \mathbb{C}$.

Then, equipped with the adjoint (49), this Banach algebra $\mathfrak{L}_p$ of (48) forms a *Banach $*$-algebra* inside $B(\mathfrak{S}_p)$.

Definition 10. *Let $\mathfrak{L}_p$ be a Banach $*$-algebra (48) in the operator space $B(\mathfrak{S}_p)$ for $p \in \mathcal{P}$. We call it the p-radial (Banach-$*$-)algebra on $\mathfrak{S}_p$.*

Let $\mathfrak{L}_p$ be the p-radial algebra (48) on $\mathfrak{S}_p$. Construct now the tensor product Banach $*$-algebra $\mathfrak{L}\mathfrak{S}_p$ by:

$$\mathfrak{L}\mathfrak{S}_p = \mathfrak{L}_p \otimes_{\mathbb{C}} \mathfrak{S}_p, \tag{50}$$

where $\otimes_{\mathbb{C}}$ is the *tensor product of Banach $*$-algebras* (Remark that $\mathfrak{S}_p$ is a C^*-algebra and $\mathfrak{L}_p$ is a Banach $*$-algebra; and hence, the tensor product Banach $*$-algebra $\mathfrak{L}\mathfrak{S}_p$ of (50) is well-defined.).

Take now a generating element $l_p^k \otimes P_{p,j}$, for some $k \in \mathbb{N}_0 = \mathbb{N} \cup \{0\}$, and $j \in \mathbb{Z}$, where $P_{p,j}$ is in the sense of (30) in $\mathfrak{S}_p$, with axiomatization:

$$l_p^0 = 1_{\mathfrak{S}_p}, \text{ the identity operator on } \mathfrak{S}_p,$$

in $B(\mathfrak{S}_p)$, satisfying:

$$1_{\mathfrak{S}_p}\left(P_{p,j}\right) = P_{p,j}, \text{ for all } P_{p,j} \in \mathfrak{S}_p,$$

for all $j \in \mathbb{Z}$.

By (50) and (32), the elements $l_p^k \otimes P_{p,j}$ indeed generate $\mathfrak{L}\mathfrak{S}_p$ under linearity, because:

$$\left(l_p \otimes P_{p,j}\right)^k = l_p^k \otimes P_{p,j},$$

for all $k \in \mathbb{N}_0$, and $j \in \mathbb{Z}$, for $p \in \mathcal{P}$, and their self-adjointness. We now focus on such generating operators of $\mathfrak{L}\mathfrak{S}_p$.

Define a linear morphism:

$$E_p : \mathfrak{L}\mathfrak{S}_p \to \mathfrak{S}_p$$

by a linear transformation satisfying that:

$$E_p\left(l_p^k \otimes P_{p,j}\right) = \frac{\left(p^{j+1}\right)^{k+1}}{\left[\frac{k}{2}\right]+1} l_p^k(P_{p,j}), \tag{51}$$

for all $k \in \mathbb{N}_0$, $j \in \mathbb{Z}$, where $\left[\frac{k}{2}\right]$ is the *minimal integer greater than or equal to* $\frac{k}{2}$, for all $k \in \mathbb{N}_0$; for example,

$$\left[\tfrac{3}{2}\right] = 2 = \left[\tfrac{4}{2}\right].$$

By the cyclicity (48) of the tensor factor $\mathfrak{L}_p$ of $\mathfrak{L}\mathfrak{S}_p$, and by the structure theorem (32) of the other tensor factor $\mathfrak{S}_p$ of $\mathfrak{L}\mathfrak{S}_p$, the above morphism E_p of (51) is a well-defined bounded surjective linear transformation.

Now, consider how our p-radial operator l_p of (47) works on $\mathfrak{S}_p$. Observe first that: if $\mathbf{c}_p$ and $\mathbf{a}_p$ are the p-creation, respectively, the p-annihilation on $\mathfrak{S}_p$, then:

$$\mathbf{c}_p\mathbf{a}_p\left(P_{p,j}\right) = P_{p,j} = \mathbf{a}_p\mathbf{c}_p\left(P_{p,j}\right),$$

for all $j \in \mathbb{Z}$, $p \in \mathcal{P}$, and hence:

$$\mathbf{c}_p\mathbf{a}_p = 1_{\mathfrak{S}_p} = \mathbf{a}_p\mathbf{c}_p on \mathfrak{S}_p. \tag{52}$$

Lemma 1. *Let $\mathbf{c}_p$, $\mathbf{a}_p$ be the p-creation, respectively, the p-annihilation on $\mathfrak{S}_p$. Then:*

$$\mathbf{c}_p^n\mathbf{a}_p^n = \left(\mathbf{c}_p\mathbf{a}_p\right)^n = 1_{\mathfrak{S}_p} = \left(\mathbf{a}_p\mathbf{c}_p\right)^n = \mathbf{a}_p\mathbf{c}_p,$$

and:

$$\mathbf{c}_p^{n_1}\mathbf{a}_p^{n_2} = \mathbf{a}_p^{n_2}\mathbf{c}_p^{n_1} on \mathfrak{S}_p, \tag{53}$$

for all $n, n_1, n_2 \in \mathbb{N}_0$.

Proof. The formula (53) holds by (52). □

By (53), one can get that:

$$l_p^n = (\mathbf{c}_p + \mathbf{a}_p)^n = \sum_{k=0}^{n} \begin{pmatrix} n \\ k \end{pmatrix} \mathbf{c}_p^k \mathbf{a}_p^{n-k} \text{ on } \mathfrak{S}_p,$$

with identities;

$$\mathbf{c}_p^0 = 1_{\mathfrak{S}_p} = \mathbf{a}_p^0, \tag{54}$$

for all $n \in \mathbb{N}$, where:

$$\begin{pmatrix} n \\ k \end{pmatrix} = \frac{n!}{k!(n-k)!}, \forall k \leq n \in \mathbb{N}_0.$$

Thus, one obtains the following proposition.

Proposition 8. *Let $l_p \in \mathfrak{L}_p$ be the p-radial operator on $\mathfrak{S}_p$. Then:*

$$l_p^{2m-1} \textit{ does not contain } 1_{\mathfrak{S}_p} - \textit{term, and} \tag{55}$$

$$l_p^{2m} \textit{ contains its } 1_{\mathfrak{S}_p} - \textit{term}, \begin{pmatrix} 2m \\ m \end{pmatrix} \cdot 1_{\mathfrak{S}_p}, \tag{56}$$

for all $m \in \mathbb{N}$.

Proof. The proofs of (55) and (56) are done by straightforward computations by (53) and (54). See [1] for more details. □

7.3. *Weighted-Semicircular Elements $Q_{p,j}$ in $\mathfrak{L}\mathfrak{S}_p$*

Fix $p \in \mathcal{P}$, and let $\mathfrak{L}\mathfrak{S}_p = \mathfrak{L}_p \otimes_{\mathbb{C}} \mathfrak{S}_p$ be the tensor product Banach $*$-algebra (50) and E_p be the linear transformation (51) from $\mathfrak{L}\mathfrak{S}_p$ onto $\mathfrak{S}_p$. Throughout this section, fix a generating element:

$$Q_{p,j} = l_p \otimes P_{p,j} of \mathfrak{L}\mathfrak{S}_p, \tag{57}$$

for $j \in \mathbb{Z}$, where $P_{p,j}$ is a projection (30) generating $\mathfrak{S}_p$. Observe that:

$$Q_{p,j}^n = (l_p \otimes P_{p,j})^n = l_p^n \otimes P_{p,j}, \tag{58}$$

for all $n \in \mathbb{N}$, for all $j \in \mathbb{Z}$.

If $Q_{p,j} \in \mathfrak{L}\mathfrak{S}_p$ is in the sense of (57) for $j \in \mathbb{Z}$, then:

$$E_p\left(Q_{p,j}^n\right) = E_p\left(l_p^n \otimes P_{p,j}\right) = \frac{\left(p^{j+1}\right)^{n+1}}{\left[\frac{n}{2}\right]+1} l_p^n\left(P_{p,j}\right), \tag{59}$$

by (51) and (58), for all $n \in \mathbb{N}$.

Now, for a fixed $j \in \mathbb{Z}$, define a linear functional $\tau_{p,j}^0$ on $\mathfrak{L}\mathfrak{S}_p$ by:

$$\tau_{p,j}^0 = \tau_j^p \circ E_p on \mathfrak{L}\mathfrak{S}_p, \tag{60}$$

where $\tau_j^p = \frac{1}{\phi(p)} \varphi_j^{(p)}$ is the linear functional (38) on $\mathfrak{S}_p$.

By the bounded-linearity of both τ_j^p and E_p, the morphism $\tau_{p,j}^0$ of (60) is a bounded linear functional on $\mathfrak{L}\mathfrak{S}_p$.

By (59) and (60), if $Q_{p,j}$ is in the sense of (57), then:

$$\tau_{p,j}^0\left(Q_{p,j}^n\right) = \frac{\left(p^{j+1}\right)^{n+1}}{\left[\frac{n}{2}\right]+1} \tau_j^p\left(l_p^n(P_{p,j})\right), \tag{61}$$

for all $n \in \mathbb{N}$.

Theorem 1. *Let $Q_{p,j} = l_p \otimes P_{p,j} \in \left(\mathfrak{LS}_p, \tau^0_{p,j}\right)$, for a fixed $j \in \mathbb{Z}$. Then, $Q_{p,j}$ is $p^{2(j+1)}$-semicircular in $\left(\mathfrak{LS}_p,\ \tau^0_{p,j}\right)$. More precisely,*

$$\tau^0_{p,j}\left(Q^n_{p,j}\right) = \omega_n \left(p^{2(j+1)}\right)^{\frac{n}{2}} c_{\frac{n}{2}}, \tag{62}$$

for all $n \in \mathbb{N}$, where ω_n are in the sense of (45). Equivalently, if $k_n^{0,p,j}(...)$ is the free cumulant on $\mathfrak{LS}_p$ in terms of the linear functional $\tau^0_{p,j}$ of (61) on $\mathfrak{LS}_p$, then:

$$k_n^{0,p,j}\left(\underbrace{Q_{p,j},\ Q_{p,j},\ ...,\ Q_{p,j}}_{n\text{-times}}\right) = \begin{cases} p^{2(j+1)} & \text{if } n = 2 \\ 0 & \text{otherwise,} \end{cases} \tag{63}$$

for all $n \in \mathbb{N}$.

Proof. The free-moment formula (62) is obtained by (55), (56) and (61). The free-cumulant formula (63) is obtained by (62) under the Möbius inversion of [14]. See [1] for details. □

8. Semicircularity on $\mathfrak{LS}$

For all $p \in \mathcal{P}, j \in \mathbb{Z}$, let:

$$\mathfrak{LS}_p(j) = \left(\mathfrak{LS}_p,\ \tau^0_{p,j}\right) \tag{64}$$

be a Banach $*$-probabilistic model of the Banach $*$-algebra $\mathfrak{LS}_p$ of (50), where $\tau^0_{p,j}$ is the linear functional (60).

Definition 11. *We call the pairs $\mathfrak{LS}_p(j)$ of (64) the j^{textth}* p-adic filters, *for all $p \in \mathcal{P}, j \in \mathbb{Z}$.*

Let $Q_{p,k} = l_p \otimes P_{p,k}$ be the kth generating elements of the jth p-adic filter $\mathfrak{LS}_p(j)$ of (64), for all $k \in \mathbb{Z}$, for fixed $p \in \mathcal{P}, j \in \mathbb{Z}$. Then, the generating elements $\{Q_{p,k}\}_{k \in \mathbb{Z}}$ of the jth p-adic filter $\mathfrak{LS}_p(j)$ satisfy that:

$$k_n^{0,p,j}\left(Q_{p,k},\ ...,\ Q_{p,k}\right) = \begin{cases} \delta_{j,k}\ p^{2(j+1)} & \text{if } n = 2 \\ 0 & \text{otherwise,} \end{cases}$$

and:

$$\tau^0_{p,j}\left(Q^n_{p,k}\right) = \delta_{j,k}\left(\omega_n \left(p^{2(j+1)}\right)^{\frac{n}{2}} c_{\frac{n}{2}}\right), \tag{65}$$

for all $p \in \mathcal{P}, j \in \mathbb{Z}$, for all $n \in \mathbb{N}$, by (62) and (63), where:

$$\omega_n = \begin{cases} 1 & \text{if } n \text{ is even} \\ 0 & \text{if } n \text{ is odd,} \end{cases}$$

for all $n \in \mathbb{N}$.

For the family:

$$\left\{\mathfrak{LS}_p(j) = \left(\mathfrak{LS}_p,\ \tau^0_{p,j}\right) : p \in \mathcal{P},\ j \in \mathbb{Z}\right\}$$

of jth p-adic filters of (64), one can define the *free product Banach $*$-probability space,*

$$\mathfrak{LS} \overset{denote}{=} \left(\mathfrak{LS},\ \tau^0\right) \overset{def}{=} \underset{p \in \mathcal{P},\ j \in \mathbb{Z}}{\star} \mathfrak{LS}_p(j). \tag{66}$$

as in [14,15], with:

$$\mathfrak{LS} = \underset{p \in \mathcal{P},\, j \in \mathbb{Z}}{\star} \mathfrak{LS}_p, \text{ and } \tau^0 = \underset{p \in \mathcal{P},\, j \in \mathbb{Z}}{\star} \tau^0_{p,j}.$$

Note that the pair $\mathbb{LS} = (\mathfrak{LS}, \tau^0)$ of (66) is a well-defined "traditional or noncommutative" Banach $*$-probability space. For more about the (free-probabilistic) *free product* of free probability spaces, see [14,15].

Definition 12. *The Banach $*$-probability space $\mathbb{LS} = (\mathfrak{LS}, \tau^0)$ of (66) is called the free Adelic filtration.*

Let $\mathbb{LS}$ be the free Adelic filtration (66). Then, by (65), one can take a subset:

$$\mathcal{Q} = \left\{ Q_{p,j} = l_p \otimes P_{p,j} \in \mathbb{LS}_p(j) \right\}_{p \in \mathcal{P},\, j \in \mathbb{Z}}$$

of $\mathbb{LS}$, consisting of "j^{th}" generating elements $Q_{p,j}$ of the "j^{th}" p-adic filters $\mathbb{LS}_p(j)$, which are the free blocks of $\mathbb{LS}$, for all $j \in \mathbb{Z}$, for all $p \in \mathcal{P}$.

Lemma 2. *Let $\mathcal{Q}$ be the above family in the free Adelic filtration $\mathbb{LS}$. Then, all elements $Q_{p,j}$ of $\mathcal{Q}$ are $p^{2(j+1)}$-semicircular in the free Adelic filtration $\mathbb{LS}$.*

Proof. Since all self-adjoint elements $Q_{p,j}$ of the family $\mathcal{Q}$ are chosen from mutually-distinct *free blocks* $\mathbb{LS}_p(j)$ of $\mathbb{LS}$, they are $p^{2(j+1)}$-semicircular in $\mathbb{LS}_p(j)$. Indeed, since every element $Q_{p,j} \in \mathcal{Q}$ is from a free block $\mathbb{LS}_p(j)$, the powers $Q^n_{p,j}$ are free reduced words with their lengths-N in $\mathbb{LS}_p(j)$ in $\mathbb{LS}$. Therefore, each element $Q_{p,j} \in \mathcal{Q}$ satisfies that:

$$\tau^0\left(Q^n_{p,j}\right) = \tau^0_{p,j}\left(Q^n_{p,j}\right) = \omega_n p^{n(j+1)} c_{\frac{n}{2}},$$

equivalently,

$$\begin{aligned} k^0_n\left(Q_{p,j}, \ldots, Q_{p,j}\right) &= k^{0,p,j}_n\left(Q_{p,j}, \ldots, Q_{p,j}\right) \\ &= \begin{cases} p^{2(j+1)} & \text{if } n = 2 \\ 0 & \text{otherwise,} \end{cases} \end{aligned}$$

for all $n \in \mathbb{N}$, by (62) and (63), where $k^0_n(\ldots)$ is the free cumulant on $\mathbb{LS}$ in terms of τ^0. Therefore, all elements $Q_{p,j} \in \mathcal{Q}$ are $p^{2(j+1)}$-semicircular in $\mathbb{LS}$, for all $p \in \mathcal{P}, j \in \mathbb{Z}$. □

Furthermore, since all $p^{2(j+1)}$-semicircular elements $Q_{p,j} \in \mathcal{Q}$ are taken from the mutually-distinct free blocks $\mathbb{LS}_p(j)$ of $\mathbb{LS}$, they are mutually free from each other in the free Adelic filtration $\mathbb{LS}$ of (66), for all $p \in \mathcal{P}, j \in \mathbb{Z}$.

Recall that a subset $S = \{a_t\}_{t \in \Delta}$ of an arbitrary (topological or pure-algebraic) $*$-probability space (A, φ) is said to be a *free family*, if, for any pair $(t_1, t_2) \in \Delta^2$ of $t_1 \neq t_2$ in a countable (finite or infinite) index set Δ, the corresponding free random variables a_{t_1} and a_{t_2} are free in (A, φ) (e.g., [7,14]).

Definition 13. *Let $S = \{a_t\}_{t \in \Delta}$ be a free family in an arbitrary topological $*$-probability space (A, φ). This family S is said to be a free (weighted-)semicircular family, if it is not only a free family, but also a set consisting of all (weighted-)semicircular elements a_t in (A, φ), for all $t \in \Delta$.*

Therefore, by the construction (66) of the free Adelic filtration $\mathbb{LS}$, we obtain the following result.

Theorem 2. *Let $\mathbb{LS}$ be the free Adelic filtration (66), and let:*

$$\mathcal{Q} = \{Q_{p,j} \in \mathbb{LS}_p(j)\}_{p \in \mathcal{P},\, j \in \mathbb{Z}} \subset \mathbb{LS}, \tag{67}$$

where $\mathbb{LS}_p(j)$ are the jth p-adic filters, the free blocks of $\mathbb{LS}$. Then, this family $\mathcal{Q}$ of (67) is a free weighted-semicircular family in $\mathbb{LS}$.

Proof. Let $\mathcal{Q}$ be a subset (67) in $\mathfrak{LG}$. Then, all elements $Q_{p,j}$ of $\mathcal{Q}$ are $p^{2(j+1)}$-semicircular in $\mathfrak{LG}$ by the above lemma, for all $p \in \mathcal{P}$, $j \in \mathbb{Z}$. Furthermore, all elements $Q_{p,j}$ of $\mathcal{Q}$ are mutually free from each other in $\mathfrak{LG}$, because they are contained in the mutually-distinct free blocks $\mathfrak{LG}_p(j)$ of $\mathfrak{LG}$, for all $p \in \mathcal{P}$, $j \in \mathbb{Z}$. Therefore, the family $\mathcal{Q}$ of (67) is a free weighted-semicircular family in $\mathfrak{LG}$. □

Now, take elements:

$$\Theta_{p,j} \overset{def}{=} \frac{1}{p^{j+1}} Q_{p,j}, \forall p \in \mathcal{P}, j \in \mathbb{Z}, \tag{68}$$

in $\mathfrak{LG}$, where $Q_{p,j} \in \mathcal{Q}$, where $\mathcal{Q}$ is the free weighted-semicircular family (67) in the free Adelic filtration $\mathfrak{LG}$.

Then, by the self-adjointness of $Q_{p,j}$, these operators $\Theta_{p,j}$ of (68) are self-adjoint in $\mathfrak{LG}$, as well, because:

$$p^{j+1} \in \mathbb{Q} \subset \mathbb{R} \text{ in } \mathbb{C},$$

satisfying $\overline{p^{j+1}} = p^{j+1}$, for all $p \in \mathcal{P}$, $j \in \mathbb{Z}$.

Furthermore, one obtains the following free-cumulant computation; if $k_n^0(...)$ is the free cumulant on $\mathfrak{LG}$ in terms of τ^0, then:

$$\begin{aligned} k_n^0\left(\Theta_{p,j}, ..., \Theta_{p,j}\right) &= k_n^{0,p,j}\left(\tfrac{1}{p^{j+1}} Q_{p,j}, ..., \tfrac{1}{p^{j+1}} Q_{p,j}\right) \\ &= \left(\tfrac{1}{p^{j+1}}\right)^n k_n^{0,p,j}\left(Q_{p,j}, ..., Q_{p,j}\right), \end{aligned} \tag{69}$$

by the *bimodule-map property* of the free cumulant (e.g., see [14]), for all $n \in \mathbb{N}$, where $k_n^{0,p,j}(...)$ are the free cumulants (63) on the free blocks $\mathfrak{LG}_p(j)$ in terms of the linear functionals $\tau_{p,j}^0$ of (60) on $\mathfrak{LG}_p$, for all $p \in \mathcal{P}$, $j \in \mathbb{Z}$.

Theorem 3. *Let $\Theta_{p,j} = \frac{1}{p^{j+1}} Q_{p,j}$ be free random variables (68) of the free Adelic filtration $\mathfrak{LG}$, for $Q_{p,j} \in \mathcal{Q}$. Then, the family:*

$$\Theta = \left\{\Theta_{p,j} \in \mathfrak{LG}_p(j) : p \in \mathcal{P},\ j \in \mathbb{Z}\right\} \tag{70}$$

forms a free semicircular family in $\mathfrak{LG}$.

Proof. Consider that:

$$\begin{aligned} k_n^0\left(\Theta_{p,j}, ..., \Theta_{p,j}\right) &= \left(\tfrac{1}{p^{j+1}}\right)^n k_n^{0,p,j}\left(Q_{p,j}, ..., Q_{p,j}\right) \text{ by (69)} \\ &= \begin{cases} \left(\tfrac{1}{p^{j+1}}\right)^2 k_2^{0,p,j}\left(Q_{p,j}, Q_{p,j}\right) & \text{if } n = 2 \\ \left(\tfrac{1}{p^{j+1}}\right)^n k_n^{0,p,j}\left(Q_{p,j}, ..., Q_{p,j}\right) = 0 & \text{otherwise,} \end{cases} \end{aligned}$$

by the $p^{2(j+1)}$-semicircularity of $Q_{p,j} \in \mathcal{Q}$ in $\mathfrak{LG}$:

$$= \begin{cases} \left(\tfrac{1}{p^{j+1}}\right)^2 \left(p^{j+1}\right)^2 = 1 & \text{if } n = 2 \\ 0 & \text{otherwise,} \end{cases} \tag{71}$$

for all $n \in \mathbb{N}$.

By the free-cumulant computation (71), these self-adjoint free random variables $\Theta_{p,j} \in \mathfrak{LG}_p(j)$ are semicircular in $\mathfrak{LG}$ by (43), for all $p \in \mathcal{P}$, $j \in \mathbb{Z}$.

Furthermore, the family Θ of (70) forms a free family in $\mathfrak{LG}$, because all elements $\Theta_{p,j}$ are the scalar-multiples of $Q_{p,j} \in \mathcal{Q}$, contained in mutually-distinct free blocks $\mathfrak{LG}_p(j)$ of $\mathfrak{LG}$, for all $j \in \mathbb{Z}$, $p \in \mathcal{P}$.

Therefore, this family Θ of (70) is a free semicircular family in $\mathfrak{LG}$. □

Now, define a Banach $*$-subalgebra $\mathbb{LS}$ of $\mathfrak{LG}$ by:

$$\mathbb{LS} \overset{def}{=} \overline{\mathbb{C}[\mathcal{Q}]} in \mathfrak{LG}, \tag{72}$$

where $\mathcal{Q}$ is the free weighted-semicircular family (67) and $\overline{Y}$ means the Banach-topology closures of subsets Y of $\mathfrak{LG}$.

Then, one can obtain the following structure theorem for the Banach $*$-algebra $\mathbb{LS}$ of (72) in $\mathfrak{LG}$.

Theorem 4. *Let $\mathbb{LS}$ be the Banach $*$-subalgebra (72) of the free Adelic filtration $\mathfrak{LG}$ generated by the free weighted-semicircular family $\mathcal{Q}$ of (67). Then:*

$$\mathbb{LS} = \overline{\mathbb{C}\left[\Theta\right]} in \mathfrak{LG}, \tag{73}$$

where Θ is the free semicircular family (70) and where "$=$" means "being identically same as sets." Moreover,

$$\mathbb{LS} \overset{*\text{-iso}}{=} \underset{p\in\mathcal{P},\, j\in\mathbb{Z}}{\star} \overline{\mathbb{C}\left[\{Q_{p,j}\}\right]} \overset{*\text{-iso}}{=} \overline{\mathbb{C}\left[\underset{p\in\mathcal{P},\, j\in\mathbb{Z}}{\star} \{Q_{p,j}\}\right]}, \tag{74}$$

in $\mathfrak{LG}$, where "$\overset{*\text{-iso}}{=}$" means "being Banach-$*$-isomorphic," and:

$$\overline{\mathbb{C}\left[\{Q_{p,j}\}\right]} \text{ are Banach } *\text{-subalgebras of } \mathfrak{LG}_p(j),$$

for all $p \in \mathcal{P}$, $j \in \mathbb{Z}$, in $\mathfrak{LG}$.

Here, $(\star)$ in the first $*$-isomorphic relation of (74) is the (free-probability-theoretic) free product of [14,15], and $(\star)$ in the second $*$-isomorphic relation of (74) is the (pure-algebraic) free product (generating noncommutative algebraic free words in the family $\mathcal{Q}$).

Proof. Let $\mathbb{LS}$ be the Banach $*$-subalgebra (72) of $\mathfrak{LG}$. Then, all generating operators $Q_{p,j} \in \mathcal{Q}$ of $\mathbb{LS}$ are contained in mutually-distinct free blocks $\mathfrak{LG}_p(j)$ of $\mathfrak{LG}$, and hence, the Banach $*$-subalgebras $\overline{\mathbb{C}\left[\{Q_{p,j}\}\right]}$ of $\mathfrak{LG}$ are contained in the free blocks $\mathfrak{LG}_p(j)$, for all $p \in \mathcal{P}$, $j \in \mathbb{Z}$. Therefore, as embedded sub-structures of $\mathfrak{LG}$, they are free from each other. Equivalently,

$$\mathbb{LS} \overset{*\text{-iso}}{=} \underset{p\in\mathcal{P},\, j\in\mathbb{Z}}{\star} \overline{\mathbb{C}\left[\{Q_{p,j}\}\right]} in \mathfrak{LG}, \tag{75}$$

by (66).

Since every free block $\overline{\mathbb{C}\left[\{Q_{p,j}\}\right]}$ of the Banach $*$-algebra $\mathbb{LS}$ of (75) is generated by a single self-adjoint (weighted-semicircular) element, every operator T of $\mathbb{LS}$ is a limit of linear combinations of free words in the free family $\mathcal{Q}$ of (67), which form noncommutative free "reduced" words (in the sense of [14,15]), as operators in $\mathbb{LS}$ of (75). Note that every (pure-algebraic) free word in $\mathcal{Q}$ has a unique free reduced word in $\mathbb{LS}$, under operator-multiplication on $\mathfrak{LG}$ (and hence, on $\mathbb{LS}$). Therefore, the $*$-isomorphic relation (75) guarantees that:

$$\mathbb{LS} \overset{*\text{-iso}}{=} \overline{\mathbb{C}\left[\underset{p\in\mathcal{P},\, j\in\mathbb{Z}}{\star} \{Q_{p,j}\}\right]}, \tag{76}$$

where the free product $(\star)$ in (76) is pure-algebraic.

Remark that, indeed, the relation (76) holds well, because all weighted-semicircular elements of $\mathcal{Q}$ are self-adjoint; if:

$$T = \prod_{l=1}^{N} Q_{p_l, j_l}^{n_l} \in \mathbb{LS}$$

is a free (reduced) word (as an operator), then:

$$T^{*} = \prod_{l=1}^{N} Q_{p_{N-l+1}, j_{N-l+1}}^{n_{N-l+1}} \in \mathbb{LS}$$

is a free word of $\mathbb{LS}$ in $\mathcal{Q}$, as well. Therefore, by (75) and (76), the structure theorem (74) holds true.

Note now that $Q_{p,j} \in \mathcal{Q}$ satisfy:

$$Q_{p,j} = p^{j+1}\Theta_{p,j}, \text{ for all } p \in \mathcal{P}, j \in \mathbb{Z},$$

where $\Theta_{p,j}$ are semicircular elements in the family Θ of (70). Therefore, the free blocks of (75) satisfy that:

$$\overline{\mathbb{C}\left[\{Q_{p,j}\}\right]} = \overline{\mathbb{C}\left[\{p^{j+1}\Theta_{p,j}\}\right]} = \overline{\mathbb{C}\left[\{\Theta_{p,j}\}\right]}, \tag{77}$$

for all $p \in \mathcal{P}, j \in \mathbb{Z}$.

Thus, one can get that:

$$\mathbb{LS} \overset{*\text{-iso}}{=} \underset{p \in \mathcal{P}, j \in \mathbb{Z}}{\star} \overline{\mathbb{C}\left[\{\Theta_{p,j}\}\right]}, \tag{78}$$

by (75) and (77).

With similar arguments of (75), we have:

$$\mathbb{LS} = \overline{\mathbb{C}[\Theta]}, set - theoretically, \tag{79}$$

by (78).

Therefore, the identity (73) holds true by (79). □

As a sub-structure, one can restrict the linear functional τ^0 of (66) on $\mathfrak{LS}$ to that on $\mathbb{LS}$, i.e., one can obtain the Banach $*$-probability space,

$$\left(\mathbb{LS}, \ \tau^0 \overset{denote}{=} \tau^0 \mid_{\mathbb{LC}}\right). \tag{80}$$

Definition 14. *Let* $(\mathbb{LS}, \tau^0)$ *be the Banach $*$-probability space* (80). *Then, we call* $(\mathbb{LS}, \tau^0)$ *the semicircular (free Adelic sub-)filtration of* $\mathfrak{LS}$.

Note that, by (66), all elements of the semicircular filtration $(\mathbb{LS}, \tau^0)$ provide "possible" non-vanishing free distributions in the free Adelic filtration $\mathfrak{LS}$. Especially, all free reduced words of $\mathfrak{LS}$ in the generator set $\{Q_{p,j}\}_{p \in P, j \in \mathbb{Z}}$ have non-zero free distributions only if they are contained in $(\mathbb{LS}, \tau^0)$. Therefore, studying free-distributional data on $(\mathbb{LS}, \tau^0)$ is to study possible non-zero free-distributional data on $\mathfrak{LS}$.

9. Truncated Linear Functionals on $\mathbb{LS}$

In *number theory*, one of the most interesting, but difficult topics is to find a number of primes or a density of primes contained in closed intervals $[t_1, t_2]$ of the real numbers $\mathbb{R}$ (e.g., [3,6,21,22]). Since the theory is deep, we will not discuss more about it here. Hhowever, motivated by the theory, we consider certain "suitable" *truncated linear functionals* on our semicircular filtration $(\mathbb{LS}, \tau^0)$ of (80) in the free Adelic filtration $\mathfrak{LS}$ of (66).

Notation: From below, we will use the following notations to distinguish their structural differences;

$\mathbb{LS} \overset{denote}{=}$ the Banach $*$-subalgebra (72) of $\mathfrak{LS}$,

$\mathbb{LS}_0 \overset{denote}{=}$ the semicircular filtration $(\mathbb{LS}, \tau^0)$ of (80).

□

9.1. *Linear Functionals* $\{\tau_{(t)}\}_{t\in\mathbb{R}}$ *on* $\mathbb{LS}$

Let $\mathbb{LS}_0$ be the semicircular filtration $(\mathbb{LS}, \tau^0)$ of the free Adelic filtration $\mathfrak{LS}$. Furthermore, let $\mathcal{Q}$ and Θ be the free weighted-semicircular family (67), respectively, the free semicircular family (70) of $\mathfrak{LS}$, generating $\mathbb{LS}$ by (73) and (74). We here truncate τ^0 on $\mathbb{LS}$ for a fixed real number $t \in \mathbb{R}$.

First, recall and remark that:

$$\tau^0 = \underset{p\in\mathcal{P},\, j\in\mathbb{Z}}{\star} \tau^0_{p,j} \text{ on } \mathbb{LS},$$

by (66) and (80). Therefore, one can sectionize τ^0 over $\mathcal{P}$, as follows;

$$\tau^0 = \underset{p\in\mathcal{P}}{\star} \tau^0_p \text{ on } \mathbb{LS},$$

with:

$$\tau^0_p = \underset{j\in\mathbb{Z}}{\star} \tau^0_{p,j} on \mathbb{LS}_p, for all p \in \mathcal{P}, \tag{81}$$

where:

$$\mathbb{LS}_p \overset{def}{=} \underset{j\in\mathbb{Z}}{\star} \overline{\mathbb{C}[\{\Theta_{p,j}\}]} \subset \mathbb{LS} \subset \mathfrak{LS}, \tag{82}$$

for each $p \in \mathcal{P}$, under (74).

From below, we understand the Banach $*$-subalgebras $\mathbb{LS}_p$ of $\mathbb{LS}$ as free-probabilistic sub-structures,

$$\mathbb{LS}_{(p)} \overset{denote}{=} \left(\mathbb{LS}_p,\ \tau^0_p\right), for all p \in \mathcal{P}. \tag{83}$$

Lemma 3. *Let* $\mathbb{LS}_{p_l}$ *be in the sense of* (82) *in the semicircular filtration* $\mathbb{LS}_0$*, for* $l = 1, 2$*. Then,* $\mathbb{LS}_{p_1}$ *and* $\mathbb{LS}_{p_2}$ *are free in* $\mathbb{LS}_0$*, if and only if* $p_1 \neq p_2$ *in* $\mathcal{P}$*.*

Proof. The proof is directly done by (81) and (82). Indeed,

$$\begin{aligned}\mathbb{LS} &= \underset{p\in\mathcal{P},\, j\in\mathbb{Z}}{\star} \overline{\mathbb{C}\left[\{\Theta_{p,j}\}\right]} \\ &= \underset{p\in\mathcal{P}}{\star}\left(\underset{j\in\mathbb{Z}}{\star} \overline{\mathbb{C}\left[\{\Theta_{p,j}\}\right]}\right) = \underset{p\in\mathcal{P}}{\star} \mathbb{LS}_p,\end{aligned}$$

by (80) and (82).

Therefore, $\mathbb{LS}_{p_1}$ and $\mathbb{LS}_{p_2}$ are free in $\mathbb{LS}_0$, if and only if $p_1 \neq p_2$ in $\mathcal{P}$. □

Fix now $t \in \mathbb{R}$, and define a new linear functional $\tau_{(t)}$ on $\mathbb{LS}$ by:

$$\tau_{(t)} \overset{def}{=} \begin{cases} \underset{p\le t}{\star} \tau^0_p & \text{on } \underset{p\le t}{\star} \mathbb{LS}_p \subset \mathbb{LS} \\ \\ O & \text{on } \mathbb{LS} \setminus \left(\underset{p\le t}{\star} \mathbb{LS}_p\right), \end{cases} \tag{84}$$

where τ^0_p are the linear functionals (81) on the Banach $*$-subalgebras $\mathbb{LS}_p$ of (82) in $\mathbb{LS}_0$, for all $p \in \mathcal{P}$, and O means the *zero linear functional* on $\mathbb{LS}$, satisfying that:

$$O(T) = 0, \text{ for all } T \in \mathbb{LS}.$$

For convenience, if there is no confusion, we simply write the definition (84) as:

$$\tau_{(t)} \overset{denote}{=} \underset{p\le t}{\star} \tau^0_p. \tag{85}$$

By the definition (84) (with a simpler expression (85)), one can easily verify that, if $t < 2$ in $\mathbb{R}$, then the corresponding linear functional $\tau_{(t)}$ is identical to the zero linear functional O on $\mathbb{LS}$. To avoid such triviality, one may refine $\tau_{(t)}$ of (84) by:

$$\tau_{(t)} \overset{def}{=} \begin{cases} \tau_{(t)} \text{ of } (84) & \text{if } t \geq 2 \\ O & \text{if } t < 2, \end{cases} \tag{86}$$

for all $t \in \mathbb{R}$.

In the following text, $\tau_{(t)}$ mean the linear functionals in (86), satisfying (84) whenever $t \geq 2$, for all $t \in \mathbb{R}$. In fact, we are not interested in the cases where $t < 2$.

For example,

$$\tau_{(\frac{\sqrt{3}}{2})} = O, \tau_{(2.1003)} = \tau_2^0, and \tau_{(6)} = \tau_2^0 \star \tau_3^0 \star \tau_5^0,$$

on $\mathbb{LS}$, under (85), etc.

Theorem 5. *Let $Q_{p,j} \in \mathcal{Q}$ and $\Theta_{p,j} \in \Theta$ in the semicircular filtration $\mathbb{LS}_0$, for $p \in \mathcal{P}$, $j \in \mathbb{Z}$, and let $t \in \mathbb{R}$ and $\tau_{(t)}$, the corresponding linear functional (86) on $\mathbb{LS}$. Then:*

$$\tau_{(t)}\left(Q_{p,j}^n\right) = \begin{cases} \omega_n p^{2(j+1)} c_{\frac{n}{2}} & \text{if } t \geq p \\ 0 & \text{if } t < p, \end{cases}$$

and:

$$\tau_{(t)}\left(\Theta_{p,j}^n\right) = \begin{cases} \omega_n c_{\frac{n}{2}} & \text{if } t \geq p \\ 0 & \text{if } t < p, \end{cases} \tag{87}$$

for all $n \in \mathbb{N}$.

Proof. By the $p^{2(j+1)}$-semicircularity of $Q_{p,j} \in \mathcal{Q}$, the semicircularity of $\Theta_{p,j} \in \Theta$ in the semicircular filtration $\mathbb{LS}_0$, and by the definition (86), if $t \geq p$ in $\mathbb{R}$, then:

$$\begin{aligned} \tau_{(t)}\left(Q_{p,j}^n\right) &= \tau_p^0\left(Q_{p,j}^n\right) = \tau_{p,j}^0\left(Q_{p,j}^n\right) \\ &= \omega_n p^{2(j+1)} c_{\frac{n}{2}}, \end{aligned}$$

and:

$$\begin{aligned} \tau_{(t)}\left(\Theta_{p,j}^n\right) &= \tau_p^0\left(\Theta_{p,j}^n\right) = \tau_{p,j}^0\left(\Theta_{p,j}^n\right) \\ &= \omega_n c_{\frac{n}{2}}, \end{aligned}$$

by (62), (71), and (81), for all $n \in \mathbb{N}$.

If $t < p$, then:

$$\tau_{(t)} = \underset{2 \leq q < t < p \text{ in } \mathcal{P}}{\star} \tau_q^0 \text{ or } O, \text{ on } \mathbb{LS}.$$

Therefore, in such cases,

$$\tau_{(t)}\left(Q_{p,j}^n\right) = \tau_{(t)}\left(\Theta_{p,j}^n\right) = 0, \text{ for all } n \in \mathbb{N},$$

by (84), (85), and (86).

Therefore, the free-distributional data (87) for the linear functional $\tau_{(t)}$ hold on $\mathbb{LS}$. □

The above theorem shows how the original free-probabilistic information on the semicircular filtration $\mathbb{LS}_0$ is affected by the new free-probabilistic models on $\mathbb{LS}$, under "truncated" linear functionals $\tau_{(t)}$ of τ^0 on $\mathbb{LS}$, for $t \in \mathbb{R}$.

Definition 15. *Let $\tau_{(t)}$ be the linear functionals (86) on $\mathbb{LS}$, for $t \in \mathbb{R}$. Then, the corresponding new Banach $*$-probability spaces,*

$$\mathbb{LS}_{(t)} \overset{denote}{=} \left(\mathbb{LS},\ \tau_{(t)}\right), \tag{88}$$

are called the semicircular t-(truncated-)filtrations of $\mathbb{LS}$ (or, of $\mathbb{LS}_0$).

Note that if t is "suitable" in the sense that "$\tau_{(t)} \neq O$ on $\mathbb{LS}$," then the free-probabilistic structure $\mathbb{LS}_{(t)}$ of (88) is meaningful.

Notation and Assumption 9.1 (NA 9.1, from below): In the following, we will say "$t \in \mathbb{R}$ is suitable," if the semicircular t-filtration "$\mathbb{LS}_{(t)}$ of (88) is meaningful," in the sense that: $\tau_{(t)} \neq O$ fully on $\mathbb{LS}$. □

Now, let us consider the following concepts.

Definition 16. *Let (A_k, φ_k) be Banach $*$-probability spaces (or C^*-probability spaces, or W^*-probability spaces, etc.), for $k = 1, 2$. A Banach $*$-probability space (A_1, φ_1) is said to be free-homomorphic to a Banach $*$-probability space (A_2, φ_2), if there exists a bounded $*$-homomorphism:*

$$\Phi : A_1 \to A_2,$$

such that:

$$\varphi_2\left(\Phi(a)\right) = \varphi_1\left(a\right),$$

for all $a \in A_1$. Such a $*$-homomorphism Φ is called a free-homomorphism.

If Φ is both a $*$-isomorphism and a free-homomorphism, then Φ is said to be a free-isomorphism, and we say that (A_1, φ_1) and (A_2, φ_2) are free-isomorphic. Such a free-isomorphic relation is nothing but the equivalence in the sense of Voiculescu (e.g., [15]).

By (87), we obtain the following free-probabilistic-structural theorem.

Theorem 6. *Let $\mathbb{LS}_q = \underset{j\in\mathbb{Z}}{\star} \overline{\mathbb{C}\left[\{Q_{q,j}\}\right]}$ be Banach $*$-subalgebras (82) of $\mathbb{LS}$, for all $q \in \mathcal{P}$. Let $t \in \mathbb{R}$ be suitable in the sense of* **NA 9.1** *and $\mathbb{LS}_{(t)}$ be the corresponding semicircular t-filtration (88). Construct a Banach $*$-probability space $\mathbb{LS}^t$ by a Banach $*$-probabilistic sub-structure of the semicircular filtration $\mathbb{LS}_0$,*

$$\mathbb{LS}^t \overset{def}{=} \underset{p\leq t}{\star} \left(\mathbb{LS}_p,\ \tau_p^0\right) = \left(\underset{p\leq t}{\star} \mathbb{LS}_p,\ \underset{p\leq t}{\star} \tau_p^0\right), \tag{89}$$

where $\tau_p^0 = \underset{j\in\mathbb{Z}}{\star} \tau_{p,j}^0$ are in the sense of (81). Then:

$$\mathbb{LS}^t is free-homomorphic to \mathbb{LS}_{(t)}. \tag{90}$$

Proof. Let $\mathbb{LS}_{(t)}$ be the semicircular t-filtration (88) of $\mathbb{LS}$, and let $\mathbb{LS}^t$ be a Banach $*$-probability space (89), for a suitably fixed $t \in \mathbb{R}$.

Define a bounded linear morphism:

$$\Phi_t : \mathbb{LS}^t \to \mathbb{LS}_{(t)},$$

by the natural embedding map,

$$\Phi_t\left(T\right) = T in \mathbb{LS}_{(t)}, for all T \in \mathbb{LS}^t. \tag{91}$$

Then, this morphism Φ_t is an injective bounded $*$-homomorphism from $\mathbb{LS}^t$ into $\mathbb{LS}_{(t)}$, by (72), (75), (82), (89), and (91).

Therefore, one obtains that:

$$\tau_{(t)}\left(\Phi(T)\right) = \tau_{(t)}(T) = \left(\underset{p \le t \text{ in } \mathcal{P}}{\star} \tau_p^0\right)(T) = \tau^t\left(T\right),$$

for all $T \in \mathbb{LS}^t$, by (87).

It shows that the Banach $*$-probability space $\mathbb{LS}^t$ of (89) is free-homomorphic to the semicircular t-filtration $\mathbb{LS}_{(t)}$ of (88). Therefore, the statement (90) holds under the free-homomorphism Φ_t of (91). □

The above theorem shows that the Banach $*$-probability spaces $\mathbb{LS}^t$ of (89) are free-homomorphic to the semicircular t-filtrations $\mathbb{LS}_{(t)}$ of (88), for all $t \in \mathbb{R}$. Note that it "seems" they are not free-isomorphic, because:

$$\left(\underset{q \le t \text{ in } \mathcal{P}}{\star} \mathbb{LS}_q\right) \subsetneqq \left(\underset{p \in \mathcal{P}}{\star} \mathbb{LS}_p\right) = \mathbb{LS},$$

set-theoretically, for $t \in \mathbb{R}$. However, we are not sure at this moment that they are free-isomorphic or not, because we have the similar difficulties discussed in [19].

Remark 3. *The famous main result of [19] says that: if $L(F_n)$ are the free group factors (group von Neumann algebras) of the free groups F_n with n-generators, for all:*

$$n \in \mathbb{N}_{>1}^{\infty} = (\mathbb{N} \setminus \{1\}) \cup \{\infty\},$$

then either (I) or (II) holds true, where:

(I) $L(F_n) \overset{*\text{-iso}}{=} L(F_\infty)$, *for all* $n \in \mathbb{N}_{>1}^{\infty}$,

(II) $L(F_{n_1}) \overset{*\text{-iso}}{\neq} L(F_{n_2})$, *if and only if* $n_1 \neq n_2 \in \mathbb{N}_{>1}^{\infty}$,

where "$\overset{*\text{-iso}}{=}$" means "being W-isomorphic." Depending on the author's knowledge, he does not know which one is true at this moment.*

We here have similar troubles. Under the similar difficulties, we are not sure at this moment that $\mathbb{LS}^t$ and $\mathbb{LS}_{(t)}$ (or $\mathbb{LS}^t$ and $\mathbb{LS}$) are $*$-isomorphic or not (and hence, free-isomorphic or not).

However, definitely, $\mathbb{LS}^t$ is free-homomorphic "into" $\mathbb{LS}_{(t)}$ in the semicircular filtration $\mathbb{LS}_0$, by the above theorem.

The above free-homomorphic relation (90) lets us understand all
"non-zero" free distributions of free reduced words of $\mathbb{LS}_{(t)}$ as those of $\mathbb{LS}^t$, for all $t \in \mathbb{R}$, by the injectivity of a free-homomorphism Φ_t of (91).

Corollary 2. *All free reduced words T of the semicircular t-filtration $\mathbb{LS}_{(t)}$ in $\mathcal{Q} \cup \Theta$, having non-zero free distributions, are contained in the Banach $*$-probability space $\mathbb{LS}^t$ of (89), whenever t is suitable. The converse holds true, as well.*

Proof. The proof of this characterization is done by (87), (89), and (90). In particular, the injectivity of the free-homomorphism Φ_t of (91) guarantees that this characterization holds. □

Therefore, whenever we consider a non-zero free-distribution having free reduced words T of semicircular t-filtrations $\mathbb{LS}_{(t)}$, they are regarded as free random variables of the Banach $*$-probability spaces $\mathbb{LS}^t$ of (89), for all suitable $t \in \mathbb{R}$.

9.2. Truncated Linear Functionals $\tau_{t_1 < t_2}$ on $\mathbb{LS}$

In this section, we generalize the semicircular t-filtrations $\mathbb{LS}_{(t)}$ by defining so-called *truncated linear functionals* on the Banach $*$-algebra $\mathbb{LS}$.

Throughout this section, let $[t_1, t_2]$ be a *closed interval* in $\mathbb{R}$, satisfying:

$$|t_1 - t_2| \neq 0, \text{ for } t_1 < t_2 \in \mathbb{R}.$$

For such a fixed closed interval $[t_1, t_2]$, define the corresponding linear functional $\tau_{t_1<t_2}$ on the semicircular filtration $\mathbb{LS}$ by:

$$\tau_{t_1<t_2} \overset{def}{=} \begin{cases} \underset{t_1 \leq p \leq t_2 \text{ in } \mathcal{P}}{\star} \tau_p^0 & \text{on } \underset{t_1 \leq p \leq t_2 \text{ in } \mathcal{P}}{\star} \mathbb{LS}_p \subset \mathbb{LS} \\ O & \text{on } \mathbb{LS} \setminus \left(\underset{t_1 \leq p \leq t_2 \text{ in } \mathcal{P}}{\star} \mathbb{LS} \right), \end{cases} \tag{92}$$

where τ_p^0 are the linear functionals (81) on the Banach $*$-subalgebras $\mathbb{LS}_p$ of (82) in $\mathbb{LS}$, for $p \in \mathcal{P}$. Similar to Section 9.1, if there is no confusion, then we simply write the definition (92) as:

$$\tau_{t_1<t_2} \overset{denote}{=} \underset{t_1 \leq p \leq t_2}{\star} \tau_p^0 on \mathbb{LS}. \tag{93}$$

To make the linear functionals $\tau_{t_1<t_2}$ of (92) be non-zero-linear functionals on $\mathbb{LS}$, the interval $[t_1, t_2]$ must be taken "suitably." For example,

$$\tau_{t_1<t_2} = O, \text{ whenever } t_2 < 2,$$

and:

$$\tau_{8<10} = O, \tau_{14<16} = O, \text{ and } \tau_{\frac{3}{7}<\frac{3}{2}} = O, \text{ etc.,}$$

but:

$$\tau_{\frac{3}{2}<8} = \tau_{(8)} = \tau_2^0 \star \tau_3^0 \star \tau_5^0 \star \tau_7^0$$

and:

$$\tau_{7<14} = \tau_7^0 \star \tau_{11}^0 \star \tau_{13}^0,$$

under (93) on $\mathbb{LS}$.

It is not difficult to check that the definition (92) of truncated linear functionals $\tau_{t_1<t_2}$ covers the definition of linear functionals $\tau_{(t)}$ of (86). In particular, $\tau_{(t)}$ is "suitable" in the sense of **NA 9.1**, then:

$$\tau_{(t)} = \tau_{2<t} = \tau_{s<t}, \text{ for all } 2 \geq s \in \mathbb{R}.$$

For our purposes, we will axiomatize:

$$\tau_{p<p} = \tau_p^0, \text{ for all } p \in \mathcal{P} \subset \mathbb{R},$$

notationally, where τ_p^0 are the linear functionals (81), for all $p \in \mathcal{P}$, under (93). Remark that the very above axiomatized notations $\tau_{p<p}$ will be used only when p are primes.

Definition 17. *Let $[t_1, t_2]$ be a given interval in $\mathbb{R}$ and $\tau_{t_1<t_2}$, the corresponding linear functional (92) on $\mathbb{LS}$. Then, we call it the $[t_1, t_2]$(-truncated)-linear functional on $\mathbb{LS}$. The corresponding Banach $*$-probability space:*

$$\mathbb{LS}_{t_1<t_2} = (\mathbb{LS}, \tau_{t_1<t_2}) \tag{94}$$

is said to be the semicircular a $[t_1, t_2]$(-truncated)-filtration.

As we discussed in the above paragraphs, the semicircular $[t_1, t_2]$-filtration $\mathbb{LS}_{t_1<t_2}$ of (94) will be "meaningful," if $t_1 < t_2$ are suitable in $\mathbb{R}$, as in **NA 9.1**.

Notation and Assumption 9.2 (**NA 9.2**, from below): In the rest of this paper, if we write "$t_1 < t_2$ are suitable," then this means "$\mathbb{LS}_{t_1<t_2}$ is meaningful," in the sense that $\tau_{t_1<t_2} \neq O$ fully on $\mathbb{LS}$, with additional axiomatization:

$$\tau_{p<p} = \tau_p^0, \text{ for } p \in \mathcal{P} \text{ in } \mathbb{R},$$

in the sense of (93). □

Theorem 7. *Let $t_1 \leq 2$ and t_2 be suitable in $\mathbb{R}$ in the sense of* **NA 9.1**.

$$\text{The semicircular}[t_1, t_2]\text{-filtration}\mathbb{LS}_{t_1<t_2}\text{is not only suitable in the sense of NA 9.2, but also, it is free-isomorphic to the semicircular } t_2\text{-filtration } \mathbb{LS}_{(t_2)} \text{ of (88).} \tag{95}$$

$$\text{The Banach } * -\text{probability space } \mathbb{LS}^{t_2} \; of \; (89) \; is \; free-homomorphic \; to \; \mathbb{LS}_{t_1<t_2}. \tag{96}$$

Proof. Suppose $t_1 \leq 2$, and t_2 are suitable in $\mathbb{R}$ in the sense of **NA 9.1**. Then, $t_1 < t_2$ are suitable in $\mathbb{R}$ in the sense of **NA 9.2**. Therefore, both the semicircular t_2-filtration $\mathbb{LS}_{(t_2)}$ and the semicircular $[t_1, t_2]$-filtration $\mathbb{LS}_{t_1<t_2}$ are meaningful.

Since t_1 is assumed to be less than or equal to two, the linear functional $\tau_{t_1<t_2} = \tau_{(t_2)}$, by (86) and (92), including the case where $\tau_{2<2} = \tau_2^0$, in the sense of (93). Therefore,

$$\mathbb{LS}_{t_1<t_2} = (\mathbb{LS}, \; \tau_{t_1<t_2}) = \left(\mathbb{LS}, \; \tau_{(t_2)}\right) = \mathbb{LS}_{(t_2)}.$$

Therefore, $\mathbb{LS}_{t_1<t_2}$ and $\mathbb{LS}_{(t_2)}$ are free-isomorphic under the identity map on $\mathbb{LS}$, acting as a free-isomorphism. Therefore, the statement (95) holds.

By (90), the Banach $*$-probability space $\mathbb{LS}^{t_2}$ of (89) is free-homomorphic to $\mathbb{LS}_{(t_2)}$. Therefore, under the hypothesis, $\mathbb{LS}^{t_2}$ is free-homomorphic to $\mathbb{LS}_{t_1<t_2}$ by (95). Equivalently, the statement (96) holds. □

The above theorem characterizes the free-probabilistic structures for semicircular $[t_1, t_2]$-filtrations $\mathbb{LS}_{t_1<t_2}$, whenever $t_1 \leq 2$, and t_2 are suitable, by (95) and (96). Therefore, we now restrict our interests to the cases where:

$$t_1 \geq 2 \text{ in } \mathbb{R}.$$

Therefore, we focus on the semicircular $[t_1, t_2]$-filtration $\mathbb{LS}_{t_1<t_2}$, where:

$$2 \leq t_1 < t_2 \text{ are suitable in } \mathbb{R},$$

in the sense of **NA 9.2**.

Theorem 8. *Let $2 \leq t_1 < t_2$ be suitable in $\mathbb{R}$, and let $\mathbb{LS}_{t_1<t_2}$ be the semicircular $[t_1, t_2]$-filtration (94). Then, the Banach $*$-probability space:*

$$\mathbb{LS}^{t_1<t_2} \overset{def}{=} \underset{t_1 \leq p \leq t_2 \text{ in } \mathcal{P}}{\star} \left(\mathbb{LS}_p, \; \tau_p^0\right), \tag{97}$$

equipped with its linear functional $\tau^{t_1<t_2} = \underset{t_1 \leq p \leq t_2}{\star} \tau_p^0$, is free-homomorphic to $\mathbb{LS}_{t_1<t_2}$ in $\mathbb{LS}$, i.e., if $2 \leq t_1 < t_2$ are suitable in $\mathbb{R}$,

$$\mathbb{LS}^{t_1<t_2} \text{ is free-homomorphic to } \mathbb{LS}_{t_1<t_2} \text{ in } \mathbb{LS}_0. \tag{98}$$

Proof. Let $\mathbb{LS}^{t_1<t_2}$ be in the sense of (97) in the semicircular filtration $\mathbb{LS}_0$, i.e.,

$$\mathbb{LS}^{t_1<t_2} = \left(\underset{t_1 \leq p \leq t_2}{\star} \mathbb{LS}_p, \; \tau^{t_1<t_2} = \underset{t_1 \leq p \leq t_2}{\star} \tau_p^0\right),$$

as a free-probabilistic sub-structure of the semicircular filtration $\mathbb{LS}_0$.

By (94), one can define the embedding map Φ from $\mathbb{LS}^{t_1<t_2}$ into $\mathbb{LS}$, satisfying;

$$\Phi(T) = T, \text{for all } T \in \mathbb{LS}^{t_1<t_2}.$$

Then, for any $T \in \mathbb{LS}^{t_1<t_2}$, one can get that:

$$\tau^{t_1<t_2}(T) = \tau_{t_1<t_2}(T) = \tau^0(T).$$

Therefore, the Banach $*$-probability space $\mathbb{LS}^{t_1<t_2}$ is free-homomorphic to $\mathbb{LS}_{t_1<t_2}$ in $\mathbb{LS}$. Therefore, the relation (98) holds. □

Remark again that we are not sure if $\mathbb{LS}^{t_1<t_2}$ and $\mathbb{LS}_{t_1<t_2}$ are free-isomorphic, or not, at this moment (see Remark 9.1 above). However, similar to (90), one can verify that all free reduced words T of $\mathbb{LS}^{t_1<t_2}$ have non-zero free distributions embedded in $\mathbb{LS}_{t_1<t_2}$, and conversely, all free reduced words of $\mathbb{LS}_{t_1<t_2}$ having non-zero free distributions are contained in $\mathbb{LS}^{t_1<t_2}$.

Corollary 3. *Let T be a free reduced word of the semicircular $[t_1, t_2]$-filtration $\mathbb{LS}_{t_1<t_2}$ in $\mathcal{Q} \cup \Theta$, and assume that the free distribution of T is non-zero for $\tau_{t_1<t_2}$. Then, T is an element of the Banach $*$-probability space $\mathbb{LS}^{t_1<t_2}$ of (97). The converse holds true.* □

9.3. More about Free-Probabilistic Information on $\mathbb{LS}_{t_1<t_2}$

In this section, we discuss more about free-probabilistic information in semicircular $[t_1, t_2]$-filtrations $\mathbb{LS}_{t_1<t_2}$, for $t_1 < t_2 \in \mathbb{R}$ (which are not necessarily suitable in the sense of **NA 9.2**).

First, let us mention about the following trivial cases.

Proposition 9. *Let $\mathbb{LS}_{t_1<t_2}$ be the semicircular $[t_1, t_2]$-filtration for $t_1 < t_2$ in $\mathbb{R}$.*

If $t_2 < 2$ in $\mathbb{R}$, then all elements of $\mathbb{LS}_{t_1<t_2}$ have the zero free distribution. (99)

Let $t_1, t_2 \geq 2$ in $\mathbb{R}$. If the closed interval $[t_1, t_2]$ does not contain a prime in $\mathbb{R}$, then all elements of $\mathbb{LS}_{t_1<t_2}$ have the zero free distribution. (100)

Proof. The proofs of the statements (99) and (100) are done immediately by (90), (95), (96), and (98). □

Even though the above results (99) and (100), themselves, are trivial, they illustrate how our original (non-zero) free-distributional data on the semicircular filtration $\mathbb{LS}_0$ are distorted under our "unsuitable" truncations.

Now, suppose $t_1 < t_2$ are suitable in $\mathbb{R}$, and:

$$t_1 \to \infty \text{ in } \mathbb{R},$$

in the sense that: t_1 is big "enough" in $\mathbb{R}$. The existence of such suitable intervals $[t_1, t_2]$ in $\mathbb{R}$ is guaranteed by the *prime number theorem* (e.g., [5,6]).

More precisely, let us collect all suitable pairs (t_1, t_2) in $\mathbb{R}^2$, i.e.,

$$\{(t_1, t_2) \in \mathbb{R}^2 : t_1 < t_2 \text{ are suitable in } \mathbb{R}\},$$

and consider its boundary.

First, consider that if $p \to \infty$ in $\mathcal{P}$ (under the usual total ordering on $\mathcal{P}$, inherited by that on $\mathbb{R}$), then:

$$\lim_{p \to \infty \text{ in } \mathcal{P}} p^{2(j+1)} = \begin{cases} 0 & \text{if } j < -1 \\ 1 & \text{if } j = -1 \\ \infty, \text{ undefined} & \text{if } j > -1, \end{cases} \tag{101}$$

for an arbitrarily-fixed $j \in \mathbb{Z}$.

Theorem 9. *Let $(t_n)_{n=1}^{\infty}$ and $(s_n)_{n=1}^{\infty}$ be monotonically "strictly"-increasing $\mathbb{R}$-sequences, satisfying:*

$$t_n < s_n \text{ are suitable in } \mathbb{R},$$

for all $n \in \mathbb{N}$. By the suitability, there exists at least one prime $p_n \in \mathcal{P}$, such that:

$$t_n \leq p_n \leq s_n, foralln \in \mathbb{N}, \tag{102}$$

where the corresponding $\mathbb{R}$-sequence $(p_n)_{n=1}^{\infty}$ is monotonically increasing.

Let $Q_{p_n,j}$ be the corresponding $p_n^{2(j+1)}$-semicircular element in the free weighted-semicircular family $\mathcal{Q}$, as a free random variable of the semicircular $[t_n, s_n]$-filtration $\mathbb{LS}_{t_n<s_n}$, where p_n are the primes of (102), for all $n \in \mathbb{N}$, for any $j \in \mathbb{Z}$. Then:

$$\lim_{n\to\infty}\left(\tau_{t_n<s_n}\left(Q_{p_n,j}^k\right)\right) = \begin{cases} 0 & \text{if } j < -1 \\ \omega_k c_{\frac{k}{2}} & \text{if } j = -1 \\ \infty & \text{if } j > -1, \end{cases} \tag{103}$$

for all $k \in \mathbb{N}$.

Proof. Suppose p_n are the primes satisfying (102) for given suitable:

$$t_n < s_n \text{ in } \mathbb{R},$$

in the sense of **NA 9.2**, for all $n \in \mathbb{N}$. Then, for the $p_n^{2(j+1)}$-semicircular elements $Q_{p_n,j} \in \mathcal{Q}$ (in $\mathbb{LS}_0$), one has that:

$$\tau_{t_n<s_n}\left(Q_{p_n,j}^k\right) = \left(\underset{t_n\le q\le s_n \text{ in } \mathcal{P}}{\star} \tau_q^0\right)\left(Q_{p_n,j}^k\right) = \tau_{p_n}^0\left(Q_{p_n,j}^k\right)$$

by (102)

$$= \tau_{p_n,j}^0\left(Q_{p_n,j}^k\right) = \omega_k p_n^{2(j+1)} c_{\frac{k}{2}}, \tag{104}$$

for all $k \in \mathbb{N}$.

Thus, we have that:

$$\lim_{n\to\infty}\left(\tau_{t_n<s_n}\left(Q_{p_n,j}^k\right)\right) = \lim_{n\to\infty}\left(\omega_k p_n^{2(j+1)} c_{\frac{k}{2}}\right)$$

by (104);

$$= \lim_{p\to\infty}\left(\omega_k p^{2(j+1)} c_{\frac{k}{2}}\right) = \left(\omega_k c_{\frac{k}{2}}\right)\left(\lim_{p\to\infty} p^{2(j+1)}\right)$$

$$= \begin{cases} 0 & \text{if } j < -1 \\ \omega_k c_{\frac{k}{2}} & \text{if } j = -1 \\ \infty & \text{if } j > -1, \end{cases}$$

by (101), for all $k \in \mathbb{N}$. Therefore, the estimation (103) holds. □

The above estimation (103) illustrates the asymptotic free-distributional data of our $p^{2(j+1)}$-semicircular elements $\{Q_{p,j} \in \mathcal{Q}\}_{p\in\mathcal{P}}$ (for a fixed $j \in \mathbb{Z}$), under our suitable truncations, as $p \to \infty$ in $\mathcal{P}$.

Corollary 4. *Let $t_1 < t_2$ be suitable in $\mathbb{R}$ under **NA 9.2**, t_1 be suitably big (i.e., $t_1 \to \infty$) in $\mathbb{R}$, and $j \le -1$ be arbitrarily fixed in $\mathbb{Z}$. Then, there exists $t_0 \in \mathbb{R}$, such that:*

$$\left|\tau_{t_1<t_2}\left(Q_{p,j}^n\right) - t_0\right| \to 0,$$

where:

$$t_0 = \begin{cases} 0 & \text{if } j < -1 \\ \omega_n c_{\frac{n}{2}} & \text{if } j = -1, \end{cases} \tag{105}$$

for all $n \in \mathbb{N}$.

Under the same hypothesis, if $j > -1$ in $\mathbb{Z}$, then:

$$\left|\tau_{t_1<t_2}\left(Q_{p,j}^n\right)\right| \to \infty, \tag{106}$$

for all $n \in \mathbb{N}$.

Proof. The estimations (105) and (106), for suitably big $t_1 \in \mathbb{R}$, are obtained by (103). □

10. Semicircularity of Certain Free Sums in $\mathbb{LS}_{t_1<t_2}$

As in Section 9, we will let $\mathbb{LS}$ be the Banach $*$-subalgebra (72) of the free Adelic filtration $\mathfrak{LS}$, and let $\mathbb{LS}_0$ be the semicircular filtration $(\mathbb{LS}, \tau^0)$ of (80).

Let (A, φ) be an arbitrary topological $*$-probability space and $a \in (A, \varphi)$. We say a free random variable a is a *free sum* in (A, φ), if:

$$a = \Sigma_{l=1}^{N} x_l, \text{ with } x_l \in (A, \varphi),$$

and the summands $x_1, ..., x_N$ of a are free from each other in (A, φ), for $N \in \mathbb{N} \setminus \{1\}$.

Let $t_1 < t_2$ be suitable in $\mathbb{R}$ in the sense of **NA 9.2**, and let $\mathbb{LS}_{t_1<t_2}$ be the corresponding semicircular $[t_1, t_2]$-filtration. Now, we define free random variables X and Y of $\mathbb{LS}$,

$$X = \sum_{l=1}^{N} Q_{p_l,j_l}^{n_l} and Y = \sum_{l=1}^{N} \Theta_{p_l,j_l}^{n_l}, \tag{107}$$

for $Q_{p_l,j_l} \in \mathcal{Q}$ and $\Theta_{p_l,j_l} \in \Theta$, for all $l = 1, ..., N$, for $N \in \mathbb{N} \setminus \{1\}$.

Remark that, the operator X (or Y) of (107) is a free sum in $\mathbb{LS}$, if and only if the summands $Q_{p_l,j_l}^{n_l}$ (resp., $\Theta_{p_l,j_l}^{n_l}$), which are the free reduced words with their lengths one, are free from each other in $\mathbb{LS}$, if and only if Q_{p_l,j_l} (resp., Θ_{p_l,j_l}) are contained in the mutually-distinct free blocks $\overline{\mathbb{C}[\{Q_{p_l,j_l}\}]}$ of $\mathbb{LS}$ by (74), if and only if the pairs (p_l, j_l) are mutually distinct from each other in the Cartesian product $\mathcal{P} \times \mathbb{Z}$, for all $l = 1, ..., N$. i.e., the given operators X and Y of (107) are free sums in $\mathbb{LS}$, if and only if:

$$(p_{l_1}, j_{l_1}) \neq (p_{l_2}, j_{l_2}) in \mathcal{P} \times \mathbb{Z}, \tag{108}$$

for all $l_1 \neq l_2$ in $\{1, ..., N\}$.

Lemma 4. *Let X and Y be in the sense of (107) in the semicircular filtration $\mathbb{LS}_0$. Assume that the pairs (p_l, j_l) are mutually distinct from each other in $\mathcal{P} \times \mathbb{Z}$, for all $l = 1, ..., N$, for $N \in \mathbb{N} \setminus \{1\}$. Then:*

$$\tau^0(X) = \Sigma_{l=1}^{N} \left(\omega_{n_l} p_l^{2(j_l+1)} c_{\frac{n_l}{2}}\right),$$

and:

$$\tau^0(Y) = \sum_{l=1}^{N} \left(\omega_{n_l} c_{\frac{n_l}{2}}\right). \tag{109}$$

Proof. Let X and Y be given as above in $\mathbb{LS}_0$. By the assumption that the pairs (p_l, j_l) are mutually distinct from each other in $\mathcal{P} \times \mathbb{Z}$, these operators X and Y satisfy the condition (108); equivalently, they are free sums in $\mathbb{LS}_0$.

Therefore, one has that:

$$\begin{aligned}\tau^0(X) &= \Sigma_{l=1}^{N} \tau^0\left(Q_{p_l,j_l}^{n_l}\right) = \Sigma_{l=1}^{N} \tau_{p_l,j_l}^0\left(Q_{p_l,j_l}^{n_l}\right) \\ &= \Sigma_{l=1}^{N} \left(\omega_{n_l} p_l^{2(j_l+1)} c_{\frac{n_l}{2}}\right),\end{aligned}$$

by the $p_l^{2(j_l+1)}$-semicircularity of $Q_{p_l,j_l} \in \mathcal{Q}$, for all $l = 1, ..., N$.

Similarly, one can get that:

$$\tau^0(Y) = \Sigma_{l=1}^{N} \tau_{p_l,j_l}^0\left(\Theta_{p_l,j_l}^{n_l}\right) = \Sigma_{l=1}^{N} \left(\omega_{n_l} c_{\frac{n_l}{2}}\right),$$

by the semicircularity of $\Theta_{p_l,j_l} \in \Theta$, for all $l = 1, ..., N$. □

Now, for the operators X and Y of (107), we consider how our truncation distorts the free-distributional data (109).

For a given closed interval $[t_1, t_2]$ in $\mathbb{R}$, where $t_1 < t_2$ are suitable in $\mathbb{R}$, we define:

$$\mathcal{P}_{[t_1,t_2]} = \{p \in \mathcal{P} : t_1 \leq p \leq t_2\} = \mathcal{P} \cap [t_1, t_2],$$

and:

$$\mathcal{P}^c_{[t_1,t_2]} = \mathcal{P} \setminus \mathcal{P}_{[t_1,t_2]}, \tag{110}$$

in $\mathcal{P}$.

By (110), the family $\{\mathcal{P}_{[t_1,t_2]}, \mathcal{P}^c_{[t_1,t_2]}\}$ forms a partition of the set $\mathcal{P}$ of all primes for the fixed interval $[t_1, t_2]$. Of course, if $t_1 < t_2$ are not suitable, then:

$$\mathcal{P}_{[t_1,t_2]} = \varnothing, \text{ and hence, } \mathcal{P} = \mathcal{P}^c_{[t_1,t_2]}.$$

Theorem 10. *Let X and Y be the operators (107), and assume they are free sums in the semicircular filtration $\mathbb{LS}_0$; and let $\mathbb{LS}_{t_1<t_2}$ be the semicircular $[t_1, t_2]$-filtration for suitable $t_1 < t_2$ in $\mathbb{R}$. Then:*

$$\tau_{t_1<t_2}(X) = \sum_{p_l \in \mathcal{P}_{[t_1,t_2]:(p_1,\ldots,p_N)}} \left(\omega_{n_l} p_l^{2(j_l+1)} c_{\frac{n_l}{2}}\right),$$

and:

$$\tau_{t_1<t_2}(Y) = \sum_{p_l \in \mathcal{P}_{[t_1<t_2]:(p_1,\ldots,p_N)}} \left(\omega_{n_l} c_{\frac{n_l}{2}}\right), \tag{111}$$

where:

$$\mathcal{P}_{[t_1,t_2]:(p_1,\ldots,p_N)} = \mathcal{P}_{[t_1,t_2]} \cap \{p_1, \ldots, p_N\} \text{ in } \mathcal{P},$$

where $\mathcal{P}_{[t_1,t_2]}$ is in the sense of (110) in $\mathcal{P}$. Clearly, if $\mathcal{P}_{[t_1,t_2]:(p_1,\ldots,p_N)}$ is empty in $\mathcal{P}$, then the formulas in (111) vanish.

Proof. The proof of (111) is done by (95), (96), (98), and (109). Indeed, if:

$$\mathcal{P}_{[t_1,t_2]:(p_1,\ldots,p_N)} = \mathcal{P}_{[t_1,t_2]} \cap \{p_1, \ldots, p_N\} \text{ in } \mathcal{P},$$

where $\mathcal{P}_{[t_1,t_2]}$ is in the sense of (110), and if:

$$\mathcal{P}_{[t_1,t_2]:(p_1,\ldots,p_N)} \neq \varnothing,$$

then:

$$\tau_{t_1<t_2}(X) = \sum_{p_l \in \mathcal{P}_{[t_1,t_2]:(p_1,\ldots,p_N)}} \tau^0_{p_l,j_l}\left(Q^{n_l}_{p_l,j_l}\right)$$

by (98)

$$= \sum_{p_l \in \mathcal{P}_{[t_1,t_2]:(p_1,\ldots,p_N)}} \left(\omega_{n_l} p_l^{2(j_l+1)} c_{\frac{n_l}{2}}\right),$$

by the $p^{2(j+1)}$-semicircularity of $Q_{p,j} \in \mathcal{Q}$.

Similarly, one can get that:

$$\tau_{t_1<t_2}(Y) = \sum_{p_l \in \mathcal{P}_{[t_1,t_2]:(p_1,\ldots,p_N)}} \left(\omega_{n_l} c_{\frac{n_l}{2}}\right),$$

by the semicircularity of $\Theta_{p,j} \in \Theta$. Therefore, the free-distributional data (111) holds, whenever:

$$\mathcal{P}_{[t_1,t_2]:(p_1,\ldots,p_N)} \neq \varnothing \text{ in } \mathcal{P}.$$

Definitely, if:

$$\mathcal{P}_{[t_1,t_2]:(p_1,\ldots,p_N)} = \varnothing,$$

then:

$$\tau_{t_1<t_2}(X) = O(X) = 0 = O(Y) = \tau_{t_1<t_2}(Y).$$

Therefore, the truncated free-distributional data (111) hold. □

Remark 4. *Let us compare the free-distributional data (109) and (111). One can check the differences between them dictated by the choices of $[t_1, t_2]$ in $\mathbb{R}$. Thus, the formula (111) also illustrates how our truncations on $\mathcal{P}$ distort the original free-probabilistic information on the semicircular filtration $\mathbb{LS}_0$.*

Let q_0 be a fixed prime in $\mathcal{P}$. Choose $t_0 < s_0 \in \mathbb{R}$ such that: (i) these quantities t_0 and s_0 satisfy:

$$t_0 \leq q_0 \leq s_0 \text{ in } \mathbb{R},$$

and (ii) q_0 is the only prime in the closed interval $[t_0, s_0]$ in $\mathbb{R}$.

By the *Archimedean property* on $\mathbb{R}$ (or *the axiom of choice*), the existence of such interval $[t_0, s_0]$, satisfying (i) and (ii) for the fixed prime q_0, is guaranteed; however, the choices of the quantities $t_0 < s_0$ are of course not unique.

Definition 18. *Let $q_0 \in \mathcal{P}$, and let $t < s \in \mathbb{R}$ be the real numbers satisfying the conditions (i) and (ii) of the above paragraph. Then, the suitable closed interval $[t, s]$ is called a q_0-neighborhood.*

Depending on prime-neighborhoods, one can obtain the following semicircularity condition on our semicircular truncated-filtrations.

Corollary 5. *Let $p \in \mathcal{P}$, $[t, s]$ be a p-neighborhood in $\mathbb{R}$, and $\mathbb{LS}_{t<s}$ be the corresponding semicircular $[t, s]$-filtration. If X and Y are free sums formed by (107) in the semicircular filtration $\mathbb{LS}_0$, then:*

$$\tau_{t<s}(X) = \Sigma_{l=1}^{N} \delta_{p,p_l} \left(\omega_{n_l} p_l^{2(j_l+1)} c_{\frac{n_l}{2}}\right),$$

and:

$$\tau_{t<s}(Y) = \sum_{l=1}^{N} \delta_{p,p_l} \left(\omega_{n_l} c_{\frac{n_l}{2}}\right), \tag{112}$$

where δ is the Kronecker delta.

Proof. The free-distributional data (112) are a special case of (111), under the prime-neighborhood condition. Indeed, in this case,

$$\mathcal{P}_{[t,s]:(p_1,\dots,p_N)} = \{p\} \cap \{p_1, \dots, p_N\} = \begin{cases} \{p\} & \text{or} \\ \varnothing, & \end{cases}$$

where $\mathcal{P}_{[t,s]:(p_1,\dots,p_N)}$ is in the sense of (111). □

More general to (112), we obtain the following result.

Proposition 10. *Let $p \in \mathcal{P}$ and $[t, s]$ be a p-neighborhood in $\mathbb{R}$, and let $\mathbb{LS}_{t<s}$ be the corresponding semicircular $[t, s]$-filtration. Then, a free random variable $T \in \mathbb{LS}_{t<s}$ has its non-zero free distribution, if and only if there exists a non-zero summand T_0 of T, such that:*

$$T_0 \in \mathbb{LS}_p in \mathbb{LS}_{t<s}, \tag{113}$$

where $\mathbb{LS}_p = \underset{j\in\mathbb{Z}}{\star} \overline{\mathbb{C}[\{\Theta_{p,j}\}]}$ is a Banach $*$-subalgebra (82) of $\mathbb{LS}$.

Proof. By (98), if $T \in \mathbb{LS}_{t<s}$ has its non-zero free distribution, then there exists a non-zero summand T_0 of T which can be a linear combination of free reduced words contained in $\underset{t \leq q \leq s \text{ in } \mathcal{P}}{\star} \mathbb{LS}_q$, and hence,

$$T_0 \in \underset{t \leq q \leq s \text{ in } \mathcal{P}}{\star} \mathbb{LS}_q, \tag{114}$$

where $\mathbb{LS}_q$ are in the sense of (82), for $q \in \mathcal{P}$.

Since $[t, s]$ is a p-neighborhood, the relation (114) is equivalent to:

$$T_0 \in \mathbb{LS}_p. \tag{115}$$

Clearly, the converse holds true as well, by (98).

Therefore, a free random variable $T \in \mathbb{LS}_{t<s}$ has its non-zero free distribution, if and only if T contains its non-zero summand $T_0 \in \mathbb{LS}_p$, by (115); equivalently, the statement (113) holds true. □

By (112) and (113), we obtain the following interesting result.

Theorem 11. *Let $X_1 = \sum_{l=1}^{N} Q_{p_l, j_l}$ and $Y_1 = \sum_{l=1}^{N} \Theta_{p_l, j_l}$ be in the sense of (107) in the semicircular filtration $\mathbb{LS}_0$, and assume that $(p_l,\ j_l)$ are mutually distinct in $\mathcal{P} \times \mathbb{Z}$, for $l = 1, ..., N$, for $N \in \mathbb{N} \setminus \{1\}$. Suppose we fix:*

$$p_{l_0} \in \{p_1, ..., p_N\},$$

and take a p_{l_0}-neighborhood $[t_0, s_0]$ in $\mathbb{R}$. Then:

$$X_1 is p_{l_0}^{2(j_{l_0}+1)} - semicircular\ in\ \mathbb{LS}_{t_0<s_0}, \tag{116}$$

$$Y_1 \text{ is semicircular in } \mathbb{LS}_{t_0<s_0}, \tag{117}$$

where $\mathbb{LS}_{t_0<s_0}$ is the semicircular $[t_0, s_0]$-filtration.

Proof. Let X_1 and Y_1 be given as above in $\mathbb{LS}$, and fix $p_{l_0} \in \{p_1, ..., p_N\}$. Note that, by the assumption, these operators X_1 and Y_1 form free sums in the semicircular filtration $\mathbb{LS}_0$, having N-many summands. Note also that they are self-adjoint in $\mathbb{LS}$ by the self-adjointness of their summands.

By (113), if an operator T has its non-zero free distribution in the semicircular $[t_0, s_0]$-filtration $\mathbb{LS}_{t_0<s_0}$, where $[t_0, s_0]$ is a p_{l_0}-neighborhood in $\mathbb{R}$, then it must have its non-zero summand T_0,

$$T_0 \in \mathbb{LS}_{p_{l_0}} \text{ in } \mathbb{LS}_{t_0<s_0}.$$

By the very construction of X_1 and Y_1, they contain their summands,

$$Q_{p_{l_0}, j_{l_0}}, \Theta_{p_{l_0}, j_{l_0}} \in \mathbb{LS}_{p_{l_0}} \text{ in } \mathbb{LS}_{t_0<s_0}.$$

Consider now that:

$$\begin{aligned} X_1^n &= \sum_{(i_1,...,i_n) \in \{1,...,N\}^n} \left(\prod_{k=1}^{N} Q_{p_{i_k}, j_{i_k}} \right) \\ &= Q_{p_{l_0}, j_{l_0}}^n + \sum_{(i_1,...,i_n) \neq (l_0,...,l_0)} \left(\prod_{k=1}^{N} Q_{p_{k_l}, j_{k_l}} \right) \\ &= Q_{p_{l_0}, j_{l_0}}^n + [\text{Rest Terms}], \end{aligned}$$

and similarly,

$$Y_1^n = \Theta_{p_{l_0}, j_{l_0}}^n + [RestTerms]', \tag{118}$$

for all $n \in \mathbb{N}$.

It is not difficult to check that:

$$\tau_{t_0<s_0}([\text{Rest Terms}]) = 0 = \tau_{t_0<s_0}([\text{Rest Terms}]'), \tag{119}$$

by (98) and (113), where [Rest Terms], and [Rest Terms]′ are from (118).

Therefore, one obtains that:

$$\begin{aligned}\tau_{t_0<s_0}\left(X_1^n\right) &= \tau_{t_0<s_0}\left(Q^n_{p_{l_0},j_{l_0}}\right) = \tau^0_{p_{l_0}}\left(Q^n_{p_{l_0},j_{l_0}}\right)\\ &= \tau^0_{p_{l_0},j_{l_0}}\left(Q^n_{p_{l_0},j_{l_0}}\right) = \omega_n p_{l_0}^{2(j_{l_0}+1)} c_{\frac{n}{2}},\end{aligned}$$

and, similarly,

$$\tau_{t_0<s_0}\left(Y_1^n\right) = \tau^0_{p_{l_0},j_{l_0}}\left(\Theta^n_{p_{l_0},j_{l_0}}\right) = \omega_n c_{\frac{n}{2}}, \tag{120}$$

for all $n \in \mathbb{N}$, by (119).

Therefore, the free sum $X_1 \in \mathbb{LS}$ is $p_{l_0}^{2(j_{l_0}+1)}$-semicircular in $\mathbb{LS}_{t_0<s_0}$; and the free sum $Y_1 \in \mathbb{LS}$ is semicircular in $\mathbb{LS}_{t_0<s_0}$, by (120). Therefore, the statements (116) and (117) hold true. □

The above theorem shows that, if there is a free sum T in the semicircular filtration $\mathbb{LS}_0$ and if we "nicely" truncate the linear functional τ^0 on $\mathbb{LS}$, then one can focus on the non-zero summand T_0 of T, whose the free distribution not only determines the truncated free distribution of T, but also follows the (weighted-)semicircular law.

11. Applications of Prime-Neighborhoods

In Section 9, we considered the semicircular truncated-filtrations $\mathbb{LS}_{t<s}$ for $t < s \in \mathbb{R}$ and studied how $[t,s]$-truncations on $\mathcal{P}$ affect, or distort, the original free-distributional data on the semicircular filtration $\mathbb{LS}_0 = \left(\mathbb{LS},\ \tau^0\right)$. As a special case, in Section 10, we introduced p-neighborhoods for primes p and considered corresponding truncated free distributions on $\mathbb{LS}$.

In this section, by using prime-neighborhoods, we provide a completely "new" model of truncated free probability on $\mathbb{LS}$ and study how the original free-distributional data on $\mathbb{LS}_0$ are distorted under this new truncation model.

Let us now regard the set $\mathcal{P}$ of all primes as a *totally ordered set* (a TOset),

$$\mathcal{P} = \{q_1 < q_2 < q_3 < q_4 < q_5 < ...\} \tag{121}$$

under the usual inequality $(\leq)$ on $\mathcal{P}$, i.e.,

$$q_1 = 2, q_2 = 3, q_3 = 5, q_4 = 7, q_5 = 11, \text{etc.}$$

For each $q_k \in \mathcal{P}$ of (121), determine a q_k-neighborhood B_k

$$B_k \overset{denote}{=} [t_k, s_k] in\mathbb{R}, \tag{122}$$

for all $k \in \mathbb{N}$.

Let τ_{B_k} be our truncated linear functionals $\tau_{t_k<s_k}$ of (92) on the Banach $*$-algebra $\mathbb{LS}$, i.e.,

$$\tau_{B_k} = \tau_{t_k<s_k}, for\ allk \in \mathbb{N}. \tag{123}$$

Then, by the truncated linear functionals of (123), one can have the corresponding semicircular B_k-filtrations,

$$\mathbb{LS}_{B_k} = \mathbb{LS}_{t_k<s_k} = \left(\mathbb{LS},\ \tau_{B_k}\right), \tag{124}$$

for all $k \in \mathbb{N}$.

We now focus on the system:

$$\mathbf{T} = \{\tau_{B_k}\}_{k=1}^{\infty} \tag{125}$$

of q_k-neighborhood-truncated linear functionals (123) for all $k \in \mathbb{N}$.

Let F be a "finite" subset of the TOset $\mathcal{P}$ of (121), and for such a set F, define a new linear functional τ_F on $\mathbb{LS}$ induced by the system $\mathbf{T}$ of (125), by:

$$\tau_F = \sum_{q_k \in F} \tau_{B_k} on \mathbb{LS}. \tag{126}$$

Before proceeding, let us consider the following result obtained from (113).

Lemma 5. *Let $p \in \mathcal{P}$ and $[t, s]$ be a p-neighborhood in $\mathbb{R}$, and let $\mathbb{LS}_{t<s}$ be the semicircular $[t, s]$-filtration. Let $\tau_p^0 = \underset{j \in \mathbb{Z}}{\star} \tau_{p,j}^0$ be the linear functional (81) on the Banach $*$-subalgebra $\mathbb{LS}_p$ of (82) in the semicircular filtration $\mathbb{LS}_0$. Define a linear functional τ^p on the Banach $*$-algebra $\mathbb{LS}$ by:*

$$\tau^p(T) \overset{def}{=} \begin{cases} \tau_p^0(T) & \text{if } T \in \mathbb{LS}_p \text{ in } \mathbb{LS} \\ O(T) = 0 & \text{otherwise,} \end{cases}$$

for all $T \in \mathbb{LS}$. Then, the Banach $*$-probability space $(\mathbb{LS}, \tau^p)$ is free-isomorphic to $\mathbb{LS}_{t<s}$, i.e.,

$$[t,\ s] \ is\ a\ p-neighborhood \Rightarrow \mathbb{LS}_{t<s}\ and\ (\mathbb{LS},\ \tau^p)\ are\ free-isomorphic. \tag{127}$$

Proof. Under the hypothesis, it is not hard to check:

$$\tau_{t<s} = \tau^p \text{ on } \mathbb{LS}.$$

Therefore, the identity map on $\mathbb{LS}$ becomes a free-isomorphism from $\mathbb{LS}_{t<s}$ onto $(\mathbb{LS},\ \tau^p)$. □

If a finite subset F is a singleton subset of $\mathcal{P}$, then the free probability on $\mathbb{LS}$ determined by the corresponding linear functional τ_F of (126) is already considered in Section 10 and in (127). Therefore, we now restrict our interests to the cases where finite subsets F have more than one element in $\mathcal{P}$.

Lemma 6. *Let F be a finite subset of the TOset $\mathcal{P}$ of (121), and let τ_F be the corresponding linear functional (126) on $\mathbb{LS}$. Then:*

$$\tau_F = \sum_{q_k \in F} \tau^{q_k} on \mathbb{LS}, \tag{128}$$

where τ^{q_k} are in the sense of (127).

Proof. The proof of (128) is done by (126) and (127) because:

$$(\mathbb{LS},\ \tau^{q_k}) \text{ and } \mathbb{LS}_{B_k}$$

are free-isomorphic for all $q_k \in F$. Therefore, the linear functional τ_F of (126) satisfies that:

$$\tau_F = \sum_{q_k \in F} \tau_{B_k} = \sum_{q_k \in F} \tau^{q_k} \text{ on } \mathbb{LS}.$$

□

By (113), (116), (117), and (128), one obtains the following result.

Theorem 12. *Let $T = \prod_{l=1}^{N} Q_{p_l, j_l}^{n_l}$ or $S = \prod_{l=1}^{N} \Theta_{p_l, j_l}^{n_l}$ be a free reduced word of $\mathbb{LS}$ with its length-N, for $N \in \mathbb{N}$. If:*

$$F \cap \{p_1, ..., p_N\} = \varnothing \text{ in } \mathcal{P},$$

then:

$$\tau_F(T) = 0 = \tau_F(S). \tag{129}$$

While, if $F \cap \{p_1, ..., p_N\} \neq \varnothing$ in $\mathcal{P}$, then:

$$\tau_F(T) = \sum_{q \in F \cap \{p_1, ..., p_N\}} \left(\omega_n q^{2(j+1)} c_{\frac{n}{2}} \right),$$

respectively,

$$\tau_F(S) = |F \cap \{p_1, ..., p_N\}| \left(\omega_n c_{\frac{n}{2}} \right), \tag{130}$$

where $|X|$ *mean the cardinalities of sets* X.

Proof. Let T and S be given free reduced words with their length-N in $\mathbb{LS}$, for $N \in \mathbb{N}$. If:

$$F \cap \{p_1, ..., p_N\} = \varnothing \text{ in } \mathcal{P},$$

then we obtain the formula (129) by (127) and (128). Indeed,

$$\tau_F(T) = \sum_{q \in F} \tau^q(T) = 0 = \sum_{q \in F} \tau^q(S) = \tau_F(S).$$

Now, assume that:

$$F \cap \{p_1, ..., p_N\} = \{p_{i_1}, ..., p_{i_k}\} \text{ in } \mathcal{P},$$

for some $k \in \mathbb{N}$, such that $1 \leq k \leq N$. Then:

$$\tau_F(T) = \left(\sum_{l=1}^{k} \tau_{p_{i_l}}^0 \right)(T) = \sum_{l=1}^{k} \tau_{p_{i_l}}^0(T)$$

by (126) and (128)

$$\begin{aligned} &= \Sigma_{l=1}^k \tau_{p_{i_l}}^0 \left(Q_{p_{i_l}, j_{i_l}}^{n_{i_l}} \right) = \Sigma_{l=1}^k \tau_{p_{i_l}, j_{i_l}}^0 \left(Q_{p_{i_l}, j_{i_l}}^{n_{i_l}} \right) \\ &= \Sigma_{l=1}^k \left(\omega_{n_l} p_{i_l}^{2(j_{i_l}+1)} c_{\frac{n_l}{2}} \right). \end{aligned} \tag{131}$$

Similarly,

$$\tau_F(S) = \Sigma_{l=1}^k \left(\omega_{n_l} c_{\frac{n_l}{2}} \right) = k \cdot \left(\omega_{n_l} c_{\frac{n_l}{2}} \right).$$

Therefore, the free-distributional data (130) hold. □

The above free-distributional data (129) and (130) characterize the free-probabilistic information on Banach $*$-probability spaces

$$(\mathbb{LS}, \tau_F),$$

for any finite subsets F of $\mathcal{P}$.

Author Contributions: The authors contributted equally in this article.

Funding: This research received no external funding.

Conflicts of Interest: The authors declare no conflict of interest.

References

1. Cho, I. Free Semicircular Families in Free Product Banach *-Algebras Induced by p-Adic Number Fields over Primes p. *Complex Anal. Oper. Theory* **2017**, *11*, 507–565. [CrossRef]
2. Cho, I.; Jorgensen, P.E.T. Semicircular Elements Induced by p-Adic Number Fields. *Opuscula Math.* **2017**, *35*, 665–703. [CrossRef]
3. Deligne, P. *Applications de la Formule des Traces Aux Sommes Trigonometriques, Cohomologies Étale (SGA 4 1/2)*; Lecture Notes in Math; Springer: Berlin, Germany, 1977; Voluеme 569.
4. Gillespie, T. Superposition of Zeroes of Automorphic L-Functions and Functoriality. Ph.D. Thesis, University of Iowa, Iowa City, IA, USA, 2010.

5. Gillespie, T. Prime Number Theorems for Rankin-Selberg L-Functions over Number Fields. *Sci. China Math.* **2011**, *54*, 35–46. [CrossRef]
6. Iwaniec, H.; Kowalski, E. *Analytic Number Theory*; Colloquium Publications: Providence, RI, USA, 2004; Volume 53.
7. Vladimirov, V.S.; Volovich, I.V.; Zelenov, E.I. *p-Adic Analysis and Mathematical Physics*; World Scientific: Singapore, 1994; Volume 1, ISBN 978-981-02-0880-6.
8. Cho, I. On Dynamical Systems Induced by p-Adic Number Fields. *Opuscula Math.* **2015**, *35*, 445–484. [CrossRef]
9. Cho, I. Dynamical Systems on Arithmetic Functions Determined by Prims. *Banach J. Math. Anal.* **2015**, *9*, 173–215. [CrossRef]
10. Cho, I. Free Distributional Data of Arithmetic Functions and Corresponding Generating Functions. *Complex Anal. Oper. Theory* **2014**, *8*, 537–570. [CrossRef]
11. Cho, I.; Jorgensen, P.E.T. Krein-Space Operators Induced by Dirichlet Characters. Available online: https://www.researchgate.net/publication/277682901_Krein-Space_Operators_Induced_by_Dirichlet_Characters (accessed on 9 Feburary 2018).
12. Cho, I. Representations and Corresponding Operators Induced by Hecke Algebras. *Complex Anal. Oper. Theory* **2016**, *10*, 437–477. [CrossRef]
13. Cho, I.; Gillespie, T. Free Probability on the Hecke Algebra. *Complex Anal. Oper. Theory* **2015**, *9*, 1491–1531. [CrossRef]
14. Speicher, R. *Combinatorial Theory of the Free Product with Amalgamation and Operator-Valued Free Probability Theory*; American Mathematical Society: Providence, RI, USA, 1998; p. 132.
15. Voiculescu, D.; Dykema, K.J.; Nica, A. *Free Random Variables*; CRM Monograph Series; American Mathematical Society: Providence, RI, USA, 1992; Volume 1.
16. Kemp, T.; Speicher, R. Strong Haagerup Inequalities for Free R-Diagonal Elements. *J. Funct. Anal.* **2007**, *251*, 141–173. [CrossRef]
17. Larsen, F. Powers of R-Diagonal Elements. *J. Oper. Theory* **2002**, *47*, 197–212.
18. Nica, A.; Shlyakhtenko, D.; Speicher, R. R-Diagonal Elements and Freeness with Amalgamation. *Can. J. Math.* **2001**, *53*, 355–381. [CrossRef]
19. Radulescu, F. Random Matrices, Amalgamated Free Products and Subfactors of the C^*-Algebra of a Free Group of Nonsingular Index. *Invent. Math.* **1994**, *115*, 347–389. [CrossRef]
20. Boedihardjo, M.; Dykema, K.J. On Algebra-Valued R-Diagonal Elements. *arXiv* **2016**, arXiv:1512.06321v2.
21. Cohen, H. *A Course in Computational Algebraic Number Theory*; Graduate Texts in Mathematic; Springer: Berlin, Germany, 1993; Volume 138.
22. Davenport, H. *Multiplicative Number Theory*, 3rd ed.; Springer: Berlin, Germany, 2000; ISBN 978-0-387-95097.

mathematics

MDPI

Article

Study on Non-Commutativity Measure of Quantum Discord

Jin Liang * **and Chengwei Zhang**

School of Mathematical Sciences, Shanghai Jiao Tong University, Shanghai 200240, China; zcw-i0@sjtu.edu.cn
* Correspondence: jinliang@sjtu.edu.cn

Received: 2 May 2019; Accepted: 11 June 2019; Published: 14 June 2019

Abstract: In this paper, we are concerned with the non-commutativity measure of quantum discord. We first present an explicit expression of the non-commutativity measure of quantum discord in the two-qubit case. Then we compare the geometric quantum discords for two dynamic models with their non-commutativity measure of quantum discords. Furthermore, we show that the results conducted by the non-commutativity measure of quantum discord are different from those conducted by both or one of the Hilbert-Schmidt distance discord and trace distance discord. These intrinsic differences indicate that the non-commutativity measure of quantum discord is incompatible with at least one of the well-known geometric quantum discords in the quantitative and qualitative representation of quantum correlations.

Keywords: quantum discord; non-commutativity measure; dynamic models

1. Introduction

In recent years, because quantum information processing is superior to classical information processing, quantum information theory and technology have developed dramatically (cf., e.g., [1–6]). As an important resource in quantum computation, quantum correlations have been investigated extensively in the last decades. So far, many forms of quantum correlations have been proposed; for example, quantum discord [7], quantum deficit [8], quantum correlation derived from the distance between the reduced states [9]. Among various quantum correlations, quantum discord and its derived measures are important (cf., e.g., [3,5–8,10–13]). Most of them are not so hard to calculate and are more robust against the effects of decoherence [10,14]. Quantum discord was initially introduced by Ollivier and Zurek and by Henderson and Vedral [7,8]. In 2010, Dakic, Vedral and Brukner [12] find a "Necessary and Sufficient Condition for Nonzero Quantum Discord" (geometric quantum discord for the Hilbert-Schmidt norm). Since then, several equivalent measures have been introduced. Recently, the non-commutativity measure of quantum discord has been discussed in [13]. In this work, we first study the problem "how to give an explicit expression of the non-commutativity measure of quantum discord in the two-qubit case?" Then we compare the geometric quantum discords (the Hilbert-Schmidt distance discord and trace distance discord) for two dynamic models of open quantum systems with their non-commutativity measure of quantum discords. We select open quantum systems as our resource quantum systems since they are significant quantum systems and they can induce occurrence of decoherence which can cause decreasing of quantum correlations and may induce failure of the algorithms.

2. Non-Commutativity Measure of Quantum Discord and Geometric Quantum Discords

Consider a composite quantum system $\mathcal{H}_{AB}$, which consists of two subsystems A and B. Quantum discord is the difference of two natural quantum extensions of the classical mutual information. In [8], the authors pointed out that the quantum discord reaches zero for and only for the classical-quantum state. So we can look at the quantum discord as the 'distance' between the state ρ and the set of classical-quantum states. The state $\rho \in \mathcal{H}_{AB}$ is classical-quantum state if and only if ρ can be written the following form (cf. [10]):

$$\rho = \sum_i |i><i|_A \otimes \rho_B^i,$$

where $\{|i>_A\}$ is any orthonormal basis of subsystem A and ρ_B is a quantum state of subsystem B.

It is known that if $\rho \in \mathcal{H}_{AB}$ is a quantum state $\rho \in \mathcal{H}_{AB}$, then (cf. [13])

$$\rho = \sum_{ij} E_{ij} \otimes B_{ij},$$

where $E_{ij} = |i><j|_A$ and $B_{ij} =< i_A|\rho|j_A >$. In [13], two non-commutativity measures are presented by:

$$D'_{N1}(\rho) = \sum_{\Omega} ||[B_{ij}, B_{i'j'}]||_2, \qquad D_{N1}(\rho) = \sum_{\Omega} ||[B_{ij}, B_{i'j'}]||_{Tr}, \tag{1}$$

where Ω is the set of all the possible pairs (regardless of the order), $[\cdot,\cdot]$ denotes the commutator, $||\cdot||_2$ is the Hilbert-Schmidt norm (i.e., $\|A\|_2 = \sqrt{Tr(A^{+}A)}$) and $||\cdot||_{Tr}$ is the trace norm (i.e., $\|A\|_{Tr} = Tr(\sqrt{A^{+}A})$).

Clearly:

(1) $D_{N1}(\rho) = 0$ and $D'_{N1}(\rho) = 0$ if and only if ρ is a quantum-classical state.
(2) D_{N1} and D'_{N1} are invariant under local unitary operations.

Moreover, we see that if $\rho = \sum_{ij} A_{ij} \otimes E'_{ij}$. According to Equation (1), we have:

$$D'_N(\rho) = \sum_{\Omega'} ||[A_{ij}, A_{i'j'}]||_2, \qquad D_N(\rho) = \sum_{\Omega'} ||[A_{ij}, A_{i'j'}]||_{Tr}, \tag{2}$$

where $E'_{ij} = |i><j|_B$, $A_{ij} =< i_B|\rho|j_B >$ and $\{|i>_B\}$ is any orthonormal basis of subsystem B, and $D_N(\rho)$ ($D'_N(\rho)$) equals zero if and only if ρ is a zero-discord state.

The Hilbert-Schmidt distance discord (cf., e.g., [12]) is defined by:

$$D_{Hs}(\rho) = \min_{\chi \in CQ} d_{Hs}(\rho, \chi), \qquad d_{Hs}(\rho, \chi) = \|\rho - \chi\|_2^2,$$

where CQ is the set of classical-quantum states. The trace distance discord (cf., e.g., [15]) is defined by:

$$D_{Tr}(\rho) = \min_{\chi \in CQ} d_{Tr}(\rho, \chi), \qquad d_{Tr}(\rho, \chi) = \|\rho - \chi\|_{Tr},$$

Both the Hilbert-Schmidt distance discord and the trace distance discord are geometric quantum discords.

3. An Explicit Expression of The Non-Commutativity Measure of Quantum Discord in the Two-Qubit Case

Since the non-commutativity measures of quantum discord are invariant under local unitary operations, every state ρ is locally unitary equivalent to:

$$\rho = \frac{1}{4}(I \otimes I + X\sigma \otimes I + I \otimes Y\sigma + \sum_i c_i \sigma_i \otimes \sigma_i), \tag{3}$$

where $X = (x_1, x_2.x_3)$, $Y = (y_1, y_2, y_3)$, $\sigma = \begin{pmatrix} \sigma_1 \\ \sigma_2 \\ \sigma_3 \end{pmatrix}$ in two qubits, and σ_i $(i = 1,2,3)$ are the three Pauli matrices (cf. [5]), which is called Bloch's representation. Therefore, if the state ρ satisfies Equation (3), we deduce that:

$$A_{00} = \frac{1}{4}\begin{pmatrix} 1 + x_3 + y_y3 + c_3 & x_1 - ix_2 \\ x_1 + ix_2 & 1 - x_3 + y_y3 - c_3 \end{pmatrix}, \quad A_{01} = \frac{1}{4}\begin{pmatrix} y_1 - iy_2 & c_1 - c_2 \\ c_1 + c_2 & y_1 - iy_2 \end{pmatrix},$$
$$A_{10} = \frac{1}{4}\begin{pmatrix} y_1 + iy_2 & c_1 + c_2 \\ c_1 - c_2 & y_1 + iy_2 \end{pmatrix}, \quad A_{11} = \frac{1}{4}\begin{pmatrix} 1 + x_3 - y_y3 - c_3 & x_1 - ix_2 \\ x_1 + ix_2 & 1 - x_3 - y_y3 + c_3 \end{pmatrix}.$$

$$[A_{00}, A_{01}]^{\dagger}[A_{00}, A_{01}] = \frac{1}{64}\begin{pmatrix} u^2 + K_+^2(c_1 + c_2)^2 & 2c_1c_2K_+(x_1 - ix_2) \\ 2c_1c_2K_+(x_1 + ix_2) & c_1^2y_2^2 + c_2^2y_1^2 + K_+^2(c_1 - c_2)^2 \end{pmatrix},$$
$$[A_{00}, A_{10}]^{\dagger}[A_{00}, A_{10}] = \frac{1}{64}\begin{pmatrix} u^2 + K_+^2(c_1 - c_2)^2 & -2c_1c_2K_+(x_1 - ix_2) \\ -2c_1c_2K_+(x_1 + ix_2) & c_1^2y_2^2 + c_2^2y_1^2 + K_+^2(c_1 + c_2)^2 \end{pmatrix},$$
$$[A_{00}, A_{11}]^{\dagger}[A_{00}, A_{11}] = \frac{1}{16}\begin{pmatrix} (x_1^2 + x_2^2)c_3^2 & 0 \\ 0 & (x_1^2 + x_2^2)c_3^2 \end{pmatrix},$$
$$[A_{01}, A_{10}]^{\dagger}[A_{01}, A_{10}] = \frac{1}{16}\begin{pmatrix} c_1^2c_2^2 & 0 \\ 0 & c_1^2c_2^2 \end{pmatrix},$$
$$[A_{01}, A_{11}]^{\dagger}[A_{01}, A_{11}] = \frac{1}{64}\begin{pmatrix} u^2 + K_-^2(c_1 + c_2)^2 & -2c_1c_2K_-(x_1 - ix_2) \\ 2c_1c_2K_-(x_1 + ix_2) & c_1^2y_2^2 + c_2^2y_1^2 + K_-^2(c_1 - c_2)^2 \end{pmatrix},$$
$$[A_{10}, A_{11}]^{\dagger}[A_{10}, A_{11}] = \frac{1}{64}\begin{pmatrix} u^2 + K_-^2(c_1 - c_2)^2 & 2c_1c_2K_-(x_1 - ix_2) \\ 2c_1c_2K_-(x_1 + ix_2) & c_1^2y_2^2 + c_2^2y_1^2 + K_-^2(c_1 + c_2)^2 \end{pmatrix}.$$

Finally, we obtain:

$$D_N(\rho) = \frac{\sum_{i=1}^4 \mu_i + 4|c_1c_2| + |c_3|\sqrt{x_1^2 + x_2^2}}{8},$$
$$D_N^{'}(\rho) = \frac{\sqrt{u^2 + K_-^2(c_1^2 + c_2^2)} + \sqrt{u^2 + K_+^2(c_1^2 + c_2^2)} + |c_1c_2| + |c_3|\sqrt{x_1^2 + x_2^2}}{2\sqrt{2}}$$

in Equation (2), where:

$$K_{\pm} = c3 \pm x3, u = \sqrt{c_1^2y_2^2 + c_2^2y_1^2},$$
$$\mu_{1,2} = \sqrt{|u^2 + K_+^2 \pm 2c_1c_2K_+\sqrt{K_+^2 + (x_1^2 + x_2^2)}|},$$
$$\mu_{3,4} = \sqrt{|u^2 + K_-^2 \pm 2c_1c_2K_-\sqrt{K_-^2 + (x_1^2 + x_2^2)}|}.$$

4. Comparisons between Two Geometric Quantum Discords for Two Dynamic Models of Open Quantum Systems and Corresponding Non-Commutativity Measure of Quantum Discords

It is known that a real physical system is never perfectly closed. Environment acting on the quantum system are regarded as noise. In mathematics, it means that there exist a series of operators (referred to as $\{ \bar{E}_u \}$) acting on the density matrix. In this way, quantum operations can be represented in an elegant form known as the operator-sum (Kraus) representation, and by [16] we see that:

$$\sum_{u,v} E_{u,v}^{\dagger} E_{u,v} = I, \qquad E_{u,v} = E_{u_A} \otimes E_{v_B}.$$

Thus, the related density matrix can be written in the following form:

$$\rho(t) = \sum\nolimits_{u,v} E_{u,v}\rho(0)E_{u,v}^{\dagger}. \quad (4)$$

Dephasing channel. It is known that the dephasing channel is the only channel that has possibility of having no energy penalty when quantum information loses. If two qubits pass through the dephasing channel respectively, the Hamiltonian ([16]):

$$H = \hbar q a^{\dagger} a(b^{\dagger} + b)$$

where a and $a^{\dagger}$, and b and $b^{\dagger}$ are the annihilation operator and creation operator of A and B respectively, and the Kraus operators are:

$$E_0 = \begin{pmatrix} 1 & 0 \\ 0 & \sqrt{1-\hbar q} \end{pmatrix} \quad E_1 = \begin{pmatrix} 0 & 0 \\ 0 & \sqrt{\hbar q}, \end{pmatrix} \quad (5)$$

$$E_0 = E_{0_A} = E_{0_B}, E_1 = E_{1_A} = E_{1_B}$$

where $q = 1 - e^{-2\gamma t}$ is photon scattering rate for the system, and γ is phase damping dissipation rate. In this paper, we suppose $\hbar = 1$.

For the Hilbert-Schmidt distance discord and trace distance discord, the dephasing channel has an important property: the freezing phenomenon (or the semi-freezing phenomenon). However, the non-commutativity measures of quantum discord is different. Actually, if we consider the initial state:

$$\rho_1(0) = \frac{1}{4}(I \otimes I + \sum_i c_i(0)\sigma_i \otimes \sigma_i)$$

with Equations (4) and (5), we have:

$$\rho_1(t) = \frac{1}{4}(I \otimes I + \sum_i c_i(t)\sigma_i \otimes \sigma_i)$$

where:

$$c_1(t) = c_1(0)e^{-2\gamma t}, \quad c_2(t) = c_2(0)e^{-2\gamma t}, \quad c_3(t) = c_3(0).$$

In Figure 1, we plot the time evolution of four different quantum discords for $c_1(0) = 0.6$, $c_2(0) = 0$, $c_3(0) = 0.2$. The plot clearly shows that the Hilbert-Schmidt distance discord and trace distance discord have the sharp transition when $t^{'} = 0.5493/\gamma$. If $t < t^{'}$, their values keep constant. If we change the initial condition (mainly about c_3), we can increase the time of the values keeping constant. However, the non-commutativity measures of quantum discord do not have this feature. They are strictly monotonous decreasing and strictly convex with time growing. Moreover, no matter what value of c_3 (except zero) could be, they still remain strictly monotonous decreasing.

Multimode vacuum field coupling the qubits. Consider a system consists of two qubits, with ω being the transition frequency and two-level energy separated by the energy gap $\hbar\omega$. The qubits are coupled to a multimode radiation field whose modes are initially in the vacuum state $|0>$. The evolution system in time is governed by the following master equation (cf., e.g., [17–19]):

$$\frac{\partial\rho}{\partial t} = -i\omega\sum_{i=1}^{2}[\sigma_3^i, \rho] - i\sum_{i\neq j}^{2}\xi_{ij}[\sigma_+^i\sigma_-^j, \rho] + \frac{1}{2}\sum_{i,j=1}^{2}\zeta_{ij}(2\sigma_-^j\rho\sigma_+^i - \{\sigma_+^j\sigma_-^i, \rho\}) \quad (6)$$

where $\sigma_{\pm}^i$ are the raising and lowering operators and σ_3^i is the energy operator (Pauli operator) of the ith qubit. The spontaneous decay rates of the qubits caused by the vacuum field $\gamma^{'} = \xi_{ii}$ coupling with the qubits. If $i \neq j$, ξ_{ij} and ζ_{ij} in Equation (6) are described by the collective damping and the "dipole-dipole" interaction, and take the forms:

$$\xi_{ij} = \frac{3}{2}\gamma'[\frac{sin(kr_{ij})}{kr_{ij}} + \frac{cos(kr_{ij})}{(kr_{ij})^2} - \frac{sin(kr_{ij})}{(kr_{ij})^3}]$$
$$\zeta_{ij} = \frac{3}{4}\gamma'[-\frac{cos(kr_{ij})}{kr_{ij}} + \frac{sin(kr_{ij})}{(kr_{ij})^2} + \frac{cos(kr_{ij})}{(kr_{ij})^3}]$$

where $k = \frac{2\pi}{\lambda}$ is the wave vector with λ being the atomic resonant wavelength and $r_{ij} = |r_i - r_j|$ is the distance between the qubits, we assume that the atomic dipole moments are parallel to each other and are polarized in the direction perpendicular to the interatomic axis.

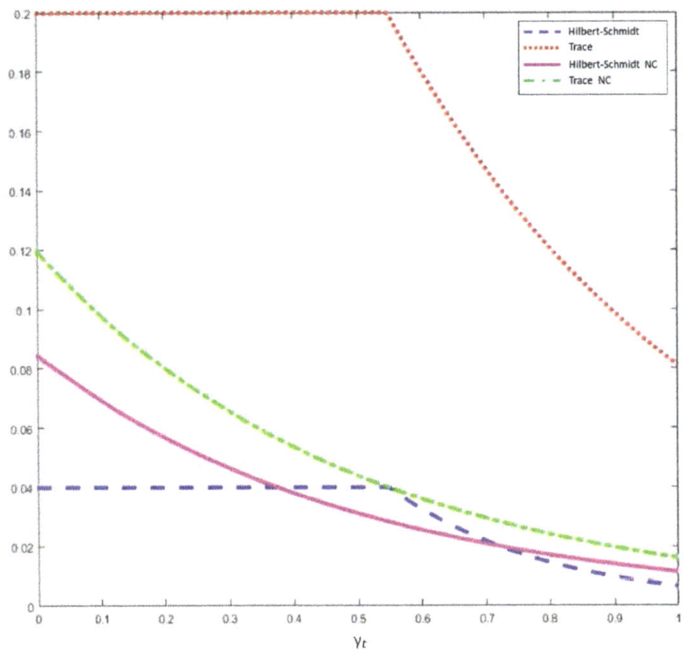

Figure 1. A two-qubit system under the operation of Dephasing channel where $c_1(0) = 0.6$, $c_2(0) = 0$, $c_3(0) = 0.2$, and NC means non-commutativity.

For this model, we consider initial state:

$$\rho_2(0) = |\Psi><\Psi|, |\Psi>= \sqrt{\alpha}|11> + \sqrt{1-\alpha}|00> .$$

Then $\rho_2(t)$ in the standard basis $\{|11>, |10>, |01>, |00>\}$ take the following forms:

$$\begin{aligned}
&\rho_{11}^{(2)}(t) = \alpha e^{-2\gamma' t} \\
&\rho_{13,41}^{(2)}(t) = \sqrt{\alpha(1-\alpha)} e^{-\gamma' t} \\
&\rho_{22,33}^{(2)}(t) = a_1[e^{-\xi_{12}^{+} t} - e^{-2\gamma' t}] + a_2[e^{-\xi_{12}^{-} t} - e^{-2\gamma' t}] \\
&\rho_{23,32}^{(2)}(t) = a_1[e^{-\xi_{12}^{+} t} - e^{-2\gamma' t}] + a_2[e^{-\xi_{12}^{-} t} - e^{-2\gamma' t}] \\
&\rho_{44}^{(2)}(t) = 1 - \rho_{11}^{2}(t) - \rho_{22}^{2}(t) - \rho_{33}^{2}(t)
\end{aligned}$$

where $\xi_{12}^{\pm} = \gamma' \pm \xi_{12}, a_{1,2} = \alpha\xi_{12}^{\pm}/\xi_{12}^{\mp}$. In Figure 2, we show the change of the discord with respect to γt and α as $\xi_{ij} = 0$, where the values are normalized.

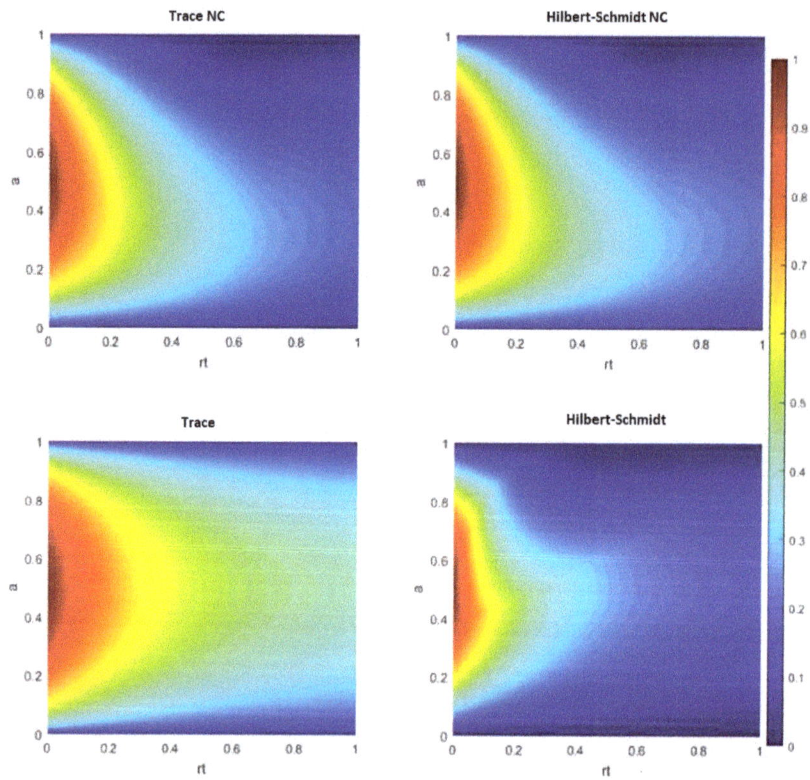

Figure 2. Change of the initial state $|\Psi>$ with respect to γt and α as the interatomic distance is $\xi_{ij} = 0$.

For this model, the trace distance discord is symmetrical about α = 0.5. Moreover, compared to the trace distance discord, we find that the non-commutativity measure of quantum discord and the Hilbert-Schmidt distance discord have more similar properties at different α values. On the other hand, the Hilbert-Schmidt distance discord will revive after α large enough. In this aspect, the non-commutativity measure of quantum discord are closer to the trace distance discord.

5. Conclusions

In conclusion, we present an explicit expression of the non-commutativity measure of quantum discord in the two-qubit case. We also compare the non-commutativity measure of quantum discord with the geometric quantum discord in the models of two qubits passing through the dephasing channel and the multimode vacuum field coupling the qubits, respectively. Our study shows that the non-commutativity measures of quantum discord lose some important features of the geometric quantum discords, such as the freezing phenomenon in the dephasing channel and the revival in the multimode vacuum field coupling the qubits.

Author Contributions: All the authors contributed equally and significantly in writing this paper. All authors read and approved the final manuscript.

Funding: The work was supported partly by the NSF of China (11571229).

Acknowledgments: The authors would like to thank the reviewers very much for valuable comments and suggestions.

Conflicts of Interest: The authors declare that they have no competing interests.

References

1. Das, D.; Bhattacharya, B.; Datta, C.; Roy, A.; Jebaratnam, C.; Majumdar, A.S.; Srikanth, R. Operational characterization of quantumness of unsteerable bipartite states. *Phys. Rev.* **2018**, *97*, 062335. [CrossRef]
2. Gharibian, S. Quantifying nonclassicality with local unitary operations. *Phys. Rev.* **2012**, *86*, 042106. [CrossRef]
3. Hosseini, S.; Rahimi-Keshari, S.; Haw, J.Y.; Assad1, S.M.; Chrzanowski, H.M.; Janousek, J.; Symul, T.; Ralph, T.C.; Lam, P.K.; Gu, M.; et al. Experimental verification of quantum discord in continuousvariable states and operational significance of discord consumption. In Proceedings of the Conference on Lasers and Electro-Optics Europe-Technical Digest 2014, San Jose, CA, USA, 3–4 January 2014.
4. Jebaratnam, C.; Aravinda, S.; Srikanth, R. Nonclassicality of local bipartite correlations. *Phys. Rev.* **2017**, *95*, 032120. [CrossRef]
5. Luo, S.L. Quantum discord for two-qubit systems. *Phys. Rev.* **2008**, *77*, 042303. [CrossRef]
6. Yune, J.; Hong, K.H.; Lim, H.T.; Lee, J.C.; Kwon, O.; Han, S.W.; Moon, S.; Kim, Y.S.; Kim, Y.H. Experimental verification of quantum discord in continuousvariable states and operational significance of discord consumption. In Proceedings of the 2015 11th Conference on Lasers and Electro-Optics Pacific Rim (CLEO-PR), Busan, Korea, 24–28 August 2015.
7. Ollivier, H.; Zurek, W.H. Quantum discord: A measure of the quantumness of correlations. *Phys. Rev. Lett.* **2002**, *88*, 017901. [CrossRef] [PubMed]
8. Henderson, L.; Vedral, V. Classical, quantum and total correlations. *J. Phys. Math. Gen.* **2001**, *34*, 6899–6905. [CrossRef]
9. Guo, Y.; Li, X.L.; Li, B.; Fan, H. Quantum correlation induced by the average distance between the reduced states. *Int. J. Theor. Phys.* **2015**, *54*, 2022–2030. [CrossRef]
10. Adesso, G.; Bromley, T.R.; Cianciaruso, M. Measures and applications of quantum correlations. *J. Phys. Math. Theor.* **2016**, *49*, 473001. [CrossRef]
11. Coecke, B.; Fritz, T.; Spekkens, R.W. A mathematical theory of resources. *Inf. Comput.* **2016**, *250*, 59–86. [CrossRef]
12. Dakic, B.; Vedral, V.; Brukner, C. Necessary and sufficient condition for nonzero quantum discord. *Phys. Rev. Lett.* **2010**, *105*, 190502. [CrossRef] [PubMed]
13. Guo, Y. Non-commutativity measure of quantum discord. *Sci. Rep.* **2016**, *6*, 25241. [CrossRef] [PubMed]
14. Mazzola, L.; Piilo, J.; Maniscalco, S. Sudden transition between classical and quantum decoherence. *Phys. Rev. Lett.* **2010**, *104*, 200401. [CrossRef] [PubMed]
15. Ciccarello, F.; Tufarelli, T.; Giovannetti, V. Toward computability of trace distance discord. *New J. Phys.* **2014**, *16*, 013038. [CrossRef]
16. Nielsen, M.A.; Chuang, I.L. *Quantum Computation and Quantum Information*; Cambridge: Cambridge University Press: London, UK, 2000.

17. Ficek, Z.; Tanas, R. Dark periods and revivals of entanglement in a two-qubit system. *Phys. Rev.* **2006,** *74*, 024304. [CrossRef]
18. Hu, M.L.; Fan, H. Robustness of quantum correlations against decoherence. *Ann. Phys.* **2012**, *327*, 851–860. [CrossRef]
19. Yu, T.; Eberly, J.H. Finite-time disentanglement via spontaneous emission. *Phys. Rev. Lett.* **2004**, *93*, 140404. [CrossRef] [PubMed]

Σ mathematics

MDPI

Article

The Prolongation Structure of the Modified Nonlinear Schrödinger Equation and Its Initial-Boundary Value Problem on the Half Line via the Riemann-Hilbert Approach

Tongshuai Liu and Huanhe Dong *

College of Mathematics and Systems Science, Shandong University of Science and Technology, Qingdao 266590, China; Liutongshuaii@163.com

* Correspondence: mathsdong@126.com

Received: 4 January 2019; Accepted: 7 February 2019; Published: 13 February 2019

Abstract: In this paper, the Lax pair of the modified nonlinear Schrödinger equation (mNLS) is derived by means of the prolongation structure theory. Based on the obtained Lax pair, the mNLS equation on the half line is analyzed with the assistance of Fokas method. A Riemann-Hilbert problem is formulated in the complex plane with respect to the spectral parameter. According to the initial-boundary values, the spectral function can be defined. Furthermore, the jump matrices and the global relations can be obtained. Finally, the potential $q(x,t)$ can be represented by the solution of this Riemann-Hilbert problem.

Keywords: prolongation structure; mNLS equation; Riemann-Hilbert problem; initial-boundary value problem

1. Introduction

In mathematics and physics, nonlinear partial differential equations play an important role due to their abundant mathematical structure and properties. Many works on nonlinear evolution equations have been studied, such as the Hamiltonian structure [1,2], the infinite conservation laws [3,4], the Bäcklund transformation [5,6] and so on [7–9]. Besides, the exact solution of these equations, which can be expressed in various forms by different methods, is also a significant subject of soliton research [10–22]. In recent years, with the development of soliton theory, more and more researchers pay attention to the Riemann-Hilbert approach. The Riemann-Hilbert approach was introduced by Fokas to analyze the initial-boundary values problem for linear and nonlinear partial differential equations [23,24]. In the past 20 years, many researchers have discussed a lot of nonlinear integrable equations for the initial-boundary values problem [25–32,32–41]. They have all made a great contribution to the development of this method. The core idea of this method is to construct the associated Riemann-Hilbert problem by the Lax pair of the integrable equation, and then in addition to the initial-boundary values problem, the long-time asymptotic behavior of the solution can be analyzed [42–46]. However, as we all know, it is difficult to determine whether a nonlinear evolution equation possesses a Lax pair or not. As far as we are concerned, the prolongation structure method is an efficient way to obtain the Lax pair, which was firstly proposed in 1975 by Wahlquist and Estabrook [47]. In recent years, a large number of scholars have improved this method, for example, Hermann deduced the prolongation structure method connection in 1976 [48], Deconinck applied the prolongation structure method to semi-discrete systems firstly [49], Wang used this approach to get the integrability of many nonlinear wave equation [50] and so on [51,52]. In this way, we can get the

Lax pair of the nonlinear evolution equation easily as long as it is integrable.

In this paper, we mainly talk about the modified nonlinear Schrödinger(mNLS) equation

$$iq_t + q_{xx} + i(|q|^2 q)_x + 2\rho|q|^2 q = 0, \tag{1}$$

which is very important in plasma physics. Recently, many properties of this equation have been studied, such as the Hamiltonian structure [53], the Darboux transformation [54], the numerical solutions [55,56] and so on [57,58]. Actually, it can become the derivative NLS equation by certain gauge transformation [59]. In this paper, we mainly discuss the mNLS equation on the half line. For simplicity, we let $\rho = 1$. Supposing that the solution $q(x,t)$ of the mNLS equation exists, and the initial-boundary values are defined as follows,
Initial values:

$$q_0(x) = q(x,0), 0 < x < \infty, \tag{2}$$

Boundary values:

$$g_0(t) = q(0,t), g_1(t) = q_x(0,t), 0 < t < T. \tag{3}$$

In order to formulate a Riemann-Hilbert problem, we need to reconstruct the Lax pair of Equation (1). Based on the initial-boundary values, the corresponding spectral functions can be defined. Eventually, the potential function $q(x,t)$ can be expressed in terms of the solution of this Riemann-Hilbert problem.

This paper is divided into four sections. The construction of the prolongation structure for the mNLS equation is in Section 2 and then in Section 3, we reconstruct the Lax pair to formulate the Riemann-Hilbert problem and some conditions and relations are derived. In the last section, we define the spectral functions according to the initial-boundary values and the Riemann-Hilbert problem is investigated.

2. The Prolongation Structures of the mNLS Equation

In order to obtain the Lax pair of the mNLS equation, we analyze the prolongation structure of this equation. This process mainly involves a fundamental theorem in Lie algebra [51].

Theorem 1. *Suppose X and Y are two elements of Lie algebra $g = sl(n+1, C)$ with $[X,Y] = aY$, $(a \neq 0)$ and $X \in$ range ad Y, it means that there exist $Z \in g$ such that $[Y,Z] = X$, so we obtain $Y = e_\pm$ and $X = \pm\frac{1}{2}ah$, where $e_\pm$ are the nilpotent and h is the neutral elements of g.*

In the beginning, we introduce these variables

$$\bar{u} = p, u_x = v, \bar{u_x} = p_x = q. \tag{4}$$

Then Equation (1) is equivalent to this set of equations as follows

$$\begin{cases} u_x - v = 0, \\ p_x - q = 0, \\ iu_t + v_x + 2iuu_x\bar{u} + iu^2\bar{u}_x + 2u^2\bar{u} = 0, \\ ip_t - q_x + 2i\bar{u}\bar{u}_x u + i\bar{u}^2 u_x - 2\bar{u}^2 u = 0. \end{cases} \tag{5}$$

We define the set of two-forms $I = \{\alpha_1, \alpha_2, \alpha_3, \alpha_4\}$,where

$$\begin{cases} \alpha_1 = du \wedge dt + vdt \wedge dx, \\ \alpha_2 = dp \wedge dt + qdt \wedge dx, \\ \alpha_3 = idu \wedge dx - dv \wedge dt + (2iuvp + iu^2q + 2u^2p)dt \wedge dx, \\ \alpha_4 = idp \wedge dx + dq \wedge dt + (ip^2v + 2ipqu - 2p^2u)dt \wedge dx. \end{cases} \tag{6}$$

It is easy to find that I is a closed ideal, actually, $dI \subset I$. After that, we define the differential one-forms

$$\omega^i = dy^i - F^i(u, v, p, q; y^i)dx - G^i(u, v, p, q; y^i)dt. \tag{7}$$

At the same time, we suppose $F^i = F^i_j y^j, G^i = F^i_j y^j$. According to the general theory of exterior differential systems, if $\tilde{I} = I \bigcup \omega^i$ is a closed ideal, it must satisfy

$$d\omega^i = \sum_{i=1}^{4}(f^i_j \alpha^j) + n^i_j \wedge \omega^j. \tag{8}$$

Combining (5)–(8), we obtain

$$\begin{cases} F_v = F_q = 0, \\ iG_v + F_u = 0, \\ iG_q - F_p = 0, \\ -G_u v - G_p q + (2iuvp + iqu^2 + 2u^2p)G_v \\ -(2ipqu + ip^2v - 2p^2u)G_q + [F, G] = 0. \end{cases} \tag{9}$$

where the bracket $[,]$ denotes the Lie bracket, namely $[F, G] = FG - GF$.

After a lengthy calculation, one solution of this set of equations can be derived

$$\begin{aligned} F &= x_0 + ux_1 + px_2, \\ G &= ix_1v - ix_2q - u^2px_1 - p^2ux_2 + iux_3 - ipx_4 - ipux_5 + x_6. \end{aligned} \tag{10}$$

with the integrability conditions

$$\begin{aligned} &2ix_1 - x_3 - i[x_1, x_5] = 0, 2ix_2 + x_4 + i[x_2, x_5] = 0, \\ &i[x_0, x_3] + [x_1, x_6] = 0, \quad i[x_0, x_4] + [x_2, x_6] = 0, \\ &[x_0, x_5] + [x_1, x_4] - [x_2, x_3] = 0, [x_1, x_3] = 0, [x_2, x_4] = 0, [x_0, x_6] = 0. \end{aligned} \tag{11}$$

where all $\{x_i\}, i = \{1, 2, ..., 6.\}$ are pending matrices. Here $\{x_1, x_2, ..., x_6\}$ depend on an incomplete Lie algebra, called prolongation algebra.

The next step is to embed the prolongation algebra in $sl(n+1, C)$. According to (11) and Theorem 1, we deduce that x_1 and x_2 is nilpotent and x_5 is neutral element. So we have

$$x_1 = \begin{pmatrix} 0 & \xi \\ 0 & 0 \end{pmatrix}, x_2 = \begin{pmatrix} 0 & 0 \\ -\xi & 0 \end{pmatrix}, x_5 = \begin{pmatrix} -\xi^2 & 0 \\ 0 & \xi^2 \end{pmatrix}. \tag{12}$$

Bringing the above results into (11), we obtain

$$\begin{aligned} x_0 &= \begin{pmatrix} -i\xi^2+i & 0 \\ 0 & i\xi^2-i \end{pmatrix}, x_3 = \begin{pmatrix} 0 & -2i\xi^3+2i\xi \\ 0 & 0 \end{pmatrix}, \\ x_4 &= \begin{pmatrix} 0 & 0 \\ -2i\xi^3+2i\xi & 0 \end{pmatrix}, x_6 = \begin{pmatrix} -2i\xi^4-2i+4i\xi^2 & 0 \\ 0 & 2i\xi^4+2i+4i\xi^2 \end{pmatrix}. \end{aligned} \tag{13}$$

where ξ is spectral parameter. Hence, the expressions of F and G can be presented eventually

$$\begin{aligned} F &= \begin{pmatrix} -i\xi^2+i & \xi q \\ -\xi\bar{q} & i\xi^2-i \end{pmatrix}, \\ G &= \begin{pmatrix} -2i\xi^4-2i+4i\xi^2+i\xi^2|q|^2 & 2\xi^3 q-2\xi q-\xi|q|^2 q+i\xi q_x \\ -2\xi^3\bar{q}+2\xi\bar{q}+\xi|q|^2\bar{q}+i\xi\bar{q}_x & 2i\xi^4+2i-4i\xi^2-i\xi^2|q|^2 \end{pmatrix}. \end{aligned} \tag{14}$$

So, the mNLS equation admits Lax pair

$$\psi_x = F\psi, \psi_t = G\psi, \tag{15}$$

where $\psi = (v_1, v_2)^T$.

3. Spectral Analysis

From the previous paragraph, we know the Lax pair of the mNLS equation. By introducing

$$Q = \begin{pmatrix} 0 & q \\ -\bar{q} & 0 \end{pmatrix}, \sigma_3 \begin{pmatrix} 1 & 0 \\ 0 & -1 \end{pmatrix}. \tag{16}$$

where the $\bar{q}$ denotes the conjugation of q, the Lax pair (15) can be rewritten in this form

$$\begin{cases} \psi_x + i\xi^2\sigma_3\psi - i\sigma_3\psi = \xi Q\psi, \\ \psi_t + 2i\xi^4\sigma_3 + 2i\sigma_3 - 4i\xi^2\sigma_3 = -i\xi^2 Q^2 + 2i\xi^3 Q - 2\xi Q + \xi Q^3 + i\xi\sigma_3 Q_x. \end{cases} \tag{17}$$

In our analysis, we assume that q decays to zero sufficiently fast as $x \to \pm\infty$. So, it is correct to extend the column vector ψ to a 2×2 matrix. For simplicity, we substitute λ^2 for $\xi^2 - 1$. Letting $\psi = \Psi e^{-i(\xi^2 x + 2\xi^4 t)\sigma_3}$, then the Lax pair (17) becomes

$$\begin{aligned} &\Psi_x + i\xi^2[\sigma_3, \Psi] = \xi Q\Psi, \\ &\Psi_t + 2i\xi^4[\sigma_3, \Psi] = (-i\xi^2 Q^2\sigma_3 + 2\xi^3 Q - 2\xi Q + \xi Q^3 + i\xi\sigma_3 Q_x)\Psi \end{aligned} \tag{18}$$

We can write (18) in full derivative form

$$d(e^{i(\xi^2 x + 2\xi^4 t)\hat{\sigma}_3}\Psi(x,t;\xi)) = e^{i(\lambda^2 x + 2\lambda^4 t)\hat{\sigma}_3} U(x,t;\xi)\Psi. \tag{19}$$

where

$$\begin{aligned} U &= U_1 dx + U_2 dt \\ &= \xi Q dx + (-i\xi^2 Q^2\sigma_3 + 2\xi^3 Q - 2\xi Q + \xi Q^3 + i\xi\sigma_3 Q_x)dt. \end{aligned} \tag{20}$$

and $\hat{\sigma}_3$ denotes the operator to matrix by $\hat{\sigma}_3 M = [\sigma_3, M]$, hence it is easy to prove $e^{\hat{\sigma}_3} M = e^{\sigma_3} M e^{-\sigma_3}$, where M is a 2×2 matrix.

3.1. The Reconstruction of Lax Pair

Expanding the solution of (19) in this way

$$\Psi = D + \frac{\Psi_1}{\xi} + \frac{\Psi_2}{\xi^2} + \frac{\Psi_3}{\xi^3} + \mathcal{O}(\frac{1}{\xi^4}), \xi \to \infty, \tag{21}$$

where $D, \Psi_1, \Psi_2, \Psi_3$ are independent of ξ. Bringing this equation into the first equation of (18), and comparing the same order of ξ's frequency, we obtain the following equations

$$\begin{aligned} &O(1) : D_x + i[\sigma_3, \Psi_2] - i[\sigma_3, D] = Q\Psi_1; \\ &O(\xi) : i[\sigma_3, \Psi_1] = QD; \\ &O(\xi^2) : i[\sigma_3, D] = 0. \end{aligned} \tag{22}$$

Using the same method, taking (21) into another equation of (18), we have

$$\begin{aligned} &O(1) : D_t - 4i[\sigma_3, \Psi_2] + 2i[\sigma_3, D] = -iQ^2\sigma_3\Psi_2 + 2Q\Psi_3 - 2Q\Psi_1 + Q^3\Psi_1 + i\sigma_3 Q_x \Psi_1; \\ &O(\xi) : 2i[\sigma_3, \Psi_3] - 4i[\sigma_3, \Psi_1] = -iQ^2\sigma_3\Psi_1 + 2Q\Psi_2 - 2QD + Q^3 D + i\sigma_3 Q_x D; \\ &O(\xi^2) : 2i[\sigma_3, \Psi_2] - 4i[\sigma_3, D] = -iQ^2\sigma_3 D + 2Q\Psi_1; \\ &O(\xi^3) : 2i[\sigma_3, \Psi_1] = 2QD; \\ &O(\xi^4) : 2i[\sigma_3, D] = 0. \end{aligned} \tag{23}$$

For (22), We find that D is a diagonal matrix from $O(\xi^2)$. Without loss of generality, we suppose

$$D = \begin{pmatrix} D_0^{11} & 0 \\ 0 & D_0^{22} \end{pmatrix}. \tag{24}$$

From $O(\xi)$ we have

$$\Psi_1^o = \frac{i}{2} Q D \sigma_3. \tag{25}$$

where Ψ_1^o denotes the off-diagonal part of Ψ_1. So, we can get D_x from $O(1)$ easily

$$D_x = \frac{i}{2} Q^2 \sigma_3 D. \tag{26}$$

For (23), after a lengthy calculation, we get

$$\begin{aligned} D_t &= \frac{3i}{4} Q^4 \sigma_3 D + \frac{1}{2} Q Q_x D - \frac{1}{2} Q_x Q D \\ &= (\frac{3i}{4}|q|^4 + \bar{q} q_x - q\bar{q}_x)\sigma_3 D. \end{aligned} \tag{27}$$

The mNLS equation admits the conservation law

$$2(|q|^2)_t = (2iq_x\bar{q} - 3|q|^4 - 2iq\bar{q}_x)_x. \tag{28}$$

From the above results, we find (26) and (27) admit the conservation law. Define

$$D(x,t) = e^{i\int_{(0,0)}^{(x,t)} \Delta(x,t)\sigma_3}. \tag{29}$$

where Δ is the differential one-form, and it is given by

$$\Delta(x,t) = \Delta_1 dx + \Delta_2 dt = -\frac{1}{2} q\bar{q} dx + [\frac{3}{4} q^2 \bar{q}^2 + \frac{i}{2}(q\bar{q}_x - q_x\bar{q})]dt. \tag{30}$$

It is not difficult to find that the integral is path independent. So, we introduce

$$\Psi(x,t;\xi) = e^{i\int_{(0,0)}^{(x,t)}\Delta\hat{\sigma}_3}\mu(x,t;\xi)D(x,t), 0<x<\infty, 0<t<T. \tag{31}$$

Then the form of the Lax pair (18) can be replaced with

$$d(e^{i(\lambda^2x+2\lambda^4t)\hat{\sigma}_3}\mu(x,t;\xi)) = W(x,t;\xi). \tag{32}$$

where

$$\begin{aligned} W(x,t;\xi) &= e^{i(\lambda^2x+2\lambda^4t)\hat{\sigma}_3}V(x,t;\xi)\mu, \\ V = V_1dx + V_2dt &= e^{-i\int_{(0,0)}^{(x,t)}\Delta\hat{\sigma}_3}(U - i\Delta\sigma_3). \end{aligned} \tag{33}$$

Considering the definitions of U and Δ, we have

$$\begin{aligned} V_1(x,t;\xi) &= \begin{pmatrix} \frac{i}{2}q\bar{q} & \xi q e^{-2i\int_{(0,0)}^{(x,t)}\Delta} \\ -\xi\bar{q}e^{2i\int_{(0,0)}^{(x,t)}\Delta} & -\frac{i}{2}q\bar{q} \end{pmatrix}, \\ V_2(x,t;\xi) &= \begin{pmatrix} V_2^{11}(x,t;\xi) & V_2^{12}(x,t;\xi) \\ V_2^{21}(x,t;\xi) & V_2^{22}(x,t;\xi) \end{pmatrix}. \end{aligned} \tag{34}$$

where

$$\begin{aligned} V_2^{11}(x,t;\xi) &= i\xi^2 q\bar{q} - \frac{3i}{4}q^2\bar{q}^2 + \frac{1}{2}(q\bar{q}_x - q_x\bar{q}), \\ V_2^{12}(x,t;\xi) &= (2\xi^3 - 2\xi q - \xi q|q|^2 + i\xi q_x)e^{-2i\int_{(0,0)}^{(x,t)}\Delta}, \\ V_2^{21}(x,t;\xi) &= (-2\xi^3\bar{q} + 2\xi\bar{q} + \xi\bar{q}|q|^2 + i\xi\bar{q}_x)e^{2i\int_{(0,0)}^{(x,t)}\Delta}, \\ V_2^{22}(x,t;\xi) &= -i\xi^2 q\bar{q} + \frac{3i}{4}q^2\bar{q}^2 - \frac{1}{2}(q\bar{q}_x - q_x\bar{q}). \end{aligned}$$

Thus, (32) changes into

$$\begin{cases} \mu_x + i\lambda^2[\sigma_3,\mu] = V_1\mu, \\ \mu_t + 2i\lambda^4[\sigma_3,\mu] = V_2\mu. \end{cases} \tag{35}$$

3.2. The Riemann-Hilbert Problem And Some Relations

Supposing that $q(x,t)$ is smooth function in the domain $D = \{0<x<\infty, 0<t<T\}$. Then we define the eigenfunctions $\mu_j(x,t,;\xi)(j=1,2,3)$ of (34) as follows

$$\mu_j(x,t;\xi) = I + \int_{(x_j,t_j)}^{(x,t)} e^{-i(\lambda x+2\lambda^2t)\hat{\sigma}_3}W(x',t';\xi), 0<x<\infty, 0<t<T. \tag{36}$$

The integral curve is from (x_j,t_j) to (x,t), where $(x_1,t_1)=(0,T),(x_2.t_2)=(0,0)$ and $(x_3,t_3)=(\infty,t)$. Furthermore, the point (x,t) is an arbitrary point in the domain D. We know that the integral of (36) is independent of the path of integration. Without loss of generality, we will consider the particular integral paths as follows, see Figure 1.

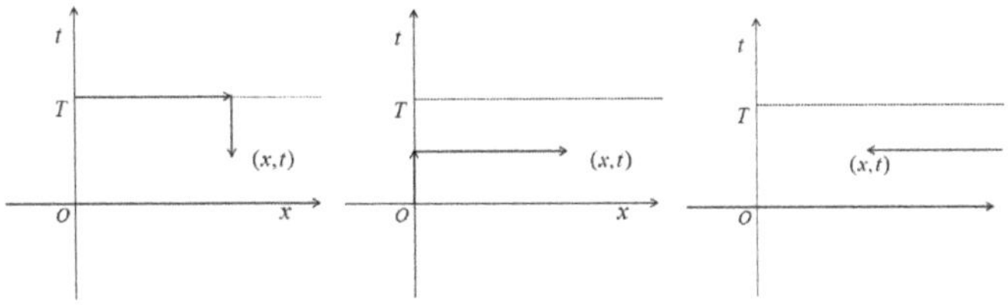

Figure 1. Integral paths.

By this method, we get

$$\begin{cases} \mu_1(x,t;\xi) = I + \int_0^x e^{i\lambda^2(x'-x)\hat{\sigma}_3}(V_1\mu_1)(x',t;\lambda)dx' \\ -e^{-i\lambda^2 x\hat{\sigma}_3}\int_t^T e^{2i\lambda^4(t'-t)\hat{\sigma}_3}(V_2\mu_1)(0,t';\xi)dt', \\ \mu_2(x,t;\xi) = I + \int_0^x e^{i\lambda^2(x'-x)\hat{\sigma}_3}(V_1\mu_2)(x',t;\xi)dx' \\ -e^{-i\lambda^2 x\hat{\sigma}_3}\int_0^t e^{2i\lambda^4(t'-t)\hat{\sigma}_3}(V_2\mu_2)(0,t';\xi)dt', \\ \mu_3(x,t,;\xi) = I - \int_\infty^x e^{i\lambda^2(x'-x)\hat{\sigma}_3}(V_1\mu_3)(x',t;\xi)dx'. \end{cases} \tag{37}$$

Noting that the first column of μ_j includes $e^{-2i[\lambda^2(x'-x)+2\lambda^4(t'-t)]}$. So, in different integral paths, we have the following inequalities

$$\begin{cases} \ell_1 : (x_1,t_1) \to (x,t) : 0 < x' < x, t < t' < T, \\ \ell_2 : (x_2,t_2) \to (x,t) : 0 < x' < x, 0 < t' < t, \\ \ell_3 : (x_3,t_3) \to (x,t) : x < x' < \infty. \end{cases} \tag{38}$$

Due to the exponential function decaying sufficiently, these inequalities imply that the first of the functions $\mu_j(x,t;\xi), (j = 1,2,3)$ are analytic if

$$\begin{aligned} &\mu_1^{(1)}(x,t;\xi) : \xi \in \{Im\xi^2 \geq 0\} \cap \{Im\xi^4 \leq 0\}, \\ &\mu_2^{(1)}(x,t;\xi) : \xi \in \{Im\xi^2 \geq 0\} \cap \{Im\xi^4 \geq 0\}, \\ &\mu_3^{(1)}(x,t;\xi) : \xi \in \{Im\xi^2 \leq 0\}. \end{aligned} \tag{39}$$

At the same time, the second column of the functions $\mu_j(x,t;\xi), (j = 1,2,3)$ are analytic if

$$\begin{aligned} &\mu_1^{(2)}(x,t;\xi) : \xi \in \{Im\xi^2 \leq 0\} \cap \{Im\xi^4 \geq 0\}, \\ &\mu_2^{(2)}(x,t;\xi) : \xi \in \{Im\xi^2 \leq 0\} \cap \{Im\xi^4 \leq 0\}, \\ &\mu_3^{(2)}(x,t;\xi) : \xi \in \{Im\xi^2 \geq 0\}. \end{aligned} \tag{40}$$

Hence, we get

$$\begin{aligned} &\mu_1(x,t;\xi) = (\mu_1^{D_2}(x,t;\xi), \mu_1^{D_3}(x,t;\xi)), \\ &\mu_2(x,t;\xi) = (\mu_2^{D_1}(x,t;\xi), \mu_2^{D_4}(x,t;\xi)), \\ &\mu_3(x,t;\xi) = (\mu_3^{D_3\cup D_4}(x,t;\xi), \mu_3^{D_1\cup D_2}(x,t;\xi)). \end{aligned} \tag{41}$$

where $\mu_j^{D_i}$ stands for μ_j is analytic if $\xi \in D_i$, where $D_i = \omega_i \cup (-\omega_i), -\omega_i = \{-\xi \in C | \xi \in \omega_i\}, \omega_i = \{\xi \in C | \frac{i-1}{4}\pi < \xi < \frac{i}{4}\pi\}$, see Figure 2.

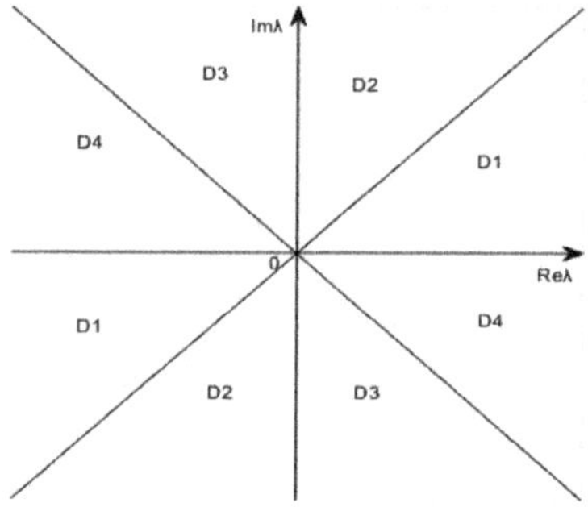

Figure 2. $D_j, j = 1,2,3,4$.

The eigenfunctions $\mu_j(j = 1,2,3)$ possess these properties.

- $\det\mu_1(x,t;\xi) = \det\mu_2(x,t;\xi) = \det\mu_3(x,t;\xi) = 1$;
- $\mu_j^{11}(x,t;\xi) = \overline{\mu_j^{22}(x,t;\bar{\xi})}, \mu_j^{12}(x,t;\xi) = \overline{\mu_j^{21}(x,t;\bar{\xi})}$;
- $\mu_j^{11}(x,t;-\xi) = \mu_j^{11}(x,t;\xi), \mu_j^{12}(x,t;-\xi) = -\mu_j^{12}(x,t;\xi)$,
  $\mu_j^{21}(x,t;-\xi) = -\mu_j^{21}(x,t;\xi), \mu_j^{22}(x,t;-\xi) = \mu_j^{22}(x,t;\xi)$.

where the eigenfunctions $\mu_j(x,t;\xi) = \begin{pmatrix} \mu_j^{11}(x,t;\xi) & \mu_j^{12}(x,t;\xi) \\ \mu_j^{21}(x,t;\xi) & \mu_j^{22}(x,t;\xi) \end{pmatrix} (j = 1,2,3)$.

For the purpose of formulating a Riemann-Hilbert problem, our main task is to find the jump matrices for every $D_i(i = 1,2,3,4)$ to any other regions. Then we define the spectral functions $s(\xi)$ and $S(\xi)$

$$\begin{cases} \mu_3(x,t;\xi) = \mu_2(x,t;\xi)e^{-i(\lambda^2x+2\lambda^4t)\hat{\sigma}_3}s(\xi), \\ \mu_1(x,t;\xi) = \mu_2(x,t;\xi)e^{-i(\lambda^2x+2\lambda^4t)\hat{\sigma}_3}S(\xi). \end{cases} \tag{42}$$

According to the above definition, we have

$$\mu_1(x,t;\xi) = \mu_3(x,t;\xi)e^{-i(\lambda^2x+2\lambda^4t)\hat{\sigma}_3}(s(\xi))^{-1}S(\xi). \tag{43}$$

Combining (37) with (42), we acquire

$$s(\xi) = \mu_3(0,0;\xi), S(\xi) = (e^{2i\lambda^4T\hat{\sigma}_3}\mu_2(0,T;\xi))^{-1}. \tag{44}$$

Owing to (37), it is clear to see that

$$\begin{aligned} \mu_1(0,t;\xi) &= I - \int_t^T e^{2i\lambda^4(t'-t)\hat{\sigma}_3}(V_2\mu_1)(0,t';\xi)dt', \\ \mu_2(0,t;\xi) &= I + \int_0^t e^{2i\lambda^4(t'-t)\hat{\sigma}_3}(V_2\mu_2)(0,t';\xi)dt', \\ \mu_3(x,0;\xi) &= I + \int_\infty^x e^{i\lambda^2(x'-x)\hat{\sigma}_3}(V_1\mu_3)(x',0;\xi)dx', \\ \mu_2(x,0;\xi) &= I + \int_0^x e^{i\lambda^2(x'-x)\hat{\sigma}_3}(V_1\mu_2)(x',0;\xi)dx', \end{aligned} \tag{45}$$

Considering the initial values $q(x,0) = u_0(x)$, $q(0,t) = g_0(t)$, boundary values $q(0,t) = g_0(t)$ and $q_x(0,t) = g_1(t)$. For convenience, the initial-boundary values of $\bar{q}(x,t)$ can be written in this form, namely $\bar{q}(x,0) = \bar{u}_0(x), \bar{q}(0,t) = \bar{g}_0(t)$, and $\bar{q}_x(0,t) = \bar{g}_1(t)$. Then $V_2(0,t;\xi)$ and $V_1(x,0;\xi)$ can be expressed with

$$V_1(x,0;\xi) = \begin{pmatrix} \frac{i}{2}|u_0|^2 & \xi u_0 e^{i\int_0^x |u_0|^2 dx'} \\ -\xi \bar{u}_0 e^{-i\int_0^x |u_0|^2 dx'} & -\frac{i}{2}|u_0|^2 \end{pmatrix},$$
$$V_2(0,t;\xi) = \begin{pmatrix} V_2^{11}(0,t;\xi) & V_2^{12}(0,t;\xi) \\ V_2^{21}(0,t;\xi) & V_2^{22}(0,t;\xi) \end{pmatrix}. \quad (46)$$

where

$$V_2^{11}(0,t;\xi) = i\xi^2|g_0|^2 - \frac{3i}{4}|g_0|^4 - \frac{1}{2}(g_0\bar{g}_1 - g_1\bar{g}_0) = -V_2^{22}(x,t;\xi),$$
$$V_2^{12}(0,t;\xi) = (2\xi^3 g_0 - 2\xi g_0 - \xi g_0|g_0|^2 + i\xi g_1)e^{-2i\int_0^t \Delta_2(0,t')dt'},$$
$$V_2^{21}(0,t;\xi) = (-2\xi^3 \bar{g}_0 + 2\xi\bar{g}_0 + \xi\bar{g}_0|g_0|^2 + i\xi\bar{g}_1)e^{2i\int_0^t \Delta_2(0,t')dt'}.$$

with

$$\Delta_2(0,t') = \tfrac{3}{4}|g_0|^4 + \tfrac{i}{2}(g_0\bar{g}_1 - g_1\bar{g}_0).$$

Due to μ_j have symmetry, the $s(\xi)$ and $S(\xi)$ also have symmetry

$$s_{11}(\xi) = \overline{s_{22}(\bar{\xi})}, s_{21}(\xi) = \overline{s_{12}(\bar{\xi})},$$
$$S_{11}(\xi) = \overline{S_{22}(\bar{\xi})}, S_{21}(\xi) = \overline{S_{12}(\bar{\xi})}.$$

Without loss of generality, we assume

$$s(\xi) = \begin{pmatrix} \overline{a(\bar{\xi})} & b(\xi) \\ \overline{b(\bar{\xi})} & a(\xi) \end{pmatrix}, S(\xi) = \begin{pmatrix} \overline{A(\bar{\xi})} & B(\xi) \\ \overline{B(\bar{\xi})} & A(\xi) \end{pmatrix}. \quad (47)$$

According to (42) and (44), we have

$$s(\xi) = I - \int_{\infty}^{0} e^{i\lambda^2(x'-x)\hat{\sigma}_3}(V_1\mu_3)(x',0;\xi)dx',$$
$$S(\xi) = (I + \int_0^T e^{2i\lambda^4 t'\hat{\sigma}_3}(V_2\mu_2)(0,t';\xi)dt')^{-1}. \quad (48)$$

The spectral functions $s(\xi), S(\xi)$ have the following properties

- $\begin{pmatrix} b(\xi) \\ a(\xi) \end{pmatrix} = \begin{pmatrix} \mu_3^{12}(0,0;\xi) \\ \mu_3^{22}(0,0;\xi) \end{pmatrix} = \mu_3^{(2)}(0,0;\xi),$
  $\begin{pmatrix} e^{-4i\lambda^4 T}B(\xi) \\ \overline{A(\bar{\xi})} \end{pmatrix} = \begin{pmatrix} \mu_2^{12}(0,T;\xi) \\ \mu_2^{22}(0,T;\xi) \end{pmatrix} = \mu_2^{(2)}(0,T;\xi).$
- $a(-\xi) = a(\xi), b(-\xi) = -b(\xi),$
  $A(-\xi) = A(\xi), B(-\xi) = -B(\xi).$
- $\det s(\xi) = \det S(\xi) = 1.$
- $a(\xi) = 1 + O(\frac{1}{\xi}), b(\xi) = O(\frac{1}{\xi}), \xi \to \infty, Im\xi^2 \geq 0,$
  $A(\xi) = 1 + O(\frac{1}{\xi}), B(\xi) = O(\frac{1}{\xi}), \xi \to \infty, Im\xi^4 \geq 0.$

These spectral functions do not exist independently, but depend on each other and satisfy certain relationships, we call it global relation

$$B(\xi)a(\xi) - A(\xi)b(\xi) = e^{4i\lambda^4 T}c^+(\xi), Im\xi^2 \geq 0. \quad (49)$$

where

$$c^+(\xi) = \int_0^\infty e^{2i\lambda^2 x'}(V_1\mu_3)(x',T;\xi)dx'.$$

For simplicity, we define $M(x,t;\xi)$

$$
\begin{aligned}
M_+ &= (\frac{\mu_2^{D_1}}{a(\xi)}, \mu_3^{D_1\cup D_2}), \xi \in D_1,\\
M_- &= (\frac{\mu_1^{D_2}}{d(\xi)}, \mu_3^{D_1\cup D_2}), \xi \in D_2,\\
M_+ &= (\mu_3^{D_3\cup D_4}, \frac{\mu_1^{D_3}}{\overline{d(\bar{\xi})}}), \xi \in D_3,\\
M_- &= (\mu_3^{D_3\cup D_4}, \frac{\mu_2^{D_4}}{\overline{a(\bar{\xi})}}), \xi \in D_4.
\end{aligned}
\tag{50}
$$

where

$$
d(\xi) = a((\xi)\overline{A(\bar{\xi})} - b(\xi)\overline{B(\bar{\xi})}, \xi \in \bar{D}_2. \tag{51}
$$

Synthesizing the above definitions, we can get

$$
detM(x,t;\xi) = 1, \tag{52}
$$

and

$$
M(x,t;\xi) = I + O(\frac{1}{\xi}), \xi \to \infty. \tag{53}
$$

Theorem 2. *Given a smooth function $q(x,t)$. Define $M(x,t;\xi)$ as (50), and define $\mu_j(x,t;\xi)(j = 1,2,3)$ like (37). Then the jump matrices can be derived through*

$$
M_+(x,t;\xi) = M_-(x,t;\xi)J(x,t;\xi), \xi^4 \in R, \tag{54}
$$

where

$$
J = \begin{cases} J_1(x,t;\xi), & arg\xi^2 = 0,\\ J_2(x,t;\xi), & arg\xi^2 = \frac{\pi}{2},\\ J_3(x,t;\xi) = J_2 J_1^{-1} J_4, & arg\xi^2 = \pi,\\ J_4(x,t;\xi), & arg\xi^2 = \frac{3}{2}\pi. \end{cases} \tag{55}
$$

and

$$
\begin{aligned}
J_1 &= \begin{pmatrix} \frac{1}{a\bar{a}} & \frac{b}{a}e^{-2i\theta(\xi)} \\ \frac{\bar{b}}{\bar{a}}e^{2i\theta(\xi)} & 1 \end{pmatrix},\\
J_2 &= \begin{pmatrix} 1 & 0 \\ -\Gamma(\xi)e^{2i\theta(\xi)} & 1 \end{pmatrix},\\
J_4 &= \begin{pmatrix} 1 & \overline{\Gamma(\bar{\xi})e^{2i\theta(\xi)}} \\ 0 & 1 \end{pmatrix}.
\end{aligned}
\tag{56}
$$

with

$$
\begin{aligned}
\theta(\xi) &= \lambda^2 x + 2\lambda^4 t = (\xi^2-1)x + 2(\xi^2-1)^2 t,\\
\Gamma(\xi) &= \frac{\overline{B(\bar{\xi})}}{a(\xi)d(\xi)}.
\end{aligned}
\tag{57}
$$

According to definition, we have to consider the residue conditions of $M(x,t;\xi)$. By analyzing, we can know that both $a(\xi)$ and $d(\xi)$ have an even zero. Hence, we suppose that

1. $a(\xi)$ has $2n$ simple zeros $\{\varepsilon_j\}_{j=1}^{2n}$, $2n = 2n_1 + 2n_2$. Furthermore, $\varepsilon_j(j = 1,2,...,2n_1)$ lie in D_1, $\bar{\varepsilon}_j(j = 1,2,...,2n_2)$ lie in D_2.

2. $d(\xi)$ has $2N$ simple zeros $\{\gamma_j\}_{j=1}^{2N}$, $2N = 2N_1 + 2N_2$. In addition, $\gamma_j (j = 1, 2, ..., 2N_1)$ lie in D_3, $\bar{\gamma}_j (j = 1, 2, ..., 2N_2)$ lie in D_4.
3. $a(\xi)$ and $d(\xi)$ do not have any of the same zeros.

Theorem 3. *For convenience, the mark $[M(x,t;\xi)]_1$ denotes the first column of $M(x,t;\xi)$. Similarly, $[M(x,t;\xi)]_2$ denotes the second column. At the same time, we let $\dot{a}(\xi) = \frac{da}{d\xi}$. Then, we get the residue condition as follows:*

$$\begin{aligned}
&(i) \quad Res\{[M(x,t;\xi)]_1, \varepsilon_j\} = \frac{1}{\dot{a}(\varepsilon_j)b(\varepsilon_j)} e^{2i\theta(\varepsilon_j)}[M(x,t;\varepsilon_j)]_2, j = 1, 2, ..., 2n_1, \\
&(ii) \quad Res\{[M(x,t;\xi)]_2, \bar{\varepsilon}_j\} = \frac{1}{\dot{a}(\bar{\varepsilon}_j)b(\bar{\varepsilon}_j)} e^{-2i\theta(\bar{\varepsilon}_j)}[M(x,t;\bar{\varepsilon}_j)]_1, j = 1, 2, ..., 2n_2, \\
&(iii) \quad Res\{[M(x,t;\xi)]_1, \gamma_j\} = \frac{\overline{B(\bar{\gamma}_j)}}{a(\gamma_j)\dot{d}(\gamma_j)} e^{2i\theta(\gamma_j)}[M(x,t;\gamma_j)]_2, j = 1, 2, ..., 2N_1, \\
&(iv) \quad Res\{[M(x,t;\xi)]_2, \bar{\gamma}_j\} = \frac{B(\bar{\gamma}_j)}{a(\bar{\gamma}_j)\dot{d}(\bar{\gamma}_j)} e^{2i\theta(\bar{\gamma}_j)}[M(x,t;\bar{\gamma}_j)]_1, j = 1, 2, ..., 2N_2.
\end{aligned} \tag{58}$$

Proof. Just prove (i), and the other proof can be proved in the same way. Firstly, we take account of $M(x,t;\xi) = (\frac{\mu_2^{D_1}a(\xi)}{,} \mu_3^{D_1 \cup D_2})$, the simple zeros $\varepsilon_j (j = 1, 2, ..., 2n_1)$ of $a(\xi)$ are the simple poles of $\frac{\mu_2^{D_1}}{a(\xi)}$. Then we get

$$Res\{\frac{\mu_2^{D_1}(x,t;\xi)}{a(\xi)}, \varepsilon_j\} = \lim_{\xi \to \varepsilon_j} (\xi - \varepsilon_j) \frac{\mu_2^{D_1}(x,t;\xi)}{a(\xi)} = \lim_{\xi \to \varepsilon_j} \frac{\mu_2^{D_1}(x,t;\varepsilon_j)}{\frac{a(\xi)-a(\varepsilon_j)}{\xi - \varepsilon_j}} = \frac{\mu_2^{D_1}(x,t;\varepsilon_j)}{\dot{a}(\varepsilon_j)},$$

Then taking $\xi = \varepsilon_j$ into the equation

$$\mu_3^{D_1 \cup D_2} = e^{-2i\theta(\xi)}b(\xi)\mu_2^{D_1} + a(\xi)\mu_2^{D_4},$$

we obtain

$$\mu_3^{D_1 \cup D_2}(x,t;\varepsilon_j) = e^{-2i\theta(\varepsilon_j)}b(\varepsilon_j)\mu_2^{D_1}(x,t;\varepsilon_j) + a(\varepsilon_j)\mu_2^{D_4}(x,t;\varepsilon_j),$$

Finally,

$$Res\{\frac{\mu_2^{D_1}}{a(\xi)}, \varepsilon_j\} = \frac{e^{2i\theta(\varepsilon_j)}}{\dot{a}(\varepsilon_j)b(\varepsilon_j)} \mu_3^{D_1 \cup D_2}(x,t;\varepsilon_j).$$

□

Now, we discuss how to derive the potential $q(x,t)$ from the spectral functions $\mu_j(x,t;\xi)$ $(j = 1, 2, 3)$. Reviewing what we did before, when (21) is a solution of (19), we have $\Upsilon_1^o = \frac{i}{2}QD\sigma_3$. Suppose

$$\mu = I + \frac{m^{(1)}}{\xi} + \frac{m^{(2)}}{\xi^2} + \frac{m^{(3)}}{\xi^3} + O(\xi^4), \xi \to \infty,$$

is a solution of (32).

As $\xi \to \infty$, letting $m(x,t) = m_{12}^{(1)}(x,t)$, namely

$$m(x,t) = \lim_{\xi \to \infty} (\xi \mu_j(x,t;\xi))_{12}.$$

By direct calculation, we have

$$q(x,t) = 2im(x,t)e^{2i\int_{(0,0)}^{(x,t)} \Delta} m(x,t), \tag{59}$$

After that, it is clear to find that

$$q\bar{q} = 4|m|^2, q\bar{q}_x - q_x\bar{q} = 4(m\bar{m}_x - m_x\bar{m}) - 32i|m|^4. \tag{60}$$

and

$$\Delta = -2|m|^2dx + [2i(m\bar{m}_x - m_x\bar{m}) + 28|m|^4]dt. \tag{61}$$

Eventually, we can get the final form of the potential $q(x,t)$.

4. The Spectral Map and the Regular Riemann-Hilbert Problem

4.1. The Spectral Map

Definition 1. *For initial values $q_0(x) = q(x,0)$, the map $\mathbb{S}$ can be defined by*

$$\mathbb{S} : \{q_0(x)\} \to \{a(\xi), b(\xi)\}$$

with

$$\begin{pmatrix} b(\xi) \\ a(\xi) \end{pmatrix} = \mu_3^{(2)}(x,0;\xi), Im\xi^2 \geq 0,$$

where $\mu_3(x,0;\xi)$ is the unique solution of the Volterra linear integral equation

$$\mu_3(x,0;\xi) = I + \int_{\infty}^{x} e^{i\lambda^2(x'-x)\hat{\sigma}_3} V_1(x',0;\xi)\mu_3(x',0;\xi)dx'$$

and $V_1(x,0;\xi)$ is given by Equation (46).

Proposition 1. *$a(\xi)$ and $b(\xi)$ possess these properties.*

(i) $a(\xi)$ *and* $b(\xi)$ *are analytic for* $\{\xi \in C | Im\xi^2 > 0\}$ *and continuous for* $\{\xi \in C | Im\xi^2 \geq 0\}$,

(ii) $a(\xi)\overline{a(\bar{\xi})} - b(\xi)\overline{b(\bar{\xi})} = 1, \xi^2 \in R$,

(iii) $a(\xi) = 1 + O(\frac{1}{\xi}), b(\xi) = O(\frac{1}{\xi}), \xi \to \infty, Im\xi^2 \geq 0$,

(iv) $a(-\xi) = a(\xi), b(-\xi) = -b(\xi), Im\xi^2 \geq 0$,

(v) *We define* $\mathbb{Q} : \{a(\xi), b(\xi)\} \to \{q_0(x)\}$, *as the inverse of map* $\mathbb{S}$, *with*

$$q_0(x) = 2im(x)e^{4i\int_0^x |m(x')|^2dx'}, m(x) = \lim_{\xi\to\infty}(\xi M^{(x)}(x,\xi))_{12}. \tag{62}$$

where $M^{(x)}(x,\xi)$ is the unique solution of the following Riemann-Hilbert problem.

- $M^{(x)}(x,\xi) = \begin{cases} M_-^{(x)}(x,\xi), & Im\xi^2 \leq 0 \\ M_+^{(x)}(x,\xi), & Im\xi^2 \geq 0 \end{cases}$ *is a meromorphic function.*
- $M_+^{(x)}(x,\xi) = M_-^{(x)}(x,\xi)J^{(x)}(x,\xi), \xi^2 \in R$,
  where

$$J^{(x)}(x,\xi) = \begin{pmatrix} \frac{1}{a(\xi)\overline{a(\bar{\xi})}} & \frac{b(\xi)}{a(\xi)}e^{-2i\lambda^2 x} \\ -\frac{\overline{b(\bar{\xi})}}{\overline{a(\bar{\xi})}}e^{2i\lambda^2 x} & 1 \end{pmatrix}, \xi^2 \in R. \tag{63}$$

- $M^{(x)}(x,\xi) = I + O(\frac{1}{\xi}), \xi \to \infty$.
- *$a(\xi)$ has $2n$ simple zeros $\{\varepsilon_j\}_{j=1}^{2n}$, $2n = 2n_1 + 2n_2$, such that ε_j, $(j = 1,2,...,2n_1)$ lie in D_1, and $\bar{\varepsilon}_j$, $(j = 1,2,...,2n_2)$ lie in D_2.*
- *The first column of $M_+^{(x)}$ has simple poles at $\xi = \varepsilon_j, j = 1,2,...,2n_1$. Furthermore, the second column of $M_-^{(x)}$ has simple poles at $\xi = \bar{\varepsilon}_j, j = 1,2,...,2n_2$. The relevant residues are given by*

$$Res\{[M^{(x)}(x,\xi)]_1, \varepsilon_j\} = \frac{e^{2i(\varepsilon_j^2-1)}x}{\dot{a}(\varepsilon_j)b(\varepsilon_j)}[M^{(x)}(x,\varepsilon_j)]_2, j = 1,2,...,2n_1, \tag{64}$$

$$Res\{[M^{(x)}(x,\xi)]_2, \bar{\varepsilon}_j\} = \frac{e^{-2i(\bar{\varepsilon}_j^2-1)}x}{\overline{\dot{a}(\bar{\varepsilon}_j)b(\bar{\varepsilon}_j)}}[M^{(x)}(x,\bar{\varepsilon}_j)]_1, j = 1,2,...,2n_2. \tag{65}$$

Definition 2. *For boundary values $g_0(t) = q(0,t), g_1(t) = q_x(0,t)$, the map $\tilde{\mathbb{S}}$ can be defined by*

$$\tilde{\mathbb{S}} : \{g_0(t), g_1(t)\} \to \{A(\xi), B(\xi)\}$$

with

$$\begin{pmatrix} B(\xi) \\ A(\xi) \end{pmatrix} = \mu_1^{(2)}(0,t;\xi), Im\xi^2 \geq 0,$$

where $\mu_1(0,t;\xi)$ is the unique solution of the Volterra linear integral equation

$$\mu_1(0,t;\xi) = I - \int_t^T e^{2i\lambda^4(t'-t)\hat{\sigma}_3} V_2(0,t';\xi)\mu_1(0,t',\xi)dt'$$

and $V_2(0,t;\xi)$ is given by (46).

Proposition 2. *$A(\xi)$ and $B(\xi)$ possess these properties.*

(i) $A(\xi)$ *and* $B(\xi)$ *are analytic for* $\{\xi \in C | Im\xi^4 > 0\}$ *and continuous* $\{\xi \in C | Im\xi^4 \geq 0\}$,

(ii) $A(\xi)\overline{A(\bar{\xi})} - B(\xi)\overline{B(\bar{\xi})} = 1, \xi^4 \in R,$

(iii) $A(\xi) = 1 + O(\frac{1}{\xi}), B(\xi) = O(\frac{1}{\xi}), \xi \to \infty, Im\xi^4 \geq 0,$

(iv) $A(-\xi) = A(\xi), B(-\xi) = -B(\xi), Im\xi^4 \geq 0,$

(v) *We define* $\tilde{\mathbb{Q}} : \{A(\xi), B(\xi)\} \to \{g_0(t), g_1(t)\}$*, as the inverse of map* $\tilde{\mathbb{S}}$*, with*

$$\begin{aligned} g_0(t) &= 2im_{12}^{(1)}(t)e^{2i\int_0^t \Delta_2(t')dt'}, \\ g_1(t) &= (4m_{12}^{(2)} + |g_0(t)|^2 m_{12}^{(1)}(t))e^{2i\int_0^t \Delta_2(t')dt'} + ig_0(t)(2m_{22}^{(2)}(t) + |g_0(t)|^2), \end{aligned} \tag{66}$$

where

$$\Delta_2(t) = 4|m_{12}^{(1)}|^4 + 8(Re[m_{12}^{(1)}\bar{m}_{12}^{(3)}] - |m_{12}^{(1)}|^2 Re[m_{22}^{(2)}]),$$

with the functions $m^{(i)}(t)(i = 1,2,3.)$ are depend on

$$M^{(t)}(t,\xi) = I + \frac{m^{(1)}(t)}{\xi} + \frac{m^{(2)}(t)}{\xi^2} + \frac{m^{(3)}(t)}{\xi^3} + O(\frac{1}{\xi^4}), \xi \to \infty,$$

where $M^t(t,\xi)$ is the unique solution of the following Riemann-Hilbert problem

- $M^{(t)}(t,\xi) = \begin{cases} M_-^{(t)}(t,\xi), & Im\xi^4 \leq 0 \\ M_+^{(t)}(t,\xi), & Im\xi^4 \geq 0 \end{cases}$ *is a meromorphic function.*
- $M_+^{(t)}(t,\xi) = M_-^{(t)}(t,\xi)J^{(t)}(t,\xi), \xi^4 \in R,$ *where*

$$J^{(t)}(t,\xi) = \begin{pmatrix} \frac{1}{A(\xi)\overline{A(\bar{\xi})}} & \frac{B(\xi)}{\overline{A(\bar{\xi})}} e^{-4i\lambda^4 t} \\ -\frac{\overline{B(\bar{\xi})}}{A(\xi)} e^{4i\lambda^4 t} & 1 \end{pmatrix}, \xi^4 \in R. \tag{67}$$

- $M^{(t)}(t,\xi) = I + O(\frac{1}{\xi}), \xi \to \infty.$
- *$A(\xi)$ has $2N$ simple zeros $\{\gamma_j\}_{j=1}^{2N}$, $2N = 2N_1 + 2N_2$, such that $\gamma_j (j = 1,2,...,2N_1)$ lie in D_3, and $\bar{\gamma}_j (j = 1,2,...,2N_2)$ lie in D_4.*
- *The first column of $M_+^{(t)}$ has simple poles at $\xi = \gamma_j, j = 1,2,...,2N_1$. And the second column of $M_-^{(t)}$ has simple poles at $\xi = \bar{\gamma}_j, j = 1,2,...,2N_2$. The relevant residues are given by*

$$Res\{[M^{(t)}(t,\xi)]_1, \gamma_j\} = \frac{e^{4i(\gamma_j^2-1)^2 t}}{\dot{A}(\gamma_j)B(\gamma_j)}[m^{(t)}(t,\gamma_j)]_2, j = 1,2,...,2N_1, \tag{68}$$

$$Res\{[M^{(t)}(t,\xi)]_2, \bar{\gamma}_j\} = \frac{e^{-4i(\bar{\gamma}_j^2-1)^2 t}}{\overline{\dot{A}(\bar{\gamma}_j)b(\bar{\gamma}_j)}}[M^{(t)}(t,\bar{\gamma}_j)]_1, j = 1,2,...,2N_2. \tag{69}$$

4.2. The Regular Riemann-Hilbert Problem

Theorem 4. *Given the smooth function $q_0(x)$, which is compatible with $g_0(t)$ and $g_1(t)$. The spectral functions $a(\xi)$,$b(\xi)$,$A(\xi)$, and $B(\xi)$ are defined according to the previous definitions. Furthermore, they satisfy the global relation (49). Clearly, it becomes $B(\xi)a(\xi) - A(\xi)b(\xi) = 0$ when $\xi \to \infty$. Define the $M(x,t;\xi)$ as the solution of this following Riemann-Hilbert problem.*

- *$M(x,t;\xi)$ is a sectionally meromorphic function in $\{\xi \in C | \xi^4 \in R\}$.*
- *The residue condition of $M(x,t;\xi)$ satisfies Theorem 3*
- *$M(x,t;\xi)$ satisfies the jump condition*

$$M_+(x,t;\xi) = M_-(x,t;\xi)J(x,t;\xi), \xi^4 \in R,$$

where the jump matrices are defined by (55)–(57).

- *$M(x,t;\xi) = I + O(\frac{1}{\xi}), \xi \to \infty$.*

Then, $M(x,t;\xi)$ not only exists but is unique. In this way, the solution of the mNLS equation can be derived, which can be defined by

$$\begin{aligned} q(x,t) &= 2im(x,t)e^{2i\int_{(0,0)}^{(x,t)} \Delta}, \\ m(x,t) &= \lim_{\xi \to \infty} (\xi \mu_j(x,t;\xi))_{12}, \\ \Delta &= -2|m|^2 dx + [2i(m\bar{m}_x - m_x m) + 28|m|^4]dt. \end{aligned} \tag{70}$$

Besides, $q(x,t)$ also satisfies the initial-boundary values condition

$$q(x,0) = q_0(x), q(0,t) = g_0(t), \text{ and } q_x(0,t) = g_1(t).$$

Proof. Actually, if there are no zeros of $a(\xi)$ and $d(\xi)$, then the 2×2 function $M(x,t;\xi)$ satisfies a non-singular Riemann-Hilbert problem. Due to the jump matrices $J(x,t;\xi)$ possessing symmetry, we can find that this problem has a unique solution. On the other hand, when $a(\xi)$ and $d(\xi)$ have a certain number of zeros, by specific mapping, the singular Riemann-Hilbert problem can become no zeros with a system of algebraic equations; the unique solvability can be proved by the following theorem. □

Theorem 5. *The Riemann-Hilbert problem in Theorem 4 with the vanishing boundary condition*

$$M(x,t;\xi) \to 0, \xi \to \infty,$$

has only the zero solution.

Proof. Firstly, we suppose that the matrix function $M(x,t;\xi)$ is a solution of the Riemann-Hilbert problem in Theorem 4. At the same time, $A^\dagger$ means the complex conjugate transpose of A, where A is a 2×2 matrix. We define

$$\begin{aligned} H_+(\xi) &= M_+(\xi)M_-^\dagger(-\bar{\xi}), Im\xi^4 \geq 0, \\ H_-(\xi) &= M_-(\xi)M_+^\dagger(-\bar{\xi}), Im\xi^4 \leq 0, \end{aligned} \tag{71}$$

where the x and t are dependent with each other. $H_+(\xi)$ and $H_-(\xi)$ are analytic in $\{\xi \in C | Im\xi^4 > 0\}$ and $\{\xi \in C | Im\xi^4 < 0\}$, respectively. Due to the symmetry, we can obtain from (54) and (55)

$$J_1^\dagger(-\bar{\xi}) = J_1(\xi), J_3^\dagger(-\bar{\xi}) = J_3(\xi), J_2^\dagger(-\bar{\xi}) = J_4(\xi). \tag{72}$$

Then

$$\begin{aligned} H_+(\xi) &= M_-(\xi)J(\xi)M_-^\dagger(-\bar{\xi}), Im\xi^4 \in R, \\ H_-(\xi) &= M_-(\xi)J^\dagger(-\bar{\xi})M_-^\dagger(-\bar{\xi}), Im\xi^4 \in R. \end{aligned} \tag{73}$$

From the above two equations, it is easy to find that $H_+(\xi) = H_-(\xi)$. This means that $H_+(\xi)$ and $H_-(\xi)$ define an entire function decaying at infinity, hence the $H_+(\xi) \equiv 0$ and $H_-(\xi) \equiv 0$. Finding $J_3(i\hbar)(\hbar \in R)$ is a 2×2 unit Hermitian matrix for any $\hbar \in R$. It is not difficult to see that $J_3(i\hbar)(\hbar \in R)$ is a positive definite matrix. Now that $H_-(\hbar) = 0$ for $\hbar \in iR$, we have

$$M_+(i\hbar)J_3(i\hbar)M_+^\dagger(i\hbar) = 0. \tag{74}$$

After simple calculation, we have $M_+(i\hbar) = 0$ for $\hbar \in R$. Therefore, $M_+(\xi) = 0$, $M_-(\xi) = 0$. □

Remark 1. *$q(x,t)$ satisfies the mNLS equation.*

In fact, if $M(x,t;\xi)$ is the solution of the Riemann-Hilbert problem defined by Theorem 4 and $q(x,t)$ is defined as the previous definition, with the help of the dressing method [45], we can find that $q(x,t)$ satisfies the Lax pair (18). Hence, $q(x,t)$ satisfies the mNLS equation.

Remark 2. *Using the same proof method in Reference [32] can we prove that $q(x,t)$ satisfies the initial values $q(x,0) = q_0(x)$ and boundary values $q(0,t) = g_0(t), q_x(0,t) = g_1(t)$, so in this paper, we leave this proof out.*

5. Conclusions

In this paper, we mainly studied the initial-boundary values problem for the mNLS equation on the half line. Before we did this, with the help of prolongation structure theory, the Lax pair of this equation was derived. Then we reconstructed the Lax pair to obtain a Riemann-Hilbert problem, and therefore, the potential function has been represented by its solution. In future work, the long time asymptotic behavior for the solutions will be analyzed.

Author Contributions: All authors contributed equally to the writing of this paper. All authors read and approved the final manuscript.

Funding: This work was supported by Nature Science Foundation of Shandong Province of China (No.ZR2018MA017), China Postdoctoral Science Foundation funded project (No.2017M610436), Open Fund of the Key Laboratory of Meteorological Disaster of Ministry of Education (Nanjing University of Information Science and Technology) (No.KLME201801).

Conflicts of Interest: The authors declare no conflict of interest.

References

1. Xu, X.X.; Sun, Y.P. Two symmetry constraints for a generalized Dirac integrable hierarchy. *J. Math. Anal. Appl.* **2018**, *458*, 1073–1090. [CrossRef]
2. Mcanally, M.; Ma, W.X. An integrable generalization of the D-Kaup-Newell soliton hierarchy and its bi-Hamiltonian reduced hierarchy. *Appl. Math. Comput.* **2018**, *323*, 220–227. [CrossRef]
3. Ma, W.X. Conservation laws by symmetries and adjoint symmetries. *Discret. Contin. Dynam. Syst. Ser. S* **2017**, *11*, 707–721. [CrossRef]
4. Xu, X.X. A deformed reduced semi-discrete Kaup-Newell equation, the related integrable family and Darboux transformation. *Appl. Math. Comput.* **2015**, *251*, 275–283. [CrossRef]
5. Hirota, R. A new form of Bäcklund transformations and its relation to the inverse scattering problem. *Progr. Theor. Phys.* **1974**, *52*, 1498–1512. [CrossRef]
6. Xu, X.X.; Sun, Y.P. An integrable coupling hierarchy of Dirac integrable hierarchy, its Liouville integrability and Darboux transformation. *J. Nonlinear Sci. Appl.* **2017**, *10*, 3328–3343. [CrossRef]
7. Zhang, L.J.; Khalique, M. Classification and bifurcation of a class of second-order ODEs and its application to nonlinear PDEs. *Discret. Contin. Dynam. Syst. Ser. S* **2017**, *11*, 777–790. [CrossRef]
8. Lou, S.Y.; Ruan, H.Y.; Chen, D.F.; Chen, W.Z. Similarity reductions of the KP equation by a direct method. *J. Phys. A Gen. Phys.* **1999**, *24*, 1455. [CrossRef]
9. Ablowitz, M.J.; Segur, H. *Solitons and the Inverse Scattering Transform*; Siam: Philadelphia, PA, USA, 1981.
10. Gu, J.Y.; Zhang, Y.; Dong, H.H. Dynamic behaviors of interaction solutions of (3+1)-dimensional Shallow Water wave equation. *Comput. Math. Appl.* **2018**, *76*, 1408–1419. [CrossRef]

11. Liu, Y.; Dong, H.H.; Zhang, Y. Solutions of a discrete integrable hierarchy by straightening out of its continuous and discrete constrained flows. *Anal. Math. Phys.* **2018**. [CrossRef]
12. Ma, W.X.; Zhou, Y. Lump solutions to nonlinear partial differential equations via Hirota bilinear forms. *J. Differ. Eq.* **2018**, *264*, 2633–2659. [CrossRef]
13. Zhang, Y.; Dong, H.H.; Zhang, X.E.; Yang, H.W. Rational solutions and lump solutions to the generalized (3+1)-dimensional Shallow Water-like equation. *Comput. Math. Appl.* **2016**, *73*, 246–252. [CrossRef]
14. Wang, H.; Wang, Y.H.; Dong, H.H. Interaction solutions of a (2+1)-dimensional dispersive long wave system. *Comput. Math. Appl.* **2018**, *75*, 2625–2628. [CrossRef]
15. Dong, H.Y.; Lu, C.N.; Yang, H.W. The finite volume WENO with Lax-Wendroff scheme for nonlinear system of Euler equations. *Mathematics* **2018**, *6*, 211. [CrossRef]
16. Lu, C.N.; Gao, Q.Q.; Fu, C.; Yang, H.W. Finite element method of BBM-Burgers equation with dissipative term based on Adaptive Moving Mesh. *Discret. Dynam. Nat. Soc.* **2017**, *2017*, 3427376. [CrossRef]
17. Tao, M.S.; Dong, H.H. Algebro-Geometric solutions for a discrete integrable equation. *Discret. Dynam. Nat. Soc.* **2017**, *2017*, 5258375. [CrossRef]
18. Hirota, R. Exact N-soliton solutions of the wave equation of long waves in Shallow-Water and in nonlinear lattices. *J. Math. Phys.* **1973**, *14*, 180–184. [CrossRef]
19. Feng, B.F.; Maruno, K.; Ohta, Y. A two-component generalization of the reduced Ostrovsky equation and its integrable semi-discrete analogue. *J. Phys. A Math. Theor.* **2016**, *50*, 055201. [CrossRef]
20. Feng, B.F.; Kawahara, T. Stationary travelling-wave solutions of an unstable KdV-Burgers equation. *Phys. D Nonlinear Phenom.* **2000**, *137*, 228–236. [CrossRef]
21. Tao, M.S.; Zhang, N.; Gao, D.Z.; Yang, H.W. Symmetry analysis for three-dimensional dissipation Rossby waves. *Adv. Differ. Eq.* **2018**, *2018*, 300. [CrossRef]
22. Ma, W.X.; Li, J.; Khalique, C.M. A Study on Lump Solutions to a Generalized Hirota-Satsuma-Ito Equation in (2+1)-Dimensions. *Complexity* **2018**, *2018*, 9059858. [CrossRef]
23. Fokas, A.S. A unified transform method for solving linear and certain nonlinear PDEs. *Proc. R. Soc. Math. Phys. Eng. Sci.* **1997**, *453*, 1411–1443. [CrossRef]
24. Fokas, A.S. On a class of physically important integrable equations. *Phys. D Nonlinear Phenom.* **1995**, *87*, 145–150. [CrossRef]
25. Fokas, A.S. *A Unified Approach to Boundary Value Problems*; Society for Industrial and Applied Mathematics: Philadelphia, PA, USA, 2008.
26. Fokas, A.S. Integrable nonlinear evolution equations on the half-Line. *Commun. Math. Phys.* **2002**, *230*, 1–39. [CrossRef]
27. Lenells, J. The derivative nonlinear Schrödinger equation on the half-line. *Phys. D Nonlinear Phenom.* **2009**, *237*, 3008–3019. [CrossRef]
28. Lenells, J. Initial-boundary value problems for integrable evolution equations with 3×3 Lax pairs. *Phys. D Nonlinear Phenom.* **2012**, *241*, 857–875. [CrossRef]
29. Lenells, J. The Degasperis-Procesi equation on the half-line. *Nonlinear Anal. Theory Methods Appl.* **2013**, *76*, 122–139. [CrossRef]
30. Lenells, J.; Fokas, A. Boundary-value problems for the stationary axisymmetric Einstein equations: A rotating disc. *Nonlinearity* **2009**, *24*, 177–206. [CrossRef]
31. Lenells, J. An integrable generalization of the sine-Gordon equation on the half-line. *Ima J. Appl. Math.* **2012**, *76*, 554–572. [CrossRef]
32. Xu, J.; Fan, E.G. A Riemann-Hilbert approach to the initial-boundary problem for derivative nonlinear Schröingder equation. *Acta Math. Sci.* **2014**, *34*, 973–994. [CrossRef]
33. Xu, J.; Fan, E.G. The initial-boundary value problem for the Ostrovsky-Vakhnenko equation on the half-line. *Math. Phys. Anal. Geom.* **2016**, *19*, 20. [CrossRef]
34. Xu, J.; Fan, E.G. Initial-boundary value problem for integrable nonlinear evolution equation with 3×3 Lax pairs on the interval. *Stud. Appl. Math.* **2016**, *136*, 321–354. [CrossRef]
35. Xu, J.; Fan, E.G. The three-wave equation on the half-line. *Phys. Lett. A* **2014**, *378*, 26–33. [CrossRef]
36. Geng, X.G.; Liu, H.; Zhu, J.Y. Initial-boundary value problems for the coupled nonlinear Schrödinger equation on the half-Line. *Stud. Appl. Math.* **2015**, *135*, 310–346. [CrossRef]
37. Tian, S.F. Initial-boundary value problems for the general coupled nonlinear Schrödinger equation on the interval via the Fokas method. *J. Differ. Eq.* **2017**, *262*, 506–558. [CrossRef]

38. Xu, J.; Fan, E.G. The unified transform method for the Sasa-Satsuma equation on the half-line. *Physics* **2013**, *17*, 445–452. [CrossRef] [PubMed]
39. Zhang, N.; Xia, T.C.; Fan, E.G. A Riemann-Hilbert approach to the Chen-Lee-Liu equation on the half line. *Acta Math. Appl. Sin. Engl. Ser.* **2018**, *34*, 493–515. [CrossRef]
40. Ma, W.X. Application of the Riemann-Hilbert approach to the multicomponent AKNS integrable hierarchies. *Nonlinear Anal. Real World Appl.* **2019**, *47*, 1–17. [CrossRef]
41. Hu, B.B.; Xia, T.C.; Ma, W.X. Riemann-Hilbert approach for an initial-boundary value problem of the two-component modified Korteweg-de Vries equation on the half-line. *Appl. Math. Comput.* **2018**, *332*, 148–159. [CrossRef]
42. Deift, P.; Zhou, X. A steepest descent method for oscillatory Riemann-Hilbert problems. Asymptotics for the mKdV equation. *Ann. Math.* **1993**, *137*, 295–368. [CrossRef]
43. Lenells, J. The nonlinear steepest descent method: Asymptotics for initial-boundary value problems. *SIAM J. Math. Anal.* **2015**, *48*, 273–311. [CrossRef]
44. Xu, J.; Fan, E.G. Long-time asymptotic for the derivative nonlinear Schrödinger equation with step-like initial value. *Math. Phys. Anal. Geom.* **2013**, *16*, 253–288. [CrossRef]
45. Lenells, J. Dressing for a novel integrable generalization of the nonlinear Schrödinger equation. *J. Nonlinear Sci.* **2009**, *20*, 709–722. [CrossRef]
46. Deift, P.; Zhou, X. Long-time asymptotics for solutions of the NLS equation with initial data in a weighted Sobolev space. *Commun. Pure Appl. Math.* **2010**, *56*, 1029–1077. [CrossRef]
47. Wahlquist, H.D.; Estabrook, F.B. Prolongation structures of nonlinear evolution equations. *J. Math. Phys.* **2008**, *16*, 1293–1297. [CrossRef]
48. Hermann, R. Pseudopotentials of Estabrook and Wahlquist, the Geometry of Solitons, and the Theory of Connections. *Phys. Rev. Lett.* **1976**, *36*, 791–795. [CrossRef]
49. Deconinck, B. A Constructive Test for Integrability of Semi-Discrete Systems. *Phys. Lett. A* **1996**, *223*, 45–54. [CrossRef]
50. Wang, D.S. Integrability of the coupled KdV equations derived from two-layer fluids: Prolongation structures and Miura transformations. *Nonlinear Anal.* **2010**, *73*, 270–281. [CrossRef]
51. Humphreys, J.E. *Introduction to Lie Algebras and Representation Theory*; Springer: New York, NY, USA, 1972.
52. Morris, H.C. Prolongation structures and nonlinear evolution equations in two spatial dimensions. II. A generalized nonlinear Schrödinger equation. *J. Math. Phys.* **1977**, *18*, 285–288. [CrossRef]
53. Karsten, T. Hamiltonian form of the modified nonlinear Schrödinger equation for gravity waves on arbitrary depth. *J. Fluid Mech.* **2011**, *670*, 404–426.
54. Wen, X.Y.; Yang, Y.Q.; Yan, Z.Y. Generalized perturbation (n, M)-fold Darboux transformations and multi-rogue-wave structures for the modified self-steepening nonlinear Schrödinger equation. *Phys. Rev. E Stat. Nonlinear Soft Matter Phys.* **2015**, *92*, 012917. [CrossRef] [PubMed]
55. Brizhik, L.; Eremko, A.; Piette, B.; Zakrzewski, W. Static solutions of a D-dimensional modified nonlinear Schrödinger equation. *Nonlinearity* **2003**, *16*, 1481–1497. [CrossRef]
56. Liu, J.Q.; Wang, Z.Q. Symmetric solutions to a modified nonlinear Schrödinger equation. *Nonlinearity* **2007**, *21*, 121. [CrossRef]
57. Strampp, W. Lax-pairs, spectral problems, and recursion operators. *J. Math. Phys.* **1984**, *25*, 2905–2909. [CrossRef]
58. Maimistov, A.I. Evolution of solitary waves which are approximately solitons of a nonlinear Schrödinger equation. *Soviet J. Exp. Theor. Phys.* **1993**, *77*, 727–731.
59. Mihalache, D.; Truta, N.; Panoiu, N.; Baboiu, D. Analytic method for solving the modified nonlinear Schrödinger equation describing soliton propagation along optical fibers. *Phys. Rev. A* **1993**, *47*, 3190–3194. [CrossRef] [PubMed]

mathematics

MDPI

Article

On the Connection between Spherical Laplace Transform and Non-Euclidean Fourier Analysis

Enrico De Micheli †

IBF–Consiglio Nazionale delle Ricerche, Via De Marini, 6-16149 Genova, Italy; enrico.demicheli@cnr.it
† Dedicated to the memory of Professor Giovanni Alberto Viano.

Received: 17 January 2020; Accepted: 17 February 2020; Published: 20 February 2020

Abstract: We prove that, if the coefficients of a Fourier–Legendre expansion satisfy a suitable Hausdorff-type condition, then the series converges to a function which admits a holomorphic extension to a cut-plane. Next, we introduce a Laplace-type transform (the so-called *Spherical Laplace Transform*) of the jump function across the cut. The main result of this paper is to establish the connection between the *Spherical Laplace Transform* and the *Non-Euclidean Fourier Transform* in the sense of Helgason. In this way, we find a connection between the unitary representation of $\mathrm{SO}(3)$ and the principal series of the unitary representation of $\mathrm{SU}(1,1)$.

Keywords: holomorphic extension; spherical Laplace transform; non-Euclidean Fourier transform; Fourier–Legendre expansion

MSC: 42A38; 44A10; 44A12; 42C10

1. Introduction

It is well known that the classical Fourier transform refers to the decomposition of a function belonging to an appropriate space into exponentials, which can be viewed as the irreducible unitary representations of the additive group of the real numbers. However, in the current most popular interpretation, particularly in connection with non-commutative groups, the phrase *harmonic analysis* has lost its original function-theoretic meaning and now it generally refers not to functions but to representations. It thus becomes natural to regard irreducible representations as the basic building blocks of the theory in place of exponential functions [1]. There are however examples in the theory of the semi-simple non-compact Lie groups where the classical setup prevails, in the sense that one can find a class of functions which play a role similar to that played by the exponentials on the real line. A typical example thereof is the group $\mathrm{SU}(1,1)$: hereafter, we shall work at the level of homogeneous spaces associated with this group and, accordingly, we shall study the spherical functions that can be constructed in these spaces.

Within this framework, a fundamental role is played by the Helgason construction of the so-called *non-Euclidean Fourier analysis* [2]. The working ambient is the symmetric space $\mathrm{SU}(1,1)/\mathrm{SO}(2)$, i.e., the non-Euclidean disk. By the use of the Poisson kernel, the analog of the plane waves are constructed in the case of the hyperbolic disk and, successively, an integral representation of the conical functions $P_{-\frac{1}{2}+i\mu}(\cosh r)$ (i.e., the Legendre functions of the first kind with index $(-\frac{1}{2}+i\mu)$, $\mu \in \mathbb{R}$) is derived in terms of these *hyperbolic waves*. It can thus be proved for these functions a product formula, which corresponds to the classical product formula of the exponentials. Finally, a Fourier transform on the non-Euclidean disk can be set up, which is exactly the tool analogous to the classical Fourier transform. In this connection, it is worth recalling that the conical functions can be associated with the principal series of the irreducible unitary representation of the group $SU(1,1)$, which acts transitively on the hyperbolic disk.

Consider now an isotropic cone in $\mathbb{R}^3$. We focus on the two-sheeted hyperboloid interior to the cone. By stereographic projection, the upper sheet of the two-sheeted hyperboloid can be mapped into the interior of the unit disk. Since the connected component of the two-sheeted hyperboloid is the homogeneous space $SO_0(1,2)/SO(2)$, the non-Euclidean Fourier analysis coincides with the harmonic analysis on $SO_0(1,2)/SO(2)$. We can thus define and study the Fourier transform on the two-sheeted hyperboloid.

Harmonic analysis can be studied also on the one-sheeted hyperboloid, which is a pseudo-Riemannian symmetric space $SO_0(1,2)/SO_0(1,1)$. It is then possible to construct on this symmetric space spherical functions, which turn out to be the Legendre functions of the second kind. A peculiar feature of this space is that it can be equipped with the partial ordering associated with the light cone in $\mathbb{R}^3$: namely, $x \geqslant y \Leftrightarrow (x-y)$ belongs to the closed future cone of $\mathbb{R}^3$. The one-sheeted hyperboloid equipped with this ordering relation is a *causal symmetric space* [3]. Accordingly, we can introduce a *Volterra algebra* of kernels, i.e., kernels $K(x,y)$ whose support is contained in the set $\Gamma \doteq \{(x,y) \in X_2 \times X_2 : x \geqslant y\}$, where X_2 is the one-sheteed hyperboloid. A kernel K is said to be invariant under $G \equiv SO_0(1,2)$ if for any $g \in G$: $K(gx,gy) = K(x,y)$, $(x,y) \in \Gamma$. An invariant Volterra kernel K can be identified with a function f on G through the equality: $K(ge_2,e_2) = f(g)$, where $g \in G$ and $e_2 = (0,0,1)$ (see next Figure 1) is the point which features $he_2 = e_2$, with $h \in SO_0(1,1)$ [4–7]. Successively, the *spherical Laplace transform* for this class of functions can be defined. This transform, as the ordinary Laplace transform, is holomorphic in a half-plane and, in the specific case of the spherical Laplace transform, this analyticity property follows from the analyticity property of the Legendre functions of the second kind $Q_\lambda(\cdot)$, which are holomorphic in the half-plane $\mathbb{C}^{(+)}_{(-1)} \doteq \{\lambda \in \mathbb{C} : \mathrm{Re}\,\lambda > -1\}$. These latter functions are indeed the spherical functions on ordered symmetric spaces in the sense of Faraut et al. [7].

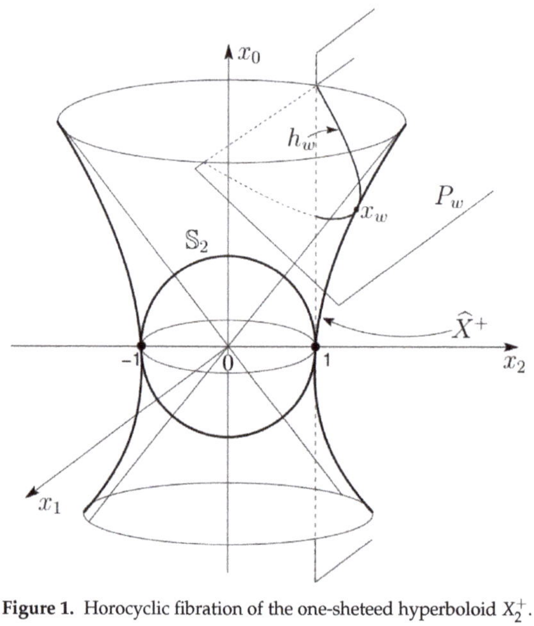

Figure 1. Horocyclic fibration of the one-sheteed hyperboloid X_2^+.

Another question closely related to this type of problems is the holomorphic extension associated with Fourier–Legendre expansions of the type: $\frac{1}{4\pi}\sum_{n=0}^{\infty}(2n+1)a_n P_n(\cos\theta)$, where $P_n(\cdot)$ denotes the Legendre polynomials. In this paper, we prove that, if the sequence $\{(n+1)^2 a_n\}_{n=0}^{\infty}$ satisfies a suitable Hausdorff-type condition [8], then the function $f(\cos\theta)$, to which the Fourier–Legendre series converges in the interval $(-1,1)$, admits a holomorphic extension to the complex $\cos\theta$-plane cut along the semi-axis $[1,+\infty)$. The first result we prove is a basic feature of the holomorphic extension associated with Fourier–Legendre series: the *dual analyticity property*. To classes of functions which are holomorphic in the $\cos\theta$-plane cut along $[1,+\infty)$, there correspond classes of functions, denoted $\tilde{a}(\lambda)$ ($\lambda \in \mathbb{C}$), which are expressed as a spherical Laplace transform, holomorphic in the half-plane $\operatorname{Re}\lambda > -\frac{1}{2}$, of Carlsonian-type (i.e., of suitable exponential growth; see next Theorem 1) and, therefore, unique interpolants of the coefficients a_n of the Fourier-Legendre series, i.e., $\tilde{a}(\lambda)|_{\lambda=n} = a_n$ $(n = 0,1,2,\ldots)$.

At this point, two strictly related problems emerge. On the one hand, the possibility of connecting, via analytic continuation, the completeness of the spherical functions for SO(3) to the completeness of the corresponding expansion of $\mathrm{SL}(2,\mathbb{R})$, which is the group acting transitively on the upper half-plane, where the model of the non-Euclidean geometry can be realized, while, on the other hand, continuing the unitary representations of SO(3) to give the unitary representations of $\mathrm{SL}(2,\mathbb{R})$. The first problem has been initially tackled by Stein and Wainger [9], while, for what concerns the second one, the reader is referred to the work of R. Hermann [10]. However, both of these problems require linking the spherical Laplace transform to the non-Euclidean Fourier analysis. When we consider the ordinary Laplace transform of a function $f \in L^1(0,+\infty) \cap L^2(0,+\infty)$,

$$\widetilde{f}(\lambda) = \int_0^{+\infty} e^{-\lambda v} f(v)\,\mathrm{d}v \qquad (\lambda \in \mathbb{C}, \operatorname{Re}\lambda > 0),$$

we obtain for $\lambda = \mathrm{i}\mu$ ($\mu \in \mathbb{R}$):

$$\widetilde{f}(\mathrm{i}\mu) - \int_{-\infty}^{+\infty} e^{-\mathrm{i}\mu v} f(v)\,\mathrm{d}v \qquad (\mu \subset \mathbb{R}),$$

which is precisely the Fourier transform of $f(v)$, it being well-defined in view of the assumption $f \in L^1(0,+\infty)$. Since $f \in L^1(0,+\infty) \cap L^2(0,+\infty)$, the Fourier transform can be inverted to recover $f(v)$ in terms of $\widetilde{f}(\mathrm{i}\mu)$ through the *inverse Fourier transform*, which converges to $f(v)$ as a limit in the mean order two. Correspondingly, we pass from the non-unitary representation $e^{-\lambda v}$ ($\lambda \in \mathbb{C}, \operatorname{Re}\lambda > 0, \mu \equiv \operatorname{Im}\lambda \in \mathbb{R}$) to the unitary irreducible representation $e^{-\mathrm{i}\mu v}$ ($\mu, v \in \mathbb{R}$) that we mentioned at the beginning of the Introduction. This connection can be extended. One of our results consists indeed of proving that the spherical Laplace transform reduces to the non-Euclidean Fourier transform at $\operatorname{Re}\lambda = -\frac{1}{2}$, which is precisely the value corresponding to the principal series of the unitary representations of the group SU(1,1), which acts transitively on the hyperbolic disk. This result has been obtained by establishing a bridge between the harmonic analysis on the one-sheeted hyperboloid and the harmonic analysis on the two-sheeted hyperboloid.

The harmonic analysis in causal symmetric spaces has been a subject of growing interest in the last three decades, and the research on these topics has flowed in various directions. Some papers have been devoted to the proof of the Paley–Wiener theorem for spherical Laplace transform [11,12] and others have treated the inversion problem extended up to the so-called Θ-transform [13]. In [14], Bertram has studied the compact symmetric space and its non-compact dual, both realized as real forms of their common complexification; this analysis has been performed by exploring the Ramanujan garden and by the use of the so-called *Master Theorem*. Working in a different direction, Gindinkin and Krötz [15] have studied the complex crown of Riemannian symmetric spaces and non-compactly causal symmetric spaces.

This allowed them to prove the conjecture that every non-compactly causal symmetric space occurs as a component of a distinguished boundary of some complex crown [15]. Finally, it is worth recalling that the question of relating the harmonic analysis of different real forms of a complex symmetric space has been studied also in the context of scattering theory and resonances [16–20].

The paper is organized as follows. In Section 2, we study the holomorphic extension associated with the Fourier–Legendre expansion by means of the spherical Laplace transform. We are thus led to develop the harmonic analysis on the complex one-sheeted hyperboloid $X_2^{(c)}$, which contains as submanifolds either the Euclidean sphere $(\mathrm{i}\mathbb{R}\times\mathbb{R}^2)\cap X_2^{(c)}$ on which the Fourier–Legendre expansion can be developed, and the real one-sheeted hyperboloid, which contains the support of the cut. In Section 3, we analyze the relationship between the spherical Laplace transform and the *non-Euclidean Fourier transform*. For this purpose, we consider a real two-sheeted hyperboloid, and, in this geometrical setting, we can recover the non-Euclidean Fourier transform in the sense of Helgason. Finally, in Section 4, some conclusions will be drawn.

2. Holomorphic Extension Associated with the Fourier–Legendre Expansion and the Spherical Laplace Transform

2.1. The Complex One-Sheeted Hyperboloid

In the space $\mathbb{C}^3$ of the variable $z = (z_0, z_1, z_2)$, we consider the complex quadric $X_2^{(c)}$ with equation

$$-z_0^2 + z_1^2 + z_2^2 = 1,$$

which is a one-sheeted complex hyperboloid. Next, we introduce two systems of coordinates: polar and horocyclic coordinates.

Polar Coordinates:

$$z_0 = -\mathrm{i}\sin\theta\cosh\varphi, \tag{1a}$$
$$z_1 = -\mathrm{i}\sin\theta\sinh\varphi, \qquad (\theta,\varphi\in\mathbb{C}), \tag{1b}$$
$$z_2 = \cos\theta. \tag{1c}$$

If $\theta = u$ $(u\in\mathbb{R})$ and $\varphi = \mathrm{i}\psi$ $(\psi\in\mathbb{R})$, (1) reads

$$x_0 = -\mathrm{i}\sin u\cos\psi, \tag{2a}$$
$$x_1 = \sin u\sin\psi, \qquad (u,\psi\in\mathbb{R}), \tag{2b}$$
$$x_2 = \cos u. \tag{2c}$$

It can be easily verified that: $-x_0^2 + x_1^2 + x_2^2 = \sin^2 u + \cos^2 u = 1$ that is, one obtains as a real submanifold of the complex one-sheeted hyperboloid $X_2^{(c)}$ the *Euclidean sphere* $\mathbb{S}_2 = (\mathrm{i}\mathbb{R}\times\mathbb{R}^2)\cap X_2^{(c)}$ ($\mathrm{i}\mathbb{R}$ referring to the coordinate z_0 of z).

Similarly, if $\theta = \mathrm{i}v$ $(v\in\mathbb{R})$ and $\varphi = \psi$ $(\psi\in\mathbb{R})$, (1) becomes

$$x_0 = \sinh v\cosh\psi, \tag{3a}$$
$$x_1 = \sinh v\sinh\psi, \qquad (v,\psi\in\mathbb{R}), \tag{3b}$$
$$x_2 = \cosh v. \tag{3c}$$

In this case, it is easily verified that: $-x_0^2 + x_1^2 + x_2^2 = \cosh^2 v - \sinh^2 v = 1$. Equation (3) describes only the subset $X_2^+ \doteq \{x \in X_2 : x_2 \geqslant 1\}$ of the real one-sheeted hyperboloid X_2, which is precisely what we need for the current analysis. Finally, let $\widehat{X}^+$ denote the (real) meridian lying in the plane $x_1 = 0$ with equation: $x_0^2 - x_2^2 = -1$.

Horocyclic Coordinates:

We now introduce another system of coordinates (τ, ζ) as follows:

$$z_0 = -\mathrm{i}\sin\tau + \frac{1}{2}\zeta^2 e^{-\mathrm{i}\tau}, \tag{4a}$$

$$z_1 = \zeta e^{-\mathrm{i}u}, \qquad (\tau, \zeta \in \mathbb{C}), \tag{4b}$$

$$z_2 = \cos\tau - \frac{1}{2}\zeta^2 e^{-\mathrm{i}\tau}. \tag{4c}$$

For $\tau = \mathrm{i}w$ ($w \in \mathbb{R}$) and $\zeta \in \mathbb{R}$, Equation (4) reads

$$x_0 = \sinh w + \frac{1}{2}\zeta^2 e^{w}, \tag{5a}$$

$$x_1 = \zeta e^{w}, \qquad (w, \zeta \in \mathbb{R}), \tag{5b}$$

$$x_2 = \cosh w - \frac{1}{2}\zeta^2 e^{w}. \tag{5c}$$

Then, we consider the intersection points x_w of $\hat{X}^+$ with the family of planes $P_w : x_0 + x_2 = e^w$ ($w \in \mathbb{R}$) (see Equation (5)), i.e., the points $x_w = (\sinh w, 0, \cosh w)$. The sections of X_2 by these planes are the (real) parabolae Π_w^+ (except in the case $x_0 + x_2 = 0$). When $\zeta = 0$ in (5), we obtain the point x_w, which is the apex of the corresponding parabola Π_w^+ (see Figure 1).

This geometrical construction can now be extended to the complex one-sheeted hyperboloid $X_2^{(c)}$. We consider the (complex) hyperbola $\widehat{X}^{(c)}$ lying in the plane $z_1 = 0$, whose equation is: $z_0^2 - z_2^2 = -1$ (see Equation (4)), and its intersections with the family of planes P_τ with equation $z_0 + z_2 = e^{-\mathrm{i}\tau}$ ($\tau \in \dot{\mathbb{C}} \doteq \mathbb{C}/2\pi\mathbb{Z}$). Each plane P_τ intersects the hyperbola $\widehat{X}^{(c)}$ at the (unique) point $z_\tau = (z_0 = -\mathrm{i}\sin\tau, z_1 = 0, z_2 = \cos\tau)$, thus defining a bijection from the set of planes $\mathcal{P} \doteq \{P_\tau : \tau \in \dot{\mathbb{C}}\}$ onto $\widehat{X}^{(c)} = \{z_\tau : \tau \in \dot{\mathbb{C}}\}$. For each $\tau \in \dot{\mathbb{C}}$, the manifold $P_\tau \cap X_2^{(c)}$ is a complex parabola Π_τ, which we call a *complex horocycle*. The set of horocycles $\{\Pi_\tau : \tau \in \dot{\mathbb{C}}\}$ defines a fibration with basis $\widehat{X}^{(c)}$ on the dense domain ${X'_2}^{(c)} \doteq \{z \in X_2^{(c)} : z_0 + z_2 \neq 0\}$ of $X_2^{(c)}$. We can therefore associate with this fibration the parametric representation of ${X'_2}^{(c)}$ given by Equation (4), (τ, ζ) being the horocyclic coordinates of the point $z \in {X'_2}^{(c)}$.

2.2. The Radon Transform

We now define a Radon-type transformation in X_2, where the horocycles defined above play the same role as the planes do in the ordinary Radon transformation. We introduce the following integral:

$$\int_{h_w} \underline{F}\left(\cosh w - \frac{1}{2}\zeta^2 e^{w}\right) \mathrm{d}\zeta = \widehat{F}(w), \tag{6}$$

where h_w is the oriented segment of horocycle, which is represented by the arc of parabola whose apex is obtained by setting $\zeta = 0$ in (5) (i.e., its coordinates are: $x_0 = \sinh w$, $x_1 = 0$, $x_2 = \cosh w$; $w \in \mathbb{R}$), and whose endpoints lie on the plane $x_2 = 1$. The function $\underline{F}$ is assumed to satisfy the regularity conditions that make integral (6) convergent.

Remark 1. *Legendre expansions, which will be our concern later, involve functions depending only on* $\cos u = x_2$ *(see (2c)). This is why, in the integral (6), we limit ourselves to consider functions only of the form* $\underline{F} = \underline{F}(\cosh w - \frac{1}{2}\zeta^2 e^w)$ *(see (5c)).*

Since the integrand in (6) is an even function of ζ, the integration domain can be restricted to the part h_w^+ of h_w with $x_1 \geqslant 0$, which can be parametrized as follows (see (5)):

$$h_w^+ = \{x \in \Pi_w^+; x = x(\zeta, w);\ \zeta(\delta) = \left[2e^{-w}(1-\delta)(\cosh w - 1)\right]^{\frac{1}{2}};\ 0 \leqslant \delta \leqslant 1\} \qquad (w \in \mathbb{R}), \tag{7}$$

positively oriented from the apex to the endpoint lying on $x_2 = 1$. For $\delta = 1$, we have $\zeta = 0$, which yields the apex of the parabola representing the horocycle; for $\delta = 0$, we have $\zeta = [2e^{-w}(\cosh w - 1)]^{1/2}$, which gives the intersection of the horocycle with the plane $x_2 = 1$.

We may now introduce another parametrization of the segment of horocycle h_w^+, which is obtained by setting $x_2 = \cosh v$ (see (3c)):

$$h_w^+ = \{x \in \Pi_w^+; x = x(\zeta, w);\ \zeta(v) = \left[2e^{-w}(\cosh w - \cosh v)\right]^{\frac{1}{2}};\ 0 \leqslant v \leqslant w\} \qquad (w \in \mathbb{R}). \tag{8}$$

Indeed, we have $\zeta = 0$ for $v = w$ (the apex of the parabola) and $\zeta = [2e^{-w}(\cosh w - 1)]^{1/2}$ for $v = 0$ (endpoint of the parabola: $x_2 = \cosh v = 1$). Since $\mathrm{d}\zeta/\mathrm{d}v = -e^{-w/2}\sinh v[2(\cosh w - \cosh v)]^{-1/2}$, integral (6) takes the form:

$$\widehat{F}(w) = 2e^{-w/2}\int_0^w \underline{F}(\cosh v)\frac{\sinh v}{[2(\cosh w - \cosh v)]^{\frac{1}{2}}}\,\mathrm{d}v \doteq e^{-w/2}\,(\mathcal{A}F)\,(w) \qquad (w \in \mathbb{R}), \tag{9}$$

where $(\mathcal{A}F)\,(w)$ is an Abel-type integral.

The fibration realized by the horocycles Π_w^+ can now be extended to the complex domain by using the complex horocycles Π_τ ($\tau = t + \mathrm{i}w; t, w \in \mathbb{R}$), whose intersection with the (complex) meridian hyperbola $\widehat{X}^{(c)}$ is the point $z_\tau = (-\mathrm{i}\sin\tau, 0, \cos\tau)$. Accordingly, we introduce the following integral:

$$2\int_{h_\tau^+} \underline{f}\left(\cos\tau - \frac{1}{2}\zeta^2 e^{-\mathrm{i}\tau}\right)\mathrm{d}\zeta = \widehat{f}(\tau), \tag{10}$$

h_τ^+ being the (oriented) arc of the complex horocycle defined by (see (4) and (7)):

$$h_\tau^+ = \{z \in \Pi_\tau; z = z(\zeta, \tau);\ \zeta(\delta) = [2e^{\mathrm{i}\tau}(1-\delta)(\cos\tau - 1)]^{\frac{1}{2}};\ 0 \leqslant \delta \leqslant 1\} \qquad (\tau \in \mathbb{C}).$$

For $\delta = 1$, we have $\zeta = 0$, i.e., the point z_τ belonging to $\widehat{X}^{(c)}$, while for $\delta = 0$ we have $\zeta = [2e^{\mathrm{i}\tau}(\cos\tau - 1)]^{1/2}$, which is the intersection of h_τ^+ with the plane $z_2 = 1$. Similarly to what done before, we may now introduce the following parametrization of h_τ^+, obtained by setting $z_2 = \cos\theta$ ($\theta \in \mathbb{C}, \theta = u + \mathrm{i}v$; see (1) and (8)):

$$h_\tau^+ = \{z \in \Pi_\tau; z = z(\zeta, \tau);\ \zeta(\theta) = [2e^{\mathrm{i}\tau}(\cos\tau - \cos\theta)]^{\frac{1}{2}};\ \theta \in \gamma_\tau\} \qquad (\tau \in \mathbb{C}),$$

where γ_τ denotes the ray in the θ-plane oriented from 0 to τ:

$$\gamma_\tau = \{\theta = \theta(\delta);\ \cos\theta(\delta) - 1 = \delta(\cos\tau - 1);\ 0 \leqslant \delta \leqslant 1\} \qquad (\tau \in \mathbb{C}).$$

Now, since $\mathrm{d}\zeta/\mathrm{d}\theta = e^{\mathrm{i}\tau/2}[2(\cos\tau - \cos\theta)]^{-1/2}\sin\theta$, integral (10) can be rewritten in the form:

$$\widehat{f}(\tau) = -2e^{\mathrm{i}\tau/2}\int_{\gamma_\tau} f(\theta)\,\frac{\sin\theta}{[2(\cos\tau - \cos\theta)]^{\frac{1}{2}}}\,\mathrm{d}\theta, \tag{11}$$

where $f(\theta) \equiv \underline{f}(\cos\theta)$. The relevant branch of the function $[2(\cos\tau - \cos\theta)]^{-1/2}$ is specified by the condition that, for $\tau = \mathrm{i}w$ and $\theta = \mathrm{i}v$ (with $w > v$), it takes the value $[2(\cosh w - \cosh v)]^{-1/2} \geqslant 0$. Putting in (11) $\tau = \mathrm{i}w$, $\theta = \mathrm{i}v$, we re-obtain precisely the r.h.s. of formula (9), once $\underline{f}$ is identified with $\underline{F}$.

Restricting formula (11) to the set of real values for the variables τ and θ, namely, $\tau = t$ and $\theta = u$ ($t, u \in \mathbb{R}$), from (11), we obtain

$$\widehat{f}(t) = -2e^{\mathrm{i}t/2}\int_0^t f(u)\,\frac{\sin u}{[2(\cos t - \cos u)]^{\frac{1}{2}}}\,\mathrm{d}u. \tag{12}$$

Accounting for the relevant branch of the factor $[2(\cos\tau - \cos\theta)]^{-1/2}$ in (11), formula (12) can be written in the following more precise form (involving a positive bracket):

$$\widehat{f}(t) = -2\mathrm{i}\,\varepsilon(t)\,e^{\mathrm{i}t/2}\int_0^t f(u)\,\frac{\sin u}{[2(\cos u - \cos t)]^{\frac{1}{2}}}\,\mathrm{d}u, \tag{13}$$

where $\varepsilon(t)$ denotes the sign function.

2.3. *Holomorphic Extension Associated with Trigonometric Series*

2.3.1. Fourier–Legendre Expansions as Trigonometric Series

Consider the following Legendre series

$$\frac{1}{4\pi}\sum_{n=0}^{\infty}(2n+1)\,a_n\,P_n(\cos u), \tag{14}$$

where $P_n(\cdot)$ denotes the Legendre polynomials, which satisfy the following integral representation:

$$P_n(\cos u) = \frac{1}{\pi}\int_0^{\pi}(\cos u + \mathrm{i}\sin u\cos\eta)^n\,\mathrm{d}\eta.$$

Suppose that expansion (14) converges to a function $\underline{f}(\cos u)$ absolutely integrable in the interval $u \in [0, \pi]$. Then, the Legendre coefficients a_n can be written as

$$a_n = 2\pi\int_0^{\pi}\underline{f}(\cos u)\,P_n(\cos u)\,\sin u\,\mathrm{d}u.$$

In Ref. [21], we proved the following proposition.

Proposition 1. *The Legendre coefficients $\{a_n\}_{n=0}^{\infty}$ coincide with the Fourier coefficients of the following form:*

$$a_n = \int_{-\pi}^{\pi}\widehat{f}(t)\,e^{\mathrm{i}nt}\,\mathrm{d}t \qquad (n = 0, 1, 2, \ldots), \tag{15}$$

where

$$\widehat{f}(t) = -2\mathrm{i}\,\varepsilon(t)\,e^{\mathrm{i}t/2}\int_0^t f(u)\,\frac{\sin u}{[2(\cos u - \cos t)]^{\frac{1}{2}}}\,\mathrm{d}u, \tag{16}$$

with $f(u) \equiv \underline{f}(\cos u)$, *and* $\varepsilon(t)$ *being the sign function.*

Proof. See Proposition 3.1 of Ref. [21]. □

Note that (16) coincides with (13). From (16), it is easy to verify that

$$\widehat{f}(t) = -e^{\mathrm{i}t}\widehat{f}(-t), \tag{17}$$

which, through (15), yields the following symmetry relation for the Legendre coefficients:

$$a_n = -a_{-n-1} \qquad (n \in \mathbb{Z}). \tag{18}$$

Now, we can introduce the following trigonometric series, assuming (18) to hold:

$$\begin{aligned} \frac{1}{2\pi}\sum_{n=-\infty}^{+\infty} a_n e^{-\mathrm{i}nt} &= \frac{1}{2\pi}\left[\sum_{n=0}^{+\infty} a_n e^{-\mathrm{i}nt} - e^{\mathrm{i}t}\sum_{n=0}^{+\infty} a_n e^{\mathrm{i}nt}\right] \\ &= \frac{1}{2\pi}e^{\mathrm{i}(t-\pi)/2}\sum_{n=-\infty}^{+\infty}(-1)^n a_n \cos\left[\left(n+\tfrac{1}{2}\right)(t-\pi)\right] = \frac{1}{2\pi}e^{\mathrm{i}(t-\pi)/2}\sum_{n=-\infty}^{+\infty} a_n \sin\left[\left(n+\tfrac{1}{2}\right)t\right], \end{aligned} \tag{19}$$

and study the associated holomorphic extension.

2.3.2. Holomorphic Extension Associated with Trigonometric Series

Consider a sequence $\{f_n\}_{n=0}^{\infty}$ of (real) numbers, and denote by Δ the difference operator:

$$\Delta f_n = f_{n+1} - f_n.$$

We have:

$$\Delta^k f_n = \underbrace{\Delta \times \Delta \times \cdots \Delta}_{k-\text{times}} f_n = \sum_{m=0}^{k}(-1)^m \binom{k}{m} f_{n+k-m}$$

(for any integer $k \geqslant 0$); Δ^0 is the identity operator, by definition. Suppose that there exists a positive constant M such that

$$(n+1)^{(1+\varepsilon)}\sum_{i=0}^{n}\binom{n}{i}^{(2+\varepsilon)}\left|\Delta^i f_{(n-i)}\right|^{(2+\varepsilon)} < M \qquad (n = 0,1,2,\ldots;\ \varepsilon > 0). \tag{20}$$

We shall refer to (20) as the *Hausdorff condition* for its relevance in the solution of the Hausdorff moment problem [8]. The tool we use to guarantee uniqueness of the interpolation of a sequence of numbers $\{f_n\}_{n=0}^{\infty}$ is Carlson's theorem [22]. Essentially, it gives growth conditions under which a function is uniquely determined by its values on non-negative integers. Let us recall that an entire function $f(z)$ is of exponential type $\tau < \infty$ if

$$\limsup_{r\to\infty}\frac{M_f(r))}{r} = \tau, \tag{21}$$

where $M_f(r)$ denotes the maximum modulus of $f(z)$ for $|z| = r$. The rate of growth of entire functions can be specified along different directions by the Phragmén-Lindelöf indicator function,

$$h_f(\theta) = \limsup_{r\to\infty}\frac{\log\left|f(re^{\mathrm{i}\theta})\right|}{r}. \tag{22}$$

Note that the preceding definitions can be extended to functions which are not entire, but regular (that is, analytic and single-valued) in a sector with vertex at the origin [22].

Theorem 1 (Carlson's theorem (Section 9.2, p. 153, [22])). *Let $f(z)$ be regular in the half-plane* $\mathrm{Re}\, z \geqslant 0$ *and*

(i) $f(z)$ *is of exponential type* $\tau < \infty$,
(ii) $h_f(\pi/2) + h_f(-\pi/2) < 2\pi$,
(iii) $f(n) = 0$ *for* $n = 0, 1, 2, \ldots$,

then $f(z)$ vanishes identically.

Among the functions of exponential type, condition (ii) requires $f(z)$ to be of exponential type less than π on the imaginary axis that is: $f(iy) = O(1)\,\exp(c|y|)$ for some $c < \pi$. We shall refer to conditions (i) and (ii) as *Carlson's bound*. Moreover, an analytic function which interpolates a sequence of numbers $\{f_n\}_{n=0}^{\infty}$ and satisfy conditions (i) and (ii) above will be called a *Carlsonian interpolant*, in view of the fact that the uniqueness of the interpolation is guaranteed by Carlson's theorem.

Proposition 2. *Suppose that the set of numbers* $\{f_n\}_{n=0}^{\infty}$, *with* $f_n \doteq (n+1)^2 a_n$ *(a_n being the Fourier–Legendre coefficients of the series (14)), satisfies condition (20). Then:*

(i) There exists a unique Carlsonian interpolant $\widetilde{a}(\lambda)$ $(\lambda \in \mathbb{C}, \mathrm{Re}\,\lambda \geqslant -\frac{1}{2})$ *of the coefficients* $\{a_n\}_{n=0}^{\infty}$, *which is holomorphic in the half-plane* $\mathrm{Re}\,\lambda > -\frac{1}{2}$. *Moreover,* $\widetilde{a}(\lambda)$ *belongs to the Hardy space* $H^2(\mathbb{C}_{-1/2}^{(+)})$.
(ii) $\lambda^2 \widetilde{a}(\lambda) \in L^2(-\infty, +\infty)$ *for any fixed value of* $\mathrm{Re}\,\lambda \geqslant -\frac{1}{2}$.
(iii) $\lambda^2 \widetilde{a}(\lambda)$ *tends uniformly to zero as* $\lambda \to \infty$ *inside any fixed half-plane* $\mathrm{Re}\,\lambda \geqslant \delta > -\frac{1}{2}$.
(iv) $\lambda \widetilde{a}(\lambda) \in L^1(-\infty, +\infty)$ *for any fixed value of* $\mathrm{Re}\,\lambda \geqslant -\frac{1}{2}$.
(v) $\widetilde{a}(-\frac{1}{2} + i\mu)$ $(\mu \in \mathbb{R})$ *is a continuous function which tends to zero as* $\mu \to \pm\infty$.
(vi) $\sup_{\substack{\sigma > -1/2 \\ \mu \in \mathbb{R}}} |\widetilde{a}(\sigma + i\mu)| = \left|\widetilde{a}\left(-\frac{1}{2} + i\mu\right)\right|$;

Proof. (i) Since the sequence $\{f_n\}_{n=0}^{\infty}$ satisfies condition (20), the numbers f_n are moments of a suitable function, that is, the following representation holds [8]:

$$f_n = \int_0^1 x^n \, \varphi(x) \, dx \qquad (n = 0, 1, 2, \ldots), \tag{23}$$

where $\varphi \in L^{(2+\varepsilon)}(0, 1)$. Let $x = e^{-s}$ in formula (23):

$$f_n = \int_0^{+\infty} e^{-(n+\frac{1}{2})s} \, e^{-s/2} \, \varphi(e^{-s}) \, ds \qquad (n = 0, 1, 2, \ldots).$$

Then, the numbers $\{f_n\}_{n=0}^{\infty}$ can be formally regarded as the restriction to the (non-negative) integers of the following Laplace transform:

$$\widetilde{f}(\lambda) = \int_0^{+\infty} e^{-(\lambda+\frac{1}{2})s} \, e^{-s/2} \, \varphi(e^{-s}) \, ds \qquad \left(\mathrm{Re}\,\lambda > -\tfrac{1}{2}\right), \tag{24}$$

being $\widetilde{f}(\lambda)|_{\lambda=n} = f_n$. We see that $e^{-s/2}\varphi(e^{-s}) \in L^2(0, +\infty)$: in fact, $\int_0^{+\infty} |e^{-s/2}\varphi(e^{-s})|^2 \, ds = \int_0^1 |\varphi(x)|^2 \, dx < \infty$, since $\varphi \in L^{(2+\varepsilon)}(0, 1)$ and, a fortiori, $\varphi(x) \in L^2(0, 1)$. Then, by the Paley–Wiener theorem [23], we have $\widetilde{f}(\lambda) \in H^2(\mathbb{C}_{-1/2}^{(+)})$, which is the Hardy space whose norm is: $\|\widetilde{f}\|_2 \doteq \sup_{\sigma > -1/2} (\int_{-\infty}^{+\infty} |\widetilde{f}(\sigma + i\mu)|^2 \, d\mu)^{1/2}$, and $\mathbb{C}_{-1/2}^{(+)} \doteq \{\lambda \in \mathbb{C}, \lambda = \sigma + i\mu; \sigma > -\frac{1}{2}, \mu \in \mathbb{R}\}$. Hence, use can be made of Carlson's theorem, which guarantees that $\widetilde{f}(\lambda)$ is the unique Carlsonian interpolant of the sequence $\{f_n\}_{n=0}^{\infty}$. Let $\widetilde{a}(\lambda) \doteq \widetilde{f}(\lambda)/(\lambda+1)^2$. $\widetilde{a}(\lambda)$ is a function holomorphic for $\operatorname{Re}\lambda > -\frac{1}{2}$, which satisfies Carlson's bound since $\widetilde{f}(\lambda)$ does. Then, $\widetilde{a}(\lambda)$ is the unique Carlsonian interpolant of the set of Fourier–Legendre coefficients $\{a_n\}_{n=0}^{\infty}$ since $\widetilde{a}(\lambda)|_{\lambda=n} = \left.\frac{\widetilde{f}(\lambda)}{(\lambda+1)^2}\right|_{\lambda=n} = \frac{f_n}{(n+1)^2} = a_n$. Finally, it is readily shown that $\widetilde{a}(\lambda) \in H^2(\mathbb{C}_{-1/2}^{(+)})$:

$$\|\widetilde{a}\|_2^2 = \sup_{\sigma>-\frac{1}{2}} \int_{-\infty}^{+\infty} \left|\frac{\widetilde{f}(\sigma+\mathrm{i}\mu)}{(\sigma+1+\mathrm{i}\mu)^2}\right|^2 \mathrm{d}\mu \leqslant 16 \sup_{\sigma>-\frac{1}{2}} \int_{-\infty}^{+\infty} \left|\widetilde{f}(\sigma+\mathrm{i}\mu)\right|^2 \mathrm{d}\mu < \infty, \tag{25}$$

and statement (i) is then proved. Regarding point (ii), since $\widetilde{f}(\lambda) \in H^2(\mathbb{C}_{-1/2}^{(+)})$, then $\widetilde{f}(\sigma+\mathrm{i}\mu) \in L^2(-\infty,+\infty)$ for any fixed value of $\sigma \geqslant -\frac{1}{2}$. Regarding point (iii), since $\widetilde{f}(\lambda) \in H^2(\mathbb{C}_{-1/2}^{(+)})$, then $\widetilde{f}(\lambda)$ tends to zero as $\lambda \to \infty$ inside any fixed half-plane $\operatorname{Re}\lambda \geqslant \delta > -\frac{1}{2}$ [23]. By Schwarz's inequality, we have for $\sigma \geqslant -\frac{1}{2}$,

$$\int_{-\infty}^{+\infty} |(\sigma+\mathrm{i}\mu)\,\widetilde{a}(\sigma+\mathrm{i}\mu)|\,\mathrm{d}\mu \leqslant \left(\int_{-\infty}^{+\infty}\left|\frac{\sigma+\mathrm{i}\mu}{(\sigma+1+\mathrm{i}\mu)^2}\right|^2 \mathrm{d}\mu\right)^{\frac{1}{2}} \left(\int_{-\infty}^{+\infty}\left|\widetilde{f}(\sigma+\mathrm{i}\mu)\right|^2 \mathrm{d}\mu\right)^{\frac{1}{2}} < \infty,$$

which proves statement (iv). Concerning point (v), return to the Laplace integral representation of $\widetilde{f}(\lambda)$ in (24). First, we want to prove that $e^{-s/2}\varphi(e^{-s}) \in L^1(0,+\infty)$, which amounts to showing that $\int_0^1 \left|\frac{\varphi(x)}{\sqrt{x}}\right| \mathrm{d}x < \infty$. By Hölder's inequality,

$$\int_0^1 \frac{|\varphi(x)|}{\sqrt{x}}\,\mathrm{d}x \leqslant \left(\int_0^1 |\varphi(x)|^{(2+\varepsilon)}\,\mathrm{d}x\right)^{\frac{1}{2+\varepsilon}} \cdot \left(\int_0^1 x^{-\frac{2+\varepsilon}{2+2\varepsilon}}\,\mathrm{d}x\right)^{\frac{1+\varepsilon}{2+\varepsilon}} < \infty,$$

where the rightmost integral converges since $\frac{2+\varepsilon}{2+2\varepsilon} < 1$ for $\varepsilon > 0$, and $\varphi \in L^{(2+\varepsilon)}(0,1)$. Then, from representation (24):

$$\left(\tfrac{1}{2}+\mathrm{i}\mu\right)^2 \widetilde{a}\left(-\tfrac{1}{2}+\mathrm{i}\mu\right) = \widetilde{f}\left(-\tfrac{1}{2}+\mathrm{i}\mu\right) = \int_0^{+\infty} e^{-\mathrm{i}\mu s} e^{-s/2}\varphi(e^{-s})\,\mathrm{d}s = \mathcal{F}\left\{h(s)e^{-s/2}\varphi(e^{-s})\right\},$$

where $\mathcal{F}$ denotes the Fourier integral operator, and $h(s)$ is the Heaviside step function. The Riemann–Lebesgue theorem guarantees that $\widetilde{a}(-\frac{1}{2}+\mathrm{i}\mu)$ is a continuous function tending to zero as $\mu \to \pm\infty$, and statement (v) is proved. Finally, in order to prove statement (vi), we note that the Laplace transform (24) holds also for $\operatorname{Re}\lambda = -\frac{1}{2}$ since $e^{-s/2}\varphi(e^{-s}) \in L^1(0,+\infty) \cap L^2(0,+\infty)$. It follows that $\sup_{\substack{\sigma>-1/2\\ \mu\in\mathbb{R}}} |\widetilde{f}(\sigma+\mathrm{i}\mu)| = |\widetilde{f}\left(-\frac{1}{2}+\mathrm{i}\mu\right)|$. Therefore, recalling that $\widetilde{a}(\lambda) = \widetilde{f}(\lambda)/(\lambda+1)^2$:

$$\sup_{\substack{\sigma>-1/2\\ \mu\in\mathbb{R}}} |\widetilde{a}(\sigma+\mathrm{i}\mu)| = \sup_{\substack{\sigma>-1/2\\ \mu\in\mathbb{R}}} \frac{|\widetilde{f}(\sigma+\mathrm{i}\mu)|}{|\sigma+1+\mathrm{i}\mu|^2} = \frac{|\widetilde{f}(-\frac{1}{2}+\mathrm{i}\mu)|}{\mu^2+1/4} = \left|\widetilde{a}\left(-\tfrac{1}{2}+\mathrm{i}\mu\right)\right|,$$

where $\widetilde{a}(-\frac{1}{2}+\mathrm{i}\mu) \in L^1(-\infty,+\infty)$ (see statements (iv) and (v)). □

Let $\xi_0 \geqslant 0$. We introduce in the complex plane $\mathbb{C}$ of the variable $\tau = t + \mathrm{i}w$ $(t, w \in \mathbb{R})$ the following domains: ${}^{\tau}\mathcal{I}_{+}^{(\pm\xi_0)} \doteq \{\tau \in \mathbb{C} : \operatorname{Im}\tau > \pm\xi_0\}$ and ${}^{\tau}\mathcal{I}_{-}^{(\pm\xi_0)} \doteq \{\tau \in \mathbb{C} : \operatorname{Im}\tau < \pm\xi_0\}$. Correspondingly, we introduce the following cut-domains: ${}^{\tau}\mathcal{I}_{+}^{(\xi_0)} \setminus {}^{\tau}\Xi_{+}^{(\xi_0)}$, where ${}^{\tau}\Xi_{+}^{(\xi_0)} \doteq \{\tau \in \mathbb{C} : \tau = 2k\pi + \mathrm{i}w,\ w > \xi_0,\ k \in$

$\mathbb{Z}\}$ (see Figure 2a) and ${}^{\tau}\mathcal{I}_{-}^{(-\xi_0)} \setminus {}^{\tau}\Xi_{-}^{(-\xi_0)}$, where ${}^{\tau}\Xi_{-}^{(-\xi_0)} \doteq \{\tau \in \mathbb{C} : \tau = 2k\pi + \mathrm{i}w,\, w < -\xi_0,\, k \in \mathbb{Z}\}$ (see Figure 2b). Finally, we denote by $\dot{A} \doteq A/2\pi\mathbb{Z}$ any subset A of $\mathbb{C}$, which is invariant under the translation group $2\pi\mathbb{Z}$. We are now ready to state the following proposition.

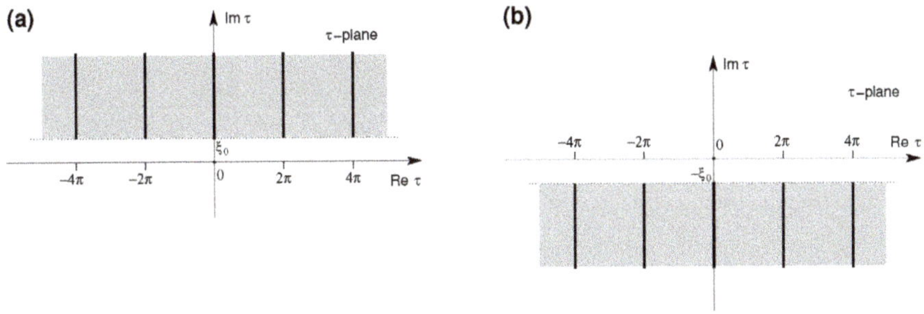

Figure 2. In the complex τ-plane, the grey regions represent: (**a**) the cut-domain ${}^{\tau}\mathcal{I}_{+}^{(\xi_0)} \setminus {}^{\tau}\Xi_{+}^{(\xi_0)}$, (**b**) the cut-domain ${}^{\tau}\mathcal{I}_{-}^{(-\xi_0)} \setminus {}^{\tau}\Xi_{-}^{(-\xi_0)}$. The cuts (thick lines) are located at $\mathrm{Re}\,\tau = 2k\pi$, $k \in \mathbb{Z}$.

Proposition 3. *Consider the trigonometric series*

$$\frac{1}{2\pi}\sum_{n=0}^{+\infty} a_n\, e^{-\mathrm{i}n\tau} \qquad (\tau = t + \mathrm{i}w; t, w \in \mathbb{R}), \tag{26}$$

and suppose that the set of numbers $\{f_n\}_{n=0}^{\infty}$, *with* $f_n \doteq (n+1)^2\, a_n$, *satisfies condition* (20). *Then:*

(i) *The series* (26) *converges uniformly on any compact subdomain of* ${}^{\tau}\mathcal{I}_{-}^{(0)}$ *to a function* $\widehat{f}^{(+)}(\tau)$ *holomorphic in* ${}^{\tau}\mathcal{I}_{-}^{(0)}$, *continuous on the axis* $\mathrm{Im}\,\tau = 0$.

(ii) *The function* $\widehat{f}^{(+)}(\tau)$ *admits a holomorphic extension to the cut-domain* ${}^{\tau}\mathcal{I}_{+}^{(0)} \setminus {}^{\tau}\dot{\Xi}_{+}^{(0)}$*: i.e., it is analytic in* $\mathbb{C} \setminus \{\tau \in \mathbb{C} : \tau = 2k\pi + \mathrm{i}w; k \in \mathbb{Z}, w > 0\}$.

(iii) *The jump function* $\widehat{F}^{(+)}(w)$*, which is the discontinuity of* $(\mathrm{i}\widehat{f}^{(+)}(\tau))$ *across the cut* ${}^{\tau}\dot{\Xi}_{+}^{(0)}$*, is a function of class* C^1 *that satisfies the bound*

$$\left|\widehat{F}^{(+)}(w)\right| \leqslant \|\widetilde{a}_\sigma\|_1\, c^{\sigma w} \qquad \left(\sigma \geqslant \tfrac{1}{2}, w \subset \mathbb{R}^+\right),$$

where $\widetilde{a}(\sigma + \mathrm{i}\mu)$ $(\mu \in \mathbb{R})$ *is the unique Carlsonian interpolant of the coefficients* a_n*, and*

$$\|\widetilde{a}_\sigma\|_1 \doteq \frac{1}{2\pi}\int_{-\infty}^{+\infty} |\widetilde{a}(\sigma + \mathrm{i}\mu)|\,\mathrm{d}\mu < \infty \qquad \left(\sigma \geqslant -\tfrac{1}{2}\right). \tag{27}$$

(iv) $\widehat{F}^{(+)}(w) = o\left(e^{-w/2}\right)$ *for* $w \to +\infty$ *and* $\widehat{F}^{(+)}(0) = 0$.

(v) $\widetilde{a}(\sigma + \mathrm{i}\mu)$ *is the Laplace transform of the jump function* $\widehat{F}^{(+)}(w)$:

$$\widetilde{a}(\sigma + \mathrm{i}\mu) = \int_0^{+\infty} \widehat{F}^{(+)}(w)\, e^{-(\sigma+\mathrm{i}\mu)w}\,\mathrm{d}w \qquad \left(\sigma > -\tfrac{1}{2}\right), \tag{28}$$

holomorphic in the half-plane $\sigma > -\frac{1}{2}$.

(vi) *The following Plancherel equality holds:*

$$\int_{-\infty}^{+\infty} |\widetilde{a}(\sigma + \mathrm{i}\mu)|^2\,\mathrm{d}\mu = 2\pi\int_{-\infty}^{+\infty} \left|\widehat{F}^{(+)}(w)e^{-\sigma w}\right|^2 \mathrm{d}w \qquad \left(\sigma \geqslant -\tfrac{1}{2}\right).$$

Proof. Since the sequence $\{f_n\}_{n=0}^{\infty}$ satisfies condition (20), then, given an arbitrary constant C, there exists an integer n_0 such that $|a_n| \leqslant C$ for $n > n_0$ and, accordingly,

$$\left|\frac{1}{2\pi}\sum_{n=0}^{\infty} a_n\, e^{-in\tau}\right| \leqslant \frac{C}{2\pi}\sum_{n=n_0+1}^{\infty} e^{nw} \qquad (w \doteq \operatorname{Im}\tau,\ C = \text{constant}). \tag{29}$$

The series on the r.h.s. of (29) converges uniformly on any compact subdomain contained in the half-plane $w < 0$. Since

$$\frac{1}{2\pi}\sum_{n=0}^{\infty} a_n\, e^{-in\tau} = \frac{1}{2\pi}\sum_{n=n_0+1}^{\infty} a_n\, e^{-in\tau} + T_{n_0}(\tau),$$

where $T_{n_0}(\tau)$ is a trigonometric polynomial, by the Weierstrass theorem on the uniformly convergent series of analytic functions, series (26) converges uniformly on any compact subdomain of ${}^{\tau}\mathcal{I}_{-}^{(0)}$ to a function $\widehat{f}^{(+)}(\tau)$ holomorphic in ${}^{\tau}\mathcal{I}_{-}^{(0)}$. Furthermore, since $(n+1)^2 a_n \xrightarrow[n\to\infty]{} 0$, given an arbitrary constant C', there exists an integer n_1 such that for $w = 0$:

$$\left|\frac{1}{2\pi}\sum_{n=n_1}^{\infty} a_n\, e^{-int}\right| \leqslant \frac{1}{2\pi}\sum_{n=n_1}^{\infty} |a_n| \leqslant \frac{C'}{2\pi}\sum_{n=n_1}^{\infty}\frac{1}{n^2} \leqslant C'\frac{\pi}{12}.$$

Then, applying once again the Weierstrass theorem on the uniformly convergent series of continuous functions, the series $\frac{1}{2\pi}\sum_{n=0}^{\infty} a_n\, e^{-int}$ converges to a continuous function $\widehat{f}^{(+)}(t)$, and statement (i) is proved. Regarding statement (ii), we write the following integral:

$$\widehat{f}_{\eta}^{(+)}(t) = \frac{\mathrm{i}}{4\pi}\int_C \tilde{a}(\lambda)\frac{e^{-\mathrm{i}\lambda(t-\eta\pi)}}{\sin\pi\lambda}\,\mathrm{d}\lambda \qquad (\eta = \pm), \tag{30}$$

where $\tilde{a}(\lambda)$ ($\operatorname{Re}\lambda \geqslant -\frac{1}{2}$) is the unique Carlsonian interpolant of the coefficients $\{a_n\}_{n=0}^{\infty}$, which is holomorphic in the half-plane $\operatorname{Re}\lambda > -\frac{1}{2}$ (see statement (i) of Proposition 2). The contour C is contained in the half-plane $\mathbb{C}_{-1/2}^{(+)}$, which encircles the semi-axis $\operatorname{Re}\lambda > -\frac{1}{2}$, and is chosen to cross the latter at a point $\sigma > -\frac{1}{2}$, $\sigma \notin \mathbb{N}$ (see Figure 3). Consider now the term $\frac{\exp[-\mathrm{i}\lambda(t-\eta\pi)]}{\sin\pi\lambda}$; the following inequalities hold ($\lambda = \sigma + \mathrm{i}\mu$):

$$\left|e^{-\mathrm{i}(\sigma+\mathrm{i}\mu)(t-\eta\pi)}\right| \leqslant 2\cosh\pi\mu \quad \text{for} \quad \begin{cases} 0 \leqslant t \leqslant 2\pi & \text{if} \quad \eta = +, \\ -2\pi \leqslant t \leqslant 0 & \text{if} \quad \eta = -, \end{cases} \tag{31}$$

$$|\sin\pi(\sigma+\mathrm{i}\mu)| \geqslant \sinh\pi\mu, \tag{32}$$

$$|\sin\pi(\sigma+\mathrm{i}\mu)| \geqslant |\sin\pi\sigma|\,\cosh\pi\mu. \tag{33}$$

From (31) and (32):

$$\left|\frac{e^{-\mathrm{i}(\sigma+\mathrm{i}\mu)(t-\eta\pi)}}{\sin\pi(\sigma+\mathrm{i}\mu)}\right| \leqslant 2\left|\frac{\cosh\pi\mu}{\sinh\pi\mu}\right| \quad \text{for} \quad \begin{cases} 0 \leqslant t \leqslant 2\pi & \text{if} \quad \eta = +, \\ -2\pi \leqslant t \leqslant 0 & \text{if} \quad \eta = -, \end{cases} \tag{34}$$

while, combining (31) and (33):

$$\left|\frac{e^{-\mathrm{i}(\sigma+\mathrm{i}\mu)(t-\eta\pi)}}{\sin\pi(\sigma+\mathrm{i}\mu)}\right| \leqslant \frac{2}{|\sin\pi\sigma|} < \infty \quad \text{for} \begin{cases} 0 \leqslant t \leqslant 2\pi & \text{if} \quad \eta = +, \\ -2\pi \leqslant t \leqslant 0 & \text{if} \quad \eta = -, \end{cases} \text{and } \sigma \notin \mathbb{N}. \tag{35}$$

Integral (30) converges since $\lambda^2\widetilde{a}(\lambda)$ tends uniformly to zero as $\lambda \to \infty$ in any fixed half-plane $\mathrm{Re}\,\lambda \geqslant \delta > -\frac{1}{2}$ (statement (iii) of Proposition 2); the term $\left|\frac{\exp[-\mathrm{i}(\sigma+\mathrm{i}\mu)(t-\eta\pi)]}{\sin\pi(\sigma+\mathrm{i}\mu)}\right|$ is bounded by a constant for $\sigma \notin \mathbb{N}$ (see (35)), and is bounded by 2 as $\mu \to \pm\infty$ in view of inequality (34). The contour $\mathcal{C}$ can be distorted and replaced by a line L_σ parallel to the imaginary axis and crossing the real axis at $\mathrm{Re}\,\lambda = \sigma$ with $\sigma \geqslant -\frac{1}{2}$ ($\sigma \notin \mathbb{N}$) (see Figure 3) provided the real variable t is kept in $[0, 2\pi]$ for $\widehat{f}_+^{(+)}(t)$, and in $[-2\pi, 0]$ for $\widehat{f}_-^{(+)}(t)$, respectively. Note that the integral along L_σ converges since $\lambda\widetilde{a}(\lambda) \in L^1(-\infty, +\infty)$ for any fixed value of $\mathrm{Re}\,\lambda \geqslant -\frac{1}{2}$ (statement (iv) of Proposition 2) and by inequality (35). We may now apply the Watson resummation to integral (30). For $t \in [0, 2\pi]$, we obtain:

$$\frac{\mathrm{i}}{4\pi}\int_{\mathcal{C}} \widetilde{a}(\lambda)\,\frac{e^{-\mathrm{i}\lambda(t-\pi)}}{\sin\pi\lambda}\,\mathrm{d}\lambda = \frac{1}{2\pi}\sum_{n=0}^{\infty} a_n\,e^{-\mathrm{i}nt} \qquad (0 \leqslant t \leqslant 2\pi),$$

where the contour $\mathcal{C}$ encircles the semi-axis $\mathrm{Re}\,\lambda > -\frac{1}{2}$ and crosses it at a point $-\frac{1}{2} \leqslant \sigma < 0$. Then, distorting the contour $\mathcal{C}$ into the line L_σ ($-\frac{1}{2} \leqslant \sigma < 0$), which is admissible as explained above, we obtain for $-\frac{1}{2} \leqslant \sigma < 0$:

$$\widehat{f}_+^{(+)}(t) = -\frac{1}{4\pi}\int_{-\infty}^{+\infty} \widetilde{a}(\sigma+\mathrm{i}\mu)\,\frac{e^{-\mathrm{i}(\sigma+\mathrm{i}\mu)(t-\pi)}}{\sin\pi(\sigma+\mathrm{i}\mu)}\,\mathrm{d}\mu = \frac{1}{2\pi}\sum_{n=0}^{\infty} a_n\,e^{-\mathrm{i}nt} \quad (0 \leqslant t \leqslant 2\pi), \tag{36}$$

and, analogously, for $t \in [-2\pi, 0]$:

$$\widehat{f}_-^{(+)}(t) = -\frac{1}{4\pi}\int_{-\infty}^{+\infty} \widetilde{a}(\sigma+\mathrm{i}\mu)\,\frac{e^{-\mathrm{i}(\sigma+\mathrm{i}\mu)(t+\pi)}}{\sin\pi(\sigma+\mathrm{i}\mu)}\,\mathrm{d}\mu = \frac{1}{2\pi}\sum_{n=0}^{\infty} a_n\,e^{-\mathrm{i}nt} \quad (-2\pi \leqslant t \leqslant 0). \tag{37}$$

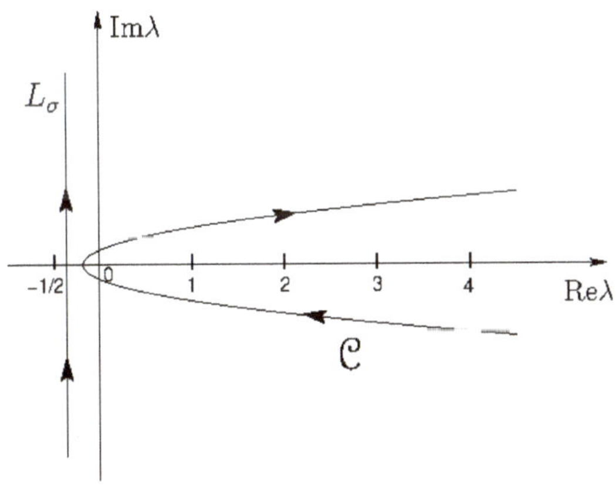

Figure 3. Integration path of integral (30).

Now, substitute into integral (36) the real variable t with the complex variable $\tau = t + \mathrm{i}w$. The resulting integral can be proved to provide an analytic continuation of $\widehat{f}_+^{(+)}(t)$ in the strip $0 < t < 2\pi$, $w \in \mathbb{R}^+$, continuous in the closure of the latter. In fact, from the first equality in (36), we formally obtain:

$$\widehat{f}_+^{(+)}(t+\mathrm{i}w) = \frac{1}{2\pi}\,e^{\sigma w}\int_{-\infty}^{+\infty} H_\sigma^t(\mu)\,e^{\mathrm{i}\mu w}\,\mathrm{d}\mu \qquad \left(0 \leqslant t \leqslant 2\pi;\, w \in \mathbb{R}^+, -\tfrac{1}{2} \leqslant \sigma < 0\right), \tag{38}$$

where

$$H^t_\sigma(\mu) \doteq -\frac{\widetilde{a}(\sigma+\mathrm{i}\mu)e^{-\mathrm{i}(\sigma+\mathrm{i}\mu)(t-\pi)}}{2\sin\pi(\sigma+\mathrm{i}\mu)}. \tag{39}$$

By inequality (35) and statement (vi) of Proposition 2:

$$\left|H^t_\sigma(\mu)\right| \leqslant \frac{|\widetilde{a}(-\frac{1}{2}+\mathrm{i}\mu)|}{|\sin\pi\sigma|} \qquad \left(0\leqslant t\leqslant 2\pi;\sigma\geqslant-\tfrac{1}{2},\sigma\notin\mathbb{N}\right), \tag{40}$$

which, along with statements (iv) and (v) of Proposition 2, guarantees that $H^t_\sigma(\mu)\in L^1(-\infty,+\infty)$ for $0\leqslant t\leqslant 2\pi,\sigma>-\frac{1}{2},\sigma\notin\mathbb{N}$. Therefore, formulas (38), (39), and (40) define $\widehat{f}^{(+)}_+(\tau)$ ($\tau=t+\mathrm{i}w$) as an analytic continuation of $\widehat{f}^{(+)}_+(t)$ in the strip $\{\tau=t+\mathrm{i}w,\ 0<t<2\pi,w\in\mathbb{R}^+\}$, continuous in the closure of the latter in view of the Riemann–Lebesgue theorem. Proceeding analogously, we can obtain an analytic continuation of $\widehat{f}^{(+)}_-(t)$ in the strip $\{\tau=t+\mathrm{i}w,\ -2\pi<t<0,w\in\mathbb{R}^+\}$, continuous on the closure of the latter. It then follows that the function $\widehat{f}^{(+)}(\tau)$ admits a holomorphic extension to the cut-domain ${}^\tau\mathcal{I}^{(0)}_+\setminus{}^\tau\dot{\Xi}^{(0)}_+$, and statement (ii) is proved.

The discontinuity $\widehat{F}^{(+)}(w)$ of $(\mathrm{i}\widehat{f}^{(+)}(\tau))$ across the cut at $t=0$ equals $\mathrm{i}[\widehat{f}^{(+)}_+(\mathrm{i}w)-\widehat{f}^{(+)}_-(\mathrm{i}w)]$ ($w\in\mathbb{R}^+$) (for the 2π-periodicity of $\widehat{f}^{(+)}(\tau)$, we may consider only the cut at $t=0$). It can be computed by replacing t by $\mathrm{i}w$ into integrals (36) and (37) and then subtracting Equation (37) from Equation (36):

$$\widehat{F}^{(+)}(w)\doteq\mathrm{i}\left[\widehat{f}^{(+)}_+(\mathrm{i}w)-\widehat{f}^{(+)}_-(\mathrm{i}w)\right]=\frac{1}{2\pi}\int_{-\infty}^{+\infty}\widetilde{a}(\sigma+\mathrm{i}\mu)e^{(\sigma+\mathrm{i}\mu)w}\,\mathrm{d}\mu \qquad \left(w\in\mathbb{R}^+,\sigma\geqslant-\tfrac{1}{2}\right), \tag{41}$$

which yields:

$$\left|\widehat{F}^{(+)}(w)\right|\leqslant\|\widetilde{a}_\sigma\|_1\,e^{\sigma w} \qquad \left(w\in\mathbb{R}^+,\sigma\geqslant-\tfrac{1}{2}\right),$$

where $\|\widetilde{a}_\sigma\|_1$, defined in (27), is guaranteed to be finite by statement (iv) of Proposition 2. Rewrite (41) as follows:

$$\widehat{F}^{(+)}(w)e^{-\sigma w}=\frac{1}{2\pi}\int_{-\infty}^{+\infty}\widetilde{a}(\sigma+\mathrm{i}\mu)\,e^{\mathrm{i}\mu w}\,\mathrm{d}\mu \qquad \left(w\in\mathbb{R}^+,\sigma\geqslant-\tfrac{1}{2}\right). \tag{42}$$

Since for any fixed $\operatorname{Re}\lambda\geqslant-\frac{1}{2}$, $\lambda\widetilde{a}(\lambda)$ and $\widetilde{a}(\lambda)$ belong to $L^1(-\infty,+\infty)$, the Riemann–Lebesgue theorem guarantees that $\widehat{F}^{(+)}(w)e^{-\sigma w}$ is a function of class C^1 tending to zero as $w\to+\infty$, and statement (iii) is proved. Hence, $\widehat{F}^{(+)}(w)$ is a continuous function of w ($w\in\mathbb{R}^+$) and $\widehat{F}^{(+)}(w)=o(e^{-w/2})$ as $w\to+\infty$. Moreover, $\widehat{F}^{(+)}(0)=0$ for the continuity of $\widehat{f}^{(+)}(\tau)$ on the real axis, and statement (iv) is proved. Inverting (42), we have:

$$\widetilde{a}(\sigma+\mathrm{i}\mu)=\int_0^{+\infty}\widehat{F}^{(+)}(w)\,e^{-(\sigma+\mathrm{i}\mu)w}\,\mathrm{d}w \qquad \left(\sigma>-\tfrac{1}{2}\right),$$

where the integral on the r.h.s. converges for $\sigma>-\frac{1}{2}$. It defines the Laplace transform of $\widehat{F}^{(+)}(w)$, holomorphic in the half-plane $\operatorname{Re}\lambda>-\frac{1}{2}$, and statement (v) follows. Finally, recalling that $\widetilde{a}(\sigma+\mathrm{i}\mu)$ ($\mu\in\mathbb{R}$) belongs to $L^2(-\infty,+\infty)$ for any fixed value $\sigma\geqslant-\frac{1}{2}$ (statement (ii) of Proposition 2 and inequality (25)), we obtain the Plancherel equality:

$$\int_{-\infty}^{+\infty}|\widetilde{a}\,(\sigma+\mathrm{i}\mu)|^2\,\mathrm{d}\mu=2\pi\int_0^{+\infty}\left|\widehat{F}^{(+)}(w)\,e^{-\sigma w}\right|^2\mathrm{d}w \qquad \left(\sigma\geqslant-\tfrac{1}{2}\right),$$

proving statement (vi). □

Proposition 4. *If in the trigonometric series (see (19))*

$$\frac{1}{2\pi}\left[\sum_{n=0}^{\infty} a_n e^{-int} - e^{it}\sum_{n=0}^{\infty} a_n e^{int}\right] \qquad (t \in \mathbb{R}), \tag{43}$$

the coefficients a_n satisfy the assumptions required by Propositions 2 and 3, i.e., the set of numbers $f_n = (n+1)^2 a_n$ satisfies condition (20), then:

(i) The series (43) converges to a continuous function $\widehat{f}(t)$ ($t \in \mathbb{R}$).
(ii) The function $\widehat{f}(t)$ admits a holomorphic extension to the cut-domain $({}^{\tau}\mathcal{I}_{+}^{(0)} \setminus {}^{\tau}\dot{\Xi}_{+}^{(0)}) \cup ({}^{\tau}\mathcal{I}_{-}^{(0)} \setminus {}^{\tau}\dot{\Xi}_{-}^{(0)})$, i.e., it is analytic in $\mathbb{C} \setminus \{\tau = 2k\pi + iw; k \in \mathbb{Z}, w \in \mathbb{R}, w \neq 0\}$.
(iii) The jump function across the cuts ${}^{\tau}\dot{\Xi}_{\pm}^{(0)}$ enjoys properties, mutatis mutandis, *analogous to conditions (iii)–(vi) of Proposition 3.*

Proof. These statements can be proved by using obvious extensions of the arguments used in the proof of Proposition 3. □

Through the holomorphic extension associated with the trigonometric series (43), we obtain a function, denoted by $\widehat{f}(\tau)$ ($\tau = t + iw; t, w \in \mathbb{R}$), which is the analytic continuation of $\widehat{f}(t)$ ($t \in \mathbb{R}$) from the real axis to the domain ${}^{\tau}\dot{\mathcal{I}} \doteq ({}^{\tau}\mathcal{I}_{+}^{(0)} \setminus {}^{\tau}\dot{\Xi}_{+}^{(0)}) \cup ({}^{\tau}\mathcal{I}_{-}^{(0)} \setminus {}^{\tau}\dot{\Xi}_{-}^{(0)})$. We can then prove the following corollary to Proposition 4.

Corollary 1. *The function $\widehat{f}(\tau)$ is holomorphic in the 2π-periodic strips $\Sigma_\tau \doteq \{\tau \in \mathbb{C} : \tau = t + iw, 2\pi k < t < 2\pi(k+1), k \in \mathbb{Z}, w \in \mathbb{R}\}$.*

Proof. We start from statement (ii) of Proposition 4. Next, by applying the Schwarz reflection principle and taking into account that the function $\widehat{f}(t)$ (i.e., the restriction of $\widehat{f}(\tau)$ to the real axis) is continuous (statement (i) of Proposition 4), the statement of the corollary follows. □

Remark 2. *In Ref. [21], the Hausdorff condition (20) was assumed to hold for the set of numbers $f_n = n^p a_n$ ($p \geqslant 2$). However, this condition does not guarantee obtaining a unique Carlsonian interpolation of the whole sequence a_n, including the first coefficient a_0. Therefore, it must be replaced by the same condition on the numbers $f_n = (n+1)^p a_n$ ($p \geqslant 2$). However, for the purpose of the current analysis, it is sufficient to take $p = 2$. Let us note that the results of Ref. [21] hold true and, in particular, the proofs of Propositions 5.1 and 6.1 of that paper, which will be used below, are correct modulo the following change: $f_n = n^p a_n \to f_n = (n+1)^p a_n$.*

2.4. Inversion of the Radon–Abel Transformation and Holomorphic Extension Associated with the Legendre Series

The first step consists of determining the inversion of the Radon–Abel transformation (12). We can prove the following proposition.

Proposition 5. *Suppose that the sequence $f_n = (n+1)^2 a_n$ (a_n being the coefficients of the Legendre expansion (14)) satisfies the Hausdorff condition (20). Then, the Radon–Abel transformation (see (12))*

$$\widehat{f}(t) = -2\,e^{it/2}\int_0^t f(u)\,\frac{\sin u}{[2(\cos t - \cos u)]^{\frac{1}{2}}}\,du \tag{44}$$

admits the following inversion:

$$f(u) = \frac{1}{\pi \sin u} \frac{\mathrm{d}}{\mathrm{d}u} \int_0^u e^{-\mathrm{i}t/2} \widehat{f}(t) \frac{\sin t}{[2(\cos u - \cos t)]^{\frac{1}{2}}} \, \mathrm{d}t.$$

Proof. See Proposition 5.1 of Ref. [21]. □

By following the same procedure used in the proof of Proposition 5 (i.e., introducing Riemann–Liouville integrals and related properties), Proposition 5 can be extended to give the inversion of the Radon–Abel transformation (11). We can state, without proof, the following proposition.

Proposition 6. *Assume that the coefficients a_n (see (14)) satisfy the conditions of Proposition 5; then, the function $\widehat{f}(\tau)$, i.e., the analytic continuation of the function $\widehat{f}(t)$ given in (44), is holomorphic in the strips Σ_τ and can be represented by the following Radon–Abel transformation (see (11)):*

$$\widehat{f}(\tau) = -2\, e^{\mathrm{i}\tau/2} \int_{\gamma_\tau} f(\theta) \frac{\sin \theta}{[2(\cos \tau - \cos \theta)]^{\frac{1}{2}}} \, \mathrm{d}\theta,$$

which admits the following inversion:

$$f(\theta) = \frac{1}{\pi \sin \theta} \frac{\mathrm{d}}{\mathrm{d}\theta} \int_{\gamma_\theta} e^{-\mathrm{i}\tau/2} \widehat{f}(\tau) \frac{\sin \tau}{[2(\cos \theta - \cos \tau)]^{\frac{1}{2}}} \, \mathrm{d}\tau, \tag{45}$$

γ_τ and γ_θ denoting the rays from zero to τ and from zero to θ, respectively.

We can now prove the following proposition.

Proposition 7. *Suppose that the sequence $f_n = (n+1)^2 a_n$ $(n = 0, 1, 2, \ldots)$ satisfies the Hausdorff condition (20), then the function $f(\theta)$ $(\theta \in \mathbb{C})$, represented by formula (45), is even, 2π-periodic, and holomorphic in ${}^{\theta}\dot{\mathcal{I}} = ({}^{\theta}\mathcal{I}_+^{(0)} \setminus {}^{\theta}\dot{\Xi}_+^{(0)}) \cup ({}^{\theta}\mathcal{I}_-^{(0)} \setminus {}^{\theta}\dot{\Xi}_-^{(0)})$.*

Proof. The assumptions on the Legendre coefficients a_n allow us to state that $\widehat{f}(\tau)$ is a 2π-periodic function, holomorphic in the domain ${}^{\tau}\dot{\mathcal{I}}$ (see Corollary 1). Moreover, it enjoys the symmetry property:

$$\widehat{f}(\tau) = -e^{\mathrm{i}\tau} \widehat{f}(-\tau),$$

which follows from (17) and from the uniqueness of the analytic continuation. The properties mentioned above imply that $\widehat{f}(\tau)$ is of the form: $\widehat{f}(\tau) = e^{\mathrm{i}\tau/2}(1 - \cos \tau)^{\frac{1}{2}}\, b(\cos \tau)$, with $b(\cos \tau)$ analytic in ${}^{\tau}\underline{D} = \{\cos \tau \in \mathbb{C}, \tau \in {}^{\tau}\dot{\mathcal{I}}\}$. Through the following parametrization of γ_θ: $\cos \tau = 1 + \delta(\cos \theta - 1)$, $(0 \leqslant \delta \leqslant 1)$, formula (45) can be rewritten as:

$$f(\theta) = \frac{\mathrm{i}}{\sqrt{2}\pi} \frac{\mathrm{d}}{\mathrm{d}(\cos \theta)} \left[(\cos \theta - 1) \int_0^1 b(1 + \delta(\cos \theta - 1)) \delta^{\frac{1}{2}} (1 - \delta)^{-\frac{1}{2}} \, \mathrm{d}\delta \right],$$

which represents a function holomorphic in ${}^{\theta}\underline{D} = \{\cos \theta \in \mathbb{C}, \theta \in {}^{\theta}\dot{\mathcal{I}}\}$. Accordingly, regarded as a function of θ, it represents an even function, 2π-periodic, and holomorphic in ${}^{\theta}\dot{\mathcal{I}}$. □

From the previous proposition and Corollary 1, it derives the following corollary.

Corollary 2. *If the sequence $\{a_n\}_{n=0}^{\infty}$ of the Legendre coefficients satisfies the conditions of Proposition 7, then $f(\theta)$ is a function analytic in the 2π-periodic strips $\Sigma_\theta \doteq \{\theta \in \mathbb{C} : \theta = u + \mathrm{i}v, 2\pi k < u < 2\pi(k+1), k \in \mathbb{Z}, v \in \mathbb{R}\}$.*

Formula (45) allows us to compute the boundary values $f_{\pm}^{(+)}(\mathrm{i}v)$, which are defined by $f_{\eta}^{(+)}(v) \doteq \lim_{u\to 0^+} f^{(+)}(\eta\, u + \mathrm{i}v), \eta = \pm, v \geqslant 0$, on the semi-axis $\theta = \mathrm{i}v$, $v \geqslant 0$, with $\gamma_{\mathrm{i}v} = \{\tau = \mathrm{i}w, 0 \leqslant w \leqslant v\}$, in terms of the corresponding boundary values $\widehat{f}_{\pm}^{(+)}(\mathrm{i}w)$, provided $\widehat{f}_{\pm}^{(+)}(\mathrm{i}w)$ satisfies a C^1-type regularity condition; the latter condition is necessary in order to perform the inversion of the Radon–Abel transform at the boundary. The C^1-continuity of the boundary values follows from the fact that the sequence $f_n = (n+1)^2 a_n$ is required to satisfy the Hausdorff condition (20) (see Propositions 2 and 3). We thus obtain:

$$F^{(+)}(v) \doteq \mathrm{i}[f_{+}^{(+)}(\mathrm{i}v) - f_{-}^{(+)}(\mathrm{i}v)] = \frac{1}{\pi \sinh v}\frac{\mathrm{d}}{\mathrm{d}v}\int_0^v e^{w/2}\widehat{F}^{(+)}(w)\,\frac{\sinh w}{[2(\cosh v - \cosh w)]^{\frac{1}{2}}}\,\mathrm{d}w, \qquad (46)$$

where $\widehat{F}^{(+)}(w) \doteq \mathrm{i}[\widehat{f}_{+}^{(+)}(\mathrm{i}w) - \widehat{f}_{-}^{(+)}(\mathrm{i}w)]$, $\widehat{f}_{\eta}^{(+)}(\mathrm{i}w) = \lim_{t\to 0^+} \widehat{f}^{(+)}(\eta\, t + \mathrm{i}w), \eta = \pm$. At this point, let us note that, for the current analysis, it is sufficient to consider the cuts in the τ- and θ-planes at $\tau = 2k\pi + \mathrm{i}w$ ($k \in \mathbb{Z}, w > 0$) and $\theta = 2k\pi + \mathrm{i}v$ ($k \in \mathbb{Z}, v > 0$), respectively. Therefore, we can limit ourselves to consider the functions: $\widehat{F}^{(+)}(w) = e^{-v/2}(\mathcal{A}F)(w)$ ($w \geqslant 0$) and, correspondingly, $F^{(+)}(v) \equiv \underline{F}^{(+)}(\cosh v)$ ($v \geqslant 0$). For simplicity, hereafter we shall omit the superscript $(+)$ in these notations.

Next, we can apply the inverse Radon–Abel operator (defined by (45)) to the series on the r.h.s. of formula (19), i.e.,

$$\widehat{f}(t) = \frac{1}{2\pi}e^{\mathrm{i}(t-\pi)/2}\sum_{n=-\infty}^{+\infty}(-1)^n a_n \cos\left[\left(n+\tfrac{1}{2}\right)(t-\pi)\right],$$

whose term by term integration is legitimate for the uniform convergence of the series, which follows from the Hausdorff conditions satisfied by the coefficients $\{a_n\}$. We now introduce the functions

$$\psi_n(\cos u) \doteq -\frac{\mathrm{i}}{\pi \sin u}\frac{\mathrm{d}}{\mathrm{d}u}\int_0^u \cos\left[\left(n+\tfrac{1}{2}\right)(t-\pi)\right]\frac{\sin t}{[2(\cos u - \cos t)]^{\frac{1}{2}}}\,\mathrm{d}t \qquad (0 < u < 2\pi),$$

which are related to the Legendre polynomials $P_n(\cos u)$ by [21]:

$$\psi_n(\cos u) = \frac{(-1)^n}{4}(2n+1)\,P_n(\cos u).$$

Recalling that $a_n = -a_{-n-1}$ ($n \in \mathbb{Z}$) (formula (18)), we finally obtain the original Legendre expansion

$$f(u) = \underline{f}(\cos u) = \frac{1}{\pi}\sum_{n=0}^{\infty}(-1)^n a_n \psi_n(\cos u) = \frac{1}{4\pi}\sum_{n=0}^{\infty}(2n+1)\,a_n\,P_n(\cos u).$$

All the results obtained for the function $\underline{f}(\cos\theta)$ in the $\cos\theta$-plane can be summarized in the following theorem.

Theorem 2. *If the sequence $f_n = (n+1)^2 a_n$ $(n = 0, 1, 2, \ldots)$ satisfies the Hausdorff condition (20), then:*

(i) Series (14) converges to a function $\underline{f}(\cos u)$ ($u \equiv \mathrm{Re}\,\theta$), analytic in the interval $-1 < \cos u < 1$.
(ii) The function $\underline{f}(\cos u)$ admits a holomorphic extension to the complex $\cos\theta$-plane ($\theta = u + \mathrm{i}v$) cut along the semi-axis $[1, +\infty)$.

2.5. Spherical Laplace Transform and Analyticity Properties in the Complex λ-Plane

A basic feature of the holomorphic extensions associated with the series expansions is the *dual analyticity* that we shall illustrate in the specific case that we have considered here, of the Legendre series (14). To classes of functions $\underline{f}(\cos\theta)$ holomorphic in the $\cos\theta$-plane cut along the semi-axis $[1,+\infty)$ (see Theorem 2), there correspond classes of analytic functions $\widetilde{a}(\lambda)$ ($\lambda \in \mathbb{C}$), which enjoy the following properties:

(a) $\widetilde{a}(\lambda)$ is the composition of the standard Laplace transform with the Abel transform (multiplied by the factor $e^{-w/2}$) of the jump function of $[\mathrm{i}\widehat{f}(\theta)]$ across the cut ${}^{\theta}\dot{\Xi}^{(0)}_{+}$, that is, the spherical Laplace transform.
(b) $\widetilde{a}(\lambda)$ is holomorphic in the half-plane $\mathrm{Re}\,\lambda > -\frac{1}{2}$.
(c) $\widetilde{a}(\lambda)$ is the (unique) Carlsonian interpolant of the Fourier–Legendre coefficients: $\widetilde{a}(\lambda)|_{\lambda=n} = a_n$.

Formulas (9) and (28) give

$$\widetilde{a}(\lambda) = \int_0^{+\infty} e^{-(\lambda+\frac{1}{2})w}\,(\mathcal{A}F)(w)\,dw \qquad \left(\mathrm{Re}\,\lambda > -\tfrac{1}{2}\right),$$

which is precisely the spherical Laplace transform, holomorphic in the half-plane $\mathrm{Re}\,\lambda > -\frac{1}{2}$ (statement (v) of Proposition 3). Moreover, $\widetilde{a}(\lambda)|_{\lambda=n} = a_n$ (statement (i) of Proposition 2). Writing explicitly the Abel transform $(\mathcal{A}F)(w)$ (see (9)), we obtain

$$\widetilde{a}(\lambda) = 2\int_0^{+\infty} e^{-(\lambda+\frac{1}{2})w}\left\{\int_0^{w} \underline{F}(\cosh v)\,\frac{\sinh v}{[2(\cosh w-\cosh v)]^{\frac{1}{2}}}\,dv\right\} dw \quad \left(\mathrm{Re}\,\lambda > -\tfrac{1}{2}\right),$$

which, interchanging the order of integration, becomes

$$\widetilde{a}(\lambda) = 2\int_0^{+\infty} \underline{F}(\cosh v)\sinh v\left\{\int_v^{+\infty} \frac{e^{-(\lambda+\frac{1}{2})w}}{[2(\cosh w-\cosh v)]^{\frac{1}{2}}}\,dw\right\} dv \quad \left(\mathrm{Re}\,\lambda > -\tfrac{1}{2}\right). \tag{47}$$

Using the integral representation of the Legendre functions of the second kind [24]

$$Q_\lambda(\cosh v) = \int_v^{+\infty} \frac{e^{-(\lambda+\frac{1}{2})w}}{[2(\cosh w-\cosh v)]^{\frac{1}{2}}}\,dw \qquad (\mathrm{Re}\,\lambda > -1, v > 0), \tag{48}$$

formula (47) can be written as follows:

$$\widetilde{a}(\lambda) = 2\int_0^{+\infty} \underline{F}(\cosh v)\,Q_\lambda(\cosh v)\,\sinh v\,dv \qquad \left(\mathrm{Re}\,\lambda > -\tfrac{1}{2}\right). \tag{49}$$

Remark 3. *The Legendre function of the second kind has a logarithmic singularity at $v = 0$, then the integral representation (48) holds if $v > 0$; nevertheless, the integral in (49) converges if $\underline{F}(\cosh v)$ is regular at $v = 0$.*

We can now state the following theorem.

Theorem 3. *If the sequence $f_n = (n+1)^2 a_n$ $(n = 0,1,2,\ldots)$, a_n being the Legendre coefficients of expansion (14), satisfies the Hausdorff condition (20), then the jump function $F(v) = \underline{F}(\cosh v)$ (defined in (46)) admits the following integral representation:*

$$F(v) = \underline{F}(\cosh v) = \frac{1}{2\pi}\int_{-\infty}^{+\infty} \widetilde{a}(\sigma + \mathrm{i}\mu)\,\left(\sigma + \tfrac{1}{2} + \mathrm{i}\mu\right)\, P_{\sigma+\mathrm{i}\mu}(\cosh v)\,\mathrm{d}\mu$$
$$= \frac{1}{2\pi\mathrm{i}}\int_{\sigma-\mathrm{i}\infty}^{\sigma+\mathrm{i}\infty} \widetilde{a}(\lambda)\,\left(\lambda + \tfrac{1}{2}\right) P_\lambda(\cosh v)\,\mathrm{d}\lambda \qquad \left(\lambda = \sigma + \mathrm{i}\mu; \sigma \geqslant -\tfrac{1}{2}\right), \tag{50}$$

where $P_\lambda(\cdot)$ denotes the Legendre function of the first kind.

Proof. See Proposition 6.1 of Ref. [21]. □

3. From Spherical Laplace Transform to Non-Euclidean Fourier Transform

3.1. Formal Derivation of Mehler's Transform from the Spherical Laplace Transform

Let us begin by considering the integral representation (48) of the Legendre function of the second kind $Q_\lambda(\cosh v)$. If $\mathrm{Re}\,\lambda = -\frac{1}{2}$, the function $Q_{-\frac{1}{2}+\mathrm{i}\mu}(\cosh v)$ can be written as the sum of its even (in μ) and odd parts, which are defined as follows:

$$Q^{(\mathrm{E})}_{-\frac{1}{2}+\mathrm{i}\mu}(\cosh v) = \int_v^{+\infty} \frac{\cos \mu w}{[2(\cosh w - \cosh v)]^{\frac{1}{2}}}\,\mathrm{d}w,$$
$$Q^{(\mathrm{O})}_{-\frac{1}{2}+\mathrm{i}\mu}(\cosh v) = -\,\mathrm{i}\int_v^{+\infty} \frac{\sin \mu w}{[2(\cosh w - \cosh v)]^{\frac{1}{2}}}\,\mathrm{d}w.$$

Recalling the relation between the Legendre functions of first and second kind (p. 140, [24])

$$P_\lambda(z) = \frac{\tan \pi\lambda}{\pi}\,\left[Q_\lambda(z) - Q_{-\lambda-1}(z)\right],$$

and exploiting the evenness (in μ) of the conical Legendre functions, i.e., $P_{-\frac{1}{2}+\mathrm{i}\mu}(\cosh v) = P_{-\frac{1}{2}-\mathrm{i}\mu}(\cosh v)$ $(\mu \in \mathbb{R})$, we obtain:

$$P_{-\frac{1}{2}+\mathrm{i}\mu}(\cosh v) = P_{-\frac{1}{2}-\mathrm{i}\mu}(\cosh v) = \frac{2}{\pi}\tan\left[\pi\left(-\tfrac{1}{2} + \mathrm{i}\mu\right)\right] Q^{(\mathrm{O})}_{-\frac{1}{2}+\mathrm{i}\mu}(\cosh v). \tag{51}$$

Let us now return to Theorem 3 and to formula (50) in the specific case $\sigma = -\frac{1}{2}$. By the μ-evenness of $P_{-\frac{1}{2}+\mathrm{i}\mu}(\cosh v)$, it follows that only the odd component $\widetilde{a}^{(\mathrm{O})}(-\frac{1}{2}+\mathrm{i}\mu)$ of $\widetilde{a}(-\frac{1}{2}+\mathrm{i}\mu)$ $(\mu \in \mathbb{R})$ contributes to the integral in formula (50):

$$F(v) = \underline{F}(\cosh v) = \frac{\mathrm{i}}{\pi}\int_0^{+\infty} \widetilde{a}^{(\mathrm{O})}\left(-\tfrac{1}{2} + \mathrm{i}\mu\right) P_{-\frac{1}{2}+\mathrm{i}\mu}(\cosh v)\,\mu\,\mathrm{d}\mu. \tag{52}$$

Moving back to formula (49), which for the moment we assume to hold also for $\mathrm{Re}\,\lambda = -\frac{1}{2}$ (this statement will be proved later in Section 3.4.3), and accounting for relationship (51), we can *formally* rewrite the odd component of $\widetilde{a}(-\frac{1}{2}+\mathrm{i}\mu)$ $(\mu \in \mathbb{R})$ as follows:

$$\widetilde{a}^{(\mathrm{O})}\left(-\tfrac{1}{2} + \mathrm{i}\mu\right) = \frac{\pi}{\tan\left[\pi\left(-\frac{1}{2} + \mathrm{i}\mu\right)\right]}\int_0^{+\infty} \underline{F}(\cosh v)\, P_{-\frac{1}{2}+\mathrm{i}\mu}(\cosh v)\sinh v\,\mathrm{d}v \qquad (\mu \in \mathbb{R}). \tag{53}$$

Noting that $(\tan[\pi(-\frac{1}{2}+\mathrm{i}\mu)])^{-1} = -\mathrm{i}\tanh\pi\mu$, we may introduce the following function: $\widetilde{F}(\mu) \doteq \widetilde{a}(-\frac{1}{2}+\mathrm{i}\mu)/(-\mathrm{i}\pi\tanh\pi\mu)$ (we omit here the superscript "(O)" in $\widetilde{a}^{(\mathrm{O})}\left(-\frac{1}{2}+\mathrm{i}\mu\right)$ since only the odd part of $\widetilde{a}(-\frac{1}{2}+\mathrm{i}\mu)$ plays a role in these transformations). Then, formulas (52) and (53) can be rewritten as

$$F(v) = \underline{F}(\cosh v) = \int_0^{+\infty} \widetilde{F}(\mu)\, P_{-\frac{1}{2}+\mathrm{i}\mu}(\cosh v)\, \tanh(\pi\mu)\, \mu\, \mathrm{d}\mu, \tag{54a}$$

$$\widetilde{F}(\mu) = \int_0^{+\infty} \underline{F}(\cosh v)\, P_{-\frac{1}{2}+\mathrm{i}\mu}(\cosh v) \sinh v\, \mathrm{d}v, \tag{54b}$$

which coincide with the Mehler transform pair, indeed (see (p. 175, [24])).

Remark 4. *(i) As we have already remarked above, we do not consider the whole spherical Laplace transform but only its odd (with respect to μ) component since only this component plays a role in the transformations being treated here.*

(ii) The class of functions $F(v)$ which we are led to consider here are of the form $\underline{F} = \underline{F}(\cosh v)$. Accordingly, the connection between spherical Laplace transform and non-Euclidean Fourier transform can be limited to this class of functions. In the non-Euclidean geometry, these functions belong to the class of radial functions *in a sense that will be clarified in what follows (see Remarks 5 and 6).*

3.2. Geometry of the Two-Sheeted Hyperboloid: Polar and Horocyclic Coordinates

Let us now return to the geometrical representation of the real one-sheeted hyperboloid X_2. It can be easily noted that, by simply swapping two coordinate axes (see Figures 1 and 4), the real meridian $\widehat{X}^+$ can be regarded as the real meridian of one of the sheets of a suitable real two-sheeted hyperboloid. It is therefore reasonable to expect that an integral representation of $\underline{F}(\cosh v)$ can be obtained also from the geometry of the two-sheeted hyperboloid.

In the space $\mathbb{R}^3$ of variables $x = (x_0, x_1, x_2)$, we consider the two-sheeted hyperboloid with equation:

$$x_0^2 - x_1^2 - x_2^2 = 1. \tag{55}$$

In the present analysis, we can limit ourselves to consider the upper sheet of this two-sheeted hyperboloid, the one with $x_0 \geqslant 1$ that will be denoted by ${}^2X_2^+$. By analogy with what we have done is Section 2, we consider two systems of coordinates: polar and horocyclic coordinates.

Polar Coordinates:

$$x_0 = \cosh v, \tag{56a}$$

$$x_1 = \sinh v \sin\psi, \qquad (v \geqslant 0, \psi \in [0, 2\pi)), \tag{56b}$$

$$x_2 = \sinh v \cos\psi. \tag{56c}$$

It is easy to verify that $x_0^2 - x_1^2 - x_2^2 = \cosh^2 v - \sinh^2 v = 1$, and $x_0 \geqslant 1$.

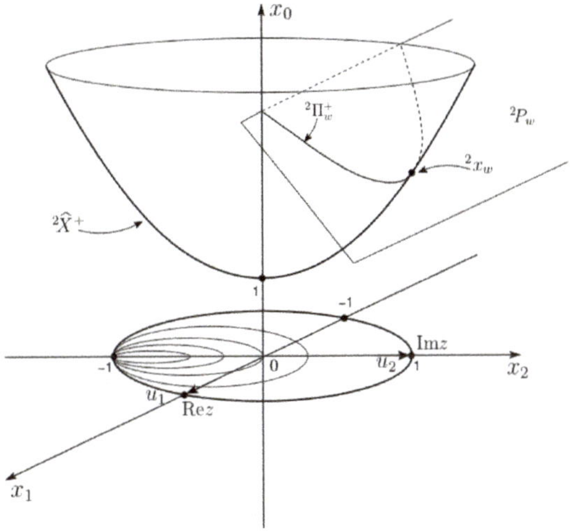

Figure 4. Upper sheet ${}^{2}\mathrm{X}_2^+$ of the real two-sheteed hyperboloid. Re z is directed along u_1 and Im z along u_2.

Horocyclic Coordinates:

$$x_0 = \cosh w + \frac{1}{2}\zeta^2 e^w, \tag{57a}$$

$$x_1 = \zeta e^w, \qquad (w, \zeta \in \mathbb{R}), \tag{57b}$$

$$x_2 = \sinh w - \frac{1}{2}\zeta^2 e^w. \tag{57c}$$

Even in this case, it is straightforward to verify that $x_0^2 - x_1^2 - x_2^2 = 1$, and $x_0 \geqslant 1$; accordingly, the horocyclic coordinates (57) are appropriate for describing the upper sheet ${}^{2}\mathrm{X}_2^+$ of the real two-sheeted hyperboloid. In particular, we focus our attention on the meridian section ${}^{2}\widehat{\mathrm{X}}^+$ of the upper sheet ${}^{2}\mathrm{X}_2^+$, which lies in the plane $x_1 = 0$ and whose equation is $x_0^2 - x_2^2 = 1$. We consider the intersections of ${}^{2}\widehat{\mathrm{X}}^+$ with the family of planes ${}^{2}P_w$ with equation $x_0 + x_2 = e^w$ $(w \in \mathbb{R})$, i.e., the points ${}^{2}x_w$, whose coordinates are $x_0 = \cosh w$, $x_1 = 0$, $x_2 = \sinh w$. The sections of ${}^{2}\mathrm{X}_2^+$ by these planes are the (real) parabolae ${}^{2}\Pi_w^+$. Setting $\zeta = 0$ in (57), we obtain the point ${}^{2}x_w$ $(w \in \mathbb{R})$, which is the apex of the corresponding parabola ${}^{2}\Pi_w^+$ (see Figure 4).

Remark 5. *As we shall see in the next section, the parabolae ${}^{2}\Pi_w^+$ generate (through a stereographic projection on the non-Euclidean disk) a family of horocycles, represented by the Euclidean circles (illustrated in Figure 4) which are tangent to the boundary of the unit disk at the point $x_1 = 0, x_2 = -1$. It is worth emphasizing that this is the sole family of horocycles which is generated by the fibration illustrated above. On the other hand, the non-Euclidean Fourier transform, which will be studied in the next section, demands considering in the non-Euclidean disk all the families of horocycles which are obtained by moving the point of tangency along the entire horizon from 0 to 2π. We shall see in Lemma 2 of Section 3.4.1 how this difficulty will be overcome (see also Remark 6).*

3.3. Stereographic Projection from the Upper Sheet of the Two-Sheeted Hyperboloid to the Non-Euclidean Unit Disk

Let $u = (u_1, u_2)$ denote the Cartesian coordinate system in the open unit disk $D \doteq \{u : u_1^1 + u_2^2 < 1\}$, which lies in the plane $x_0 = 0$:

$$u_1 = \tanh\left(\frac{v}{2}\right)\sin\psi \qquad (v \geqslant 0, \psi \in [0, 2\pi)), \tag{58a}$$

$$u_2 = \tanh\left(\frac{v}{2}\right)\cos\psi. \tag{58b}$$

Then: $|u|^2 = u_1^2 + u_2^2 = \tanh^2(v/2)$. Let $B \doteq \{u : u_1^1 + u_2^2 = 1\}$ denote the boundary of D, i.e., the horizon. Consider the Riemannian structure

$$\mathrm{d}s^2 = \frac{4(\mathrm{d}u_1^2 + \mathrm{d}u_2^2)}{[1 - (u_1^2 + u_2^2)]^2} = \mathrm{d}v^2 + \sinh^2 v\,\mathrm{d}\psi^2, \tag{59}$$

where $\mathrm{d}v^2 + \sinh^2 v\,\mathrm{d}\psi^2$ can also be obtained from (56) through the equality: $\mathrm{d}s^2 = -\mathrm{d}x_0^2 + \mathrm{d}x_1^2 + \mathrm{d}x_2^2$. Let us embed the non-Euclidean disk in the complex z-plane so that each point of the unit disk can be represented either by the coordinates (u_1, u_2) in (58) or through the polar representation of z, which in the present case reads: $z = |z|\exp[\mathrm{i}(\frac{\pi}{2} - \psi)]$, $|z| = \tanh(\frac{v}{2})$. Note that in this latter representation we are forced to write $\exp[\mathrm{i}(\frac{\pi}{2} - \psi)]$ with ψ induced by (56) (instead of the standard expression $\exp(\mathrm{i}\psi)$) in order to have the correct correspondence between Cartesian and polar representations. Accordingly, the points b of the boundary B of the non-Euclidean disk will be described by $b = \exp[\mathrm{i}(\frac{\pi}{2} - \phi)]$, ϕ being defined analogously to ψ that is, measured from the positive u_2-axis and increasing toward the positive u_1-axis. The Riemannian structure (59) induces the usual non-Euclidean distance on D:

$$d(0, z) = \ln\frac{1 + |z|}{1 - |z|} \qquad z = |z|\,e^{\mathrm{i}(\frac{\pi}{2} - \psi)},$$

where $|z| = \tanh(v/2)$ and 0 is the center of the unit disk. In D, the geodesics are circular arcs intersecting the unit circle at right angles. In particular, all diameters of the unit circle are straight lines since these diameters can be considered as arcs of infinite large radius. A pencil of parallel straight lines is given by arcs of Euclidean circles orthogonal to the unit circle, lying in its interior and intersecting the boundary B at a common point b. The lines orthogonal to this pencil of parallel geodesics are the circles tangent from within to the horizon at the point b. Since these circles are the Euclidean images of the horocycles, we shall refer to the point of contact b as the normal to the horocycle.

The coordinates (x_0, x_1, x_2) that describe the point P varying on the upper sheet ${}^2\mathrm{X}_2^+$ of the two-sheeted hyperboloid are related to the coordinates (u_1, u_2) in the open non-Euclidean unit disk D through the following formulas (Lemma 1, p. 49, [25]):

$$x_0 = \frac{1 + |u|^2}{1 - |u|^2}, \qquad x_1 = \frac{2u_1}{1 - |u|^2}, \qquad x_2 = \frac{2u_2}{1 - |u|^2}. \tag{60}$$

Each line of intersection of ${}^2\mathrm{X}_2^+$ with the plane $ax_0 + bx_1 + cx_2 = 0$ $(a, b, c \in \mathbb{R})$ is mapped by the transformation (60) into a circular arc intersecting the horizon at right angles (Lemma 3, p. 51, [25]). In order to find the curve into which a straight line on ${}^2\mathrm{X}_2^+$ is mapped, it is sufficient to substitute into the equation of the plane $ax_0 + bx_1 + cx_2 = 0$ the explicit expression (60) of the variables x_i $(i = 0, 1, 2)$ in terms of u_1, u_2. Then, the equation

$$a\frac{1+|u|^2}{1-|u|^2} + \frac{2bu_1}{1-|u|^2} + \frac{2cu_2}{1-|u|^2} = 0 \tag{61}$$

is reduced after simple algebra to the equation of the circle

$$\left(u_1 + \frac{b}{a}\right)^2 + \left(u_2 + \frac{c}{a}\right)^2 = \frac{b^2 + c^2 - a^2}{a^2} \qquad (a \neq 0), \tag{62}$$

with radius $r = \sqrt{b^2 + c^2 - a^2}/|a|$ and center at the point $(-b/a, -c/a)$ if $a \neq 0$, and to the line $bu_1 + cu_2 = 0$ if $a = 0$. Note that the image into D of a straight line on ${}^2X_2^+$ is not the entire circle (62) but only its part contained in the unit disk.

Let us now map the hyperbola ${}^2\widehat{X}^+ = {}^2X_2^+ \cap \{x_1 = 0\}$, namely, the meridian of the upper sheet of the two-sheeted hyperboloid, into the non-Euclidean unit disk. By setting $a = c = 0$ and $b = 1$ in the equation of the planes given above, from (61), we see that ${}^2\widehat{X}^+$ is mapped into the diameter $u_1 = 0$ of the non-Euclidean disk: precisely (see (58)), we have the diameter $u_1 = 0$, $u_2 = \tanh(v/2)$, which tends to $u_2 = +1$ (i.e., $z = +\mathrm{i}$) for $v \to +\infty$ and $\psi = 0$, and to $u_2 = -1$ (i.e., $z = -\mathrm{i}$) for $v \to +\infty$ and $\psi = \pi$ (see Figure 4).

Next, we map the parabola ${}^2\Pi_w^+$, with apex ${}^2x_w = (\cosh w, 0, \sinh w)$ lying on the meridian ${}^2\widehat{X}^+$, which is represented in horocyclic coordinates by Equation (57). The apex ${}^2x_w$ of the parabola lies on the right branch $x_2 > 0$ of the meridian ${}^2\widehat{X}^+$ for $w > 0$, and on the left branch $x_2 < 0$ for $w < 0$; for $w = 0$, the apex of the parabola has coordinates $(1, 0, 0)$. These parabolae are generated by the intersection of ${}^2X_2^+$ with the plane ${}^2P_w$ whose equation is: $x_0 + x_2 = e^w$. By substituting in this latter equation the expressions of x_0 and x_2 given in (60), we obtain:

$$u_1^2 + \left(u_2 + \frac{1}{1+e^w}\right)^2 = \frac{e^{2w}}{(1+e^w)^2}, \tag{63}$$

which represents a circle. In view of (63), we can thus say that the mapping of the parabolae ${}^2\Pi_w^+$ into the unit non-Euclidean disk are the Euclidean circles (images of horocycles) with center in $(u_1 = 0, u_2 = -(1+e^w)^{-1})$, radius $r_w = (1+e^{-w})^{-1}$, and tangent from within to the horizon at the point $(u_1 = 0, u_2 = -1)$ (i.e., $z = -\mathrm{i}$; see Figure 4). Moreover, these circles cut orthogonally the diameter $u_1 = 0$ of D in the point $(u_1 = 0, u_2 = \tanh(w/2))$, which lies above the center of D when $w > 0$ and below the center of D when $w < 0$.

3.4. Connection between Spherical Laplace Transform and Non-Euclidean Fourier Transform

3.4.1. Preparatory Lemmas

Lemma 1. *(i) The Poisson kernel*

$$\mathsf{P}(z,b) = \frac{1-|z|^2}{1+|z|^2 - 2|z|\cos(\psi - \phi)} \qquad \left(z = |z|e^{\mathrm{i}\psi'}, \psi' = \frac{\pi}{2} - \psi; b = e^{\mathrm{i}\phi'}, \phi' = \frac{\pi}{2} - \phi\right),$$

is constant on each horocycle $H_b(z)$ with normal b and passing through $z \in D$.

(ii) The function

$$[\mathsf{P}(z,b)]^\lambda = \left[\frac{1-|z|^2}{1+|z|^2 - 2|z|\cos(\psi - \phi)}\right]^\lambda \qquad (\lambda \in \mathbb{C}), \tag{64}$$

is an eigenfunction of the Laplace–Beltrami operator on the hyperbolic disk D, corresponding to the eigenvalue $\lambda(\lambda-1)$.

(iii) The hyperbolic waves (horocyclic waves) are represented by the following expression:

$$e^{\lambda\langle z,b\rangle} = \left[\frac{1-|z|^2}{1+|z|^2-2|z|\cos(\psi-\phi)}\right]^{\lambda} \qquad (\lambda \in \mathbb{C}), \tag{65}$$

where $\langle z,b\rangle$ *is the signed non-Euclidean distance between the center of the unit disk D and the horocycle* $H_b(z)$ *with normal b that passes through* $z \in D$.

(iv) The following equality holds:

$$\left[\frac{1-|z|^2}{1+|z|^2-2|z|\cos(\psi-\phi)}\right]^{\lambda} = \frac{1}{[\cosh v - \sinh v \cos(\psi-\phi)]^{\lambda}} \qquad (\lambda \in \mathbb{C}). \tag{66}$$

Proof. (i) The level lines of the Poisson kernel $\mathsf{P}(z,b)$ are the circles tangent from within to the unit circle at the point $b = e^{i\phi'}$; when interpreted in a non-Euclidean fashion, they represent horocycles $H_b(z)$ with normal b (p. 7, [26]).

(ii) The Laplace–Beltrami operator Δ_D on the non-Euclidean unit disk D is given by [2,27]

$$\Delta_D = \frac{1}{4}\left[1-(u_1^2+u_2^2)\right]^2 \left(\frac{\partial^2}{\partial u_1^2}+\frac{\partial^2}{\partial u_2^2}\right).$$

For $\lambda \in \mathbb{C}$, a direct computation gives [2,27]

$$\Delta_D\,[\mathsf{P}(z,b)]^{\lambda} = \lambda(\lambda-1)\,[\mathsf{P}(z,b)]^{\lambda} \qquad (\lambda \in \mathbb{C}). \tag{67}$$

(iii) In the Euclidean case, the function $x \mapsto e^{ik(x,\omega)}$, $k \in \mathbb{R}$, $\omega \in \mathbb{S}^{(n-1)}$, $x \in \mathbb{R}^n$, represents a plane wave with normal ω: it is an eigenfunction of the Laplacian in $\mathbb{R}^n$ and is constant on every hyperplane perpendicular to ω. In the case of the non-Euclidean disk D, the geometric analog of the plane wave is the function represented by Equation (65) [2,27]. In fact, it is an eigenfunction of the Laplace–Beltrami operator on D, as proved by statement (ii) (see (67)). Let $z_{\min} = |z_{\min}|e^{i\chi}$ denote the point on the horocycle $H_b(z)$ (i.e., the one with normal $b = e^{i\phi'}$ and passing through z) that is closest to the center of D. Then, $\chi = \phi'$ or $\chi = \phi' + \pi$ depending on whether the origin of D lies, respectively, outside or inside the horocycle $H_b(z)$. Therefore, from the definition of $\langle z,b\rangle$, we have:

$$|\langle z,b\rangle| \doteq \ln\frac{1+|z_{\min}|}{1-|z_{\min}|} = |\ln \mathsf{P}(z_{\min},b)| = |\ln \mathsf{P}(z,b)|,$$

the last equality following from statement (i). If we define $\langle z,b\rangle \doteq \ln \mathsf{P}(z,b)$, then $\langle z,b\rangle$ is indeed constant on each horocycle $H_b(z)$ and represents the hyperbolic analog of (x,ω). Moreover, $\langle z,b\rangle$ is positive if z is such that the origin of D lies outside the horocycle $H_b(z)$, whereas it is negative if the origin falls inside.

(iv) Equality (66) follows by plugging $|z| = \tanh(v/2)$ in formula (64). □

We now introduce the Spherical Functions $\Phi_\lambda(g)$ on G/K ($g \in G = \mathrm{SU}(1,1)$, $K = \mathrm{SO}(2)$).

Definition 1. *The* Spherical Functions *on G/K are defined by [28]*

$$\Phi_\lambda(g) \doteq \int_B \left|\frac{\mathrm{d}(g^{-1}\cdot b)}{\mathrm{d}b}\right|^{\lambda} \mathrm{d}b \qquad (g \in G, \lambda \in \mathbb{C}),$$

where $B = \{z : |z| = 1\}$ is the boundary of the non-Euclidean disk D.

Lemma 2 (Eymard [28]). *The functions $\Phi_\lambda(g)$ satisfy the following properties:*

(i) $\Phi_\lambda(g) = P_{-\lambda}(\cosh v)$, $(g \in G, \lambda \in \mathbb{C}, v \in \mathbb{R})$, *where $P_{-\lambda}(\cdot)$ are the Legendre functions of the first kind.*
(ii) $\Delta_D \Phi_\lambda(g) = \lambda(\lambda - 1)\Phi_\lambda(g)$, $(g \in G, \lambda \in \mathbb{C})$, *where Δ_D is the non-Euclidean Laplace–Beltrami operator.*
(iii) *For $\lambda = -\frac{1}{2} + i\mu$ ($\mu \in \mathbb{R}$):*

$$
\begin{aligned}
&(iii.a) \quad P_{-\frac{1}{2}+i\mu}(\cosh v) = P_{-\frac{1}{2}-i\mu}(\cosh v), \\
&(iii.b) \quad P_{-\frac{1}{2}+i\mu}(\cosh v) = \frac{1}{2\pi}\int_0^{2\pi} \left(\frac{1}{\cosh v - \sinh v \cos(\psi - \phi)}\right)^{\frac{1}{2}-i\mu} d\phi \\
&\qquad = \frac{1}{2\pi}\int_0^{2\pi} \left(\frac{1 - |z|^2}{1 + |z|^2 - 2|z|\cos(\psi - \phi)}\right)^{\frac{1}{2}-i\mu} d\phi \\
&\qquad = \frac{1}{2\pi}\int_B e^{(\frac{1}{2}-i\mu)\langle z,b\rangle}\, db \qquad (z \in D).
\end{aligned}
\tag{68}
$$

Proof. See Ref. [28]. □

Remark 6. *Representation (68) of the conical function is attained by varying within the integral the angle ϕ (or, equivalently, the normal b) in such a way as to span the entire horizon ($\phi \in [0, 2\pi]$). This amounts to considering in the non-Euclidean disk all the families of horocycles which are obtained by varying the normal b on the horizon. On the other hand, we have seen in Section 3.2 that only one family of horocycles (the one with tangency point at $z = -i$) results from the fibration of the upper sheet ${}^2X_2^+$ of the real two-sheeted hyperboloid. However, from formula (68), we see that the same representation of the conical function can be obtained by keeping ϕ fixed and varying ψ in the range $[0, 2\pi]$. Let us recall once again that here we consider only functions of the form $\underline{F} = \underline{F}(\cosh v)$ and therefore not depending on b (see Remark 4(ii)). Consequently, in the non-Euclidean transform (that will be analyzed in the next subsection), the b-dependence of the integrand derives solely from the integral representation (68) of the conical function.*

3.4.2. Non-Euclidean Fourier Transform

It is well known that the classical Fourier transform refers to the decomposition of a function, belonging to an appropriate space, into exponentials of the form e^{ikr} (k real), which can also be viewed as the irreducible unitary representations of the additive group of the real numbers. However, the non-Euclidean disk is not a group. Therefore, a straightforward generalization of this viewpoint is not applicable here. Nevertheless, in view of the fact that the functions $P_{-\lambda}(\cosh v)$ correspond for $\lambda = \frac{1}{2} - i\mu$ to the principal series of the irreducible unitary representations of the group $SU(1,1)$, the exponentials $e^{(\frac{1}{2}-i\mu)\langle z,b\rangle}$ ($\mu \in \mathbb{R}$) (see statement (iii.b) of Lemma 2) represent the analog of the Euclidean exponentials and play the same role in the non-Euclidean Fourier analysis. We can now state the following classical theorem due to Helgason.

Theorem 4 (Helgason [2,27]). *For $f \in C_c^\infty(D)$, let $\widetilde{f}$ denote the Fourier transform*

$$
\widetilde{f}(\mu, b) = \int_D e^{(\frac{1}{2}-i\mu)\langle z,b\rangle} f(z)\, dz \qquad (\mu \in \mathbb{R}, b \in B), \tag{69}
$$

where dz is the invariant surface element on D. Then:

(i) The inverse Fourier transform is given by

$$f(z) = \frac{1}{(2\pi)^2} \int_{\mathbb{R}^+} \int_B e^{(\frac{1}{2}+\mathrm{i}\mu)\langle z,b\rangle} \widetilde{f}(\mu,b)\, \mu \tanh \pi\mu \, \mathrm{d}\mu \, \mathrm{d}b, \tag{70}$$

where $\mathrm{d}b$ *is the angular measure on* B.

(ii) The mapping $f \to \widetilde{f}$ *extends to an isometry of the space* $L^2(D, \mathrm{d}z)$ *onto the space* $L^2(\mathbb{R}^+ \times B, \frac{1}{2\pi}\mu \tanh \pi\mu \, \mathrm{d}\mu \, \mathrm{d}b)$.

Proof. See Refs. [2,27]. □

Remark 7. *(i) Since we are considering functions of the form* $\underline{F} = \underline{F}(\cosh v)$*, we can restrict the integrals* (69) *and* (70) *to the case of radial functions [27] that is, to functions* $f(z) = F(d(0,z))$*,* F *even. In fact, since* $d(0,z) = \ln \frac{1+|z|}{1-|z|}$ *and* $|z| = \tanh(v/2)$*, then* $d(0,z) = v$ *and* $f(z) = F(v) = \underline{F}(\cosh v)$*. Hence,* $\widetilde{f}(\mu,b)$ *is an even function of* μ *alone. In view of statements (iii.a) and (iii.b) of Lemma 2 (see also Remark 6), noting that the expression of the invariant surface element* $\mathrm{d}z$ *in the coordinates* (v,ψ) *is* $\mathrm{d}z = \sinh v \, \mathrm{d}v \, \mathrm{d}\psi$*, and writing* $\widetilde{f}(\mu,b) = 2\pi\widetilde{F}(\mu)$*, then formulas* (69) *and* (70) *read*

$$\widetilde{F}(\mu) = \int_0^{+\infty} \underline{F}(\cosh v)\, P_{-\frac{1}{2}+\mathrm{i}\mu}(\cosh v) \sinh v \, \mathrm{d}v, \tag{71a}$$

$$F(v) = \underline{F}(\cosh v) = \int_0^{+\infty} \widetilde{F}(\mu)\, P_{-\frac{1}{2}+\mathrm{i}\mu}(\cosh v)\, \mu \tanh \pi\mu \, \mathrm{d}\mu, \tag{71b}$$

which coincide with formulas (54)*, i.e., the Mehler transform pair.*

(ii) It is worth noting the close analogy between the non-Euclidean Fourier transform pair (69) *and* (70) *and the Euclidean Fourier transform in the plane* $\mathbb{R}^2$*. The Fourier transform* $\widetilde{f}$ *of a function* f *on* $\mathbb{R}^2$ *is given by*

$$\widetilde{f}(\mu\omega) = \int_{\mathbb{R}^2} f(x) e^{-\mathrm{i}\mu\,(x,\omega)} \, \mathrm{d}x.$$

Then, the Fourier inversion formula, valid, for example, if $f \in C_c^\infty(\mathbb{R}^2)$*, reads*

$$f(x) = \frac{1}{(2\pi)^2} \int_{\mathbb{R}^+} \int_{\mathbb{S}^1} \widetilde{f}(\mu\omega)\, e^{\mathrm{i}\mu\,(x,\omega)}\, \mu \, \mathrm{d}\omega \, \mathrm{d}\mu,$$

$\mathrm{d}\omega$ *denoting the circular measure on* $\mathbb{S}^1$ *(p. 4, [2]).*

3.4.3. Connection between the Spherical Laplace Transform at $\mathrm{Re}\,\lambda = -\frac{1}{2}$ and the Non-Euclidean Fourier Transform

We can now prove the following theorem.

Theorem 5. *Suppose that the sequence* $\{f_n\}_{n=0}^{\infty}$*, where* $f_n = (n+1)^2 a_n$ *(*a_n *being the Legendre coefficients of expansion* (14)*) satisfies the Hausdorff condition* (20)*. Then:*

(i) The spherical Laplace transform at $\mathrm{Re}\,\lambda = -\frac{1}{2}$*, restricted to the odd component (in* μ*) of* $\widetilde{a}(\sigma + \mathrm{i}\mu)$ *(see Section 3.1) reads*

$$\widetilde{F}(\mu) = \int_0^{+\infty} \underline{F}(\cosh v) P_{-\frac{1}{2}+\mathrm{i}\mu}(\cosh v) \sinh v \, \mathrm{d}v, \tag{72}$$

where the equality holds in the sense of the L^2*-norm. Precisely,* $\underline{F}(\cosh v) \in L^2(1, +\infty)$ *and* $\widetilde{F}(\mu)$ *is an even function of* μ *which belongs to* $L^2(\mathbb{R}^+, \mu \tanh \pi\mu \, \mathrm{d}\mu)$.

(ii) The inverse of formula (72) *reads*

$$F(v) = \underline{F}(\cosh v) = \int_0^{+\infty} \widetilde{F}(\mu) P_{-\frac{1}{2}+\mathrm{i}\mu}(\cosh v)\, \mu \tanh \pi\mu \,\mathrm{d}\mu. \tag{73}$$

(iii) The following equality holds:

$$\int_0^{+\infty} \left|\widetilde{F}(\mu)\right|^2 \mu \tanh \pi\mu \,\mathrm{d}\mu = \int_1^{+\infty} |\underline{F}(\cosh v)|^2 \,\mathrm{d}(\cosh v). \tag{74}$$

(iv) The spherical Laplace transform coincides at $\mathrm{Re}\,\lambda = -\frac{1}{2}$ *with the non-Euclidean Fourier transform.*

Proof. We start by proving statements (ii) and (iii) and, successively, we shall prove (i) and (iv).

(ii) In view of the assumptions on $\{f_n\}_{n=0}^{\infty}$, we may use the results of Theorem 3 and from formula (50) with $\sigma \equiv \mathrm{Re}\,\lambda = -\frac{1}{2}$:

$$F(v) = \underline{F}(\cosh v) = \frac{1}{4\pi} \int_{-\infty}^{+\infty} \widetilde{a}\left(-\tfrac{1}{2} + \mathrm{i}\mu\right) (2\mathrm{i}\mu) P_{-\frac{1}{2}+\mathrm{i}\mu}(\cosh v) \,\mathrm{d}\mu. \tag{75}$$

Let $\widetilde{F}(\mu) \doteq \frac{\widetilde{a}(-1/2+\mathrm{i}\mu)}{-\mathrm{i}\pi \tanh \pi\mu}$, as we already did when passing from (53) to (54b). Formula (75) can be rewritten as

$$F(v) = \underline{F}(\cosh v) = \int_0^{+\infty} \widetilde{F}(\mu)\, P_{-\frac{1}{2}+\mathrm{i}\mu}(\cosh v)\, \mu \tanh \pi\mu \,\mathrm{d}\mu, \tag{76}$$

which coincides with the inverse spherical Laplace transform (73). Note that formula (76) coincides with the inversion of the non-Euclidean Fourier transform (70) in the specific case of radial functions.

(iii) First, we prove that the l.h.s. of (74) is convergent. The integral $\int_0^{+\infty} |\widetilde{F}(\mu)|^2 \mu \tanh \pi\mu \,\mathrm{d}\mu$ is split into the integral over $[0, 1]$ plus the integral over $[1, +\infty)$. We have $\int_0^1 |\widetilde{F}(\mu)|^2 \mu \tanh \pi\mu \,\mathrm{d}\mu = \frac{1}{\pi^2} \int_0^1 |\widetilde{a}(-\frac{1}{2} + \mathrm{i}\mu)|^2 \frac{\mu}{\tanh \pi\mu} \,\mathrm{d}\mu$, which is convergent since $\lim_{\mu\to 0}(\mu/\tanh \pi\mu) = 1/\pi$, and $\widetilde{a}(-\frac{1}{2} + \mathrm{i}\mu)$ belongs to $L^2(0, 1)$ for inequality (25). The integral $\int_1^{+\infty} |\widetilde{F}(\mu)|^2 \mu \tanh \pi\mu \,\mathrm{d}\mu$ is convergent since $|\tanh \pi\mu|^{-1} \leqslant M$ (M constant) for $\mu \in [1, +\infty)$ and $\int_1^{+\infty} |\sqrt{\mu}\, \widetilde{a}(-\frac{1}{2} + \mathrm{i}\mu)|^2 \,\mathrm{d}\mu < \int_1^{+\infty} |\mu^2 \widetilde{a}(-\frac{1}{2} + \mathrm{i}\mu)|^2 \,\mathrm{d}\mu < \infty$ in view of statement (ii) of Proposition 2. We have proved above that (76) coincides with the inverse non-Euclidean Fourier transform (70). From statement (ii) of Theorem 4, we know that the mapping $f \to \widetilde{f}$ (which, in our case, corresponds to the mapping $F(v) = \underline{F}(\cosh v) \to \widetilde{F}(\mu)$) is an isometry of $L^2(D, \mathrm{d}z)$ onto $L^2(\mathbb{R}^+ \times B, \frac{1}{2\pi}\mu \tanh \pi\mu \,\mathrm{d}\mu \mathrm{d}b)$. Therefore, the convergence of integral $\int_0^{+\infty} |\widetilde{F}(\mu)|^2 \mu \tanh \pi\mu \,\mathrm{d}\mu$ along with the above mentioned isometry allows us to state equality (74) and that $\underline{F}(\cosh v) \in L^2(1, +\infty)$.

(i) Consider formula (54b), which is the formal expression of the spherical Laplace transform at $\mathrm{Re}\,\lambda = -\frac{1}{2}$ (restricted to its odd component with respect to the variable $\mu \in \mathbb{R}$). We note that it coincides with formula (72) which, by formula (74) proved above, holds as an equality in the sense of the L^2-norm, as specified by statement (i).

(iv) We have indeed proved that formulas (72) and (73) coincide, respectively, with formulas (71a) and (71b), which are the Mehler reduction of the non-Euclidean Fourier transform pair (69) and (70). □

4. Conclusions

In summary, we have proved the following results. Let the Fourier–Legendre expansion $\frac{1}{4\pi}\sum_{n=0}^{\infty}(2n + 1)a_n P_n(\cos\theta)$, and assume the numbers $\{f_n\}_{n=0}^{\infty}$ ($f_n = (n+1)^2 a_n$) to satisfy the Hausdorff-type condition (20). Then:

1. (a) The Fourier–Legendre expansion converges to a function $\underline{f}(\cos u)$ analytic in the interval $-1 < \cos u < 1$.
(b) The function $\underline{f}(\cos u)$ admits a holomorphic extension to the complex $(\cos\theta)$-plane ($\theta = u + iv$) cut along the semi-axis $[1, +\infty)$.
2. (a) The Fourier–Legendre coefficients $\{a_n\}_{n=0}^{\infty}$ are the restrictions to non-negative integers of a transform, called spherical Laplace transform, which is the composition of the ordinary Laplace transform with the Abel transform of $F(v) \equiv \underline{F}(\cosh v)$, which is the jump function across the cut $[1, +\infty)$. Namely, for $\operatorname{Re}\lambda > -\frac{1}{2}$,

$$\widetilde{a}(\lambda) = \int_0^{+\infty} e^{-(\lambda+\frac{1}{2})w}\,(\mathcal{A}F)(w)\,\mathrm{d}w = 2\int_0^{+\infty} \underline{F}(\cosh v)\,Q_\lambda(\cosh v)\sinh v\,\mathrm{d}v, \tag{77}$$

$Q_\lambda(\cdot)$ being the Legendre function of the second kind.
(b) The function $\widetilde{a}(\lambda)$, holomorphic in the half-plane $\operatorname{Re}\lambda > -\frac{1}{2}$, satisfies Carlson's bound and interpolates uniquely the coefficients $\{a_n\}_{n=0}^{\infty}$, that is, $a_n = \widetilde{a}(\lambda)|_{\lambda=n}$.
3. The jump function across the cut admits the following integral representation:

$$\begin{aligned} F(v) = \underline{F}(\cosh v) &= \frac{1}{2\pi}\int_{-\infty}^{+\infty} \widetilde{a}(\sigma + \mathrm{i}\mu)\left(\sigma + \tfrac{1}{2} + \mathrm{i}\mu\right) P_{\sigma+\mathrm{i}\mu}(\cosh v)\,\mathrm{d}\mu \\ &= \frac{1}{2\pi\mathrm{i}}\int_{\sigma-\mathrm{i}\infty}^{\sigma+\mathrm{i}\infty} \widetilde{a}(\lambda)\left(\lambda + \tfrac{1}{2}\right) P_\lambda(\cosh v)\,\mathrm{d}\lambda \qquad \left(\sigma \geqslant -\tfrac{1}{2}\right), \end{aligned} \tag{78}$$

where $P_\lambda(\cdot)$ denotes the Legendre function of the first kind. Representation (78) is the inverse of the spherical Laplace transform (77).
4. (a) For $\operatorname{Re}\lambda = -\frac{1}{2}$ the spherical Laplace transform, restricted to the odd component in μ of $\widetilde{a}(\sigma + \mathrm{i}\mu)$, reads

$$\widetilde{F}(\mu) = \int_0^{+\infty} \underline{F}(\cosh v) P_{-\frac{1}{2}+\mathrm{i}\mu}(\cosh v)\sinh v\,\mathrm{d}v, \tag{79}$$

where $\widetilde{F}(\mu) = \frac{\widetilde{a}^{(\mathrm{O})}(-\frac{1}{2}+\mathrm{i}\mu)}{-\mathrm{i}\pi\tanh\pi\mu}$ ($\widetilde{a}^{(\mathrm{O})}(-\frac{1}{2} + \mathrm{i}\mu)$ denoting the odd component of $\widetilde{a}(-\frac{1}{2} + \mathrm{i}\mu)$). Equality (79) holds in the sense of L^2-norm: $\underline{F}(\cosh v) \in L^2(1, +\infty)$, $\widetilde{F}(\mu) \in L^2(\mathbb{R}^+, \mu\tanh\pi\mu\,\mathrm{d}\mu)$. The inverse of formula (79) reads

$$F(v) = \underline{F}(\cosh v) = \int_0^{+\infty} \widetilde{F}(\mu) P_{-\frac{1}{2}+\mathrm{i}\mu}(\cosh v)\,\mu\tanh\pi\mu\,\mathrm{d}\mu. \tag{80}$$

(b) Formulas (79) and (80) can be written explicitly, passing through Mehler transform, in terms of non-Euclidean Fourier transform as follows:

$$\widetilde{f}(\mu, b) = \int_D e^{(\frac{1}{2}-\mathrm{i}\mu)\langle z,b\rangle} f(z)\,\mathrm{d}z \qquad (\mu \in \mathbb{R}, b \in B), \tag{81}$$

where $\widetilde{f}(\mu, b) = 2\pi\widetilde{F}(\mu)$ and $\mathrm{d}z$ is the invariant surface element on the non-Euclidean disk D. The inverse of (81) is

$$f(z) = \frac{1}{(2\pi)^2}\int_{\mathbb{R}^+}\int_B e^{(\mathrm{i}\mu+\frac{1}{2})\langle z,b\rangle}\widetilde{f}(\mu, b)\,\mu\tanh\pi\mu\,\mathrm{d}\mu\,\mathrm{d}b,$$

where $\mathrm{d}b$ is the angular measure on the boundary B of the unit disk D.
(c) The functions $P_{-\frac{1}{2}+\mathrm{i}\mu}(\cosh v)$ can be represented as follows:

$$P_{-\frac{1}{2}+\mathrm{i}\mu}(\cosh v) = \frac{1}{2\pi}\int_B e^{(\frac{1}{2}-\mathrm{i}\mu)\langle z,b\rangle}\,\mathrm{d}b \qquad (\mu \in \mathbb{R}, z \in D, |z| = \tanh(v/2)),$$

where: $\langle z, b\rangle = \ln \mathsf{P}(z, b)$ is the *signed* non-Euclidean distance from the center of D to the horocycle with normal b passing through $z \in D$, and P denotes the Poisson kernel.

(d) The conical functions $P_{-\frac{1}{2}+i\mu}(\cosh v) = P_{-\frac{1}{2}-i\mu}(\cosh v)$ correspond to the fundamental series of the unitary irreducible representation of the group $SU(1,1)$, which acts transitively on the non-Euclidean disk D.

5. Last but not least, we wish to spend a few words to emphasize the differences between the classical Stein and Wainger approach and ours. First of all, we want to stress the great relevance of the pioneering work of Stein and Wainger; nevertheless, we believe that some remarks are in order.

(a) Stein and Wainger [9] assume that the Legendre coefficients are the restriction to the integers of a function (denoted by $a(s)$ in their notation), which belongs to a space $H^2_*(\operatorname{Re} s > -\frac{1}{2})$. This latter is the space of functions $a(s)$ which belong to $H^2(\operatorname{Re} s > -\frac{1}{2})$ and, for which, in addition, the squared norm $\|a\|^2_* = \int_{-\infty}^{+\infty} |a(-\frac{1}{2}+it) - a(-\frac{1}{2}-it)|^2 \frac{t\,dt}{\tanh \pi t}$ is finite. In their approach, it remained open and rather obscure the following question: How can it be established if the coefficients $\{a_n\}$ are the restriction of a function belonging to $H^2_*(\operatorname{Re} s > -\frac{1}{2})$? Conversely, in our approach, we start directly from the Legendre coefficients, which are required to satisfy a suitable Hausdorff-type condition strictly connected with the moment theory. This second approach seems more direct, especially in the applications (e.g., scattering theory), where only the coefficients of the expansion are known.

(b) A geometrical analysis of the problem (see Section 3 of this paper) is missing in Stein and Wainger's work. Correspondingly, the remarkable results of Helgason on the non-Euclidean Fourier analysis are not mentioned.

(c) In Stein and Wainger's paper, the analytical properties of the Spherical Laplace Transform, as well as its character of being the composition of a Laplace and an Abel-type transform, do not emerge. This also makes the connection between the Mehler transform and the spherical Laplace transform not transparent.

Funding: This research received no external funding.

Conflicts of Interest: The author declares no conflict of interest.

References

1. Furstenberg, H. Translation-invariant cones of functions on semi-simple Lie groups. *Bull. Am. Math. Soc.* **1965**, *71*, 271–326. [CrossRef]
2. Helgason, S. *Groups and Geometric Analysis: Integral Geometry, Invariant Differential Operators, and Spherical Functions*; Academic Press: Orlando, FL, USA, 1984.
3. Hilgert, J.; Ólafsson, G. *Causal Symmetric Spaces: Geometry and Harmonic Analysis*; Academic Press: San Diego, CA, USA, 1997.
4. Faraut, J.; Viano, G.A. Volterra algebra and the Bethe–Salpeter equation. *J. Math. Phys.* **1986**, *27*, 840–848. [CrossRef]
5. Faraut, J. Algèbres de Volterra et Transformation de Laplace Sphérique sur Certains Espaces Symétriques Ordonnés. In *Symposia Mathematica*; Istituto Nazionale di Alta Matematica: Roma, Italy, 1987; Volume 29, pp. 183–196.
6. Faraut, J. Espaces symétriques ordonnés et algèbres de Volterra. *J. Math. Soc. Jpn.* **1991**, *43*, 133–147. [CrossRef]
7. Faraut, J.; Hilgert, J.; Ólafsson, G. Spherical functions and ordered symmetric spaces. *Ann. Inst. Fourier* **1994**, *44*, 927–966. [CrossRef]
8. Widder, D.V. *The Laplace Transform*; Princeton University Press: Princeton, NJ, USA, 1966.
9. Stein, E.M., Wainger, S. Analytic properties of expansions, and some variants of Parseval–Plancherel formulas. *Ark. Mat.* **1965**, *5*, 553–567. [CrossRef]
10. Hermann, R. *Lie Groups for Physicists*; W.A. Benjamin: New York, NY, USA, 1966.

11. Andersen, N.B.; Ólafsson, G. A Paley-Wiener theorem for the spherical Laplace transform on causal symmetric space of rank 1. *Proc. Am. Math. Soc.* **2001**, *129*, 173–179. [CrossRef]
12. Ólafsson, G.; Schlichtkrull, H. A local Paley–Wiener theorem for compact symmetric spaces. *Adv. Math.* **2008**, *129*, 202–215. [CrossRef]
13. Pasquale, A. The Θ-spherical transform and its inversion. *Math. Scand.* **2004**, *95*, 265–284. [CrossRef]
14. Bertram, W. Ramanujan's Master Theorem and duality of symmetric spaces. *J. Funct. Anal.* **1997**, *148*, 117–151. [CrossRef]
15. Gindinkin, S.; Krötz, B. Complex crowns of Riemannian symmetric spaces and non-compactly causal symmetric spaces. *Trans. Am. Math. Soc.* **2002**, *354*, 3299–3327. [CrossRef]
16. Guillopè, L.; Zworski, M. The wave trace for Riemann surfaces. *Geom. Funct. Anal.* **1999**, *9*, 1156–1168. [CrossRef]
17. Hilgert, J.; Pasquale, A.; Przebinda, T. Resonances for the Laplacian on products of two rank one Riemannian symmetric spaces. *J. Funct. Anal.* **2017**, *272*, 1477–1523. [CrossRef]
18. Mazzeo, R.; Vasy, A. Analytic continuation of the resolvent of the Laplacian on $\mathrm{SL}(3) \setminus \mathrm{SO}(3)$. *Am. J. Math* **2004**, *126*, 821–844. [CrossRef]
19. Miatello, R.; Wallach, N.R. The resolvent of the Laplacian on locally symmetric spaces. *J. Differ. Geom.* **1992**, *36*, 663–698. [CrossRef]
20. Strohmaier, A. Analytic continuation of resolvent kernels on noncompact symmetric spaces. *Math. Z.* **2005**, *250*, 411–425. [CrossRef]
21. De Micheli, E.; Viano, G.A. Holomorphic extension associated with Fourier–Legendre expansions. *J. Geom. Anal.* **2002**, *12*, 355–374. [CrossRef]
22. Boas, R.P. *Entire Functions*; Academic Press: New York, NY, USA, 1954.
23. Hoffman, K. *Banach Spaces of Analytic Functions*; Prentice–Hall Inc.: Englewood Cliffs, NJ, USA, 1962.
24. Erdélyi, A. (Ed.) Higher Transcendental Functions. In *Bateman Manuscript Project*; McGraw–Hill: New York, NY, USA, 1953; Volume 1.
25. Mishchenko, A.; Fomenko, A. *A Course of Differential Geometry and Topology*; Mir: Moscow, Russia, 1980.
26. Nevanlinna, R. *Analytic Functions*; Springer: Berlin, Germany, 1970.
27. Helgason, S. Lie Groups and Symmetric Spaces. In *Battelle Rencontres – 1967 Lectures in Mathematics and Physics*; De Witt-Morette, C., Wheeler, J.A., Eds.; Benjamin: New York, NY, USA, 1968; pp. 1–71.
28. Eymard, P. Le noyau de Poisson et la theorie des groupes. In *Symposia Mathematica*; Istituto Nazionale di Alta Matematica: Roma, Italy, 1977; Volume 22, pp. 107–132.

mathematics MDPI

Article

A Deformed Wave Equation and Huygens' Principle

Salem Ben Saïd *, Sara al-Blooshi, Maryam al-Kaabi, Aisha al-Mehrzi and Fatima al-Saeedi

Department of Mathematical Sciences, College of Science, United Arab Emirates University, P. O. Box No. 15551, Al Ain, Abu Dhabi, UAE; 201502530@uaeu.ac.ae (S.a.-B.); 201506313@uaeu.ac.ae (M.a.-K.); 201601854@uaeu.ac.ae (A.a.-M.); 201735617@uaeu.ac.ae (F.a.-S.)

* Correspondence: salem.bensaid@uaeu.ac.ae

Received: 1 October 2019; Accepted: 4 November 2019; Published: 19 December 2019

Abstract: We consider a deformed wave equation where the Laplacian operator has been replaced by a differential-difference operator. We prove that this equation does not satisfy Huygens' principle. Our approach is based on the representation theory of the Lie algebra $\mathfrak{sl}(2,\mathbb{R})$.

Keywords: generalized Fourier transform; deformed wave equation; Huygens' principle; representation of $\mathfrak{sl}(2,\mathbb{R})$

MSC: 43A30; 22E70

1. Introduction

The wave problem consists of the wave equation and some initial data,

$$\Delta u(x,t) = \partial_{tt}u(x,t), \quad u(x,0) = f(x),\ \partial_t u(x,0) = g(x), \qquad \text{for } x \in \mathbb{R}^n \text{ and } t \in \mathbb{R}.$$

This problem is certainly one of the most interesting problems of mathematical physics. Standard techniques involving the Fourier transform show that there are two distributions P_1 and P_2 on $\mathbb{R}^n \times \mathbb{R}$ such that $u = P_1 * f + P_2 * g$. Here $*$ represents the Euclidean convolution product. The distributions P_1 and P_2 are called propagators.

One of the most celebrated features of the wave equation is Huygens' principle: When the dimension n is odd and starting from 3, the propagators are supported entirely on spherical shells. This is the reason why in our three-dimensional word, transmission of signals is possible and we can hear each other. A two-dimensional world would be drastically different from this point of view.

The problem of classifying all second order differential operators which obey Huygens' principle is known as the Hadamard problem [1]. This problem has received a good deal of attention and the literature is extensive (see, for instance, [2–13]). Nevertheless, this problem is still far from being fully solved.

In this paper we will consider a deformed wave equation where the Laplacian Δ is replaced by a certain differential-difference operator. We will prove the non-existence of Huygens' principle for the deformed wave equation for all $n \geqslant 1$. The main tool is the representation theory of the Lie algebra $\mathfrak{sl}(2,\mathbb{R})$.

More precisely, we will consider the deformed wave equation $2\|x\|\Delta_k u(x,t) = \partial_{tt}u(x,t)$ with compactly supported initial data (f,g). Here Δ_k is the differential-difference Dunkl Laplacian (see (2)), where k is a multiplicity function for the Dunkl operators. The operator $\|x\|\Delta_k$ appeared in [14] and played a crucial rule in the study of the so-called $(k,1)$-generalized Fourier transform. When $k \equiv 0$, the deformed wave equation becomes $2\|x\|\Delta u(x,t) = \partial_{tt}u(x,t)$ and the $(0,1)$-generalized Fourier transform reduces to a Hankel type transform on $\mathbb{R}^n$. We refer the reader to [14] for a detailed study on the generalized Fourier transform.

We begin with a straightforward treatment of the Cauchy problem for the deformed wave equation by means of the $(k,1)$-generalized Fourier transform, and derive the existence of propagators $P_{k,1}$ and $P_{k,2}$, in terms of which, the Cauchy problem is solved. Huygens' principle for the deformed wave equation is that $P_{k,1}$ and $P_{k,2}$ are supported entirely on the set $\{(x,t) \in \mathbb{R}^n \times \mathbb{R} : \|x\| - \frac{1}{2}t^2 = 0\}$. It is not a simple task to study the support property from the precise form of the propagators. However, subtler dilatation properties of the propagators allow us to show that Huygens' principle holds true if, and only if, $P_{k,1}$ and $P_{k,2}$ generate a finite dimensional representation of the Lie algebra $\mathfrak{sl}(2,\mathbb{R})$. It is here that the construction of a representation of $\mathfrak{sl}(2,\mathbb{R})$ plays a crucial role. This construction was inspired by [14]. A closer investigation shows that $P_{k,1}$ and $P_{k,2}$ cannot generate finite dimensional representations of $\mathfrak{sl}(2,\mathbb{R})$, and therefore, Huygens' principle does not hold for the deformed wave equation for any $n \geqslant 1$ and any multiplicity function k. The strategy uses proof by contradiction. It is noteworthy mentioning that the case $k \equiv 0$ is already new.

It would be interesting to understand the interpretation(s) of the non-existence of Huygens' principle for the deformed wave equation from the physics point of view. It would also be fascinating to ask whether Huygens's principle holds for other seminal Dunkl-type equations such as the Dunkl–Dirac equation (see [15] for more details about the Dunkl–Dirac operator). For the Euclidean Dirac equation, this problem has been investigated in [16].

2. Background

Let $\langle \cdot, \cdot \rangle$ be the standard Euclidean scalar product in $\mathbb{R}^n$. For $x \in \mathbb{R}^n$, denote $\|x\| = \langle x, x\rangle^{1/2}$.

For α in $\mathbb{R}^n \backslash \{0\}$, we write r_α for the reflection with respect to the hyperplane $\alpha^\perp$ orthogonal to α defined by

$$r_\alpha(x) := x - 2\frac{\langle \alpha, x\rangle}{\langle \alpha, \alpha\rangle}\alpha, \quad x \in \mathbb{R}^n.$$

A finite set $\mathcal{R} \subset \mathbb{R}^n \backslash \{0\}$ is called a root system if $r_\alpha(\mathcal{R}) \subset \mathcal{R}$ for every $\alpha \in \mathcal{R}$. The finite group $G \subset O(n)$ generated by the reflections $\{r_\alpha : \alpha \in \mathcal{R}\}$ is called the finite Coxeter group associated with $\mathcal{R}$. A multiplicity function for G is a function $k : \mathcal{R} \to \mathbb{R}_{\geqslant 0}$ which is constant on G-orbits.

For $1 \leqslant j \leqslant n$, the Dunkl operator is defined in [17] by

$$T_j f(x) = \partial_j f(x) + \frac{1}{2}\sum_{\alpha \in \mathcal{R}} k(\alpha) \frac{f(x) - f(r_\alpha(x))}{\langle \alpha, x\rangle}\langle \alpha, e_j\rangle, \quad x \in \mathbb{R}^n,$$

where ∂_j is the standard directional derivative and $\{e_1, \ldots, e_n\}$ is the canonical orthonormal basis in $\mathbb{R}^n$. The most important property of these operators is that they commute. The operators T_j and ∂_j are intertwined by the following Laplace type operator

$$V_k f(x) = \int_{\mathbb{R}^n} f(y)\, d\mu_x^k(y), \tag{1}$$

where μ_x^k is a unique compactly supported probability measure with $\operatorname{supp}(\mu_x^k) \subset \{y \in \mathbb{R}^n : \|y\| \leqslant \|x\|\}$ (see [17,18]).

The Dunkl Laplacian, which is akin to the Euclidean Laplace operator Δ, is defined by $\Delta_k := T_1^2 + \cdots + T_n^2$ and is given explicitly, for suitable function f, by

$$\Delta_k f(x) = \Delta f(x) + \sum_{\alpha \in \mathcal{R}} k(\alpha) \left(\frac{\langle \nabla f(x), \alpha\rangle}{\langle \alpha, x\rangle} - \frac{\|\alpha\|^2}{2} \frac{f(x) - f(r_\alpha(x))}{\langle \alpha, x\rangle^2} \right), \quad x \in \mathbb{R}^n, \tag{2}$$

where ∇ is the gradient. It is worth mentioning that if $k(\alpha) = 0$ for all $\alpha \in \mathcal{R}$, then Δ_k reduces to the Euclidean Laplacian Δ. We refer the reader to [19] for the theory of Dunkl's operators. This theory, which started with the seminal paper [17], was developed extensively afterwards and continues to receive considerable attention (see, e.g., [20–32]).

Next we will introduce some definitions and results for the generalized Fourier transform; for details we refer to [14]. For $a > 0$, let

$$\Delta_{k,a} := \|x\|^{2-a}\Delta_k - \|x\|^a,$$

where $\|x\|^a$ on the right-hand side of the formula stands for the multiplication operator by $\|x\|^a$. The operator $\Delta_{k,a}$ is symmetric on the Hilbert space $L^2(\mathbb{R}^n, \vartheta_{k,a})$ consisting of square integrable functions against the measure $\vartheta_{k,a}(x)dx := \|x\|^{a-2}\prod_{\alpha\in\mathcal{R}} |\langle\alpha, x\rangle|^{k(\alpha)}dx$.

The (k,a)-generalized Fourier transform $\mathcal{F}_{k,a}$ was defined in [14] to be

$$\mathcal{F}_{k,a} := e^{i\frac{\pi}{2}\left(\frac{n+2\langle k\rangle+a-2}{a}\right)} \exp\left(i\frac{\pi}{2a}\Delta_{k,a}\right),$$

where $\langle k\rangle := \frac{1}{2}\sum_{\alpha\in\mathcal{R}} k(\alpha)$. We pin down that $\mathcal{F}_{k,a}$ is a unitary operator from $L^2(\mathbb{R}^n, \vartheta_{k,a})$ onto itself and the inversion formula is given as

$$\mathcal{F}_{k,\frac{1}{r}}^{-1} = \mathcal{F}_{k,\frac{1}{r}}, \qquad (\mathcal{F}_{k,\frac{2}{2r+1}}^{-1} f)(x) = (\mathcal{F}_{k,\frac{2}{2r+1}} f)(-x), \tag{3}$$

where r is any nonnegative integer. The transform $\mathcal{F}_{k,a}$ reduces to the Euclidean Fourier transform if $k \equiv 0$ and $a = 2$; to the Kobayashi–Mano Hankel transform [33] if $k \equiv 0$ and $a = 1$; to the Dunkl transform [34] if $k > 0$ and $a = 2$. In this paper we consider the case $k > 0$ and $a = 1$. For more details, we refer the reader to ([14] Sections 4 and 5) (see also [35–41]).

Let us collect the main properties of the $(k,1)$-transform $\mathcal{F}_{k,1} := \mathcal{F}_k$. In ([14] Theorem 4.23), the authors proved that for $n + 2\langle k\rangle > 1$, there exists a kernel $B_k(x,y)$ such that for every $f \in L^2(\mathbb{R}^n, \vartheta_{k,1})$,

$$\mathcal{F}_k f(x) = c_k^{-1}\int_{\mathbb{R}^n} f(y)B_k(x,y)\vartheta_{k,1}(y)dy, \quad x \in \mathbb{R}^n,$$

where, for $x = r\theta'$ and $y = t\theta''$, the kernel B_k is given by

$$B_k(x,y) = V_k\left(\widetilde{J}_{\frac{n-3}{2}+\langle k\rangle}\left(\sqrt{2rt(1+\langle\theta',\cdot\rangle)}\right)\right)(\theta'').$$

Here V_k is the Dunkl intertwining operator (1) and $\widetilde{J}_\nu(z)$ is the normalized Bessel function. Above,

$$c_k := \Gamma(n + 2\langle k\rangle - 1)\int_{S^{n-1}}\prod_{\alpha\in\mathcal{R}}|\langle\alpha,\eta\rangle|^{k(\alpha)}d\sigma(\eta).$$

It is noteworthy mentioning that

$$\mathcal{F}_k \circ \|x\| = -\|x\|\Delta_k\mathcal{F}_k, \qquad \mathcal{F}_k \circ (-\|x\|\Delta_k) = -\|x\| \circ \mathcal{F}_k. \tag{4}$$

Recently, in [42] the authors defined a translation operator τ_x, for $x \in \mathbb{R}^n$, on the space $L^1 \cap L^\infty(\mathbb{R}^n, \vartheta_{k,1})$ by

$$\mathcal{F}_k(\tau_x f)(\xi) = B(x,\xi)\mathcal{F}_k(f)(\xi), \qquad \xi \in \mathbb{R}^n.$$

Here are some basic properties of the translation operator:

(i) $\tau_0 = \mathrm{Id}$;
(ii) $\tau_x f(y) = \tau_y f(x)$;
(iii) $\tau_x f_\lambda = (\tau_{\lambda x} f)_\lambda$, where $f_\lambda(x) = f(\lambda x)$ for $\lambda > 0$.

By means of the translation operator, a convolution $\circledast$ on the Schwartz space $\mathcal{S}(\mathbb{R}^n)$ was defined by

$$f \circledast g(x) = c_k^{-1} \int_{\mathbb{R}^n} f(y)\tau_x g(y)\vartheta_{k,1}(y)dy, \quad x \in \mathbb{R}^n.$$

In particular, $f \circledast g = g \circledast f$ and $\mathcal{F}_k(f \circledast g) = \mathcal{F}_k f \cdot \mathcal{F}_k g$ (see [42] for more details).

Next we turn our attention to the convolution of distributions (see [42,43]). Denote by $\mathcal{S}'(\mathbb{R}^n)$ the dual of the Schwartz space $\mathcal{S}(\mathbb{R}^n)$. If $T \in \mathcal{S}'(\mathbb{R}^n)$, then $\mathcal{F}_k(T)$ is defined by

$$\langle \mathcal{F}_k(T), \varphi \rangle := \langle T, \mathcal{F}_k(\varphi) \rangle, \qquad \forall \varphi \in \mathcal{S}(\mathbb{R}^n).$$

It is worth mentioning that $\mathcal{S}(\mathbb{R}^n)$ is stable by $\mathcal{F}_k$ (see [36]). The convolution $T \circledast f$ of $T \in \mathcal{S}'(\mathbb{R}^n)$ and $f \in \mathcal{S}(\mathbb{R}^n)$ is defined in [42] by

$$T \circledast f(x) = \langle T, \tau_x f \rangle.$$

In particular, a result analogous to the Euclidean convolution shows that $T \circledast f \in \mathcal{S}'(\mathbb{R}^n) \cap C^\infty(\mathbb{R}^n)$ and $\mathcal{F}_k(T \circledast f) = \mathcal{F}_k T \cdot \mathcal{F}_k f$.

3. The Deformed Wave Equation and Huygens' Principle

For $n + 2\langle k \rangle - 1 > 0$, where $\langle k \rangle = \frac{1}{2}\sum_{\alpha \in \mathcal{R}} k(\alpha)$, we consider the following Cauchy problem for the wave equation

$$\begin{aligned} 2\|x\|\Delta_k^x u_k(x,t) &= \partial_{tt} u_k(x,t), \quad (x,t) \in \mathbb{R}^n \times \mathbb{R}, \\ u_k(x,0) &= f(x), \ \ \partial_t u_k(x,0) = g(x), \end{aligned} \tag{5}$$

where the functions f and g belong to the space $\mathcal{D}(\mathbb{R}^n)$ of smooth functions with compact support. Here the superscript in Δ_k^x indicates the relevant variable, while the subscript t indicates differentiation in the t-variable. Next, we will prove the following statements:

($\mathcal{S}$) Let $u_k(x,t)$, $x \in \mathbb{R}^n$ and $t \in \mathbb{R}$, satisfy $2\|x\|\Delta_k^x u_k(x,t) - \partial_{tt} u_k(x,t) = 0$, then u_k does not satisfy Huygen's principle. In other words, the solution u_k is expressed as a sum of $\circledast$-convolution of f and g with distributions that are not supported entirely on the set $\{(x,t) \in \mathbb{R}^n \times \mathbb{R} : \|x\| = \frac{1}{2}t^2\}$.

For $t \in \mathbb{R}$, denote by $P_{k,t}$ the 2×2 matrix of tempered distributions on $\mathbb{R}^n$

$$P_{k,t} = \begin{pmatrix} P_{k,t}^{11} & P_{k,t}^{12} \\ P_{k,t}^{21} & P_{k,t}^{22} \end{pmatrix} := \begin{pmatrix} \mathcal{F}_k\big(\cos(t\sqrt{2\|\cdot\|})\big) & \mathcal{F}_k\big(\sin(t\sqrt{2\|\cdot\|})/\sqrt{2\|\cdot\|}\big) \\ \mathcal{F}_k\big(-\sqrt{2\|\cdot\|}\sin(t\sqrt{2\|\cdot\|})\big) & \mathcal{F}_k\big(\cos(t\sqrt{2\|\cdot\|})\big) \end{pmatrix}. \tag{6}$$

Set $U_k(x,0) := \begin{pmatrix} f(x) \\ g(x) \end{pmatrix}$, where the initial data $(f,g) \in \mathcal{D}(\mathbb{R}^n) \times \mathcal{D}(\mathbb{R}^n)$. Thus, we may define the vector column $U_k(x,t)$ by

$$\begin{aligned} U_k(x,t) &:= \left\{P_{k,t} \circledast U_k(\cdot,0)\right\}(x) \\ &= \left\{\begin{pmatrix} P_{k,t}^{11} & P_{k,t}^{12} \\ P_{k,t}^{21} & P_{k,t}^{22} \end{pmatrix} \circledast \begin{pmatrix} f \\ g \end{pmatrix}\right\}(x). \end{aligned} \tag{7}$$

By applying the Fourier transform $\mathcal{F}_k$ to (7), in the x-variable, we get

$$\mathcal{F}_k(U_k(\cdot,t))(\xi) = e^{t\mathbb{A}}\mathcal{F}_k(U_k(\cdot,0))(\xi), \tag{8}$$

where

$$\mathbb{A} := \begin{pmatrix} 0 & 1 \\ -2\|\xi\| & 0 \end{pmatrix}. \tag{9}$$

Above we have used the fact that $\mathcal{F}_k^{-1} = \mathcal{F}_k$ (see (3)). That is $\mathcal{F}_k(U_k(\cdot,t))(\xi)$ is a solution to the following ordinary differential equation

$$\partial_t \mathcal{F}_k(U_k(\cdot,t))(\xi) = \mathbb{A}\mathcal{F}_k(U_k(\cdot,t))(\xi) = \begin{pmatrix} 0 & 1 \\ -2\|\xi\| & 0 \end{pmatrix} \mathcal{F}_k(U_k(\cdot,t))(\xi). \tag{10}$$

Now, recall from (4) that $-\|\xi\|\mathcal{F}_k(f)(\xi) = \mathcal{F}_k(\|x\|\Delta_k f)(\xi)$, and using the injectivity of the Fourier transform $\mathcal{F}_k$, we deduce that

$$\partial_t U_k(x,t) = \begin{pmatrix} 0 & 1 \\ 2\|x\|\Delta_k & 0 \end{pmatrix} U_k(x,t). \tag{11}$$

Hence, if we write $U_k(x,t) = \begin{pmatrix} u_k(x,t) \\ v_k(x,t) \end{pmatrix}$, then $u_k(x,t)$ satisfies the following equation

$$\partial_{tt} u_k(x,t) = 2\|x\|\Delta_k u_k(x,t).$$

Moreover, since $f, g \in \mathcal{D}(\mathbb{R}^n)$, it follows from (7) and the properties of the $\circledast$-convolution that $u_k(\cdot,t) \in C^\infty(\mathbb{R}^n)$ for all $t \in \mathbb{R}$.

Furthermore, $u_k(x,t) \to f(x)$ as $t \to 0$. Indeed, if δ denotes the Dirac functional, then, as $t \to 0$, $\mathcal{F}_k(\cos(t\sqrt{2\|\cdot\|})) \to \delta$ in $\mathcal{S}'(\mathbb{R}^n)$ and thus in $\mathcal{D}'(\mathbb{R}^n)$. On the other hand $\mathcal{F}_k(\sin(t\sqrt{2\|\cdot\|})/\sqrt{2\|\cdot\|}) \to 0$ as $t \to 0$. Using the continuity of the convolution $\circledast$, we deduce that

$$u_k(x,t) \to (\delta \circledast f)(x) = f(x) \qquad \text{as} \quad t \to 0.$$

Similarly, one can prove that $(\partial_t u_k)(x,t) \to g(x)$ as $t \to 0$.

We mention that the solution u_k constructed above is unique. To prove this claim, we need the lemma below. Let

$$\mathcal{E}_k[u_k](t) = \int_{\mathbb{R}^n} \left\{ |\partial_t \mathcal{F}_k(u_k(\cdot,t))(\xi)|^2 + 2\|\xi\||\mathcal{F}_k(u_k(\cdot,t))(\xi)|^2 \right\} \vartheta_{k,1}(\xi) d\xi. \tag{12}$$

Lemma 1. *Assume that $n + 2\langle k\rangle - 1 > 0$ and that the initial data $f, g \in \mathcal{D}(\mathbb{R}^N)$. Then the total energy $\mathcal{E}_k[u_k]$ is independent of t.*

Proof. Since

$$\mathcal{F}_k(u_k(\cdot,t))(\xi) = \cos(t\sqrt{2\|\xi\|})\mathcal{F}_k f(\xi) + \frac{\sin(t\sqrt{2\|\xi\|})}{\sqrt{2\|\xi\|}}\mathcal{F}_k g(\xi), \qquad \text{for all} \quad t \in \mathbb{R},$$

we deduce that

$$\begin{aligned} |\mathcal{F}_k(u_k(\cdot,t))(\xi)|^2 &= \cos^2(t\sqrt{2\|\xi\|})|\mathcal{F}_k f(\xi)|^2 + \frac{\sin^2(t\sqrt{2\|\xi\|})}{2\|\xi\|}|\mathcal{F}_k g(\xi)|^2 \\ &\quad + \sqrt{2}\frac{\cos(t\sqrt{\|\xi\|})\sin(t\sqrt{\|\xi\|})}{\sqrt{\|\xi\|}}\mathrm{Re}\left(\mathcal{F}_k f(\xi)\overline{\mathcal{F}_k g(\xi)}\right), \end{aligned}$$

and

$$\begin{aligned} |\partial_t \mathcal{F}_k(u_k(\cdot,t))(\xi)|^2 &= \cos^2(t\sqrt{2\|\xi\|})|\mathcal{F}_k g(\xi)|^2 + 2\|\xi\|\sin^2(t\sqrt{2\|\xi\|})|\mathcal{F}_k f(\xi)|^2 \\ &\quad - 2\sqrt{2\|\xi\|}\cos(t\sqrt{2\|\xi\|})\sin(t\sqrt{2\|\xi\|})\mathrm{Re}\left(\mathcal{F}_k f(\xi)\overline{\mathcal{F}_k g(\xi)}\right). \end{aligned}$$

Thus we have

$$\mathcal{E}_k[u_k](t) = \int_{\mathbb{R}^n} \left\{2\|\xi\||\mathcal{F}_k f(\xi)|^2 + |\mathcal{F}_k g(\xi)|^2\right\} \vartheta_{k,1}(\xi) d\xi. \tag{13}$$

Hence, we established the lemma. □

Now let us go back to the uniqueness of the solution u_k. Assume that $u_k^{(1)}$ and $u_k^{(2)}$ are two solutions of the wave equation with the same initial data, then $u_k^{(1)} - u_k^{(2)}$ is a solution of the wave equation with zero initial data. Therefore, by (13), we have $\mathcal{E}_k[u_k^{(1)} - u_k^{(2)}](t) = 0$. Hence, (12) implies $\partial_t \mathcal{F}_k((u_k^{(1)} - u_k^{(2)})(\cdot, t))(\xi) = 0$ for every $t \in \mathbb{R}$. That is, the function $t \mapsto \mathcal{F}_k((u_k^{(1)} - u_k^{(2)})(\cdot, t))(\xi)$ is a constant, which implies $\mathcal{F}_k((u_k^{(1)} - u_k^{(2)})(\cdot, t))(\xi) = \mathcal{F}_k((u_k^{(1)} - u_k^{(2)})(\cdot, 0))(\xi) = 0$. Using the injectivity of the Fourier transform $\mathcal{F}_k$, we deduce that $(u_k^{(1)} - u_k^{(2)})(x, t) = 0$ for all $x \in \mathbb{R}^n$ and $t > 0$. This proves that the solutions of the wave equation are uniquely determined by the initial Cauchy data.

The following theorem collects all the above facts and discussions.

Theorem 1. *The solution to the Cauchy problem (5) is given uniquely by*

$$u_k(x, t) = (P_{k,t}^{11} \circledast f)(x) + (P_{k,t}^{12} \circledast g)(x),$$

where, for a fixed t, $P_{k,t}^{11}$ and $P_{k,t}^{12}$ are the tempered distributions on $\mathbb{R}^n$ given by

$$P_{k,t}^{11} = \mathcal{F}_k\big(\cos(t\sqrt{2\|\cdot\|})\big), \qquad P_{k,t}^{12} = \mathcal{F}_k\big(\sin(t\sqrt{2\|\cdot\|})/\sqrt{2\|\cdot\|}\big).$$

The distributions $P_{k,t}^{ij}$ will be called the propagators.

We shall now prove the statement $(\mathcal{S})$. To do so, we will assume that the propagators $P_{k,t}^{11}$ and $P_{k,t}^{12}$ are supported entirely on the set $\mathcal{C} = \{(x, t) \in \mathbb{R}^n \times \mathbb{R} : \|x\| = \frac{1}{2}t^2\}$ and we will show that this assumption cannot hold. Our approach uses the representation theory of the Lie algebra $\mathfrak{sl}(2, \mathbb{R})$, following [43,44].

Assume that the propagators $P_{k,t}^{11}$ and $P_{k,t}^{12}$ are supported entirely on the set $\mathcal{C} = \{(x, t) \in \mathbb{R}^n \times \mathbb{R} : \|x\| = \frac{1}{2}t^2\}$.

We start by investigating certain properties of the wave equation, which are reflected in properties of the propagators. To see this, we define the 2×2 matrix $P_k = \begin{pmatrix} P_k^{11} & P_k^{12} \\ P_k^{21} & P_k^{22} \end{pmatrix}$ of entrywise distributions on $\mathbb{R}^{n+1}$, where

$$P_k^{ij}(\psi_1 \otimes \psi_2) := \int_{\mathbb{R}} P_{k,t}^{ij}(\psi_1)\psi_2(t) dt, \qquad i, j = 1, 2,$$

for $\psi_1 \in \mathcal{S}(\mathbb{R}^n)$ and $\psi_2 \in \mathcal{S}(\mathbb{R})$. Here we used the fact that $\mathcal{S}(\mathbb{R}^{n+1}) \simeq \mathcal{S}(\mathbb{R}^n)\widehat{\otimes}\mathcal{S}(\mathbb{R})$ is the unique topological tensor product of $\mathcal{S}(\mathbb{R}^n)$ and $\mathcal{S}(\mathbb{R})$ as nuclear spaces. From the constructive proof of Theorem 1, it follows that

$$2\|x\|\Delta_k P_k^{ij} = \partial_{tt} P_k^{ij}, \qquad i, j = 1, 2.$$

Next, we will investigate the dilations of the propagators under a dilation operator. This will inform us on the degree of the "homogeneity" of the distributions P_k^{ij}, with $i, j = 1, 2$. For $\lambda > 0$ and a function ψ on $\mathbb{R}^{n+1}$, let

$$S_\lambda^x \psi(x, t) := \psi(\lambda^2 x, t), \qquad S_\lambda^t \psi(x, t) := \psi(x, \lambda t),$$

where the superscript denotes the relevant variable. Set $S_\lambda := S_\lambda^x \circ S_\lambda^t$. By duality, the operators S_λ^x, S_λ^t, and S_λ act on distributions in the standard way.

We begin by looking to the properties of $P_{k,t}^{ij}$ under the dilation S_λ. Observe that if $u_k(x,t)$ is a solution to (5) with initial data $(f(x), g(x))$, then $S_\lambda u_k(x,t)$ solves the wave equation with initial data $(S_\lambda^x f(x), \lambda S_\lambda^x g(x))$. Thus

$$S_\lambda U_k(x,t) = P_{k,t} \circledast \begin{bmatrix} S_\lambda^x f \\ \lambda S_\lambda^x g \end{bmatrix}. \tag{14}$$

On the other hand

$$\begin{aligned} S_\lambda U_k(x,t) = \begin{bmatrix} S_\lambda u_k(x,t) \\ \partial_t\{S_\lambda u_k(x,t)\} \end{bmatrix} &= \begin{bmatrix} u_k(\lambda^2 x, \lambda t) \\ \lambda\{\partial_t u_k\}(\lambda^2 x, \lambda t) \end{bmatrix} \\ &= \begin{bmatrix} u_k \\ \lambda \partial_t u_k \end{bmatrix}(\lambda^2 x, \lambda t) \\ &= \begin{bmatrix} 1 & 0 \\ 0 & \lambda \end{bmatrix}\begin{bmatrix} u_k \\ \partial_t u_k \end{bmatrix}(\lambda^2 x, \lambda t) \\ &= \begin{bmatrix} 1 & 0 \\ 0 & \lambda \end{bmatrix}\left\{P_{k,\lambda t} \circledast \begin{bmatrix} f \\ g \end{bmatrix}\right\}(\lambda^2 x) \\ &= \begin{bmatrix} 1 & 0 \\ 0 & \lambda \end{bmatrix} S_\lambda^x\left\{P_{k,\lambda t} \circledast \begin{bmatrix} f \\ g \end{bmatrix}\right\}(x). \end{aligned}$$

Using the fact that S_λ^x preserves the convolution of a distribution with a function, a fact that can be easily checked using the properties of the translation operator, we get

$$\begin{aligned} S_\lambda U_k(x,t) &= \begin{bmatrix} 1 & 0 \\ 0 & \lambda \end{bmatrix}\left\{S_\lambda^x P_{k,\lambda t} \circledast \begin{bmatrix} S_\lambda^x f \\ S_\lambda^x g \end{bmatrix}\right\}(x) \\ &= \begin{bmatrix} 1 & 0 \\ 0 & \lambda \end{bmatrix}\left\{S_\lambda^x P_{k,\lambda t} \circledast \begin{bmatrix} 1 & 0 \\ 0 & \lambda^{-1} \end{bmatrix}\begin{bmatrix} S_\lambda^x f \\ \lambda S_\lambda^x g \end{bmatrix}\right\}(x). \end{aligned} \tag{15}$$

Comparing (14) with (15) gives $S_\lambda^x P_{k,\lambda t}^{ij} = \lambda^{j-i} P_{k,t}^{ij}$, for $i,j = 1,2$. Now we can obtain the dilation properties of P_k^{ij} as follows: For $\psi_1 \in \mathcal{S}(\mathbb{R}^n)$ and $\psi_2 \in \mathcal{S}(\mathbb{R})$, we have

$$\begin{aligned} S_\lambda(P_k^{ij})(\psi_1 \otimes \psi_2) &= P_k^{ij}(S_{\lambda^{-1}}^x(\psi_1) \otimes S_{\lambda^{-1}}^t(\psi_2)) \\ &= \int_{\mathbb{R}} P_{k,t}^{ij}(S_{\lambda^{-1}}^x(\psi_1)) S_{\lambda^{-1}}^t(\psi_2)(t)dt \\ &= \lambda \int_{\mathbb{R}} P_{k,\lambda t}^{ij}(S_{\lambda^{-1}}^x(\psi_1))\psi_2(t)dt \\ &= \lambda \int_{\mathbb{R}} S_\lambda^x(P_{k,\lambda t}^{ij}(\psi_1))\psi_2(t)dt \\ &= \lambda^{1+j-i} \int_{\mathbb{R}} P_{k,t}^{ij}(\psi_1)\psi_2(t)dt \\ &= \lambda^{1+j-i} P_k^{ij}(\psi_1 \otimes \psi_2). \end{aligned}$$

We summarize the above computations.

Proposition 1. *For $n + 2\langle k\rangle - 1 > 0$, we have*

(1) *The distribution P_k^{ij} satisfies the deformed wave equation, i.e.,*

$$(\|x\|\Delta_k - \frac{1}{2}\partial_{tt})P_k^{ij} = 0, \qquad i,j = 1,2. \tag{16}$$

(2) *For $\lambda > 0$,*

$$S_\lambda P_k^{ij} = \lambda^{1+j-i} P_k^{ij}, \qquad i,j = 1,2.$$

Next we shall describe the structure of a representation of the Lie algebra $\mathfrak{sl}(2,\mathbb{R})$ on $\mathcal{S}(\mathbb{R}^{n+1})$. This structure, together with Proposition 1, will allow us to prove that the Assumption 3 does not hold true.

We take a basis for the Lie algebra $\mathfrak{sl}(2,\mathbb{R})$ as

$$\mathbf{e} := \begin{pmatrix} 0 & 1 \\ 0 & 0 \end{pmatrix}, \qquad \mathbf{f} := \begin{pmatrix} 0 & 0 \\ 1 & 0 \end{pmatrix}, \qquad \mathbf{h} := \begin{pmatrix} 1 & 0 \\ 0 & -1 \end{pmatrix}.$$

The triple $\{\mathbf{e}, \mathbf{f}, \mathbf{h}\}$ satisfies the commutation relations

$$[\mathbf{e}, \mathbf{f}] = \mathbf{h}, \qquad [\mathbf{h}, \mathbf{e}] = 2\mathbf{e}, \qquad [\mathbf{h}, \mathbf{f}] = -2\mathbf{f},$$

where $[A, B] := AB - BA$.

Choose $x_1, x_2, \ldots, x_n$ as the usual system of coordinates on $\mathbb{R}^n$. Let

$$\mathbb{E}_{n,1} := i(\|x\| - \frac{1}{2}t^2), \quad \mathbb{F}_{n,1} := i(\|x\|\Delta_k - \frac{1}{2}\partial_{tt}), \quad \mathbb{H}_{n,1} := n + 2\langle k \rangle - \frac{1}{2} + 2\sum_{\ell=1}^{n} x_\ell \partial_\ell + t\partial_t.$$

Using ([14] Theorem 3.2), the following commutation relations hold

$$[\mathbb{E}_{n,1}, \mathbb{F}_{n,1}] = \mathbb{H}_{n,1}, \quad [\mathbb{H}_{N,1}, \mathbb{E}_{N,1}] = 2\mathbb{E}_{N,1}, \quad [\mathbb{H}_{n,1}, \mathbb{F}_{n,1}] = -2\mathbb{F}_{n,1}. \tag{17}$$

These are the commutation relations of a standard basis of the Lie algebra $\mathfrak{sl}(2,\mathbb{R})$. Equation (17) gives rise to a representation ω_k of the Lie algebra $\mathfrak{sl}(2,\mathbb{R})$ on the Schwartz space $\mathcal{S}(\mathbb{R}^{n+1})$ by setting

$$\omega_k(\mathbf{h}) = \mathbb{H}_{n,1}, \qquad \omega_k(\mathbf{e}) = \mathbb{E}_{n,1}, \qquad \omega_{k,a}(\mathbf{f}) = \mathbb{F}_{n,1}. \tag{18}$$

An analogue of the representation ω_k has been intensively studied in [14].

Recall that the Huygens' principle is equivalent to the fact that the propagators P_k^{11} and P_k^{12} are supported on the set $C = \{(x,t) \in \mathbb{R}^n \times \mathbb{R} \mid \|x\| - \frac{1}{2}t^2 = 0\}$. Since C is the locus of zeros of $\|x\| - \frac{1}{2}t^2$, then, P_k^{ij} is supported on C if and only if

$$\mathbb{E}_{n,1}^m \cdot P_k^{ij} = 0 \tag{19}$$

for some positive integer m (see, for instance, ([44] p. 173)). In the light of Proposition 1(1) together with the dilatation property of P_k^{ij}, which implies that P_k^{ij} is an eigen-distribution for $\mathbb{H}_{n,1}$, Equation (19) amounts to saying the distribution P_k^{ij} generates a finite-dimensional representation ω_k^* for $\mathfrak{sl}(2,\mathbb{R})$ on $\mathcal{S}'(\mathbb{R}^{n+1})$. Thus, the qualitative part of Huygens' principle holds.

Theorem 2. *Huygens' principle holds if and only if P_k^{ij} is supported on the set C, if and only if P_k^{ij} generates a finite-dimensional representation ω_k^* for $\mathfrak{sl}(2,\mathbb{R})$ on $\mathcal{S}'(\mathbb{R}^{n+1})$.*

Theorem 3. *Huygens' principle cannot hold when*

$$n - 1 + 2\langle k \rangle \notin \mathbb{Z}.$$

Proof. In ([14] Theorem 3.21) the authors proved that the spectrum of the element $\mathbf{k} := i(\mathbf{f} - \mathbf{e})$ acting on $\mathcal{S}'(\mathbb{R}^{n+1})$ via the dual representation ω_k^* is $n - 1 + 2\langle k \rangle + 2\mathbb{Z}$, whereas, it is well known, the spectrum of $\mathbf{k}$ in finite-dimensional representations of $\mathfrak{sl}(2,\mathbb{R})$ is contained in $\mathbb{Z}$. □

The above theorem leaves the possibility that the wave equation may satisfy Huygens' principle when $n-1+2\langle k\rangle \in \mathbb{Z}$.

Now, using Proposition 1 (2), we get

$$\Big\{2\sum_{\ell=1}^{n} x_\ell \partial_\ell + t\partial_t\Big\} P_k^{ij} = (1+j-i)P_k^{ij}.$$

Therefore

$$\mathbb{H}_{n,1} P_k^{ij} = -\left(n+2\langle k\rangle - \frac{1}{2} + i - j - 1\right) P_k^{ij}, \qquad i,j = 1,2.$$

That is P_k^{ij} is an eigendistributation for $\mathbb{H}_{n,1}$ with eigenvalue $-(n+2\langle k\rangle - \frac{1}{2} + i - j - 1)$. Keeping in mind the fact that $\mathbb{F}_{n,1} \cdot P_k^{ij} = 0$, and in the light of Theorem 3, clearly each distribution P_k^{ij} cannot generate a finite-dimensional ω_k^* for $\mathfrak{sl}(2,\mathbb{R})$ on $\mathcal{S}'(\mathbb{R}^{n+1})$; otherwise $n+2\langle k\rangle - \frac{1}{2} + i - j - 1 \in \mathbb{Z}$ which is impossible in view of Theorem 3. That is our Assumption 3 does not hold true.

Theorem 4. *The solution $u_k(x,t)$ to the Cauchy problem (5) does not satisfy the Huygens' principle.*

Author Contributions: Investigation, S.a.-B., M.a.-K., A.a.-M., F.a.-S.; methodology, S.B.S.; supervision, S.B.S.; writing–original draft preparation, S.B.S.; writing–review and editing, S.B.S. All authors have read and agreed to the published version of the manuscript.

Funding: This research was funded by United Arab Emirates University (UAEU), SURE Plus 2019 grant # G00003109.

Acknowledgments: The authors would like to thankfully acknowledge the financial support awarded by UAEU through the SURE Plus 2019 grant # G00003109. We thank the referees for their valuable comments.

Conflicts of Interest: The authors declare no conflict of interest.

References

1. Hadamard, J. *Lectures on Cauchy's Problem in Linear Partial Differential Equations;* Dover Publications: New York, NY, USA, 1923.
2. Ben Saïd, S. Huygens' principle for the wave equation associated with the trigonometric Dunkl-Cherednik operators. *Math. Res. Lett.* **2006**, *13*, 43–58. [CrossRef]
3. Berest, Y.; Veselov, A.P. Huygens' principle and integrability. *Russ. Math. Surv.* **1994**, *49*, 5–77. [CrossRef]
4. Branson, T.; Ólafsson, G.; Schlichtkrull, H. Huygens' Principle in Riemannian Symmetric Spaces. *Math. Ann.* **1995**, *301*, 445-462. [CrossRef]
5. Courant, R.; Hilbert, D. *Methods of Mathematical Physics;* Interscience Publ.: New York, NY, USA, 1962; Volume II.
6. Chalykh, O.A.; Feigin, M.V.; Veselov, A.P. New integrable generalizations Calogero-Moser Quantumproblem. *J. Math. Phys.* **1998**, *39*, 695–703. [CrossRef]
7. Günther, P. *Huygens' Principle and Hyperbolic equations;* Academic Press: Boston, MA, USA, 1988.
8. Lagnese, J.E.; Stellmacher, K.L. A method of generating classes of Huygens's operators. *J. Math. Mech.* **1967**, *17*, 461–472.
9. Lax, P.D.; Phillips, R.S. *Scattering Theory;* Academic Press: New York, NY, USA; London, UK, 1967.
10. McLenaghan, R.G. Huygens' principle. *Ann. Inst. Henri Poincaré* **1982**, *37*, 211–236.
11. Ørsted, B. The conformal invariance of Huygens' principle. *J. Diff. Geom.* **1981**, *16*, 1–9. [CrossRef]
12. Stellmacher, K. Ein Beispiel einer Huygensschen Differentialgleichung. *Nachr. Akad. Wiss. Göttingen Math.* **1953**, *10*, 133–138.
13. Schimming, R. A review on Huygens' principle for linear hyperbolic differential operators. In Proceedings of the International Symposium Group-Theoretical Methods in Mechanics, Novosibirsk, Russia, 25–29 August 1978;
14. Ben Saïd, S.; Kobayashi, T.; Ørsted, B. Laguerre semigroup and Dunkl operators. *Compos. Math.* **2012**, *148*, 1265–1336. [CrossRef]

15. de Bie, H.; Genest, V.; Vinet, L. Dirac-Dunkl equation on S2 and the Bannai-Ito algebra. *Comm. Math. Phys.* **2016**, *344*, 447–464. [CrossRef]
16. Pascazio, S.; Pepe, F.V.; Pérez-Pardo, J.M. Huygens' principle and Dirac-Weyl equation. *Eur. Phys. J. Plus* **2017**, *132*, 287. [CrossRef]
17. Dunkl, C.F. Differential-difference operators associated to reflection groups. *Trans. Am. Math. Soc.* **1989**, *311*, 167–183. [CrossRef]
18. Rösler, M. Positivity of Dunkl's intertwining operator. *Duke Math. J.* **1999**, *98*, 445–463. [CrossRef]
19. Dunkl, C.F.; Xu, Y. *Orthogonal Polynomials of Several Variables*; Cambridge Univ. Press: Cambridge, UK, 2001.
20. Amri, B.; Gaidi, M. $L^p - L^q$ estimates for the solution of the Dunkl wave equation. *Manuscripta Math.* **2019**, *159*, 379–396. [CrossRef]
21. Dai, F.; Ye, W. Almost everywhere convergence of Bochner-Riesz means with critical index for Dunkl transforms. *J. Approx. Theory* **2016**, *205*, 43–59. [CrossRef]
22. Deleaval, L. On the boundedness of the Dunkl spherical maximal operator. *J. Topol. Anal.* **2016**, *8*, 475–495. [CrossRef]
23. Gorbachev, D.V.; Ivanov, V.I.; Tikhonov, S.Y. Positive L^p-bounded Dunkl-type generalized translation operator and its applications. *Constr. Approx.* **2019**, *49*, 555–605. [CrossRef]
24. De Jeu, M. The Dunkl transform. *Invent. Math.* **1993**, *113*, 147–162. [CrossRef]
25. Mejjaoli, H. Dunkl-Schrödinger semigroups and applications. *Appl. Anal.* **2013**, *92*, 1597–1626. [CrossRef]
26. Mejjaoli, H.; Trimèche, K. On a mean value property associated with the Dunkl Laplacian operator and applications. *Integral Transform. Spec. Funct.* **2001**, *12*, 279–302. [CrossRef]
27. Mejjaoli, H. Nonlinear generalized Dunkl-wave equations and applications. *J. Math. Anal. Appl.* **2011**, *375*, 118–138. [CrossRef]
28. Mejjaoli, H. Strichartz estimates for the Dunkl wave equation and application. *J. Math. Anal. Appl.* **2008**, *346*, 41–54. [CrossRef]
29. Thangavelu, S.; Xu, Y. Convolution operator and maximal function for the Dunkl transform. *J. Anal. Math.* **2005**, *97*, 25–55. [CrossRef]
30. Trimèche, K. Paley-Wiener theorems for the Dunkl transform and Dunkl translation operators. *Integral Transform. Spec. Funct.* **2002**, *13*, 17–38. [CrossRef]
31. Trimèche, K. Harmonic analysis on measures spaces attached to some Dunkl operators on $\mathbb{R}^d$ and applications. *J. Inequal. Spec. Funct.* **2017**, *8*, 104–117.
32. Gallardo, L.; Rejeb, C. A new mean value property for harmonic functions relative to the Dunkl Laplacian operator and applications. *Trans. Am. Math. Soc.* **2015**, *368*, 3727–3753. [CrossRef]
33. Kobayashi, T.; Mano, G. The inversion formula and holomorphic extension of the minimal representation of the conformal group. In *Harmonic Analysis, Group Representations, Automorphic Forms and Invariant Theory: In Honor of Roger Howe*; Li, J.S., Tan, E.C., Wallach, N., Zhu, C.B., Eds.; World Scientific: Singapore, 2007; pp. 159–223.
34. Dunkl, C.F. Hankel transforms associated to finite reflection groups. *Contemp. Math.* **1992**, *138*, 123–138.
35. Mejjaoli, H. Wavelet-multipliers analysis in the framework of the k-Laguerre theory. *Linear Multilinear Algebra* **2019**, *67*, 70–93. [CrossRef]
36. Johansen, T.R. Weighted inequalities and uncertainty principles for the (k, a)-generalized Fourier transform. *Internat. J. Math.* **2016**, *27*, 44. [CrossRef]
37. Ben Saïd, S. Strichartz estimates for Schrödinger-Laguerre operators. *Semigroup Forum* **2015**, *90*, 251–269. [CrossRef]
38. Ben Saïd, S.; Kobayashi, T.; Ørsted, B. Generalized Fourier transforms $\mathcal{F}_{k,a}$. *C. R. Math. Acad. Sci. Paris* **2009**, *347*, 1119–1124. [CrossRef]
39. de Bie, H.; Oste, R.; van der Jeugt, J. Generalized Fourier transforms arising from the enveloping algebras of (2) and $\mathfrak{osp}(1|2)$. *Int. Math. Res. Not. IMRN* **2016**, *15*, 4649–4705. [CrossRef]
40. de Bie, H.; Ørsted, B.; Somberg, P.; Soucek, V. Dunkl operators and a family of realizations of $\mathfrak{osp}(1|2)$. *Trans. Am. Math. Soc.* **2012**, *364*, 3875–3902. [CrossRef]
41. de Bie, H.; Pan, L.; Constales, D. Explicit formulas for the Dunkl dihedral kernel and the (k, a)-generalized Fourier kernel. *J. Math. Anal. Appl.* **2018**, *460*, 900–926.
42. Ben Saïd, S.; Deleaval, L. Translation operator and maximal operator for the $(k, 1)$-generalized Fourier transform. *J. Geom. Anal.* **2019**, in press. [CrossRef]

43. Ben Saïd, S.; Ørsted, B. The wave equation for Dunkl operators. *Indag. Math. (N.S.)* **2005**, *16*, 351–391. [CrossRef]
44. Howe, R.; Tan, E.C. *Nonabelian Harmonic Analysis. Applications of* SL(2, ℝ); Universitext; Springer: New York, NY, USA, 1992.

mathematics

MDPI

Article

On the Gibbs Effect Based on the Quasi-Affine Dual Tight Framelets System Generated Using the Mixed Oblique Extension Principle

Mutaz Mohammad

Department of Mathematics and Statistics, College of Natural and Health Sciences, Zayed University, 144543 Abu Dhabi, UAE; Mutaz.Mohammad@zu.ac.ae; Tel.: +971-2-599-3496

Received: 14 September 2019; Accepted: 2 October 2019; Published: 12 October 2019

Abstract: Gibbs effect represents the non-uniform convergence of the nth Fourier partial sums in approximating functions in the neighborhood of their non-removable discontinuities (jump discontinuities). The overshoots and undershoots cannot be removed by adding more terms in the series. This effect has been studied in the literature for wavelet and framelet expansions. Dual tight framelets have been proven useful in signal processing and many other applications where translation invariance, or the resulting redundancy, is very important. In this paper, we will study this effect using the dual tight framelets system. This system is generated by the mixed oblique extension principle. We investigate the existence of the Gibbs effect in the truncated expansion of a given function by using some dual tight framelets representation. We also give some examples to illustrate the results.

Keywords: Gibbs phenomenon; quasi-affine; shift-invariant system; dual tight framelets; oblique extension principle; B-splines

1. Introduction

The Gibbs effect was first recognized over a century ago by Henry Wilbraham in 1848 (see Ref. [1]). However, in 1898 Albert Michelson and Samuel Stratton (see Ref. [2]) observed it via a mechanical machine that they used to calculate the Fourier partial sums of a square wave function. Soon after, Gibbs explained this effect in two publications [3,4]. In his first short paper, Gibbs failed to notice the phenomenon and the limits of the graphs of the Fourier partial sums was inaccurate. In the second paper, he published a correction and gave the description of overshoot at the point of jump discontinuity. In fact, Gibbs did not provide a proof for his argument but only in 1906 a detailed mathematical description of the effect was introduced and named after Gibbs phenomenon by Maxime (see Ref. [5]) as he believed Gibbs to be the first person noticing it. This phenomenon has been studied extensively in Fourier series and many other situations such as the classical orthogonal expansions (see Refs. [6–8]), spline expansion (see Refs. [9,10]), wavelets and framelets series (see Refs. [11–17]), sampling approximations (see Ref. [18]), and many other theoretical investigations (see Refs. [19–23]). By considering Fourier series, it is impossible to recover accurate point values of a periodic function with many finitely jump discontinuities from its Fourier coefficients. Wavelets and their generalizations (framelets) have great success in coefficients recovering and have many applications in signal processing and numerical approximations (see Refs. [24–27]). However, many of these applications are represented by smooth functions that have jump discontinuities. However, expanding these functions will create (most often) unpleasant ringing effect near the gaps. It is the aim of this article to analyze the Gibbs effect of dual tight framelets using a different/higher order of vanishing moments.

Let us recall the preliminary background by introducing some notations (e.g., see Refs. [28–30]). Let $L_2(\mathbb{R})$ denote the space of all square integrable functions over the space $\mathbb{R}$, where

$$L_2(\mathbb{R}) = \left\{ f : \mathbb{R} \to \mathbb{R}; \int_{\mathbb{R}} |f|^2 < \infty \right\}.$$

Definition 1 ([31]). *Let $\psi \in L_2(\mathbb{R})$. For $j, k \in \mathbb{Z}$, define the function $\psi_{j,k}$ by*

$$\psi_{j,k} = 2^{j/2}\psi(2^j \cdot -k).$$

Then, we say the function ψ is a wavelet if the set $\left\{\psi_{j,k}\right\}_{j,k\in\mathbb{Z}}$ forms an orthonormal basis for $L_2(\mathbb{R})$.

Every square integrable function $f \in L_2(\mathbb{R})$ has a wavelet representation and this requires an orthonormal basis. However, the existence of such complete orthonormal basis is in general hard to construct and their representation is too restrictive and rigid. Therefore, frames were defined by the idea of an additional lower bound of the Bessel sequence which does not constitute an orthonormal set and are not linearly independent. In this paper, we will use dual tight framelets constructed by the mixed oblique extension principle (MOEP) (see Ref. [32]) which enables us to construct dual tight framelets for $L_2(\mathbb{R})$ of the form $\{\psi^\ell_{j,k}, \tilde{\psi}^\ell_{j,k}, \ell = 1, \cdots, r\}_{j,k}$. The MOEP provides an important method to construct dual framelets from refinable functions and gives us a better number of vanishing moments for ψ_ℓ and therefore a better imation orders. In fact, using the unitary extension principle UEP (see Ref. [32]), it is known that the approximation order of the system will not exceed 2, whereas the MOEP will give us a better approximation (see Ref. [33]). Please note that the MOEP is a generalization of the UEP and the oblique extension principle OEP. extension principle OEP (see Ref. [34]), which is again given to ensure that the system

$$\{\psi^\ell_{j,k}, \tilde{\psi}^\ell_{j,k}, \ell = 1, \cdots, r\}_{j,k\in\mathbb{Z}},$$

forms a dual tight framelets for $L_2(\mathbb{R})$. We refer the reader to Ref. [34] for the general setup of the MOEP.

Definition 2 ([31]). *A sequence $\Psi = \{\psi^\ell_{j,k}, \ell = 1, \cdots, r\}_{j,k}$ of elements in $L_2(\mathbb{R})$ is a framelet for $L_2(\mathbb{R})$ if there exists constants $A, B > 0$ such that*

$$A\|f\|^2 \leq \sum_{\ell=1}^{r}\sum_{j,k}^{\infty} |\langle f, \psi^\ell_{j,k}\rangle|^2 \leq B\|f\|^2, \forall f \in L_2(\mathbb{R}). \tag{1}$$

The numbers A, B are called frame bounds. If we can choose $A = B = 1$, then Ψ is called a tight framelet for $L_2(\mathbb{R})$.

Please note that we obtain a family of functions $\tilde{\Psi} = \{\tilde{\psi}^\ell_{j,k}, \ell = 1, \cdots, r\}_{j,k}$ such that

$$B^{-1}\|f\|^2 \leq \sum_{\ell=1}^{r}\sum_{j,k} |\langle f, \tilde{\psi}^\ell_{j,k}\rangle|^2 \leq A^{-1}\|f\|^2, \forall f \in L_2(\mathbb{R}). \tag{2}$$

The family $\tilde{\Psi}$ is called dual (reciprocal) framelet of the framelet Ψ. Equations (1) and (2) implies, respectively, the following equation

$$\langle f, g\rangle = \sum_{\ell=1}^{r}\sum_{j,k} \langle f, \tilde{\psi}^\ell_{j,k}\rangle\langle \psi^\ell_{j,k}, g\rangle \tag{3}$$

It follows directly from Equation (3) that any function $f \in L_2(\mathbb{R})$ has the following framelet representation

$$f = \sum_{\ell=1}^{r} \sum_{j\in\mathbb{Z}} \sum_{k\in\mathbb{Z}} \langle f, \tilde{\psi}_{j,k}^{\ell} \rangle \psi_{j,k}^{\ell} = \sum_{\ell=1}^{r} \sum_{j\in\mathbb{Z}} \sum_{k\in\mathbb{Z}} \langle f, \psi_{j,k}^{\ell} \rangle \tilde{\psi}_{j,k}^{\ell}. \tag{4}$$

The framelet constructions of Ψ and $\tilde{\Psi}$ require mother wavelets, called refinable functions ϕ and $\tilde{\phi}$, where a compactly supported function $\phi \in L_2(\mathbb{R})$ is said to be refinable if

$$\phi(x) = 2 \sum_{k\in\mathbb{Z}} h_0[k]\, \phi(2x-k), \tag{5}$$

for some finite supported sequence $h_0[k] \in \ell_2(\mathbb{Z})$. The sequence h_0 is called the *low pass filter* of ϕ. For convenience, we define $\psi_{0,k}^{\ell}(\cdot) = \phi_{0,k}(\cdot)$ and $\tilde{\psi}_{0,k}^{\ell}(\cdot) = \tilde{\phi}_{0,k}(\cdot)$. Therefore, Equation (4) can be rewritten as

$$f = \sum_{k} \langle f, \tilde{\phi}(\cdot - k) \rangle \phi(\cdot - k) + \sum_{\ell=1}^{r} \sum_{j=1}^{\infty} \sum_{k\in\mathbb{Z}} \langle f, \tilde{\psi}_{j,k}^{\ell} \rangle \psi_{j,k}^{\ell} \tag{6}$$

The above series expansion (6) can be truncated as

$$\mathcal{Q}_n f = \sum_{\ell=1}^{r} \sum_{j<n} \sum_{k\in\mathbb{Z}} \langle f, \tilde{\psi}_{j,k}^{\ell} \rangle \psi_{j,k}^{\ell}. \tag{7}$$

which is typical in kernel-based system identification approaches (see Ref. [35])

Please note that $\mathcal{Q}_n f$ can be described by a reproducing kernel Hilbert space which is given by a linear combination of its frame and dual frame product.

$$\mathcal{Q}_n f(x) = \int_{\mathbf{R}} f(y)\, \mathcal{D}_n(x,y)\, dy, \tag{8}$$

where

$$\mathcal{D}_n(x,y) = \sum_{\ell=1}^{r} \sum_{j<n} \sum_{k\in\mathbb{Z}} \tilde{\psi}_{j,k}^{\ell}(y)\, \psi_{j,k}^{\ell}(x),$$

is called the kernel of $\mathcal{Q}_n f$. Figure 1 shows the graphs of the kernel $\mathcal{D}_2(x,y)$ for different framelets.

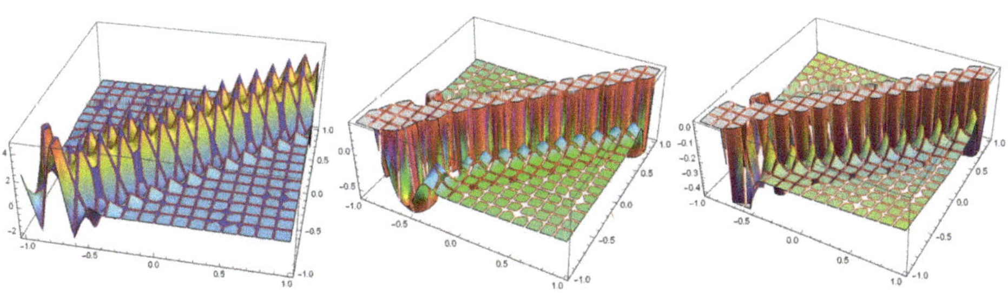

Figure 1. The graphs of the kernel, $\mathcal{D}_2(x,y)$, using the framelets of Example 2, 3 and 4, respectively.

It is known from the approximation theory, see e.g., Refs. ([28,35]), that the truncated expansion (7) is equivalent to

$$\mathcal{Q}_n f = \sum_{k\in\mathbb{Z}} \langle f, \tilde{\phi}_{n,k} \rangle \phi_{n,k}, \quad n \in \mathbb{Z}^+, \ f \in L_2(\mathbb{R}). \tag{9}$$

The general setup is to construct a set of functions as the form of Ψ, which can be summarized as follows: Let V_0 be the closed space generated by $\{\phi(\cdot - k)\}_{k\in\mathbb{Z}}$, i.e., $V_0 = \overline{span}\,\{\phi(\cdot - k)\}_{k\in\mathbb{Z}}$, and

$V_j = \{f(2^j x) : f(x) \in V_0, x \in \mathbb{R}\}$. Let $\{V_j, \phi\}_{j\in\mathbb{Z}}$ be the multiresolution analysis (MRA) generated by the function ϕ and $\Psi \subset V_1$ such that

$$\psi^\ell = 2 \sum_{k\in\mathbb{Z}} h_\ell[k]\, \phi(2\cdot -k), \tag{10}$$

where $\{h_\ell[k], k \in \mathbb{Z}\}_{\ell=1}^r$ is a finitely supported sequence called *high pass filters* of the system. Please note that from Equation (7), the functions ϕ, ψ^ℓ and $\tilde{\phi}, \tilde{\psi}^\ell$ are playing a great role. They are used for computing the coefficients of the expansion of the function f in terms of ϕ and ψ, and recovering the projection of f onto V_j from the coefficients $\langle f, \tilde{\psi}^\ell_{j,k}\rangle$. The Fourier transform of a function $f \in L_2(\mathbb{R})$ is defined to be

$$\mathcal{F}(f)(\omega) = \hat{f}(\omega) = \int_{\mathbb{R}} f(x)\, e^{-i\omega x} dx, \quad \omega \in \mathbb{R},$$

and the Fourier series of a sequence $h \in \ell_2(\mathbb{Z})$ is defined by

$$\mathcal{F}(h)(\omega) = \hat{h}(\omega) = \sum_{k\in\mathbb{Z}} h[k] e^{-i\omega k}, \quad \omega \in \mathbb{R}.$$

2. Gibbs Effect in Quasi-Affine Dual Tight Framelet Expansions

In this section, we study the Gibbs effect by using dual tight framelet in the quasi-affine tight framelet expansions generated via the MOEP. In general, and by using the expansion in Equation (7), we have $\lim_{n\to\infty} \mathcal{Q}_n f(x) = f(x)$ around x, where f is continuous except at many finite points. Hence, it is sufficient to study this effect by considering the following function

$$f(x) = \begin{cases} 1-x, & 0 < x \le 1 \\ -1-x, & -1 \le x < 0 \\ 0, & \text{else} \end{cases}.$$

In fact, this function is useful in the sense that other functions that have the same type of gaps, can be represented as expansions in terms of f plus a continuous function at $x = 0$. Please note that if we define $\mathcal{S}$ as

$$\mathcal{S}(x) = \begin{cases} \xi+1-x, & \xi < x \le \xi+1 \\ \xi-1-x, & \xi-1 \le x < \xi \\ 0, & \text{else} \end{cases},$$

then, $\mathcal{S}$ has a jump discontinuity at the point ξ and $\mathcal{S}(x) = f(x-\xi)$. Thus, we have the following result.

Theorem 1. *Any function with finitely many jump discontinuities can be written in terms of $\mathcal{S}$ plus a continuous function at the origin.*

Proof. Let g be a discontinuous function with a jump discontinuity, say at $x = \xi$, of magnitude D. We could put several of these together for g but we would likely only be looking at one such function at a time. Suppose that $\mathcal{S}$ and g are in the same direction of the needed jump (i.e., if $g(\xi^+) > g(\xi^-)$, then $J(\xi^+) > J(\xi^-)$, and similarly for $g(\xi^+) < g(\xi^-)$) or multiply $\mathcal{S}$ by (- or +)D to create the needed jump in the same direction. Define

$$F(x) = (+\ or\ -)D\mathcal{S}(x) + d,$$

so that d is a constant that makes the jump endpoints of F and g matched at $x = \xi$. Our continuous function in the neighborhood of the point ξ will then be $g(x) - F(x)$. $\square$

The definition of the Gibbs effect under the quasi-projection approximation $\mathcal{Q}_n$ is defined as follows.

Definition 3. *Suppose a function f is smooth and continuous everywhere except at $x_\circ$, i.e., limits $\lim_{x\to x_\circ^+} f(x)$ and $\lim_{x\to x_\circ^-} f(x)$ exist, and that $f(x_\circ^+)$ & $\neq$ &$f(x_\circ^-)$. Define $\mathcal{Q}_n f$ to be the truncated partial sum of Equation (7). We say that the framelet expansion of f exhibits the Gibbs effect at the right-hand side of $x_\circ$ if there is a sequence $d_s > 0$ converging to $x_\circ$, and*

$$\lim_{n,s\to\infty} \mathcal{Q}_n f(d_s) \begin{cases} > f(x_\circ^+), & if \;\; f(x_\circ^+) > f(x_\circ^-) \\ < f(x_\circ^+), & if \;\; f(x_\circ^+) < f(x_\circ^-) \end{cases}.$$

Similarly, we can define the Gibbs effect on the left-hand side of $x_\circ$.

Let Ψ to be the system defined by Definition 2. Thus, the corresponding quasi-affine system $X^J(\Psi)$ generated by Ψ is defined by a collection of translations and dilation of the elements in Ψ such that

$$X^J(\Psi) = \left\{\psi_{j,k}^\ell : 1 \le \ell \le r, j, k \in \mathbb{Z}, \right\}$$

where

$$\psi_{j,k}^\ell = \begin{cases} 2^{j/2}\psi^\ell(2^j \cdot -k), & j \ge J \\ 2^j\psi^\ell(2^j \cdot -2^j k)), & j < J \end{cases}.$$

In the study of our expansion, we consider $J = 0$. Many applications in framelet and approximation theory are modeled by non-negative functions. One family of such important functions are the B-splines, where the B-spline B_m of order m is defined by

$$B_m = B_{m-1} * B_1 = \int_{(0,1]} B_{m-1}(\cdot - x)dx$$

where

$$B_1 = \chi_{(0,1]}.$$

Figure 2 shows the graphs of the B-splines B_m for different order.

It is known that sparsity of the framelets representations is due to the vanishing moments of the underling refinable wavelet (see Ref. [29]). We say ψ has N vanishing moments if

$$\int x^r \psi_{j,k}^\ell(x)dx = 0, \; for \;\; r = 0, 1, \cdots, N-1, \tag{11}$$

which is equivalent to that $\hat{\psi}_{j,k}^{(r)}(0) = 0$, for all $r = 0, 1, \cdots, N-1$. This implies that the framelet $\psi_{j,k}^\ell(x)$ is orthogonal to the polynomials $1, x, \cdots, x^{N-1}$. The following statement is well known in the literature [13] for wavelets, but we present the proof for the reader's convenience by considering the general quasi-affine dual framelet system.

Proposition 1. *Assume that $\tilde{\Psi}, \Psi$ is a quasi-affine dual framelet system for $L_2(\mathbb{R})$ and that ψ, where $\psi \in \Psi$ has a vanishing moment of order N. Then for any polynomial $d(x)$ of degree at most $N-1$, we have*

$$\mathcal{Q}d = d,$$

where $\mathcal{Q} = \mathcal{Q}_0$ is defined by Equation (9) for $n = 0$.

Proof. From the definition of $\tilde{\Psi}, \Psi$, we know that all the generators must have a compact support. Therefore, we can find a positive integer A such that the support of all these generators lie in the interval $[-A, A]$. Define

$$C_B = \chi_{[-B,B]}, \quad where \ B \geq A. \tag{12}$$

Let $d(\cdot)$ be a polynomial of degree at most $N-1$. Then, by the vanishing moment property of $\psi_{j,k}^{\ell}$ we have

$$\int_{\mathbb{R}} C_B d(x) \psi_{j,k}^{\ell} dx = \int_{-B}^{B} d(x) \psi_{j,k}^{\ell} dx = 0, \quad where \ x < |B - A|. \tag{13}$$

Now, the proof is completed by taking $B \to \infty$ and using Equations (6) and (13). Thus, we have

$$d(\cdot) = C_B d(\cdot) = \sum_k \langle d, \tilde{\phi}_{0,k} \rangle \phi_{0,k}(\cdot) = \mathcal{Q} d(\cdot).$$

□

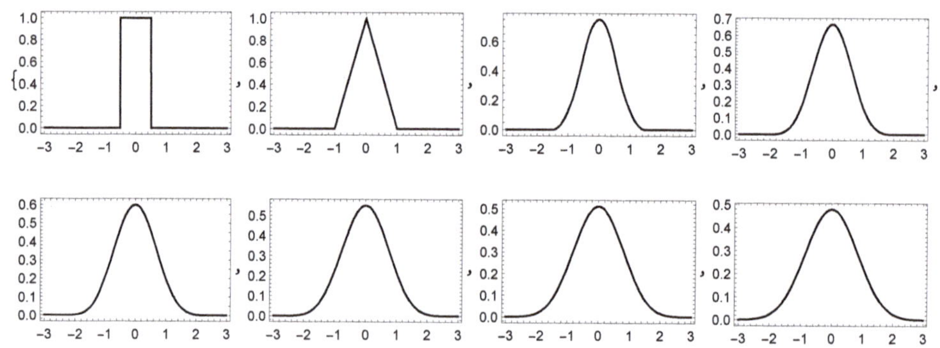

Figure 2. B-splines of order 1 through 8, respectively.

Now, we present some examples of dual tight framelets constructed by the MOEP in Ref. [34].

Example 1. *Let* $B_1 = \phi = \chi_{[0,1)}$. *Define,*

$$\begin{aligned} \psi_1(x) &= \frac{529}{1497} \chi_{[0,1)}(2x) + \frac{-173}{489} \chi_{[0,1)}(2x-1), \\ \tilde{\psi}_1(x) &= \frac{-1208}{3415} \chi_{[0,1)}(2x) + \frac{489}{1384} \chi_{[0,1)}(2x-1). \end{aligned}$$

Then, the resulting system generates a dual tight framelet for $L_2(\mathbf{R})$*. We illustrate the framelet and its dual framelet generators in Figure 3.*

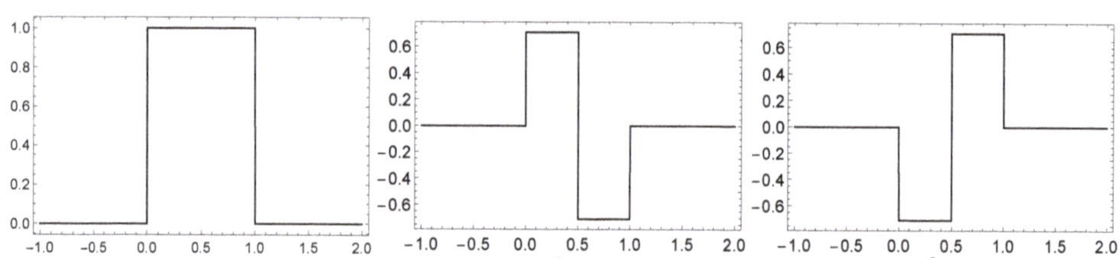

Figure 3. The graph of the Haar dual tight framelet of Example 1.

Example 2. *Let $B_2 = \phi = \tilde{\phi}$. Then,*

$$\mathcal{F}(h_0)(\xi) = \hat{h}_0(\xi) = \hat{\tilde{h}}_0(\xi) = 0.25(1 + \cos(\xi) - i\sin(\xi))^2.$$

Thus, by using the MOEP, one can find the following high pass filters,

$$\begin{aligned}
\hat{h}_1(\xi) &= -(1 - \cos(\xi) + i\sin(\xi))^2, \\
\hat{h}_2(\xi) &= -16\sin^4(\xi/2)(\cos(\xi) - i\sin(\xi)), \\
\hat{\tilde{h}}_1(\xi) &= \frac{1}{480}e^{-i\xi}\sin^4(\xi/2)(32{,}668 + 57{,}569\cos(\xi) + 39{,}422\cos(2\xi) + 21{,}191\cos(3\xi) + \\
&\quad 8428\cos(4\xi) + 2233\cos(5\xi) + 426\cos(6\xi) + 71\cos(7\xi)), \\
\hat{\tilde{h}}_2(\xi) &= -\frac{1}{480}e^{-i\xi}\sin^2(\xi/2)(-614 - 726\cos(\xi) + 85\cos(2\xi) + 412\cos(3\xi) + 458\cos(4\xi) \\
&\quad +284\cos(5\xi) + 71\cos(6\xi)).
\end{aligned}$$

Then, $\psi^\ell, \tilde{\psi}^\ell, \ell = 1, 2$ forms a dual tight framelets for $L_2(\mathbb{R})$. These functions have vanishing moments (vm) as follows, $vm(\psi_1) = vm(\tilde{\psi}_2) = 2$ while $vm(\psi_2) = vm(\tilde{\psi}_1) = 4$. See Figure 4 for their graphs.

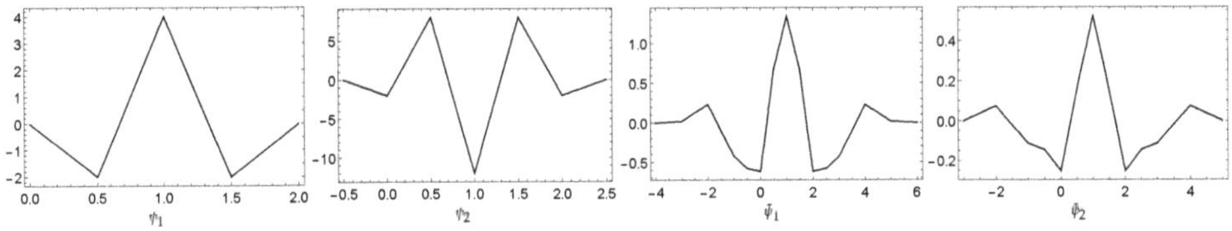

Figure 4. The graphs of the framelets and its dual framelets of Example 2.

Example 3. *Let $B_4 = \phi = \tilde{\phi}$. Then,*

$$\hat{h}_0(\xi) = \hat{\tilde{h}}_0(\xi) = 0.0625(1 + \cos(\xi) - i\sin(\xi))^4.$$

We have the following high pass filters,

$$\begin{aligned}
\hat{h}_1(\xi) &= (1 - \cos(\xi) + i\sin(\xi))^4, \\
\hat{h}_2(\xi) &= 16\sin^4(\xi/2)(\cos(3\xi) - i\sin(3\xi)).
\end{aligned}$$

The high pass filters for the dual framelets in $\ell_2(\mathbb{Z})$, where $k \in \mathbb{Z}$, is given by

$$\begin{cases}
\tilde{h}_1[k] = [\frac{311}{3{,}870{,}720}, \frac{311}{967{,}680}, \frac{865}{387{,}072}, \frac{473}{64{,}512}, \frac{1783}{4{,}83{,}840}, \frac{-19{,}967}{967{,}680}, \frac{-67{,}453}{1{,}935{,}360}, \frac{-17{,}887}{967{,}680}, \frac{233{,}473}{1{,}935{,}360}, \frac{17{,}887}{967{,}680}, \\
\frac{67{,}453}{1{,}935{,}360}, \frac{19{,}967}{967{,}680}, \frac{1783}{483{,}840}, \frac{473}{64{,}512}, \frac{865}{387{,}072}, \frac{311}{967{,}680}, \frac{311}{3{,}870{,}720}], \\
\tilde{h}_2[k] = [\frac{311}{483{,}840}, \frac{311}{120{,}960}, \frac{2119}{483{,}840}, \frac{47}{10{,}080}, \frac{-4111}{483{,}840}, \frac{-5687}{120{,}960}, \frac{-21{,}103}{483{,}840}, \frac{2627}{15{,}120}, \frac{-21{,}103}{483{,}840}, \frac{-5687}{120{,}960}, \\
\frac{-4111}{483{,}840}, \frac{47}{10{,}080}, \frac{2119}{483{,}840}, \frac{311}{120{,}960}, \frac{311}{483{,}840}].
\end{cases}$$

Then, $\psi^\ell, \tilde{\psi}^\ell, \ell = 1, 2$ forms a dual tight framelets for $L_2(\mathbb{R})$. Here we have $vm(\psi_1) = vm(\psi_2) = vm(\tilde{\psi}_1) = vm(\tilde{\psi}_2) = 4$. Their graphs are depicted in Figure 5.

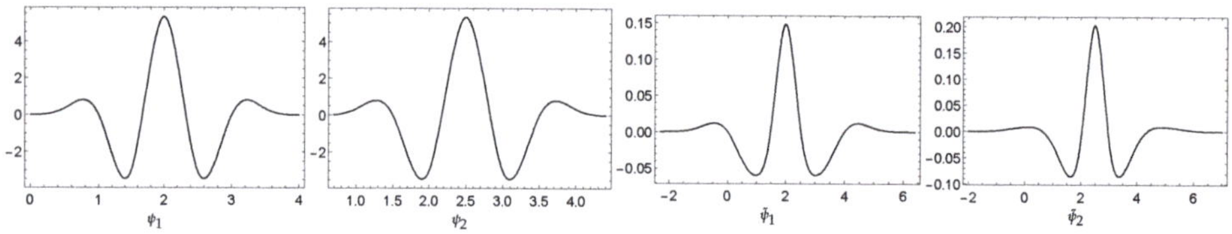

Figure 5. The graphs of the framelets and its dual framelets of Example 3.

Example 4. *Let $B_4 = \phi$, and $B_2 = \tilde{\phi}$. Thus,*

$$\hat{h}_0(\xi) = 0.0625(1 + \cos(\xi) - i\sin(\xi))^4, \text{ and } \hat{\tilde{h}}_0(\xi) = 0.25(1 + \cos(\xi) - i\sin(\xi))^2.$$

Then, we have the following tight framelets,

$$\psi_1(x) = \begin{cases} -8x^3/3 & \text{if } 0 \le x \le 1/2 \\ -2 + 12x - 24x^2 + 40x^3/3 & \text{if } 1/2 < x \le 1 \\ 38 - 108x + 96x^2 - 80x^3/3 & \text{if } 1 < x < 3/2 \\ -142 + 252x - 144x^2 + 80x^3/3 & \text{if } 3/2 \le x < 2 \\ 178 - 228x + 96x^2 - 40x^3/3 & \text{if } 2 \le x < 5/2 \\ (8/3)(-3 + x)^3 & \text{if } 5/2 \le x \le 3 \\ 0 & \text{if otherwise,} \end{cases}$$

$$\psi_2(x) = \begin{cases} -(1/3)(-1 + 2x)^3 & \text{if } 1/2 \le x \le 1 \\ -47/3 + 46x - 44x^2 + 40x^3/3 & \text{if } 1 < x < 3/2 \\ 358/3 - 224x + 136x^2 - 80x^3/3 & \text{if } 3/2 \le x < 2 \\ (2/3)(-461 + 624x - 276x^2 + 40x^3) & \text{if } 2 \le x < 5/2 \\ 953/3 - 334x + 116x^2 - 40x^3/3 & \text{if } 5/2 \le x \le 3 \\ (1/3)(-7 + 2x)^3 & \text{if } 3 < x \le 7/2 \\ 0 & \text{if otherwise} \end{cases}$$

and the high pass filters for its dual tight framelets in $\ell_2(\mathbb{Z})$, where $k \in \mathbb{Z}$, are given by:

$$\begin{cases} \tilde{h}_1[k] = [\frac{13}{15{,}360}, \frac{13}{7680}, \frac{61}{7680}, \frac{109}{7680}, \frac{-733}{15{,}360}, \frac{-35}{256}, \frac{409}{1280}, \frac{-35}{256}, \frac{-733}{15{,}360}, \frac{109}{7680}, \frac{61}{7680}, \frac{13}{7680}, \frac{13}{15{,}360},], \\ \tilde{h}_2[k] = [\frac{13}{2560}, \frac{13}{1280}, \frac{-89}{7680}, \frac{-1}{30}, \frac{-199}{3840}, \frac{313}{1920}, \frac{-199}{3840}, \frac{-1}{30}, \frac{-89}{7680}, \frac{13}{1280}, \frac{13}{2560}]. \end{cases}$$

Then, again $\psi^\ell, \tilde{\psi}^\ell, \ell = 1, 2$ forms a dual tight framelets for $L_2(\mathbb{R})$. Here we have $vm(\psi_1) = vm(\psi_2) = 2, vm(\tilde{\psi}_1) = vm(\tilde{\psi}_2) = 4$. Their graphs are depicted in Figure 6.

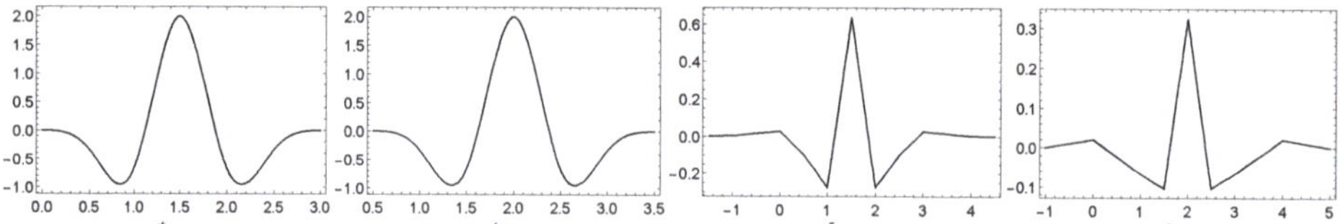

Figure 6. The graphs of the framelets and its dual framelets of Example 4.

We will use the framelet expansion defined by Equation (7) to present the numerical evidence of the Gibbs effect by determining the maximal overshoot and undershoot of the truncated expansion

$\mathcal{Q}_n f$ near the origin. The behavior of the truncated functions $\mathcal{Q}_n$ of a function with jump discontinuities is related to the existence of the Gibbs phenomenon, which is unpleasant in application, and not so easy to avoid. Therefore, examining a series of representations to avoid it or at least reduce it, is very important.

Proposition 2. *For any two refinable compactly supported functions ϕ and $\tilde{\phi}$ in $L_2(\mathbb{R})$. If*

$$\mathcal{Q}\,\mathrm{sgn} \leq 1 \text{ on } (x_\circ, \infty) \text{ and } \mathcal{Q}\,\mathrm{sgn} \geq -1 \text{ on } (-\infty, x_\circ),$$

then $\mathcal{Q}_n f$ exhibits no Gibbs effect.

Proof. Please note that $(\mathcal{Q}_n f)(2^{-n}x) = \mathcal{Q}(f(2^{-n}x))$ for all $n \in \mathbb{N}$. In particular, $(\mathcal{Q}_n\,\mathrm{sgn})(2^{-n}x) = \mathcal{Q}(\mathrm{sgn}(2^{-n}x))$. Suppose that the truncated function $\mathcal{Q}_n f$ do exhibit the Gibbs effect near $x_\circ^+$. Thus, there exists an open interval $\mathcal{U} \subseteq (x_\circ, \infty)$ such that $\mathcal{Q}f(x) > 1$, $\forall x \in \mathcal{U}$. Therefore, $\exists u > x_\circ$ such that $\max_{x \in (x_\circ, u)} \mathcal{Q}f(x) = U > 1$. Define a sequence $u_n > x_\circ$, $\forall n$ such that $u_n \to x_\circ$ as $n \to \infty$ (one can take $u_n = \epsilon_n + x_\circ$ such that $\epsilon_n \to 0$ as $n \to \infty$). Hence, $\max_{x \in (x_\circ, u_n)} \mathcal{Q}_n f(x) = \max_{x \in (x_\circ, u)} \mathcal{Q}f(x) = U > 1$, a contradiction. Similarly, we can prove the case when $x < x_\circ^-$ in the same fashion. □

Please note that it is important to use non-negative functions in framelet analysis due to its use in a variety of applications. One of those functions is the *B*-splines. The following statement will require such non-negativity to avoid the Gibbs effect.

Theorem 2. *Let ϕ and $\tilde{\phi}$ be any two non-negative refinable real valued compactly supported functions in $L_2(\mathbb{R})$ such that*

$$\int \mathrm{sgn}(x)\phi(x-k) > 0 \text{ and } \int \mathrm{sgn}(x)\tilde{\phi}(x-k) > 0 \text{ on } \mathbb{R}.$$

Assume further $\sum_k \phi(x-k) = 1$. If the vanishing moment of ϕ and $\tilde{\phi}$ is one, then $\tilde{\mathcal{Q}}f$ exhibit no Gibbs effect.

Proof. It suffices to show this for $\mathrm{sgn}(x)$ as $f(x) = \mathrm{sgn}(x) - x$ on $[-1, 1]$. Please note that Proposition 1 is held for $d = 1$, i.e.,

$$\tilde{\mathcal{Q}}1 = 1.$$

Now, for $x \in \mathbb{R}$, and since

$$\langle \mathrm{sgn}(x), \phi(x-k) \rangle = \hat{\phi}(0),$$

by assumption, we have

$$\begin{aligned} \tilde{\mathcal{Q}}\,\mathrm{sgn}(x) - 1 &= \sum_k \langle \mathrm{sgn}(x), \phi(x-k) \rangle \tilde{\phi}(x-k) \\ &= -\sum_k \left(\int_{-\infty}^{-k} \phi(x)dx \right) \tilde{\phi}(x-k) < 0. \end{aligned}$$

The other side is analogue. Thus, $-1 \leq \tilde{\mathcal{Q}}\,\mathrm{sgn} \leq 1$ for all $x \in \mathbb{R}$. □

3. Results and Discussion

We present some numerical illustration by using the dual tight framelets which will generalize the result in Ref. [14]. The results show that if the dual framelet has vanishing moments of order of at least two, then $\mathcal{Q}_n f$ must exhibit the Gibbs effect. However, $\mathcal{Q}_n f$ has no Gibbs effect by using dual tight framelets of vanishing moments of order one.

Now we present an illustration for the Gibbs effect using the above dual tight framelets by showing the maximum overshoots, undershoots of $f(x)$, and some related graphs. This is to showcase the absence of the effect in Table 1 and Figures 7 and 8.

Table 1. Approximate maximum overshoot and undershoot in neighborhoods of $x = 0$ using $\mathcal{Q}_n f$ of Example 1.

Level	Maximum	Minimum
$n = 2$	0.907276	−0.907276
$n = 3$	0.964562	−0.964562
$n = 5$	0.994801	−0.994802
$n = 10$	0.999968	−0.999972

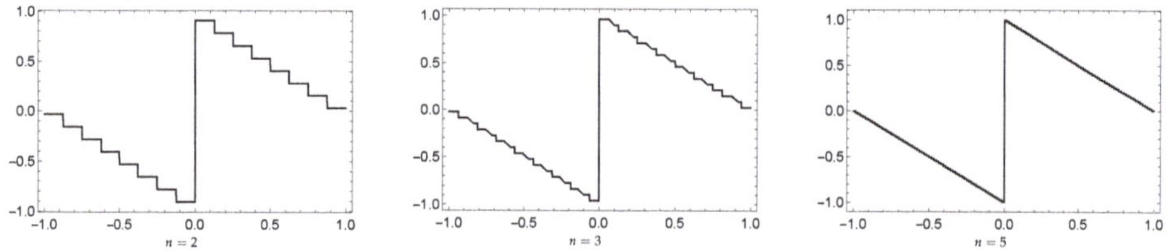

Figure 7. Illustration for the absence of the Gibbs effect using the quasi-operator $\mathcal{Q}_n f$ of Example 1.

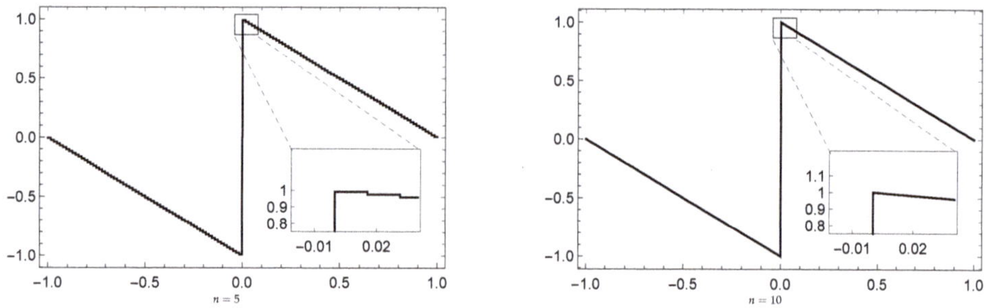

Figure 8. Illustration for the absence of the Gibbs effect using the quasi-operator $\mathcal{Q}_n f$ of Example 1 for $n = 5, 10$, respectively.

In Figures 9 and 10 we illustrate the graphs of the function $\mathcal{Q}f$ and $\mathcal{Q}\,\mathrm{sgn}$, receptively, generated using the B-splines dual tight framelets. In Tables 2–4 we show the approximated values for the overshoots and undershoots of $\mathcal{Q}_n f$. Figures 11–14 illustrate the graphs of the Gibbs effect.

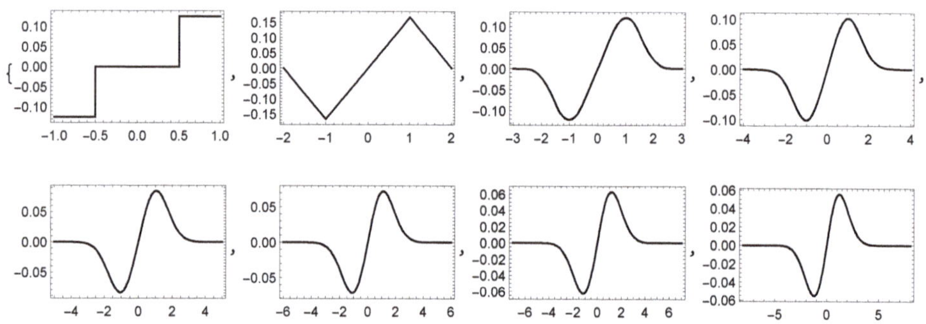

Figure 9. Graphs of $\mathcal{Q}f(x)$ by the B-splines of order 1 through 8, respectively.

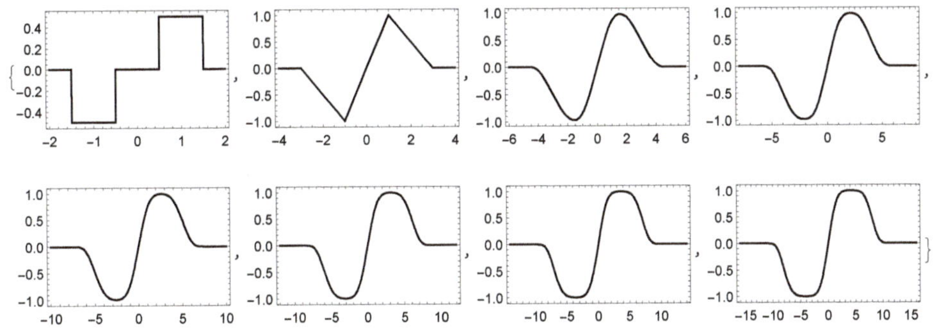

Figure 10. Graphs of $\mathcal{Q}\,\mathrm{sgn}(x)$ by the B-splines of order 1 through 8, respectively.

Table 2. Approximate maximum overshoot and undershoot in neighborhoods of $x = 0$ using $\mathcal{Q}_n f$ of Example 2 for $n = 2, 3, 5, 10$.

Level	Maximum	Minimum
$n = 2$	1.0021	−1.01204
$n = 3$	1.12708	−1.12708
$n = 5$	1.22083	−1.18958
$n = 10$	1.25111	−1.25111

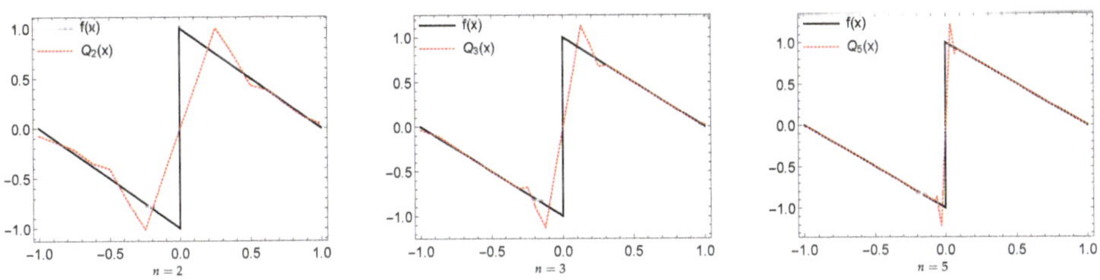

Figure 11. Illustration of the Gibbs effect using the quasi-operator $\mathcal{Q}_n f$ of Example 2.

Table 3. Approximate maximum overshoot and undershoot in neighborhoods of $x = 0$ using $\mathcal{Q}_n f$ of Example 3 for $n = 2, 3, 5, 10$.

Level	Maximum	Minimum
$n = 2$	0.997622	−0.996194
$n = 3$	1.07451	−1.07451
$n = 5$	1.13384	−1.13384
$n = 10$	1.15337	−1.15337

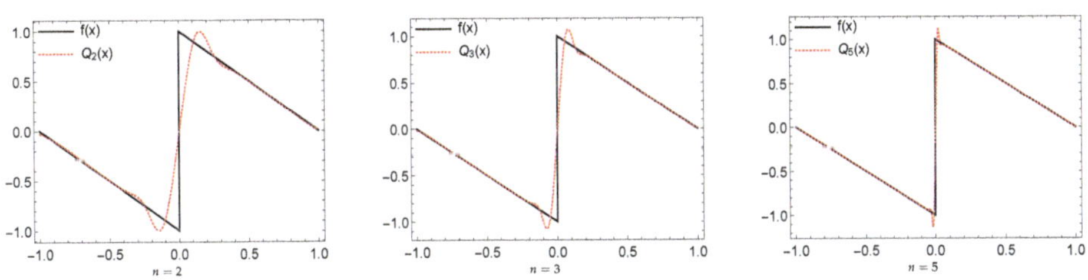

Figure 12. Illustration of the Gibbs effect using the quasi-operator $\mathcal{Q}_n f$ of Example 3.

Table 4. Approximate maximum overshoot and undershoot in neighborhoods of $x = 0$ using $\mathcal{Q}_n f$ of Example 4 for $n = 2, 3, 5, 10$.

Level	Maximum	Minimum
$n = 2$	1.08519	−1.08526
$n = 3$	1.14823	−1.14823
$n = 5$	1.19514	−1.19514
$n = 10$	1.21028	−1.21028

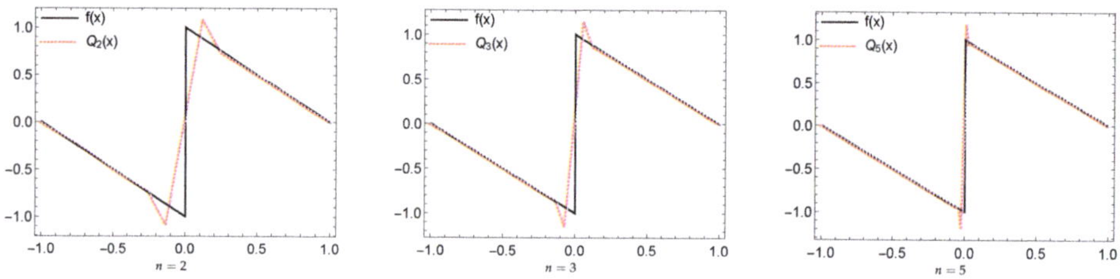

Figure 13. Illustration of the Gibbs effect using the quasi-operator $\mathcal{Q}_n f$ of Example 4.

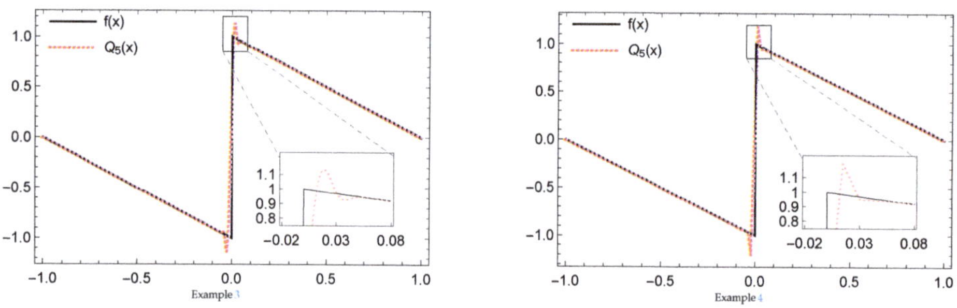

Figure 14. Gibbs effect illustrations by $\mathcal{Q}_5 f(x)$ of Example 3 and 4, respectively.

4. Conclusions

According to the above results, we show that the Gibbs effect is absent when the dual tight framelets of vanishing moments of order one are used to represent a function with jump discontinuities at the origin. Please note that increasing the vanishing moments, e.g., using the MOEP, will increase the approximation order of the framelet representation $\mathcal{Q}_n f$ that used to expand the function f; however, the Gibbs effects cannot be avoided for any level of n. Quite a few examples of dual tight framelets, numerical results, and graphical illustrations have been presented for the absence and presence of Gibbs effect.

Funding: This research was funded by Zayed University Research Fund.

Acknowledgments: I would like to thank the anonymous reviewers for their valuable comments to improve the quality of the paper.

Conflicts of Interest: The author declares no conflict of interest.

Abbreviations

The following abbreviations are used in this manuscript:

$L_2(\mathbb{R})$	The space of all square integrable functions over $\mathbb{R}$
$\psi, \tilde{\psi}$	The wavelet and its dual function
$\Psi, \tilde{\Psi}$	The framelets and its dual framelets system
UEP	The unitary extension principle
OEP	The oblique extension principle
MOEP	The mixed oblique extension principle
$\langle f, g \rangle$	The inner product of f and g
ϕ	The refinable function
$\mathcal{D}_n f$	The reproducing kernel Hilbert space of the function f
$\mathcal{Q}_n f$	The truncated partial sum of the framelet system
$h_0[k]$	Low pass filter of the framelet system
$h_\ell[k]$	High pass filters of the framelet system
V_j	The multiresolution analysis generated by the function ϕ
$\mathcal{F}(f)$	The Fourier transform of a function f
B_m	The B-spline of order m
$vm(f)$	The vanishing moments of the function f

References and Note

1. Wilbraham, H. On a certain periodic function. *Camb. Dublin Math. J.* **1848**, *3*, 198–201.
2. Michelson, A.; Stratton, S. A new harmonic analyser. *Philos. Mag.* **1898**, *45*, 85–91. [CrossRef]
3. Gibbs, J.W. Fourier's Series. *Nature* **1898**, *59*, 200. [CrossRef]
4. Gibbs, J.W. Fourier's Series. *Nature* **1899**, *59*, 606. [CrossRef]
5. Maxime, B. Introduction to the theory of Fourier's series. *Ann. Math.* **1906**, *7*, 81–152.
6. Shim, H.T. A summability for Meyer wavelets. *J. Appl. Math. Comput.* **2002**, *9*, 657–666. [CrossRef]
7. Shen, X. On Gibbs phenomenon in wavelet expansions. *J. Math. Study* **2002**, *35*, 343–357.
8. Shen, X. Gibbs Phenomenon for Orthogonal Wavelets with Compact Support. In *Advances in the Gibbs Phenomenon*; Jerri, J., Ed.; Sampling Publishing: Potsdam, Germany, 2011; pp. 337–369,
9. Foster, J.; Richards, F. The Gibbs phenomenon for piecewise-linear approximation. *Am. Math. Mon.* **1991**, *98*, 47–49. [CrossRef]
10. Richards, F.B. A Gibbs phenomenon for spline functions. *J. Approx. Theory* **1991**, *66*, 334–351. [CrossRef]
11. Han, B. Gibbs Phenomenon of Framelet Expansions and Quasi-projection Approximation. *J. Fourier Anal. Appl.* **2018**. [CrossRef]
12. Gribonval, R.; Nielsen, M. On approximation with spline generated framelets. *Constr. Approx.* **2004**, *20*, 207–232.
13. Kelly, S. Gibbs phenomenon for wavelets. *Appl. Comp. Harmon. Anal.* **1996**, *3*, 72–81. [CrossRef]
14. Mohammad, M.; Lin, E.B. Gibbs phenomenon in tight framelet expansions. *Commun. Nonlinear Sci. Numer. Simul.* **2018**, *55*, 84–92. [CrossRef]
15. Mohammad, M.; Lin, E.B. Gibbs Effects Using Daubechies and Coiflet Tight Framelet Systems. *Contemp. Math. AMS* **2018**, *706*, 271–282.
16. Mohammad, M. Special B-spline Tight Framelet and It's Applications. *J. Adv. Math. Comput. Sci.* **2018**, *29*, 1–18. [CrossRef]
17. Shim, H.T.; Volkmer, H. On the Gibbs Phenomenon for Wavelet Expansions. *J. Approx. Theory* **1996**, *84*, 74–95. [CrossRef]
18. Zygmund, A. *Trigonometric Series*, 2nd ed.; Cambridge Univ. Press: Cambridge, UK, 1959.
19. Adcock, B.; Hansen, A. Stable reconstructions in Hilbert spaces and the resolution of the Gibbs phenomenon. *Appl. Comput. Harmon. Anal.* **2012**, *32*, 357–388. [CrossRef]
20. Conditions on shape preserving of stationary polynomial reproducing subdivision schemes.
21. Aldwairi, M.; Flaifel, Y. Baeza-Yates and Navarro Approximate String Matching for Spam Filtering. In Proceedings of the Second International Conference on Innovative Computing Technology (INTECH 2012), Casablanca, Morocco, 18–20 September 2012; pp. 16–20.

22. Gottlib, D.; Shu, C. On the Gibbs phenomenon and its resolution. *SIAM Rev.* **1997**, *39*, 644–668. [CrossRef]
23. Jerri, A. *The Gibbs Phenomenon in Fourier Analysis, Splines and Wavelet Approximations*; Kluwer Academic Publishers: Dordrecht, The Netherlands, 1998.
24. Li, B.; Chen, X. Wavelet-based numerical analysis: A review and classification. *Finite Elem. Anal. Des.* **2014**, *81*, 14–31. [CrossRef]
25. Barg, A.; Glazyrin, A.; Okoudjou, K.A.; Yu, W. Finite two-distance tight frames. *Linear Algebra Appl.* **2015**, *475*, 163–175. [CrossRef]
26. Ganiou, A.; Atindehou, D.; Kouagou, Y.B.; Okoudjou, K.A. On the frame set for the 2-spline. *arXiv* **2018**, arXiv:1806.05614.
27. Hussein, R.; Shaban, K.; El-Hag, H. Energy conservation-based thresholding for effective wavelet denoising of partial discharge signals. *Sci. Meas. Technol. IET* **2016**, *10*, 813–822. [CrossRef]
28. Han, B. Framelets and wavelets: Algorithms, analysis, and applications. In *Applied and Numerical Harmonic Analysis*; Birkhauser/Springer: Cham, Switzerland, 2017; p. 724.
29. Daubechies, I. *Ten Lectures on Wavelets*; SIAM: Philadelphia, PA, USA, 1992.
30. Daubechies, I.; Grossmann, A.; Meyer, Y. Painless nonorthogonal expansions. *J. Math. Phys.* **1986**, *27*, 1271–1283. [CrossRef]
31. Christensen, O. *An Introduction to Frames and Riesz Bases*; Birkhauser: Boston, MA, USA, 2003.
32. Ron, A.; Shen, Z. Affine systems in $L^2(\mathbf{R}^d)$: the analysis of the analysis operators. *J. Funct. Anal.* **1997**, *148*, 408–447. [CrossRef]
33. Ron, A.; Shen, Z. Affine systems in $L^2(\mathbb{R}^d)$ II: dual systems. *J. Fourier Anal. Appl.* **1997**, *3*, 617–637. [CrossRef]
34. Daubechies, I.; Han, B.; Ron, A.; Shen, Z. Framelets: MRA-based constructions of wavelet frames. *Appl. Comput. Harmon.* **2003**, *14*, 1–46. [CrossRef]
35. Zorzi, M.; Chiuso, A. The harmonic analysis of kernel functions. *Automatica* **2018**, *94*, 125–137. [CrossRef]

mathematics

MDPI

Article

Preliminary Analysis of a Fully Ceramic Microencapsulated Fuel Thermal–Mechanical Performance

Ping Chen [1,2,*], Suizheng Qiu [1,*], Shichao Liu [2], Yi Zhou [2], Yong Xin [2], Shixin Gao [2], Xi Qiu [2] and Huaiyu Lu [2]

[1] Shanxi Key Lab. of Advanced Nuclear Energy and Technology, School of Nuclear Science and Technology, Xi'an Jiaotong University, Xi'an 710049, China

[2] Science and Technology on Reactor System Design Technology Laboratory, Nuclear Power Institute of China, Chengdu 610200, China; hit_lsc@163.com (S.L.); hit_yf@163.com (Y.Z.); hit_yf306@163.com (Y.X.); liushichao19861010@163.com (S.G.); qiuxi_drsi@163.com (X.Q.); luhuaiyu0424@163.com (H.L.)

* Correspondence:chenping_npic@163.com (P.C.); szqiu@mail.xjtu.edu.cn (S.Q.); Tel./Fax: +86-028-8590-8295 (P.C.)

Received: 20 March 2019; Accepted: 30 April 2019; Published: 20 May 2019

Abstract: In this paper, a two-dimensional characteristic unit was used to simulate the thermal–mechanical performance of a fully ceramic microencapsulated (FCM) fuel pellet, and the criterion of FCM structure integrity was discussed. FCM structure integrity can be reflected though the integrity of the silicon carbide (SiC) matrix or SiC layers because of the excellent fission retention capability of SiC ceramics. The maximum temperature of the SiC matrix under normal conditions of the pressure water reactor (PWR) environment was about 1390 K, which was lower than the decomposition point of SiC. The maximum hoop stress of the SiC matrix, especially the inner part, was up to about 1200 MPa, and the hoop stress of the non-fuel region part was lower than the inner part, which can be attributed to the deformation of tristructural-isotopic (TRISO) particles. The hoop stress of the SiC layers at the end of life was only about 180 MPa, which is much lower than the strength of the chemical vapor deposition (CVD)-SiC. The failure probability of the SiC layer was lower than 9×10^{-5}; thus, the integrity of SiC layers and the fission retention capability were maintained. The structure integrity of FCM fuel was broken because the SiC matrix cracked.

Keywords: FCM fuel; thermal–mechanical performance; failure probability; silicon carbide

1. Introduction

Fully ceramic microencapsulated (FCM) fuel, which can improve the accident tolerance of light water reactors (LWRs), receives much attention because of its excellent oxidation resistance, fission retention capability, high thermal conductivity, and irradiation stability [1–4]. FCM was designed to improve the fission retention capacity during accident processes; fission retention capacity may be reflected by the integrity of the silicon carbide (SiC) matrix and layers. FCM is composed of historic tri-isotropic fuel embedded in a fully dense and impermeable SiC matrix. The tristructural-isotopic (TRISO) particle consists of the fuel kernel surrounded by four successive layers [5]: the low-density carbon buffer layer, whose function is to slow down and retain the generated fission product, and the inner and outer pyroltytic graphite layers, which surround the silicon carbide (SiC) micro pressure vessel. The function of the pyrolytic carbon is severalfold, including the protection of the kernel from aggressive process gases used during TRISO processing, to protect the SiC shell from energetic fission product recoil damage, and to provide thermal–mechanical stability and toughening for this multilayer system. Historically, TRISO and its fuel forms (graphite matrix compacts in the form of the cylindrical fuel of a prismatic high-temperature gas-cooled reactor or the spherical pebble of a pebble-bed modular high-temperature gas-cooled reactor) operate at a temperature in excess of 1000 °C. As the envisioned LWR application of the FCM fuel is at a less aggressive operating temperature for the TRISO (nominally 350–500 °C), it is anticipated that higher burn-ups and significantly improved safety benefits realized by the extraordinary fission product retention may be realized [6].

Lower fissile loading was an obvious shortcoming of FCM fuel compared with uranium dioxide (UO_2) pellets. The feasibility of FCM fuel to satisfy the reasonable fuel cycle lengths in LWRs was demonstrated by Sen et al. [7]. The effect of the UO_2 kernel size, enrichment, and type on the effective full power days (EFPD) was discussed. The results indicated that uranium nitride (UN) kernel FCM fuel can meet the fissile loading requirement of LWRs by optimizing the size of the kernel and coated layers. The sizes of the UN kernel and buffer layer were 800 μm and 100 μm, respectively, and the fraction of the FCM had to be higher than 44 vol.%. The thermal–mechanical performance of the TRISO with UO_2 kernels was simulated by PARFUM, PASTA, and other models [8]. In the CO production model, internal pressure and gap heat transfer were considered. The Recoil and Booth models were used to calculate the fission gas release. The thermal and mechanical performance of the coated layer can be calculated in the above models. The performance of the UN kernel TRISO fuel in the LWR environment was reported, and the influence of temperature and kernel size on internal pressure and the survivability of TRISO fuel was studied [9]. The tangential stress of the inner pyrocarbon (IPyC) and SiC layers was calculated using COMSOL multi-physics software (COMSOL-5.2, MERCURY LEARNING AND INFORMATION LLC, Dulles, Virginia). However, the influence of the SiC matrix on TRISO particle performance was not discussed.

The thermal–mechanical performance of FCM pellets is difficult to simulate because of its complex structure and material properties. The temperature field and thermal conductivity of FCM pellets were studied previously [10–12], but thermal and mechanical coupled simulation results are rare. The temperature of the FCM pellet used in a pressure water reactor (PWR) environment was calculated by COMSOL software in the literature [10]. Both homogeneous and heterogeneous models were used in the report, and the calculated result of the homogeneous models was in good agreement with the result calculated by RELAP. The mechanical performance of FCM was not studied. Ougouag and co-workers [13] studied the influence of the SiC matrix on the performance of TRISO fuel by adding an SiC matrix on the outer surface of a single TRISO particle model. The results indicate that the thickness of the SiC matrix has an obvious influence on the stress distribution of the SiC and IPyC layers. The structure and coated layer size can affect the stress condition of the FCM. The SiC layer stress increased upon ignoring the outer pyrocarbon (OPyC) layer or increasing the SiC matrix thickness. The stress of the SiC layer with a matrix was much higher than the single particle. Schappel [14] simulated the thermal–mechanical performance of an FCM pellet and TRISO particle separately using ABAQUS software. TRISO particles were subtracted from the FCM pellet to simplify the model and decrease the computation amount. The performance of the coated layers was not detected, and the interaction between the TRISO particle and matrix was not reflected in the literature.

There is no criterion to evaluate the integrity of the FCM pellet, and the integrity of the FCM pellet was not studied in previous work as far as we know. In this paper, the integrity of the FCM pellet was defined according to fission retention capacity. The integrity of the FCM pellet under a PWR environment was discussed by investigating the mechanical performance and failure probability of the SiC matrix and SiC layers. A two-dimensional model was established, while the buffer layer and fuel kernel were subtracted from the FCM pellet. The internal pressure of TRISO particles embedded in different locations of FCM fuel was calculated using a TRISO simulation. The interaction between the TRISO particle and the SiC matrix was investigated, and the thermal–mechanical properties of the coated layers and SiC matrix were discussed. The fission production capacity of the FCM pellet was evaluated by investigating the failure probability of SiC layers and the hoop stress of the SiC matrix.

2. Governing Equation and Material Properties

2.1. Governing Equation

COMSOL multi-physics software offered the heat transfer model for the solid materials to calculate the temperature distribution in FCM and TRISO fuel. The temperature field of FCM pellet and TRISO fuel are determined by the heat conduction equation:

$$\rho C_p \frac{\partial T}{\partial t} + \nabla \cdot q - E_f \dot{F} = 0 \tag{1}$$

where T, ρ and C_p are the temperature, density (kg/m^3) and heat capacity (J/(kg·K)) of the solid material, respectively. E_f and $\dot{F}$ are the energy released in a single fission event and volumetric fission rate, respectively. The heat flux can be written as follows:

$$q = -k\nabla T. \tag{2}$$

The stress in all coated layers and the SiC matrix was caused by the accumulation of internal pressure and the deformation of other materials. The internal pressure of TRISO fuel was caused by fission gas release. The TRISO particle located in a different part of the FCM pellet possessed a different internal pressure caused by the temperature gradient of the FCM pellet [9]. The deformation of the coated layers and SiC matrix was caused by the internal and outer pressure, thermal expansion, irradiation and creep strain. The strain tensor of FCM was written as follows:

$$\varepsilon = \varepsilon^e + \varepsilon^T + \varepsilon^i + \varepsilon^c \tag{3}$$

where ε is the strain, ε^e, ε^T, ε^i and ε^c represent the elastic strain, thermal expansion strain, irradiation strain and creep strain.

2.2. Material Properties

The properties of dense pyrolytic carbon, SiC layer, and SiC matrix were used in this simulation. The properties of dense pyrolytic carbon in PASTA code used in a high temperature gas cooled reactor have been employed to analyze the performance of the FCM pellet in an LWR environment by Idaho National Laboratory (INL), and the appropriateness of the properties of dense pyrolytic carbon have been proved in previous works [15,16]. The appropriateness of the properties of SiC ceramics described in the following part used in an LWR environment, has been reported [17,18].

2.2.1. Dense Pyrolytic Carbon (PyC) Layers

The properties of the SiC layer, SiC matrix and PyC materials are taken into account in this paper. The elastic modulus of the PyC layer is anisotropic and can be calculated by the following Equation [19]:

$$\begin{aligned} E_{PyC} = 25.5(0.384 + 0.000324\rho_{PyC})(0.481 + 0.519BAF) \\ (1 + 0.23\Phi)(0.9560275 + 0.00015T) \end{aligned} \tag{4}$$

where Φ is the neutron flux ($\times 10^{25}$ n/m^2), T is the temperature (K), and BAF is the anisotropic parameters of PyC.

The radial and tangential irradiation strain of the PyC layers are expressed by the following equations [17]:

$$\dot{\varepsilon}_r = -0.077\exp(-\Phi) + 0.031 \tag{5}$$

$$\dot{\varepsilon}_\theta = -0.036\exp(-2.1\Phi) - 0.01. \tag{6}$$

The irradiation deformation of PyC (as shown in Equations (5) and (6)) was saturated when the neutron flux reached 8.14×10^{25} n/m^2 [16].

Equation (7) is given to calculate the creep strain of the PyC layers along the radial and other directions [17].

$$\dot{\varepsilon}_{cr,r} = K_{pyc}[\sigma_r - v_c(\sigma_\theta + \sigma_\phi)]\dot{\Phi} \tag{7}$$

where $\dot{\Phi}$ is the fast neutron flux rate (10^{25}n·m^{-2}·s^{-1}), v_c is the Poisson ratio (v_c = 0.23), and K_{pyc} is the temperature-dependent creep coefficient.

2.2.2. SiC Layer and SiC Matrix

The elastic modulus and thermal expansion coefficient of the SiC layer are represented as in the following equations [17]:

$$E_{SiC} = 460 - 0.04T\exp\left(-\frac{962}{T}\right) \tag{8}$$

$$\alpha_{sic}(\times 10^{-6}) = \begin{cases} -1.8267 + 0.0178T - 1.5544 \\ \times 10^{-5}T^2 + 4.5246 \times 10^{-9}T^3 & T < 1273K \\ 5.0 & T > 1273K \end{cases} \tag{9}$$

where E and α are the elastic modulus and thermal expansion coefficient, respectively.

The swelling model of the SiC matrix is the function of temperature and neutron flex, which can be written as follows [19]:

$$\dot{S} = k_s\gamma^{-1/3}\exp\left(-\frac{\gamma}{\gamma_{sc}}\right) \tag{10}$$

where S is the swelling rate (s^{-1}), K_s is the coefficient of the swelling rate (dpa$^{-2/3}$), γ is the neutron flux (dpa), and γ_{sc} is the characteristic dose for swelling saturation by the negative feedback mechanism (dpa). The swelling of the SiC can be obtained from the time integration of Equation (10):

$$S = S_s\left[1 - \exp\left(-\frac{\gamma}{\gamma_{sc}}\right)\right]^{2/3} \tag{11}$$

where S_s and γ_{sc} are the function of the temperature and can be expressed as follows:

$$S_s(T) = 0.05837 - 1.0089 \times 10^{-4}T + 6.9368 \times 10^{-8}T^2 - 1.8152 \times 10^{-11}T^3 \tag{12}$$

$$\gamma_{sc}(dpa) = -0.4603 + 2.6674 \times 10^{-3}T - 4.3176 \times 10^{-6}T^2 + 2.3803 \times 10^{-9}T^3. \tag{13}$$

Figure 1 shows the swelling of SiC materials as a function of temperature and dose calculated from Equation (11).

The thermal expansion coefficient of the SiC matrix is expressed as follows [20]:

$$\alpha(10^{-6}/K) = 0.7738 + 7.080 \times 10^{-3}T - 4.951 \times 10^{-6}T^2 + 1.372 \times 10^{-9}T^3. \tag{14}$$

The thermal conductivity of the SiC layer is different from that of the SiC matrix due to the differences in their fabrication and chemical components. Si and C exhibit similar electronegative properties; thus, SiC is mainly combined by C and Si with a covalent bond, according to the Principle of Electro negativity. Unirradiated SiC thermal resistance increases with temperature and defects due to an increase in phonon-phonon scattering. Because of irradiation can increase the quantity of defects, the thermal resistance of SiC increased. The thermal resistance of SiC can be simplified as Equation (15) by neglecting the contribution to the electron to the thermal conductivity at a high temperature [21]:

$$\frac{1}{k} = R_0 + R_{irr} \tag{15}$$

where R_0 and R_{irr} are the thermal resistance (K/W) of the prior- and post-irradiated SiC, respectively.

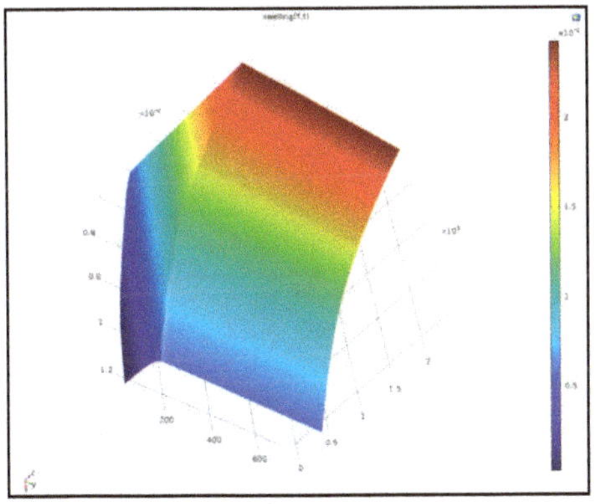

Figure 1. Variation of the swelling silicon carbide (SiC) material as functions of temperature and neutron flux.

The thermal resistance of the unirradiated SiC lay and matrix is expressed as below:

$$R_0 = \frac{1}{-3.7 \times 10^{-8} T^3 + 1.54 \times 10^{-4} T^2 - 0.214T + 153.1}. \tag{16}$$

The thermal resistance induced by the irradiation can be written:

$$R_{irr} = \frac{1}{6.08 \cdot S} \tag{17}$$

where S is the irradiation-induced volume swelling shown in Equation (11).

Failure probability of the SiC layer can be reflected by the stress distribution. The failure probability of brittleness material can be calculated by the following Equation [19]:

$$P = 1 - \exp\left[-\int_V (\sigma_p/\sigma_0)^m dV\right] \tag{18}$$

where V is the characteristic volume, m is the Weibull modulus of the SiC layer, and σ_0 is the characteristic strength, while σ_p is the true stress. The characteristic strength and the Weibull modulus of the SiC layer were 350 MPa and 5, respectively [15].

3. Geometry and Boundary Conditions

The performance of a TRISO particle with a UN kernel was simulated in our previous work [22], and the calculated internal pressure at a different temperature of TRISO was used as input parameters for the FCM pellet simulation. A 1/8 sphere unit was used to simulate the performance of the TRISO particle by defining the boundary condition. The heat source of the TRISO particle was conversed from the liner power of the FCM pellet. The surface temperature was set according to the calculated result of the FCM pellet. The symmetry boundary condition was set on the 3 side surfaces. The structure and boundary condition schematic diagram of the TRISO particle unit is shown in Figure 2. The typical size of the TRISO particle had been studied in [22], and the size of the kernel and coated layers have great influence on the performance of TRISO particles. The structural size of the TRISO particle was similar with that of our previous work [23] and was set as follows: the diameter of the UN kernel was 800 μm, while the thickness of buffer, IPyC, SiC, and OPyC layers were 100 μm, 30 μm, 40 μm, and 30 μm, respectively.

The FCM pellet with 39.8 vol.% TRISO particle loading was simulated in this work. The diameter and height of the FCM pellet was 7 mm and 7 mm, respectively. A three-dimensional model is shown in Figure 3. In order to decrease the calculated quantity of the FCM pellet, a 2-dimensional model was used in this simulation. The 2-dimensional unit was obtained from the 3-dimensional one; the cutting pattern is shown in Figure 3.

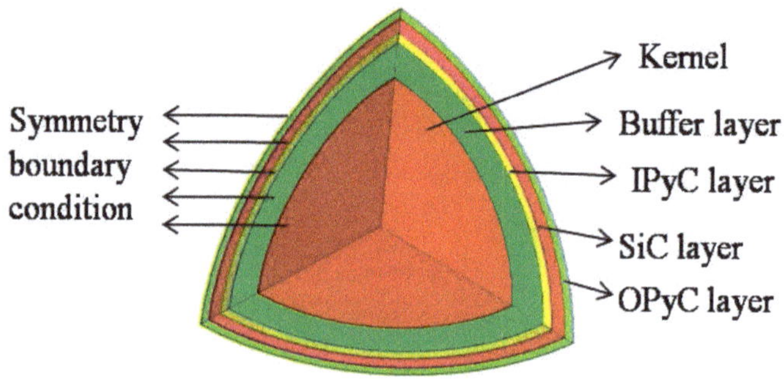

Figure 2. Schematic diagram of a tristructural-isotopic (TRISO) particle characteristic unit. IPyC, inner pyrocarbon; OPyC, outer pyrocarbon.

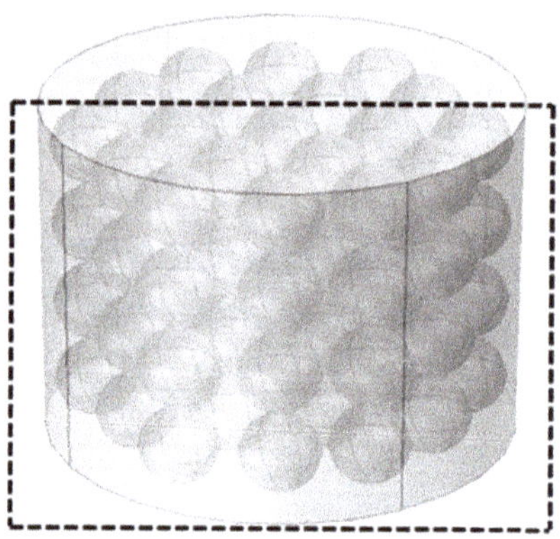

Figure 3. A three-dimensional model of a fully ceramic microencapsulated (FCM) pellet with 39.8 vol.% TRISO particle loading.

Figure 4 shows the 2-dimensional model; the structure and boundary condition was also revealed. The kernel and buffer layers were separated from the IPyC layer during the operation process and have little influence on the performance of the other coated layers and the SiC matrix; this phenomenon has been proven by previous works [19,22]. The kernel and buffer layers were subtracted from the TRISO particle, and the other coated layers, including the IPyC, SiC, and OPyC layers were reserved to reflect the SiC layer's performance and to simulate the interaction between the coated layers and the SiC matrix. The side surface temperature of the FCM pellet was set as about 800 K, which was set as a solid heat transfer boundary. Heat flux was set on the IPyC surface per particle and the value was about 3.53×10^5 W/m^2, which corresponds to an average linear power of 315 W/cm for the FCM pellet. The internal pressure has an obvious effect on the performance of the coated layers, and the internal pressure of the TRISO particle was calculated from the TRISO particle model at a different surface temperature. The bottom part of the FCM pellet was set as a spring foundation to assure simulation convergence. The pressure on the side surface was set as a constant value (P = 2 MPa), which corresponds to the helium pressure of the fuel rod.

A schematic of the calculation flow in the COMSOL software was is shown in Figure 5, and each part of the computable flow graph was described as follows. The two-dimensional model was used to simulate the performance of the FCM pellet. Solid mechanics and heat transfer modules were selected to calculate the temperature and mechanical field of the FCM under a normal condition. The three-dimensional model of the FCM pellet was established, and the two-dimensional model was obtained from the three-dimensional ones, as shown in Figure 3. The geometric model was established by set the size of pellet and coated layers. The material model and boundary condition, as mentioned above, was input, and the grid was subdivided using the native function of COMSOL, according to the physical field. The temperature and stress field of the FCM pellet was output.

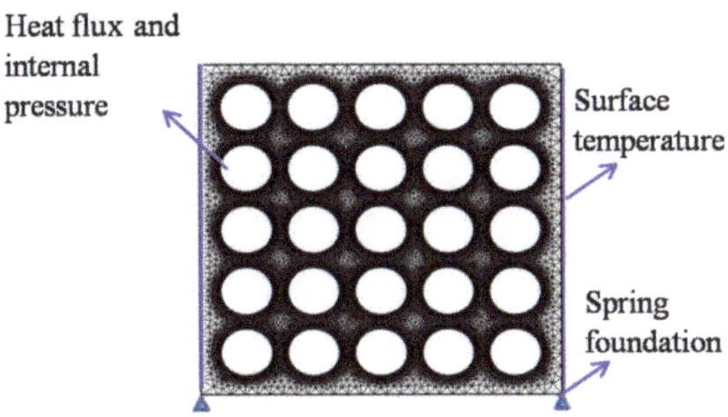

Figure 4. Two-dimensional model of the FCM pellet; the structure and boundery condition were also revealed.

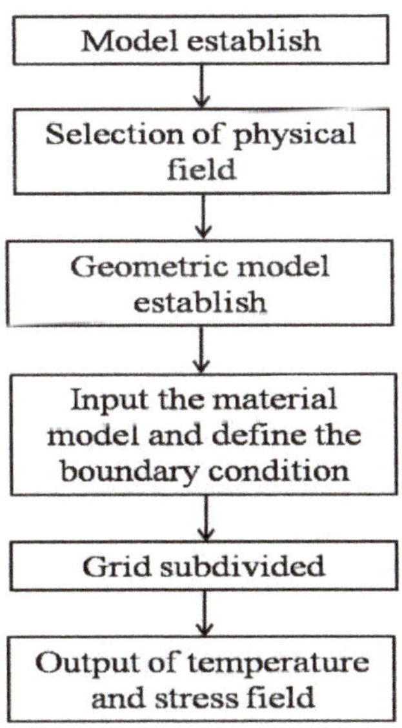

Figure 5. Computable flow graph of simulation process.

4. Results and Discussion

4.1. Temperature Distribution of FCM Pellet

The calculated temperature distribution of an FCM pellet in an LWR environment is shown in Figure 6. The highest temperature of the SiC matrix was located in the center of the FCM pellet, and the maximum temperature at the end of life was about 1390 K, which was similar with the results in the literature [14]. The maximum matrix temperature was much lower than the decomposition or molten point of SiC. The maximum temperature increased rapidly at the beginning and was followed by a slower linear increase. The swelling of the SiC matrix is the function of temperature and neutron flex, which can cause thermal conductivity degradation of SiC matrix, as discussed in the above section (shown in Equation (17)). The thermal conductivity of the SiC matrix reached an essentially steady state condition after 100 d when the SiC swelling reached saturation at about 1 dpa. The maximum temperature of the SiC matrix was much lower than that of the UO_2 pellet in the same operational environment, which was attributed to the high thermal conductivity of the SiC matrix.

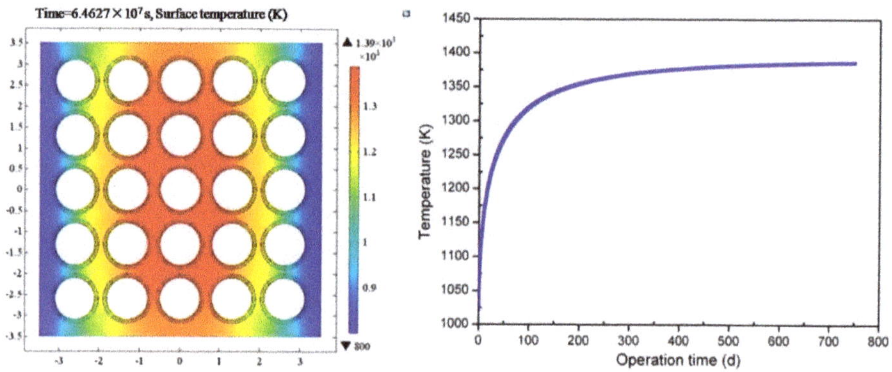

Figure 6. Matrix temperature profile at end of life (EOL) of FCM pellet and variation of the maximum matrix temperature with operation time.

4.2. Mechanical Performance of SiC Matrix

The internal pressure of TRISO particles located at different part of the FCM pellet was calculated using a 1/8 sphere unit model. Internal pressure of the TRISO particle with a different surface temperature is shown in Figure 7. The internal pressure of the TRISO particle increased with the surface temperature. No interaction occurred between the UN kernel and Buffer layers, and the internal pressure was caused by the fission gas. The calculated result was similar with the reported ones, which used UN as a TRISO kernel [23], but the internal pressure was much lower than that in the literature [14]. This result may be attributed to the different kernel, and the interaction between the UO_2 and Buffer layer may increase the internal pressure dramatically. The internal pressure increased with burn up linearly because of the increase in fission gas production. The maximum internal pressure was about 3 MPa at 1400 EFPD when the surface temperature was up to 1200 K, which was in good agreement with the literature [22]. Internal pressure has an obvious influence on the stress distribution of the TRISO particle and the FCM pellet.

Variation of hoop stress on marginal parts of the FCM pallet (non-fuel part) is shown in Figure 8. The bottom part (Parts A, B, and C) shows relatively lower stress than the side part (Parts D and E), which may be caused by the different temperature gradient. Because of the higher temperature of the inner matrix (shown in Figure 6), a higher thermal deformation was induced; thus, the hoop stress of the inner part (Parts B and C) was much higher than that of the outer part (Part A). The non-fuel part hoop stress was up to about 500 MPa at the end of life, which was confirmed by the literature [14]. Cracks may be produced because of the high hoop stress of the SiC matrix during the running operation, but the cracks may have disappeared or been deflected when the cracks extended to the interface between the matrix and the TRISO particles [6]; the formation and extension of cracks will be investigated in our future work. The hoop stress decreased at the beginning and then increased. The variation of hoop stress may be attributed to the irradiation deformation of the PyC layers; the shrinkage of PyC at the beginning may release the stress of SiC matrix, and then the swelling of PyC may cause an increase in hoop stress.

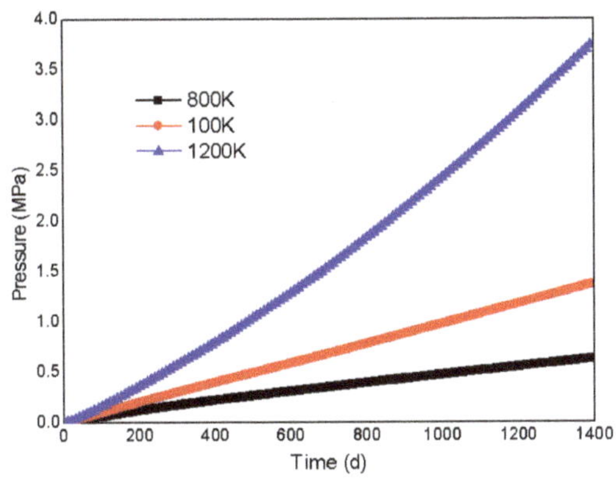

Figure 7. Fission gas pressure inside the TRISO particle throughout its life time.

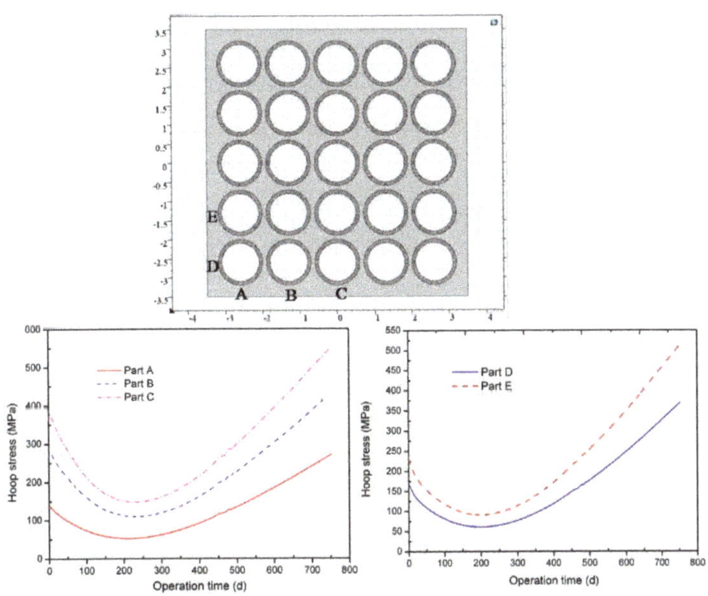

Figure 8. Variation of the hoop stress of the SiC matrix located in the marginal parts of the FCM pallet.

Figure 9 shows the hoop stress of the inner SiC matrix located between the two TRISO particles. The maximum hoop stress of the inner SiC matrix was up to 1200 MPa at the end of life, and this result was in good agreement with the literature [14] (the maximum value in the literature was about 1160 MPa). The inner SiC matrix between two TRISO particles suffered large hoop stress; the stress decreased at the beginning and then increased. This phenomenon was caused by the interaction between the TRISO particle and the SiC matrix. The hoop stress of the SiC matrix, including the non-fuel region and the inner part, was higher than the SiC's intrinsic strength, and the SiC matrix may be broken during the operation process.

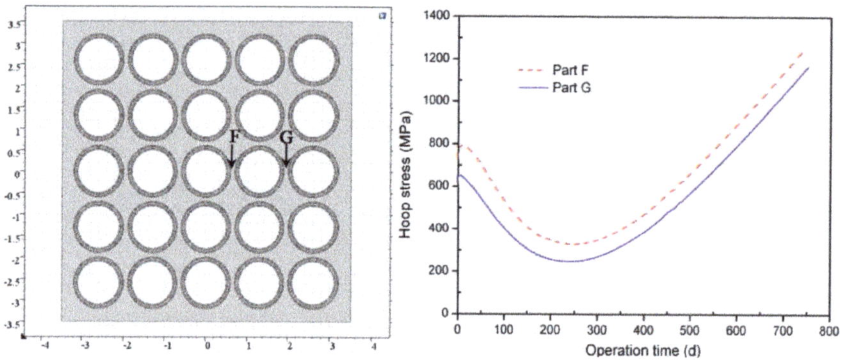

Figure 9. Variation of hoop stress of the inner SiC matrix located between two TRISO particles.

4.3. Mechanical Performance of SiC Layer

The fission product capabilities of FCM can be reflected by the structural integrity of the SiC matrix and the the SiC layers in the TRISO particles. The fission products can be effectively prevented by the SiC layers, which have been proven by an advanced gas cooled reactor [8]. The thermal-mechanical performance and failure probability of the SiC layers were important parameters to evaluate the integrity and fission product capabilities of FCM. There was no consideration of SiC layer performance on FCM pellets in previous reports as far as we know. The performance of the SiC layer located in different parts of the FCM pellet was calculated in this work.

The hoop stress of the SiC layers in TRISO particles, which embed in different parts of the FCM pellet, is shown in Figure 10, and the detected TRISO particles were labeled. SiC layers located in different parts of the FCM suffered different hoop stress. TRISO particles located in the inner part of the FCM (such as No. 1, No. 2, No. 5 and No. 8) suffered high hoop stress, which can be attributed to the high temperature gradient and thermal deformation. The maximum hoop stress of the SiC layers was about 180 MPa, which was much lower than the SiC strength (350 MPa) [15]. The maximum stress of the SiC layer was similar to the literature [19]. The hoop stress of all the SiC layers decreased at the beginning, and this result may be attributed to the irradiation deformation of the IPyC and OPyC layers, which was similar to the calculated result of the TRISO particles [22]. The irradiation deformation of the IPyC and OPyC layers can protect the SiC layers at the beginning time [19]. Then, irradiation deformation of the IPyC and OPyC layers was saturated, and the protective effect disappeared, thus the hoop stress of the SiC layer increased. The irradiation deformation of the SiC matrix may pull or push the TRISO particles located in different parts of FCM pellet, which may increase the hoop stress of the SiC layers.

The failure probability of the SiC layers located in different parts of the FCM pellet is shown in Figure 11. They are higher than the single TRISO particle with probabilities lower than 5.0×10^{-5} [22], which was confirmed by the stress distribution of the SiC layer. Similar to the single TRISO particle model, the pressure contribution to SiC failure was very limited because of the low gas release and internal pressure of the TRISO particles with the UN kernel under a PWR environment (shown in Figure 7). The maximum failure probability of the TRISO was No. 8 (showed in Figure 11), which was the inner one located on the side edge of FCM pellet. A large temperature gradient and matrix swelling deformation at a relatively low temperature (shown in Figure 1) may induce large hoop stress on the SiC layer, and the failure probability increased. As a whole, the SiC layers located in different parts of the FCM pellet suffered low stress and the integrity of the SiC layers was maintained.

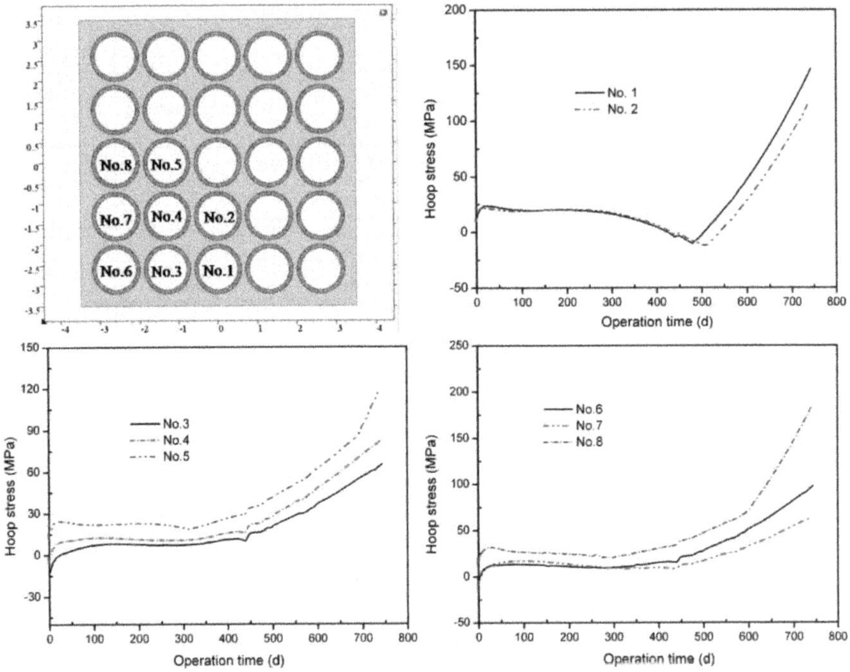

Figure 10. Variation of hoop stress on the SiC layer located in different parts in the FCM pellet.

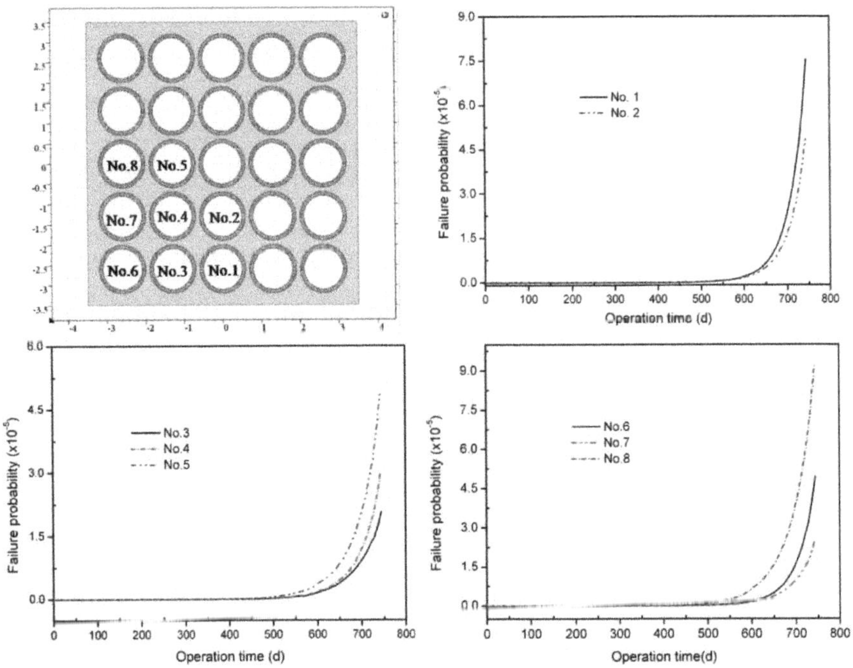

Figure 11. Failure probability of the SiC layers located in different parts of the FCM pellet.

5. Conclusions

A two-dimensional model of an FCM pellet with a homogenous distribution of TRISO particles was generated to simulate the thermal-mechanical performance of an FCM pellet. The UN kernel and buffer layers were subtracted because the kernel and buffer layer were separated from other coated layers and have little influence on the thermal-mechanical performance of the SiC layer and matrix, which has been proven by a high temperature gas cooled reactor. The interaction between the matrix and TRISO particle was considered, and the performance of SiC layer and matrix were calculated. The maximum matrix temperature increased rapidly at the beginning and was followed by a slower linear increase, which corresponded to the degradation of SiC thermal conductivity. The SiC matrix, including the non-fuel region and inner parts, suffered a large hoop, which can break the SiC matrix. The maximum hoop stress of the SiC layers located in different parts was about 180 MPa, which was much lower than the intrinsic strength of the SiC ceramics. The failure probability of the SiC layers was lower than 9×10^{-5}, which indicated that the structural integrity of the SiC layers was maintained. The integrity of the SiC layers is beneficial to the capacity of fission product retention.

Author Contributions: Conceptualization, P.C. and S.Q.; methodology, S.L.; software, S.L.; validation, Y.Z., Y.X. and S.G.; formal analysis, H.L.; investigation, X.Q.; resources, X.Q.; data curation, P.C.; writing—original draft preparation, S.L.; writing—review and editing, X.Q.; visualization, X.Q.; supervision, S.Q.; project administration, P.C.; funding acquisition, P.C.

Funding: This research received no external funding.

Conflicts of Interest: The authors declare no conflict of interest.

References

1. Chun, J.H.; Lim, S.W.; Chung, B.D. Safety evaluation of accident-tolerant FCM fueled core with SiC-coated zircalloy cladding for design-basis-accidents and beyond DBAs. *Nucl. Eng. Des.* **2015**, *289*, 287–295. [CrossRef]
2. Snead, L.L.; Terrani, K.A.; Venneri, F. Fully ceramic microencapsulated fuels: A transformational technology for present and next generation reactors-properties and fabrication of FCM fuel. *Trans. Am. Nucl. Soc.* **2011**, *104*, 668–670.
3. Snead, L.L.; Terrani, K.A.; Katoh, Y. Stability of SiC-matrix microencapsulated fuel constituents at relevant LWR conditions. *J. Nucl. Mater.* **2014**, *448*, 389–398. [CrossRef]
4. Lu, C.; Hiscox, B.D.; Terrani, K.A. Fully ceramic microencapsulated fuel in prismatic high temperature gas-cooled reactors: Analysis of reactor performance and safety Characteristics. *Ann. Nucl. Energy* **2018**, *117*, 277–287. [CrossRef]
5. Terrani, K.A.; Zinkle, S.J.; Snead, L.L. Snead, Microencapsulated fuel technology for commercial light water and advanced reactor application. *J. Nucl. Mater.* **2012**, *427*, 209–224. [CrossRef]
6. Terrani, K.A.; Kiggans, J.O.; Katoh, Y. Fabrication and characterization of fully ceramic microencapsulated fuels. *J. Nucl. Mater.* **2012**, *426*, 268–276. [CrossRef]
7. Sen, R.S.; Pope, M.A.; Ougouag, A.M. Assessment of possible cycle lengths for fully encapsulated microstructure fueled light water reactor concepts. *Nucl. Eng. Des.* **2013**, *255*, 310–320. [CrossRef]
8. Powers, J.J.; Wirth, B.D. A review of TRISO fuel performance models. *J. Nucl. Mater.* **2010**, *405*, 74–82. [CrossRef]
9. Besmann, T.M.; Ferber, M.K.; Lin, H.T. Fission product release and survivability of UN-kernel LWR TRISO fuel. *J. Nucl. Mater.* **2014**, *448*, 412–419. [CrossRef]
10. Kamalpour, S.; Salehi, A.A.; Khala, H. The potential impact of Fully Ceramic Microencapsulated (FCM) fuel on thermal hydraulic performance of SMART reactor. *Nucl. Eng. Des.* **2018**, *339*, 39–52. [CrossRef]
11. Lee, Y.; Cho, N.Z. Steady and transient-state analyses of fully ceramic microencapsulated fuel loaded reactor core via two-temperature homogenized thermal conductivity model. *Ann. Nucl. Energy* **2015**, *76*, 283–296. [CrossRef]
12. Lee, H.G.; Kim, D.; Lee, S.J. Thermal conductivity analysis of SiC ceramics and fully ceramic microencapsulated fuel composites. *Nucl. Eng. Des.* **2017**, *311*, 9–15. [CrossRef]

13. Ougouag, A.M.; Kloosterman, J.L. Investigation of bounds on particle packing in pebble-bed high temperature reactors. *Nucl. Eng. Des.* **2006**, *236*, 669–676. [CrossRef]
14. Schappel, D.; Terrani, K. *Thermo Mechanical Analysis of Fully Ceramic Microencapsulated Fuel during in-Pile Operation*; American Nuclear Society-ANS: La Grange Park, IL, USA, 2016.
15. Boer, B.; Sen, S.R.; Pope, A.M. *Material Performance of Fully-Ceramic Micro-Encapsulated Fuel under Selected LWR Design Basis Scenarios: Final Report*; INL/EXT-11-23313; Idaho National Laboratory: Idaho Falls, ID, USA, 2011.
16. Boer, B.; Ougouag, A.M. Stress analysis of coated particle fuel in graphite of High-Temperature Reactors. *Nucl. Technol.* **2008**, *162*, 276–292. [CrossRef]
17. Miller, G.K.; Petti, D.A.; Maki, J.T. *PARFUME Theory and Model Basis Report*; Idaho National Laboratory, Next Generation Nuclear Plant Project; Idaho National Laboratory: Idaho Falls, ID, USA, 2009.
18. Demancge, P.; Marian, J. TRISO fuel element thermal-mechanical performance modeling for the hybrid LIFE engine with Pu fuel blanket. *J. Nucl. Mater.* **2010**, *405*, 144–155. [CrossRef]
19. Katoh, Y.; Koyanagi, T.; Singh, G.; Terrani, K.A.; Petrie, C.M.; Snead, L.L.; Deck, C. *Irradiation-High Heat Flux Synergism in Silicon Carbide-Based Fuel Claddings for Light Water Reactors*; American Nuclear Society-ANS: La Grange Park, IL, USA, 2016.
20. Katoh, Y.; Ozawa, K.; Shih, C. Continuous SiC fiber, CVI SiC matrix composites for nuclear applications: Properties and irradiation effects. *J. Nucl. Mater.* **2014**, *448*, 448–476. [CrossRef]
21. Stone, J.G.; Schleicher, R.; Deck, C.P. Stress analysis and probalibistic assessment of multi-layer SiC-based accident tolerant nuclear fuel cladding. *J. Nucl. Mater.* **2015**, *466*, 682–697. [CrossRef]
22. Collin, B.P. Modeling and analysis of UN TRISO fuel for LWR application using the PARFUME code. *J. Nucl. Mater.* **2014**, *451*, 65–77. [CrossRef]
23. Li, W.; Wu, X.; Liu, S. Performance analysis of TRISO coated fuel particle with UN kernel. *Atom. Energy Sci. Technol.* **2018**, *52*, 283–289. (In Chinese)

mathematics

MDPI

Article

A Numerical Study on the Crack Development Behavior of Rock-Like Material Containing Two Intersecting Flaws

Bing Dai [1], Ying Chen [1,*], Guoyan Zhao [2], Weizhang Liang [2,*] and Hao Wu [2]

1 School of Resources Environment and Safety Engineering, University of South China, Hengyang 421000, China; daibingusc@usc.edu.cn

2 School of Resources and Safety Engineering, Central South University, Changsha 410000, China; gyzhao@csu.edu.cn (G.Z.); hoekwu@csu.edu.cn (H.W.)

* Correspondence: csu_chenying@csu.edu.cn (Y.C.); wzlian@csu.edu.cn (W.L.); Tel.: +86-185-6906-9231 (Y.C.); +86-151-1141-9038 (W.L.)

Received: 31 October 2019; Accepted: 10 December 2019; Published: 11 December 2019

Abstract: It is quite often that rocks contain intersecting cracks. Therefore, crack behavior cannot be completely studied by only considering several isolated, single flaws. To investigate the crack behavior of rock or rock-like material containing intersecting flaws under uniaxial loading, numerical simulations were carried out using parallel bonded-particle models containing two intersecting flaws with different inclination angles (varying β) and different intersection angles (varying αα). The crack propagation processes are analyzed and two typical patterns of linkage are observed between two intersecting flaws: (1) One-tip-linkage that contains three subtypes: Coalescence position near the tip; coalescence position at the flaw, but far away from the tip; coalescence position outside the flaw at a certain distance from the tip; and (2) two-tip-linkage with two subtypes: Straight linkage and arc linkage. The geometries of flaws influence the coalescence type. Moreover, the effects of intersection angle α and inclination angle β on the peak stress, the stress of crack initiation, and the stress of crack coalescence are also investigated in detail.

Keywords: crack growth behavior; particle model; intersecting flaws; uniaxial compression

1. Introduction

Numerous discontinuities on different scales occur in rocks and rock masses, such as flaws, joints, cracks, and faults, which are induced during the formation of rock masses and successive tectonic processes. The existence of joints, cracks, or faults in rock masses has two mechanical effects [1–4]: (1) The strength and stiffness of the rock mass will be decreased; and (2) they will generate new fractures that may further develop and connect with other flaws and then may lead to the degradation of the strength and stiffness of the rock or rock mass, respectively. Therefore, the mechanism of crack propagation and coalescence is of great interest in geomechanics [5,6].

To further study the laws of cracks and the development of single cracks, physical experiments were conducted [7–18]. The observed crack patterns under uniaxial compression can be categorized as follows: (1) Wing cracks initiated at the tips of single, pre-existing flaws propagating along the direction of axial stress; and (2) secondary shear or mixed mode tensile-shear cracks initiated later from the tips of single pre-existing flaws propagating as quasi-coplanar or oblique cracks.

Crack coalescence mechanisms were studied by other researchers with rock-like materials containing two or more parallel flaws in a compression state [1,19–27]. It was observed that flaw angle, bridging angle, non-overlapping length, and frictional coefficient have significant effects on the crack initiation modes and the patterns of crack coalescence. Moreover, it was shown that the peak strength and stiffness of rocks and rock-like material depends on the number and geometry of flaws [28].

Previous work provides a good understanding of the coalescence patterns obtained from specimens with one or several single, non-intersecting flaws. However, joint patterns with more than one joint set are common in nature. Pollard and Aydin [29] defined several types of joint intersection geometries, which can be classified as orthogonal (+intersections) and non-orthogonal (X intersections). Both types can be divided into three groups according to the continuity of the joints at the intersections: (1) "all continuous joints"; (2) "continuous and discontinuous sets"; and (3) "all discontinuous sets".

Lee and Jeon [30] used three materials (PMM, Diastone and Hwangdeung granite) to study the coalescence of a horizontal flaw with an underneath lying, inclined flaw. They stated that the geometry of a double-flaw can improve the understanding of crack propagation and coalescence because echelon type of cracks can be initiated by fracture planes that are not parallel to each other. Zhang et al. also studied the crack coalescence of gypsum with two non-parallel flaws under different flaw angles and different bridge angles. Their results showed that the stress distribution in the bridge area of the non-parallel flaws is more complicated than that of the parallel flaws. This difference affects the crack initiation stress as well as the coalescence pattern. Their studies provided fundamental understanding of the coalescence of non-parallel flaws, but observations made from two non-parallel flaws can only provide a limited understanding of the complex behavior of the rock mass with several crossing discontinuities. Besides this, numerical simulation also plays an important role in crack coalescence research [1,22,31–33]. One of the popular tools nowadays is the discrete element method [18,34–40].

In the present study, numerical simulation results are shown in respect to peak strength, crack initial stress, crack coalescence stress, and coalescence patterns of rock-like material with two intersecting flaws under uniaxial compression using a bonded-particle model (BPM).

2. Numerical Simulation Procedure

We used the bonded particle model (BPM), which is one of the contact models available in the two-dimensional particle flow code (PFC2D) software for numerical simulation. Turco [41] suggested a micro-mechanical model, which completely is described by the definition of the strain energy of the interaction between two nearby grains without consideration of the dynamical effects. This model can also effectively handle large deformation processes, even if quasistatic simulations require the solution of remarkable numerical challenges because they might present instabilities and bifurcation phenomena [42]. In the present study, numerical simulation focuses on the peak strength, crack initial stress, crack coalescence stress, and coalescence patterns of rock-like material with two intersecting flaws under uniaxial compression, thus the bonded particle model (BPM) could satisfy this study.

In PFC (such as that in BPM), the model specimens are created by combining a large number of circular particles with varying diameters. These particles are interconnected using parallel bonds, as shown in Figure 1. Tensile and shear stresses acting on the parallel bonds are calculated during simulations and if the maximum tensile stress exceeds the tensile strength of the parallel bonds or the maximum shear stress exceeds the shear strength of the parallel bonds, the parallel bonds will break and a microscopic tensile or shear crack will form (Figure 1).

To resemble the real cracking processes, discrete cracks that are close enough are connected with each other by a unique technique [18]. The symbol "a" stands for the centroids' distance form a crack to the other one nearby, and the symbol "c" stands for the length of any of these two cracks (Figure 2). If a/c is less than or equal to 1, the crack, and the other one nearby, are regarded as a single continuous crack, and then the centroids of the two micro-cracks will connect with each other as a macroscopic crack trace line [18].

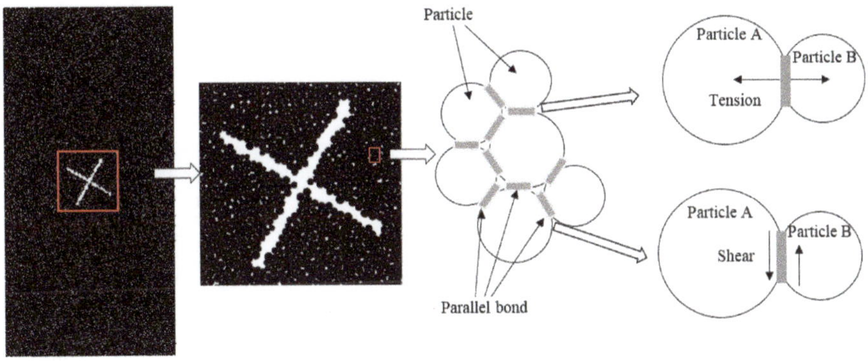

Figure 1. Illustration of the parallel bond model.

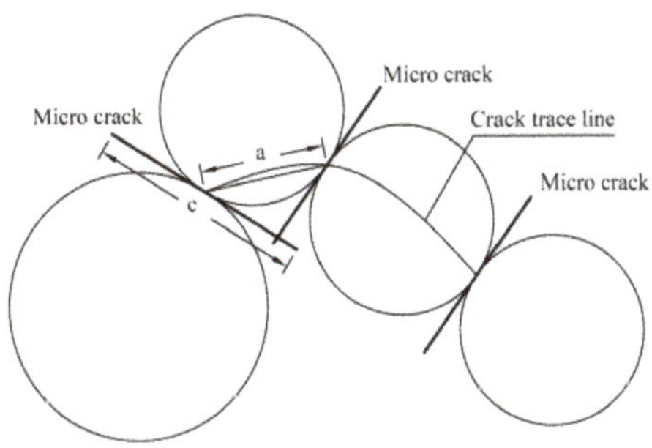

Figure 2. Constructing a macro-crack based on connecting centroids of micro-cracks [18].

The geometry and parameters of the model specimens were chosen in such a way that the model duplicates the mechanical behavior of the gypsum specimens used in the experimental investigation of Wong and his subsequent numerical simulations [43]. The model specimens were created with the dimensions 50 mm in width and 100 mm in height, which are consistent with those of the specimens used for the laboratory testing by Wong [25]. The model specimens created in this manner comprised about 19,650 particles. The porosity of the model specimens was found to be 0.01.

The simulation program consisted of two stages. First, model specimens free of pre existing flaws (i.e., intact) were simulated. Micro-parameters for the particles were selected in order to obtain a consistent macro-scale mechanical behavior with that of the gypsum specimens used by Wong [25]. Table 1 outlines these micro parameters assigned for the particles and parallel bonds of the numerical specimens independently. The next section reports the comparison of the mechanical properties of the numerical specimens of the present study and those of the specimens used for laboratory testing by Wong [25].

The second stage of the study involved a series of numerical simulations on model specimens containing two intersecting flaws with varied geometry to explore the influence of various geometries on the mechanical behavior of model specimens; flaw length L and thickness were fixed at 20 and 1 mm for both flaws, respectively, while the intersection angle α of the two flaws varied at 90°, 60°, 45°, and 30° (Figure 3). The inclination angle of one flaw (angle β of flaw B as shown in Figure 3b) was also varied in the next stage, while α was kept unchanged. To better distinguish the crack coalescence types and for the clarity of the explanations, the bridging zone (the zone surrounding the intersection of flaws) was divided into four zones, as shown in Figure 3c (I, II, III, and IV).

Table 1. Micro-parameters used for the model specimens.

Particle Parameters	Value	Parallel Bond Parameters	Value
Ec Young's modulus of the particle	2.5 GPa	$\overline{E}c$ Young's modulus of the parallel bond	2.5 GPa
kn/ks ratio of normal to shear stiffness of the particle	1	$\overline{kn}/\overline{ks}$ ratio of normal to shear stiffness of the parallel bond	1
Grain friction coefficient,	0.01	$\overline{\sigma}c$ tensile strength of the parallel bond	6 MPa
Rmax/Rmin ratio of max particle radius to min particle radius	1.66	$\overline{c}$ cohesion of the parallel bond	15 MPa
Rmin Lower bound of particle radius	0.21 mm	$\overline{\lambda}$ radius multiplier	1
ρ particle density	1.83 g/m^3	$\overline{\Phi}$ Friction angle of the parallel bond	0

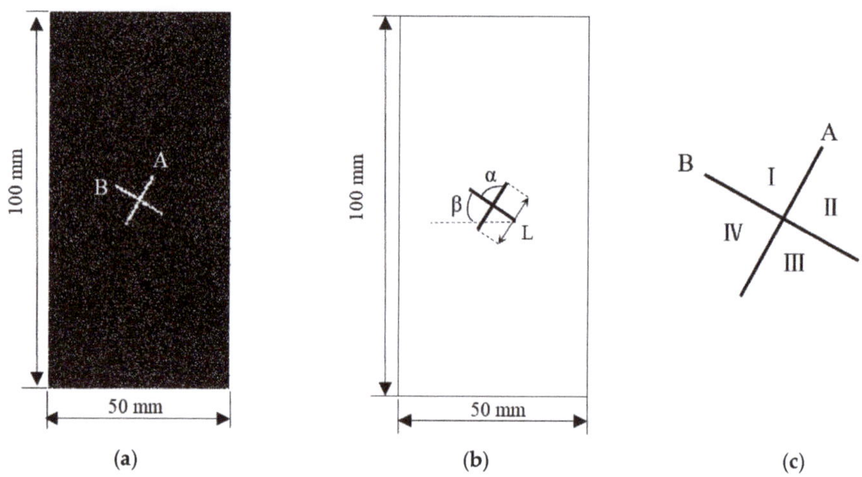

Figure 3. Geometries of the model containing two intersecting flaws: (**a**) Flaws are numbered A and B; (**b**) flaw inclination angle β, intersection angle α and flaw length L; and (**c**) four bridging zones.

For all the simulations in the present numerical study, specimens were vertically loaded in a velocity-controlled manner and no confining pressure was applied (i.e., uniaxial compression). A sufficiently low loading rate −0.05 m/s was applied to ensure that the specimens remained in a quasi-static equilibrium throughout the test.

3. Results and Discussion

3.1. Mechanical Properties of Intact Models

We first determined the macro-scale mechanical properties of the model specimens and compared them with the corresponding properties reported by Wong [43]. Table 2 and Figure 4 show this comparison. It can be seen that the numerical model has very closely reproduced the mechanical behavior of gypsum specimens used in the laboratory investigation by Wong [43]. Following this verification, we simulated the mechanical behavior of specimens containing two intersecting flaws and the forthcoming sub-sections report the results of those simulations.

Table 2. Lab test results [43] vs. numerical simulation results.

Parameter	Experimental Tests	BPM
Density (g/cm^3)	1.54	1.54
Young's modulus (GPa)	5.96	6.09
Poisson's ratio	0.15	0.15
Uniaxial compressive strength (MPa)	33.85	34.2
Tensile strength (MPa)	3.2	4.8

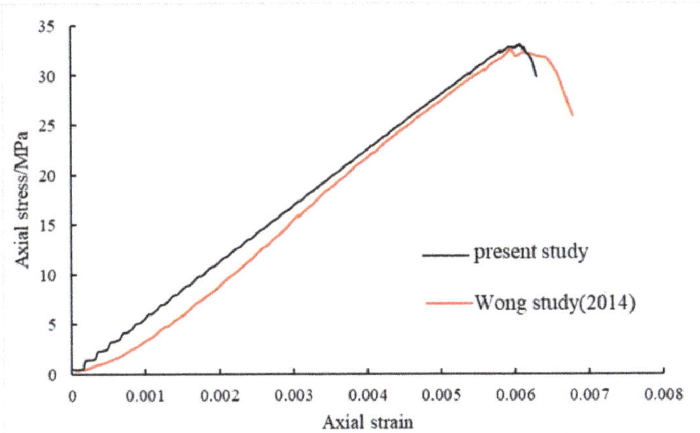

Figure 4. The stress-strain curves: Lab test vs. numerical simulation results.

3.2. *Mechanical Behavior of Specimens Containing Flaws*

3.2.1. Effect of the Flaw Geometries on Peak Strength

The influence of flaw geometry was investigated in regards to the strength of the sandstone samples, gypsum, and marble containing single crack, parallel cracks, or non-parallel cracks in uniaxial loading [20,21,44,45]. Their results revealed that flaw geometry has a key effect on the strength. Thus, variations of peak strength against β for different values of α are shown in Figure 5.

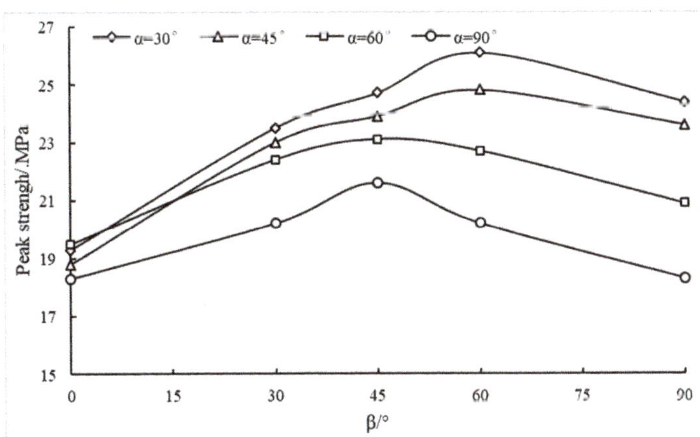

Figure 5. Variations of peak strength against β for different values of α.

From Figure 5, the following general trends can be deduced:

(1) For low values of α (below app. 50°), the peak strength first increases up to about β = 60° and then decreases.

(2) For higher values of α (above app. 50°), the peak strength first increased up to about β = 45° and then decreased.

(3) For the values of β (above app. 15°), the peak strength decreased with increasing α.

To explain this, the peak strengths of the model containing single crack against β for different initial angles are shown in Figure 6.

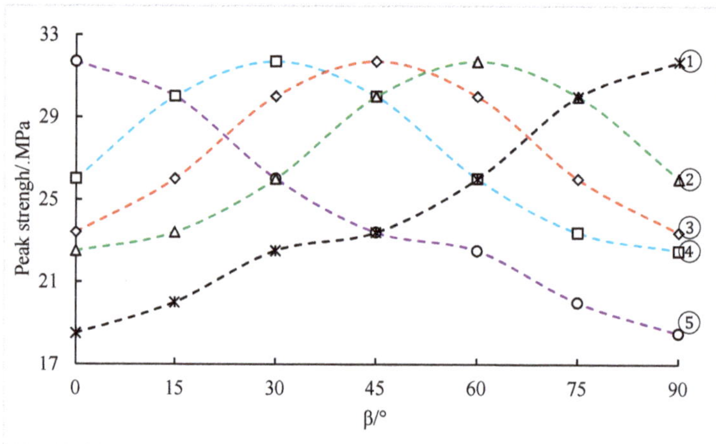

Figure 6. The peak strengths of the model containing a single flaw against β for different flaw initial inclination angles. ① ② ③ ④ ⑤ curves corresponding to five flaw initial inclination angles in Figure 7.

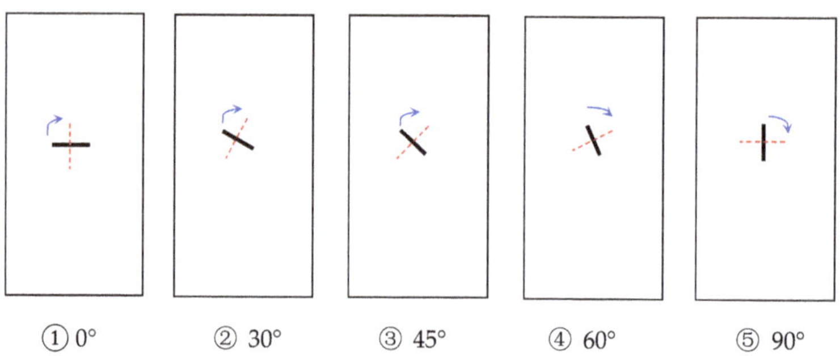

Figure 7. Five flaw initial inclination angles of the model containing a single flaw. The red dotted line represents the final state of rotation for the flaw. The blue arrow represents the direction of rotation.

The overall strength of a rock mass containing two sets of discontinues is given by the lowest strength envelope in the individual strength curves, according to Bray's suggestion [46]. It is clear from Figure 6 that the strength curves of the model containing intersect cracks had a similar trend with the lowest strength envelope in the individual strength curves. For example, the lowest strength envelope in the individual strength curves ① ⑤ show the strength of the model containing two intersect cracks (α = 90°); the peak strength first increases up to about β = 45° and then decreases with the increase of the inclination angle. It is worth noting that the lowest strength envelope in the individual strength curves ① ④ are a little bit different (when β = 60°) from the trend of α = 60° in Figure 5. One plausible explanation is that it was due to the two cracks effecting each other. Another plausible explanation is that it was due to a random distribution of particle sizes.

3.2.2. Crack Initiation and Coalescence Behavior

The crack evolution of models containing two intersecting flaws (α = 90°, 60°, 45°, and 30°) is shown in Table 3. Five stages corresponded to five points on the stress–strain curve in Figure 8: Document crack initiation, propagation, coalescence, as well as crack type at peak stress or post-peak stress level. To shorten and avoid repetition, only selected constellations of α and β are documented in detail.

Table 3. The crack evolution pattern at different stages of loading with different flaw geometry. The cracks corresponding to five points on the stress–strain curve in Figure 8. Microscopic tensile and shear cracks are shown in white and red, respectively. Original flaws are in yellow.

Flaw Geometry	Crack Evolution Pattern at Different Stages of Loading				
	(A)	(B)	I	(D)	I
α = 90° β = 30°					
α = 60° β = 0°					
α = 45° β = 45°					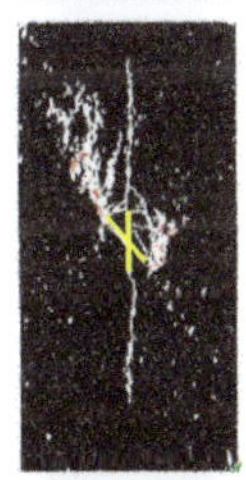
α = 30° β = 0°		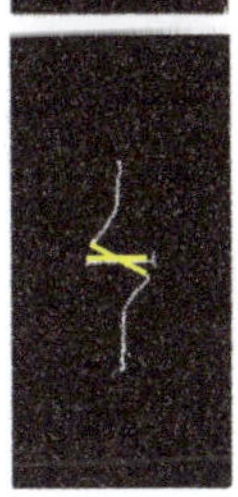	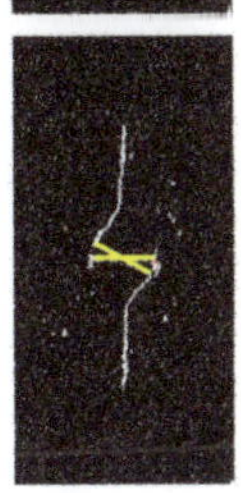	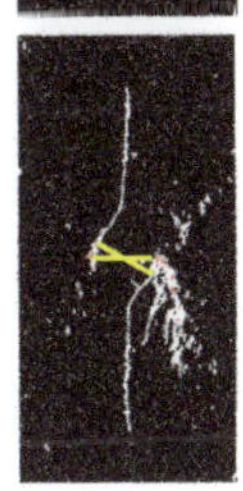	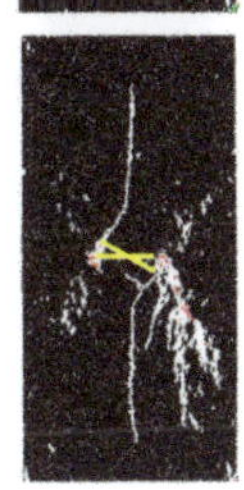

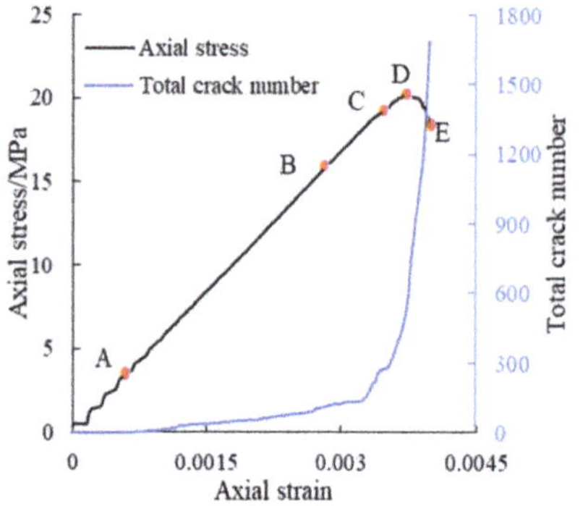

(**a**) $\alpha = 90°$, $\beta = 30°$

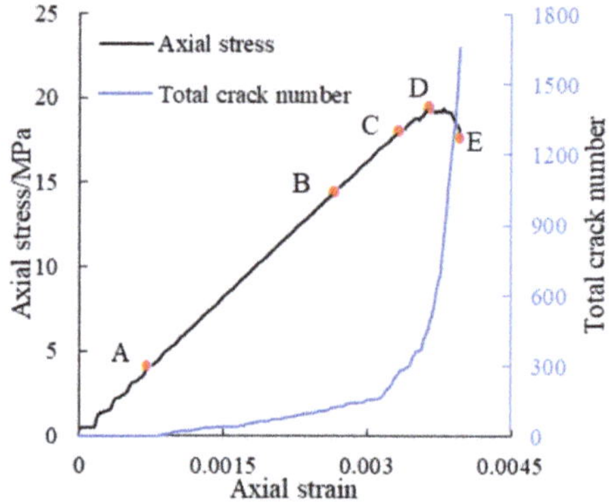

(**b**) $\alpha = 60°$, $\beta = 0°$

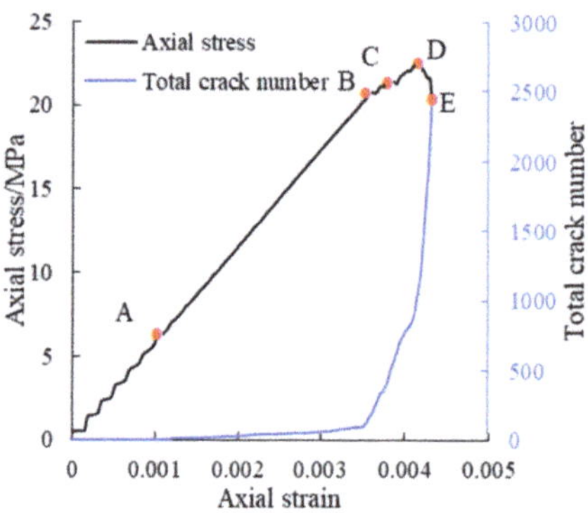

(**c**) $\alpha = 45°$, $\beta = 45°$

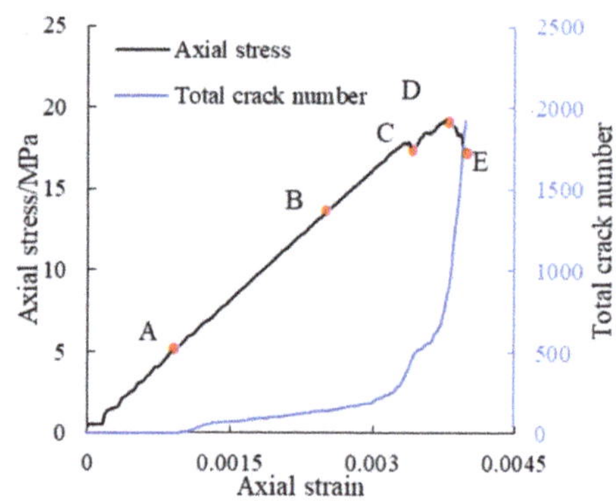

(**d**) $\alpha = 30°$, $\beta = 0°$

Figure 8. Axial stress (black line) and total crack number (blue line) versus axial strain with different values of α and β. Five points, A, B, C, D, and E, on the stress–strain curve are monitored to observe the crack initiation, propagation, and coalescence.

For flaw geometry with $\alpha = 90°$ and $\beta = 30°$, macroscopic wing cracks initiate from the tips of the A flaw. With an increasing load, the wing cracks extend, and macroscopic tensile cracks appear in the tip area of the B flaw as secondary cracks. When axial stress reaches 18.8 MPa, the first coalescence is achieved in bridge zone II by the extension of the secondary crack starting from the right tip of the B flaw. During a further load increase up to the peak strength, a downward extension of the macroscopic tensile crack, starting from the left tip of the B flaw, reaches the left tip of the A flaw, and the second coalescence occurs in bridge zone IV. The test specimen is broken by the development of cracks during the post-peak stress stage. The coalescence cracks in the bridge region become progressively wider during the cracking processes.

For flaw geometry with $\alpha = 60°$ and $\beta = 0°$, and when axial stress has reached 4.8 MPa, macroscopic wing cracks start to propagate from the tips of the A flaw. With a further load increase, the wing cracks extend and some microscopic cracks appear in the tip area of the B flaw (secondary cracks). At 18 MPa vertical stress, the first coalescence is achieved in bridge zone III by the extension of the secondary crack initiated from the right tip of the B flaw. At peak stress of 19.5 MPa, coalescence cracks become wider and some new microscopic tensile and shear cracks propagate downwards or upwards from the tips of the B flaw. Beyond peak stress, some secondary cracks extend quickly towards the edge of the model, and finally the model fails.

For flaw geometry with α = 45° and β = 45°, macroscopic tensile cracks start from the tips of the A flaw. With an increasing load, wing cracks extend and some microscopic tensile cracks are initiated at the tips of the B flaw. With a further stress increase, secondary tensile and shear cracks propagate towards the wing crack initiated at the top tip of the A flaw. At 21.1 MPa, the first coalescence occurs in bridge zone I. At peak stress of 23.9 MPa, the secondary cracks propagate further upwards from the right tip of the B flaw towards the top tip of the A flaw. Secondary coalescence occurs in bridge zone II. Additional cracks are induced close to the B flaw tip regions in the post-peak stage.

For flaw geometry with α = 30° and β = 0°, at 5.4 MPa, a macroscopic wing crack starts to propagate from the tips of the A flaw. Further load increases lead to the initiation of micro tensile cracks at the right tip of the B flaw. When the axial stress increases further up to 17.6 MPa, the first coalescence is achieved in bridge zone III by the extension of a secondary crack initiated from the right tip of the B flaw. At peak strength, coalescence occurs in bridge zone I. Meanwhile, some mixed mode tensile–shear cracks appear and extend downwards near the right tips of the A and B flaws. In the post-peak region at a stress of 17.4 MPa, some secondary cracks become wider and extend to the edge of the model and the model fails.

3.3. Crack Initiation Stress

In the numerical studies performed by Zhang et al., wing cracks start to propagate simultaneously at the tips of the flaws. However, in this study, the macro-cracks initiated at the tips of the two intersecting flaws do not occur simultaneously. Therefore, the stress state when a macro-crack starts to develop from either the A flaw or B flaw is called crack initiation (CI) stress.

Specimens with different flaw geometries have different peak strengths; therefore, CI stress cannot be compared directly. Thus, the CI stress σ_{ci} is normalized by the respective peak strength σ_{cn} (see Figure 9). From Figure 9, the following general trends can be deduced:

1. The ration σ_{ci}/σ_{cn} varies between 0.2 and 0.4.
2. For the same values of α, the minimum values of σ_{ci}/σ_{cn} are observed at β = 30°, which means that the first crack initiates more easily (except α = 60°, β = 45°). For four cases of β = 90° and one case of β = 0°, α = 90° is the maximum value of σ_{ci}/σ_{cn} observed.

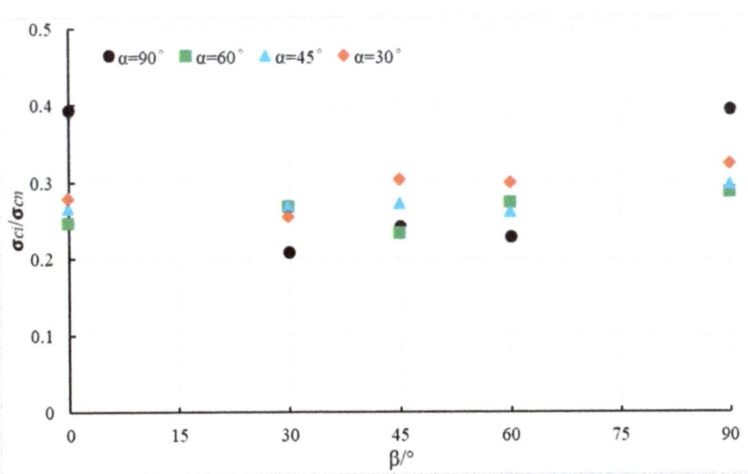

Figure 9. σ_{ci}/σ_{cn} versus flaw inclination angle β for different values of α.

To better understand the effect of α and β in the ratio σ_{ci}/σ_{cn}, the differences ($\Delta\ \sigma_{ci}/\sigma_{cn}$) between the maximum and minimum values of σ_{ci}/σ_{cn} are calculated, as shown in Figure 9. From Figure 10a, the difference of σ_{ci}/σ_{cn} is first decreased down to α = 45° and then increased with the increase of α, which indicates that the inclination angle β had less effect on the ration σ_{ci}/σ_{cn} when α = 45°.

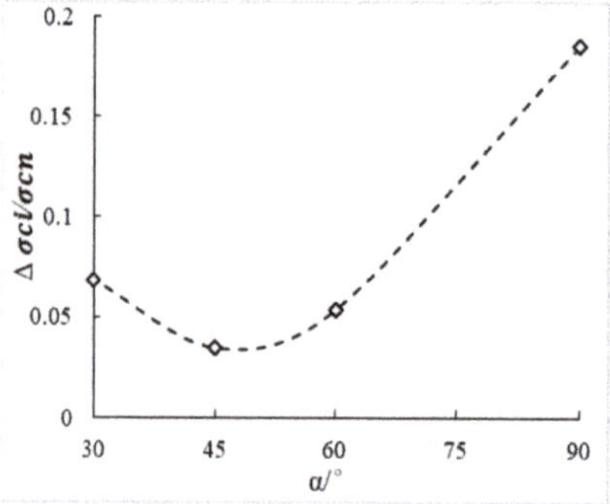

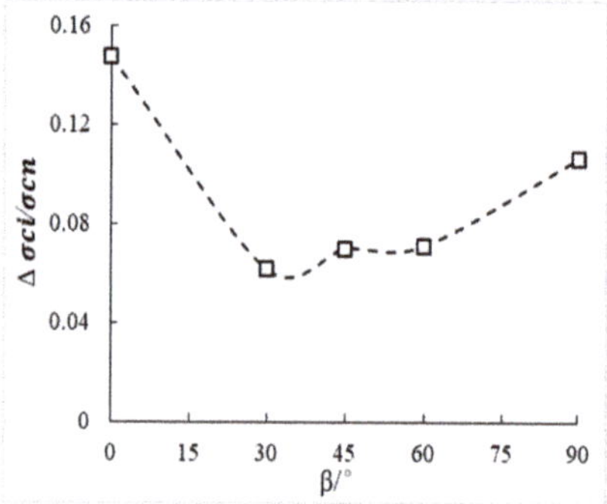

Figure 10. Variations of $\Delta\ \sigma_{ci}/\sigma_{cn}$ against α (**a**) and β (**b**).

From Figure 10b, it can be deduced that the difference of σ_{ci}/σ_{cn} also decreases down to β = 30° and then increases with the increase of β, which indicates that the intersection angle α had less of an effect on the ration σ_{ci}/σ_{cn} when β = 30°.

3.4. Crack Coalescence

Many crack coalescence types have been observed in physical experiments [1,21,26] and have been classified as seven different crack coalescence types, based on their geometry and propagation mechanism using a high-speed camera. Zhang [40] classified five types of linkages observed between two non-parallel flaws using parallel bonded-particle models. However, these crack coalescence types are based on the coalescence between two parallel or non-parallel flaws. Our study reveals that two intersect flaws show different coalescence behavior. Table 4 shows the crack coalescence patterns for two intersecting flaws with different values for α and β.

The different crack coalescence patterns observed in the case of two intersecting flaws with different geometries are shown in Figure 11 and are summarized in Table 5. Two typical patterns can be distinguished:

1. A macroscopic tensile crack initiated from the tip of one of the flaws extends upwards or downwards to the other flaw but does not reach the tip and some microscopic shear or mixed tensile-shear cracks participate at coalescence. This is classified as one-tip-linkage. According to the coalescence position, the one-tip-linkage mechanism contains three sub-types: Coalescence position near the tip (Figure 11(1),(2)), coalescence position at the flaw, but far away from the tip (Figure 11(4)), or coalescence position outside the flaw with a certain distance from the tip (Figure 11(3)).
2. A macroscopic tensile crack initiated from the tip of one flaw extends upwards or downwards to the tip of the other flaw. The coalescence occurred by the linkage of tensile cracks and is classified as two-tips-linkage. The two tips linkage has two sub-types: Straight linkage (Figure 11(5)) and arc linkage (Figure 11(6)).

Table 4. Crack coalescence patterns for two intersecting flaws with different values of α and β. Microscopic tensile and shear cracks are shown in white and red, respectively. Original flaws are in yellow.

Flaw Geometry	The Crack Coalescence Patterns of Loading				
	$\beta = 0°$	$\beta = 30°$	$\beta = 45°$	$\beta = 60°$	$\beta = 90°$
$\alpha = 90°$	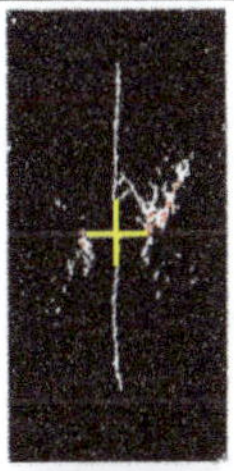		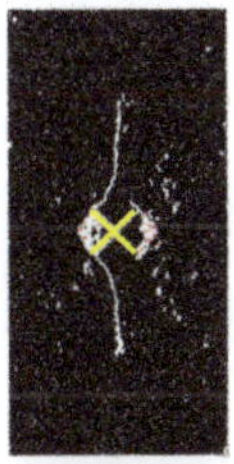		
$\alpha = 60°$	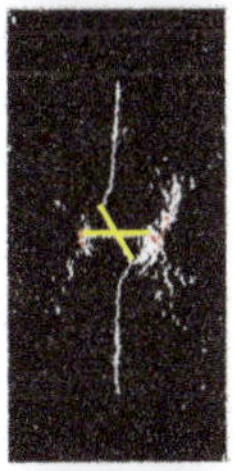		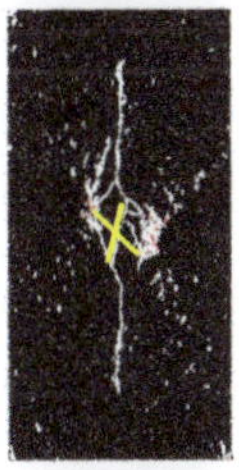		
$\alpha = 45°$	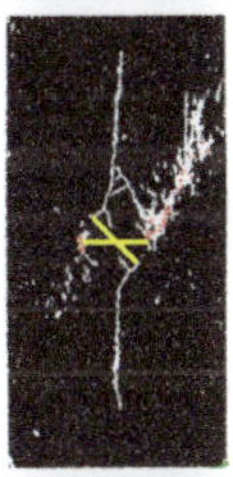				
$\alpha = 30°$	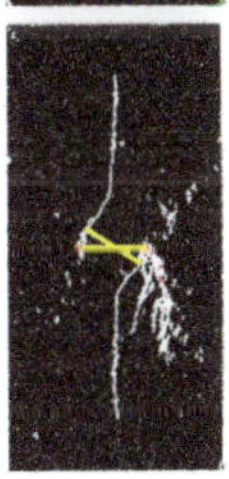	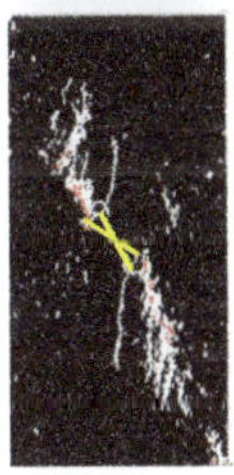		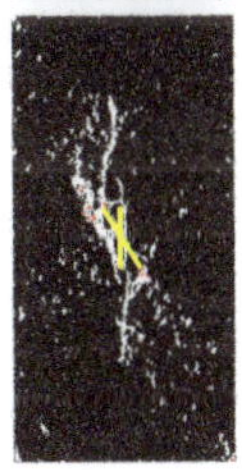	

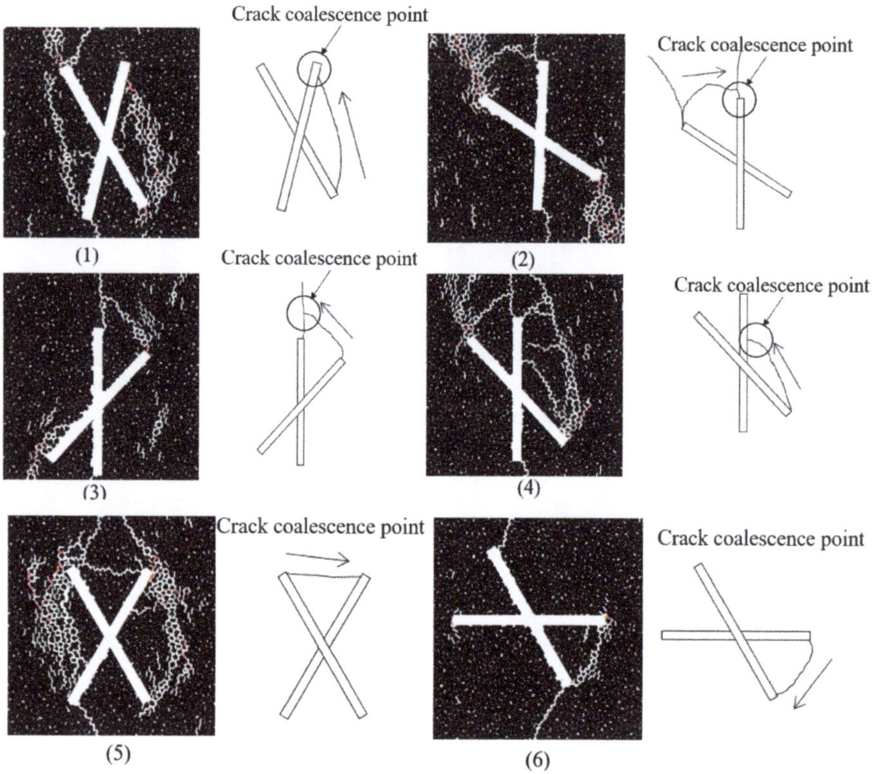

Figure 11. Observed crack coalescence patterns for two intersecting flaws (left: Simulation results, right: Simplified sketch). White and red segments represent microscopic tensile and shear cracks, respectively. The arrows represent the crack propagation direction.

Table 5. The coalescence patterns of different geometries.

Angle/°	Coalescence Types of Different Geometries									
	β = 0°		β = 30°		β = 45°		β = 60°		β = 90°	
	Type	Zone	Type	Zone	Type	Zone	Type	Zone	Type	Zone
α = 90°	(3)	II	(6)	II IV	(6)	II IV	(3) (6)	II III	(3)	II
α = 60°	(6)	III	(2) (3)	I III	(1) (2) (3) (4) (5)	I II IV	(1) (5) (6)	I II IV	(3)	I III
α = 45°	(3)	II	(2) (6)	I III	(2) (3) (4) (6)	I II	(1) (4) (6)	II IV	(3) (6)	I II
α = 30°	(5) (6)	I III	(5) (6)	I III	(2) (3)	I II	(2) (3) (4)	I III IV	(3) (4) (6)	I III

3.5. Crack Coalescence Stress

The coalescence stress σ_{co} is the vertical stress that is observed when the first coalescence of the two intersecting flaws occurs. The ratio between crack coalescence stress σ_{co} and peak strength σ_{cn} varies, as shown in Figure 12. The crack coalescence stress reaches 85%–100% of the peak stress.

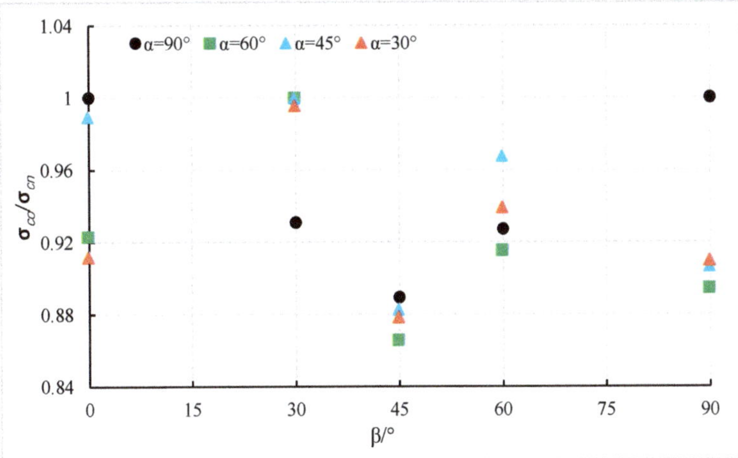

Figure 12. Ratio between coalescence stress σ_{co} and peak strength σ_{cn} for different values of α and β.

For two cases of α = 90° and one case of α = 60°, α = 45°, and α = 30°, the ratios of their cases are at or just less than 1.0, which means that their model coalesced near or at the peak stress. In contrast, for four cases of β = 45°, three cases of β = 90°, and one case of β = 0°, the ratios of their cases are significantly less than 1.0, which means that the coalescence occurs more easily. In addition, for any intersection angle α, the minimum values of σ_{co}/σ_{cn} are observed at β = 45°, which indicates that coalescence occurred more easily with that geometry.

4. Conclusions

Numerical simulations were performed on rock-like specimens containing two intersecting flaws with different inclination angle β and different intersection angle α subjected to uniaxial compressive loads. The main conclusions can be summarized as follows:

1. For any intersection angle α, the strength is increases for flaw inclination angle β, ranging from 0° to about 45°. For intersection angle α up to about 45°, the strength is further slightly increased for inclination angle β, bigger than 45°. For intersection angle α, it is bigger than about 45° and the strength is strongly decreased for inclination angle β, bigger than 45°.
2. The macro-crack initiated from the tips of two intersect flaws that did not occur simultaneously. The crack initial stress σ_ci of the first crack, obtained at the tip of flaw (A flaw or B flaw), normalized by the respective specimen peak strength σ_{cn}, which changed between 0.2 and 0.4 with different flaw geometries.
3. Two major crack coalescence patterns were observed: (a) One-tip-linkage (either near the crack tip, at the flaw but far away from the crack tip or outside the flaw at a certain distance from the tip) and (b) two-tips-linkage (straight or arc linkage). The geometry of flaws governed the coalescence type.

The ratio between coalescence stress σ_{co} and peak strength σ_{cn} varied between 0.85–1.0, which meant the first coalescence occurred near or at the peak strength. For the same values of α, the minimum values of σ_{co}/σ_{cn} were observed at β = 45°, which indicated that coalescence occurred more easily with that geometry.

Author Contributions: Conceptualization, B.D.; Methodology, Y.C.; Software, G.Z.; Validation, W.L.; Formal analysis, H.W.; Data curation, Y.C.; Writing—original draft preparation, Y.C.; Writing—review and editing, B.D.; Funding acquisition, B.D.

Funding: This study was financially supported by the national natural science foundation (51804163).

Conflicts of Interest: The authors declare no conflict of interest.

References

1. Bobet, A.; Einstein, H.H. Fracture coalescence in rock-type materials under uniaxial and biaxial compression. *Int. J. Rock Mech. Min. Sci.* **1998**, *35*, 863–888. [CrossRef]
2. Peng, K.; Zhou, J.; Zou, Q.; Yan, F. Deformation characteristics of sandstones during cyclic loading and unloading with varying lower limits of stress under different confining pressures. *Int. J. Fatigue* **2019**, *127*, 82–100. [CrossRef]
3. Hoek, E.; Martin, C.D. Fracture initiation and propagation in intact rock–a review. *J. Rock Mech. Geotech. Eng.* **2014**, *6*, 287–300. [CrossRef]
4. Peng, K.; Zhou, J.; Zou, Q.; Zhang, J.; Wu, F. Effects of stress lower limit during cyclic loading and unloading on deformation characteristics of sandstones. *Constr. Build. Mater.* **2019**, *217*, 202–215. [CrossRef]
5. Liang, W.; Zhao, G.; Wang, X.; Zhao, J.; Ma, C. Assessing the rockburst risk for deep shafts via distance-based multi-criteria decision making approaches with hesitant fuzzy information. *Eng. Geol.* **2019**, *260*, 105211. [CrossRef]
6. Wang, S.; Li, X.; Yao, J.; Gong, F.; Li, X.; Du, K.; Du, S. Experimental investigation of rock breakage by a conical pick and its application to non-explosive mechanized mining in deep hard rock. *Int. J. Rock Mech. Min. Sci.* **2019**, *122*, 104063. [CrossRef]
7. Adams, M.; Sines, G. Crack extension from flaws in a brittle material subjected to compression. *Tectonophysics* **1978**, *49*, 97–118. [CrossRef]
8. Peng, K.; Liu, Z.; Zou, Q.; Zhang, Z.; Zhou, J. Static and dynamic mechanical properties of granite from various burial depths. *Rock Mech. Rock Eng.* **2019**, *52*, 3545–3566.
9. Camones, L.A.M.; do Amaral Vargas, E.; de Figueiredo, R.P.; Velloso, R.Q. Application of the discrete element method for modeling of rock crack propagation and coalescence in the step-path failure mechanism. *Eng. Geol.* **2013**, *153*, 80–94. [CrossRef]
10. Dyskin, A.V.; Sahouryeh, E.; Jewell, R.J.; Joer, H.; Ustinov, K.B. Influence of shape and locations of initial 3-D cracks on their growth in uniaxial compression. *Eng. Fract. Mech.* **2003**, *70*, 2115–2136. [CrossRef]
11. Peng, K.; Wang, Y.; Zou, Q.; Liu, Z.; Mou, J. Effect of crack angles on energy characteristics of sandstones under a complex stress path. *Eng. Fract. Mech.* **2019**, *218*, 106577. [CrossRef]
12. Li, Y.P.; Chen, L.Z.; Wang, Y.H. Experimental research on pre-cracked marble under compression. *Int. J. Solids Struct.* **2005**, *42*, 2505–2516. [CrossRef]
13. Zhang, B.; Li, Y.; Fantuzzi, N.; Zhao, Y.; Liu, Y.-B.; Peng, B.; Chen, J. Investigation of the Flow Properties of CBM Based on Stochastic Fracture Network Modeling. *Materials* **2019**, *12*, 2387. [CrossRef]
14. Wong, L.N.Y.; Einstein, H.H. Crack coalescence in molded gypsum and Carrara marble: Part 1. Macroscopic observations and interpretation. *Rock Mech. Rock Eng.* **2009**, *42*, 475–511. [CrossRef]
15. Wong, L.N.Y.; Einstein, H.H. Crack coalescence in molded gypsum and Carrara marble: Part 2—Microscopic observations and interpretation. *Rock Mech. Rock Eng.* **2009**, *42*, 513–545. [CrossRef]
16. Wong, R.H.C.; Chau, K.T.; Tang, C.A.; Lin, P. Analysis of crack coalescence in rock-like materials containing three flaws—Part I: Experimental approach. *Int. J. Rock Mech. Min. Sci.* **2001**, *38*, 909–924. [CrossRef]
17. Yang, S.Q. Crack coalescence behavior of brittle sandstone samples containing two coplanar fissures in the process of deformation failure. *Eng. Fract. Mech.* **2011**, *78*, 3059–3081. [CrossRef]
18. Zhang, X.P.; Wong, L.N.Y. Cracking processes in rock-like material containing a single flaw under uniaxial compression: A numerical study based on parallel bonded-particle model approach. *Rock Mech. Rock Eng.* **2012**, *45*, 711–737. [CrossRef]
19. Ko, T.Y.; Einstein, H.; Kemeny, J. Crack coalescence in brittle material under cyclic loading. In Proceedings of the Golden Rocks 2006, The 41st US Symposium on Rock Mechanics (USRMS), Golden, CO, USA, 17–21 June 2006; American Rock Mechanics Association: Alexandria, VA, USA, 2006.
20. Park, C.H.; Bobet, A. Crack coalescence in specimens with open and closed flaws: A comparison. *Int. J. Rock Mech. Min. Sci.* **2009**, *46*, 819–829. [CrossRef]

21. Sagong, M.; Bobet, A. Coalescence of multiple flaws in a rock-model material in uniaxial compression. *Int. J. Rock Mech. Min. Sci.* **2002**, *39*, 229–241. [CrossRef]
22. Tang, C.A.; Lin, P.; Wong, R.H.C.; Chau, K.T. Analysis of crack coalescence in rock-like materials containing three flaws—Part II: Numerical approach. *Int. J. Rock Mech. Min. Sci.* **2001**, *38*, 925–939. [CrossRef]
23. Zhang, Z.; Cao, S.; Li, Y.; Guo, P.; Yang, H.; Yang, T. Effect of moisture content on methane adsorption- and desorption- induced deformation of tectonically deformed coal. *Adsorpt. Sci. Technol.* **2018**, *36*, 1648–1668. [CrossRef]
24. Wong, L.N.Y.; Li, H.Q. Numerical study on coalescence of two pre-existing coplanar flaws in rock. *Int. J. Solids Struct.* **2013**, *50*, 3685–3706. [CrossRef]
25. Zhao, Y.; Cao, S.; Li, Y.; Zhang, Z.; Guo, P.; Yang, H.; Zhang, S.; Pan, R. The occurrence state of moisture in coal and its influence model on pore seepage. *RSC Adv.* **2018**, *8*, 5420–5432. [CrossRef]
26. Wong, R.H.; Chau, K.T. Crack coalescence in a rock-like material containing two cracks. *Int. J. Rock Mech. Min. Sci.* **1998**, *35*, 147–164. [CrossRef]
27. Yang, S.Q.; Jing, H.W. Strength failure and crack coalescence behavior of brittle sandstone samples containing a single fissure under uniaxial compression. *Int. J. Fract.* **2011**, *168*, 227–250. [CrossRef]
28. Yang, S.Q. *Strength, Failure and Crack Evolution Behavior of Rock Material Containing Pre-Existing Fissures*; Springer: Cham, Switzerland, 2015; p. 246.
29. Pollard, D.D.; Aydin, A. Progress in understanding jointing over the past century. *GSA Bull.* **1988**, *100*, 1181–1204. [CrossRef]
30. Lee, H.; Jeon, S. An experimental and numerical study of fracture coalescence in pre-cracked specimens under uniaxial compression. *Int. J. Solids Struct.* **2011**, *48*, 979–999. [CrossRef]
31. Li, H.; Wong, L.N.Y. Influence of flaw inclination angle and loading condition on crack initiation and propagation. *Int. J. Solids Struct.* **2012**, *49*, 2482–2499. [CrossRef]
32. Ning, Y.; Yang, J.; An, X.; Ma, G. Modelling rock fracturing and blast-induced rock mass failure via advanced discretisation within the discontinuous deformation analysis framework. *Comput. Geotech.* **2011**, *38*, 40–49. [CrossRef]
33. Wu, Z.; Wong, L.N.Y. Frictional crack initiation and propagation analysis using the numerical manifold method. *Comput. Geotech.* **2012**, *39*, 38–53. [CrossRef]
34. Amarasiri, A.; Kodikara, J. Use of material interfaces in DEM to simulate soil fracture propagation in Mode I cracking. *Int. J. Geomech.* **2010**, *11*, 314–322. [CrossRef]
35. Cundall, P.A.; Strack, O.D. A discrete numerical model for granular assemblies. *Geotechnique* **1979**, *29*, 47–65. [CrossRef]
36. Chen, W.; Konietzky, H.; Tan, X.; Frühwirt, T. Pre-failure damage analysis for brittle rocks under triaxial compression. *Comput. Geotech.* **2016**, *74*, 45–55. [CrossRef]
37. Chen, W.; Konietzky, H.; Abbas, S.M. Numerical simulation of time-independent and-dependent fracturing in sandstone. *Eng. Geol.* **2015**, *193*, 118–131. [CrossRef]
38. Gui, Y.L.; Zhao, Z.Y.; Kodikara, J.; Bui, H.H.; Yang, S.Q. Numerical modelling of laboratory soil desiccation cracking using UDEC with a mix-mode cohesive fracture model. *Eng. Geol.* **2016**, *202*, 14–23. [CrossRef]
39. Gui, Y.L.; Bui, H.H.; Kodikara, J.; Zhang, Q.B.; Zhao, J.; Rabczuk, T. Modelling the dynamic failure of brittle rocks using a hybrid continuum-discrete element method with a mixed-mode cohesive fracture model. *Int. J. Impact Eng.* **2016**, *87*, 146–155. [CrossRef]
40. Zhang, X.P.; Wong, L.N.Y.; Wang, S. Effects of the ratio of flaw size to specimen size on cracking behavior. *Bull. Eng. Geol. Environ.* **2015**, *74*, 181–193. [CrossRef]
41. Turco, E. In-plane shear loading of granular membranes modeled as a Lagrangian assembly of rotating elastic particles. *Mech. Res. Commun.* **2018**, *92*, 61–66. [CrossRef]
42. Turco, E.; dell'Isola, F.; Misra, A. A nonlinear Lagrangian particle model for grains assemblies including grain relative rotations. *Int. J. Numer. Anal. Methods Geomech.* **2019**, *43*, 1051–1079. [CrossRef]
43. Wong, L.N.Y.; Zhang, X.P. Size effects on cracking behavior of flaw-containing specimens under compressive loading. *Rock Mech. Rock Eng.* **2014**, *47*, 1921–1930. [CrossRef]
44. Shen, B.; Stephansson, O.; Einstein, H.H.; Ghahreman, B. Coalescence of fractures under shear stresses in experiments. *J. Geophys. Res. Ser.* **1995**, *100*, 5975. [CrossRef]

45. Vesga, L.F.; Vallejo, L.E.; Lobo-Guerrero, S. DEM analysis of the crack propagation in brittle clays under uniaxial compression tests. *Int. J. Numer. Anal. Methods Geomech.* **2008**, *32*, 1405–1415. [CrossRef]
46. Bray, J.W. A study of jointed and fractured rock. *Rock Mech. Eng. Geol.* **1967**, *5*, 117–136.

mathematics

MDPI

Article

Cost-Based Optimum Design of Reinforced Concrete Retaining Walls Considering Different Methods of Bearing Capacity Computation

Neda Moayyeri [1], Sadjad Gharehbaghi [1] and Vagelis Plevris [2,*]

1 Department of Civil Engineering, Behbahan Khatam Alanbia University of Technology, Behbahan P.O. Box 6361647189, Iran; moayyeri@bkatu.ac.ir (N.M.); sgharehbaghi@bkatu.ac.ir (S.G.)

2 Department of Civil Engineering and Energy Technology, OsloMet—Oslo Metropolitan University, 0166 Oslo, Norway

* Correspondence: vageli@oslomet.no; Tel.: +47-67238853

Received: 30 October 2019; Accepted: 8 December 2019; Published: 12 December 2019

Abstract: This paper investigates the effect of computing the bearing capacity through different methods on the optimum construction cost of reinforced concrete retaining walls (RCRWs). Three well-known methods of Meyerhof, Hansen, and Vesic are used for the computation of the bearing capacity. In order to model and design the RCRWs, a code is developed in MATLAB. To reach a design with minimum construction cost, the design procedure is structured in the framework of an optimization problem in which the initial construction cost of the RCRW is the objective function to be minimized. The design criteria (both geotechnical and structural limitations) are considered constraints of the optimization problem. The geometrical dimensions of the wall and the amount of steel reinforcement are used as the design variables. To find the optimum solution, the particle swarm optimization (PSO) algorithm is employed. Three numerical examples with different wall heights are used to capture the effect of using different methods of bearing capacity on the optimal construction cost of the RCRWs. The results demonstrate that, in most cases, the final design based on the Meyerhof method corresponds to a lower construction cost. The research findings also reveal that the difference among the optimum costs of the methods is decreased by increasing the wall height.

Keywords: reinforced concrete; retaining wall; optimization; bearing capacity; particle swarm optimization; PSO

1. Introduction

Reinforced concrete retaining walls (RCRWs) are referred to as structures that withstand the pressure resulting from the difference in the levels caused by embankments, excavations, and/or natural processes. Such situations frequently occur in the construction of several structures, such as bridges, railways, and highways. Due to the frequent application of RCRWs in civil engineering projects, minimizing the construction cost of such structures is an issue of crucial importance.

The satisfaction of both geotechnical and structural design constraints is a key component in the design of RCRWs. In most cases, primary dimensions are initially estimated based on reasonable assumptions and the experience of the designer. Then, in order to reach a cost-effective design while satisfying the design constraints, the design variables (particularly the wall dimensions) need to be revised by using a trial-and-error process, which makes it rather grueling. On the other hand, there is no guarantee that the final design will be the best possible one. To eliminate this problem, which can hinder the designer from reaching a cost-effective solution, and by considering the advances in computational technologies during the recent decades, it makes sense to express the design in the form of a formal optimization problem.

The design optimization of RCRWs has received significant attention during the last two decades. Some of the pertinent works are briefly investigated herein. As a benchmark work, Saribas and Erbatur [1] used a nonlinear programming method and investigated the sensitivity of the optimum solutions to parameters such as backfill slope, surcharge load, internal friction angle of retained soil, and yield strength of reinforcing steel. The simulated annealing (SA) algorithm has been also applied to minimize the construction cost of RCRWs [2,3]. Camp and Akin [4] developed a procedure to design cantilever RCRWs using Big Bang–Big Crunch optimization. They captured the effects of surcharge load, backfill slope, and internal friction angle of the retained soil on the values of low-cost and low-weight designs with and without a base shear key. Khajehzadeh et al. [5] used the particle swarm optimization with passive congregation (PSOPC), claiming that the proposed algorithm was able to find an optimal solution better than the original PSO and nonlinear programming. In their work, the weight, cost, and CO_2 emissions were chosen as the three objective functions to be minimized. Gandomi et al. [6] optimized RCRWs by using swarm intelligence techniques, such as accelerated particle swarm optimization (APSO), firefly algorithm (FA), and cuckoo search (CS). They concluded that the CS algorithm outperforms the other ones. They also investigated the sensitivity of the algorithms to surcharge load, base soil friction angle, and backfill slope with respect to the geometry and design parameters. Kaveh and his colleagues (e.g., [7–10]) optimized the RCRWs using nature-inspired optimization algorithms, including charged system search (CSS), ray optimization algorithm (RO), dolphin echolocation optimization (DEO), colliding bodies of optimization (CBO), vibrating particles system (VPS), enhanced colliding bodies of optimization (ECBO), and democratic particle swarm optimization (DPSO). Temur and Bekdas [11] employed the teaching–learning-based optimization (TLBO) algorithm to find the optimum design of cantilever RCRWs. They concluded that the minimum weight of the RCRWs decreases as the internal friction angle of the retained soil increases, and increases with the values of the surcharge load. Ukritchon et al. [12] presented a framework for finding the optimum design of RCRWs, considering the slope stability. Aydogdu [13] introduced a new version of a biogeography-based optimization (BBO) algorithm with levy light distribution (LFBBO) and, by using five examples, it was shown that this algorithm outperforms some other metaheuristic algorithms. In this work, the cost of the RCRWs was used as the criterion to find the optimum design. Nandha Kumar and Suribabu [14] adopted the differential evolution (DE) algorithm to solve the design optimization problem of RCRWs. The results of sensitivity analysis showed that width and thickness of the base slab and toe width increases as the height of stem increases. Gandomi et al. [15] studied the importance of different boundary constraint handling mechanisms on the performance of the interior search algorithm (ISA). Gandomi and Kashani [16] minimized the construction cost and weight of RCRWs analyzed by the pseudo-static method. They employed three evolutionary algorithms, DE, evolutionary strategy (ES), and BBO, and concluded that BBO outperforms the others in finding the optimum design of RCRWs. More recently, Mergos and Mantoglou [17] optimized concrete retaining walls by using the flower pollination algorithm, claiming that this method outperforms PSO and GA.

By taking a look at the studies so far reported, it can be noticed that there has been no work done in assessing the effect of using different available methods of determining the bearing capacity on the optimum design of the RCRWs. The current study investigates this important issue. In order to model and design the RCRWs, a code is developed in MATLAB [18]. To reach a design with minimum construction cost, an optimization problem is defined and the construction cost is considered as the single objective function to be minimized. The design criteria, including both geotechnical and structural limitations, are considered as the optimization constraints. The wall geometrical dimensions and the amount of steel reinforcement are used as the design variables. The particle swarm optimization (PSO) [19] algorithm is used to find the optimum solution.

2. Design of Retaining Walls

The design of RCRWs includes two sets of variables. The first set consists of the geometrical dimensions of the concrete wall, namely variables X_1 to X_8, that are defined as shown in Table 1 and depicted graphically in Figure 1. It can be seen that these variables fully define the geometry of the structure.

Table 1. Description of the geometric variables and their lower and upper bounds.

Geometric Variable	Description	Lower Bound	Upper Bound
X_1	Width of the base slab	0.4*H*	0.8*H*
X_2	Toe width	0.1*H*	0.6*H*
X_3	Thickness at the bottom of the stem	0.2 m	0.5 m
X_4	Thickness at the top of the stem	0.2 m	0.4 m
X_5	Thickness of the base	0.2 m	0.3*H*
X_6	Distance from the toe to the front face of the shear key	0.5*H*	0.8*H*
X_7	Width of shear key	0.2 m	0.4 m
X_8	Depth of the shear key	0.2 m	0.9 m

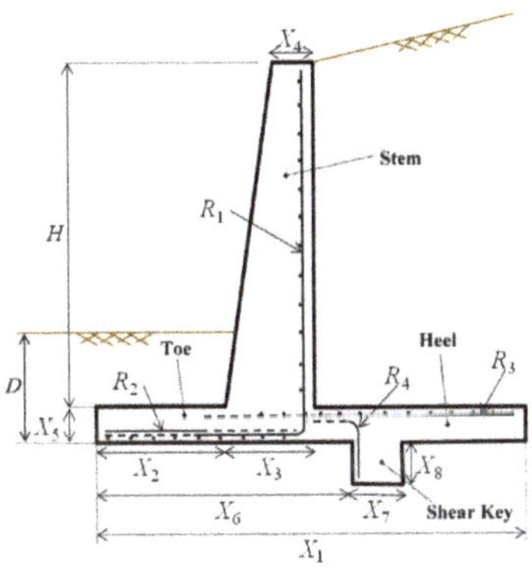

Figure 1. Design variables (X_1 to X_8 and R_1 to R_4), (redesigned based on [4]).

The second set is the amounts of steel reinforcement for four components of the wall—the stem, heel, toe, and shear key. As such, four variables, R_1 to R_4, are introduced to represent the steel reinforcement of the different components, as shown in Table 2. In this paper, a total of 223 possible reinforcement configurations were used, resulting from the combinations of using 3–28 evenly spaced bars, with varying sizes (bar diameter) from 10 to 30 mm. It is worth mentioning that the combinations used for the steel reinforcement, as listed in Table 3, are obtained such that the allowable minimum and maximum amount of steel area per unit meter length of the wall are satisfied as per ACI318-14 code [20].

Table 2. Description of the reinforcement variables.

Reinforcement Variable	Description
R_1	Vertical steel reinforcement of the stem
R_2	Horizontal steel reinforcement of the toe
R_3	Horizontal steel reinforcement of the heel
R_4	Vertical steel reinforcement of shear key

Table 3. Steel reinforcement combinations (adopted from [4]).

Index Number	Bars Quantity and Size (*R*)	Bars Cross-Sectional Area (cm^2) (in Ascending Order)
1	3ϕ10	2.356 (lower bound)
2	4ϕ10	3.141
3	3ϕ12	3.392
4	5ϕ10	3.926
5	4ϕ12	4.523
.	.	.
.	.	.
221	16ϕ30	113.097
222	17ϕ30	120.165
223	18ϕ30	127.234 (upper bound)

The allowable value of spacing between longitudinal bars (ds_{all}) is defined as follows [20]:

$$ds_{\text{all}} = \max\{25mm, d_{\text{bl}}, 1.33d_{\max}\}, \tag{1}$$

in which d_{bl} and $d_{\max}$ are the diameter of longitudinal reinforcements and diameter of greatest aggregate of concrete, respectively. For the purposes of steel reinforcement design, ACI318-14 code [20] has been considered.

In the design of RCRWs, geotechnical stability control and the satisfaction of the structural requirements are mandatory. In geotechnical design, the stability of the structure shall be controlled against possible overturning, sliding, and bearing capacity failure modes. On the other hand, in the structural design phase, each component of the structure, including the stem, toe, heel, and shear key, shall be checked against shear and moment demands [21]. Figure 2 shows all the forces acting on the retaining wall. As shown in this figure, P_{a} is the resultant force of the active pressure p_{a} per unit length of the wall; Q_{s} is the resultant force of the distributed surcharge load q; W_{C} is the weight of all sections of the reinforced concrete wall; W_{S} and W_{T} are defined as the weight of backfill behind the retaining wall and the weight of the soil on the toe, respectively; P_{p} is the resultant force due to the passive pressure (p_{p}) on the front face of the toe and shear key per unit length of the wall; P_{b} is the resultant force caused by the pressure acting on the base soil; $q_{\max}$ and $q_{\min}$ are the maximum and minimum soil pressure intensity at the toe and the heel of the retaining wall, respectively. In this paper, the water level and the seismic actions were not considered in computing the forces acting on the RCRWs. The pressure distributions on the base and retaining soil have also been illustrated in Figure 2.

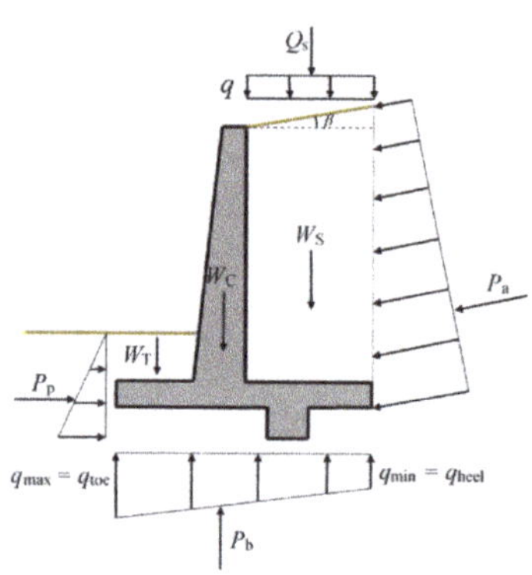

Figure 2. Forces acting on the retaining wall (redesigned based on [6]).

In this study, Rankine theory was used to evaluate the active and passive forces acting on the unit length of the retaining wall (p_a and p_p, respectively), and they are computed as follows [21]:

$$p_a = \frac{1}{2} k_a \gamma_{rs} H'^2 + k_a q H', \tag{2}$$

$$p_p = \frac{1}{2} k_p \gamma_{bs} D'^2 + 2c_{bs} \sqrt{k_p} D'^2, \tag{3}$$

where k_a and k_p are the Rankine active and passive earth pressure coefficients, respectively; D' is the buried depth of shear key; γ_{rs} is the unit weight of backfill; H' is the height of the soil located on the embedment depth of the base slab at the edge of the heel; γ_{bs} and ff_{bs} are the unit weight and the cohesion of soil in front of the toe and beneath the base slab, respectively. k_a is computed as [21]:

$$k_a = \cos\beta \frac{\cos\beta - \sqrt{\cos^2\beta - \cos^2\phi_{rs}}}{\cos\beta + \sqrt{\cos^2\beta - \cos^2\phi_{rs}}}, \tag{4}$$

in which β and ϕ_{rs} are the slope and internal friction angle of the back fill, respectively; k_p is computed as follows [21]:

$$k_p = \tan^2\left(45 + \frac{\phi_{bs}}{2}\right), \tag{5}$$

in which ϕ_{bs} (in degrees) is the internal friction angle of soil in front of the toe and beneath the base slab.

2.1. Geotechnical Stability Demands

A retaining wall may fail due to overturning about the toe, sliding along the base slab, and the loss of bearing capacity of the soil supporting the base [21]. The checks for these three failure modes are described in this section. The safety factor against overturning about the toe is defined as follows:

$$FS_O = \frac{\sum M_R}{\sum M_O}, \tag{6}$$

in which $\sum M_R$ is the sum of the moments tending to resist against overturning about the toe and $\sum M_O$ is the sum of the moments tending to overturn the wall about the toe [21]. The safety factor against sliding along the base slab is computed as follows:

$$FS_S = \frac{\sum F_R}{\sum F_D}, \tag{7}$$

where $\sum F_R$ is the sum of the resisting forces against the sliding and $\sum F_D$ is the sum of the horizontal driving forces.

The safety factor against bearing capacity failure mode is computed by:

$$FS_B = \frac{q_u}{q_{max}}, \tag{8}$$

in which q_u is the ultimate bearing capacity of the soil supporting the base slab. The ultimate bearing capacity is the load per unit area of the foundation at which shear failure occurs in the soil. For retaining walls, q_u is computed as follows [21]:

$$q_u = c_{bs} N_c F_{cs} F_{cd} F_{ci} + q N_q F_{qs} F_{qd} F_{qi} + 0.5 \gamma_{bs} B' N_\gamma F_{\gamma s} F_{\gamma d} F_{\gamma i}, \tag{9}$$

where:

$$q = \gamma_{bs} D, \tag{10}$$

in which D is the embedment depth of the toe (base slab). B' is computed as:

$$B' = B - 2e, \tag{11}$$

where B is the width of the base slab. As mentioned earlier, q_{max} and q_{min}, which are the maximum and minimum stresses occurring at the end of the toe and heel of the structure, respectively, are computed as follows [21]:

$$q_{\min}^{\max} = \frac{\sum V}{B}\left(1 \pm \frac{6e}{B}\right), \tag{12}$$

where e is defined as the eccentricity of the resultant force, which can be computed as follows:

$$e = \frac{B}{2} - \frac{\sum M_{R} - \sum M_{O}}{\sum V}. \tag{13}$$

Note that if e becomes greater than $B/6$, then some tensile stress will be applied at the soil located under the heel section. In such a case, since the tensile strength of soil is negligible, design and calculations should be repeated.

In Equation (9), the N coefficients are used for the modification of the bearing capacity value and the F coefficients are used for the modification of the shape, depth, and inclination factors. The bearing capacity factors N_c, N_q, and N_γ are, respectively, the contributions of cohesion, surcharge, and unit weight of soil to the ultimate load-bearing capacity [21]. Some of these coefficients vary in accordance with the method used, resulting in different values for q_u. Hence, this issue could affect the final design of the RCRWs. In this paper, the effects of three methods of Meyerhof [22], Hansen [23], and Vesic [24] in computing q_u were investigated, particularly their effect on the optimized construction cost of the RCRWs. The N and F coefficients corresponding to each of the three design methods are listed in Appendix A.

2.2. Structural Requirements

The moment and shear capacity of all components of the retaining wall must be greater than their corresponding demands. The flexural strength of each component can be computed as follows [20]:

$$M_n = \phi_m A_s f_y\left(d - \frac{a}{2}\right), \tag{14}$$

in which ϕ_m is known as the nominal strength coefficient (equal to 0.9 [20]); A_s is the cross-sectional area of the steel reinforcement; f_y is the yield strength of steel; d is the effective depth of the cross section; and a is the depth of the compressive stress block. The shear strength is estimated by [20]:

$$V_n = 0.17\phi_v \sqrt{f_c} bd, \tag{15}$$

where ϕ_v is the nominal strength coefficient (equal to 0.75 [20]); f_c is the specified compressive strength of concrete; and b is the width of the cross section.

2.2.1. Flexural Moment and Shear Force Demands of Stem

As shown in Figure 2, by considering the active force acting on the unit length of the wall due to surcharge load and the weight of the backfill on the stem, the critical section for flexural moment is at the intersection of the stem with the base slab. In addition, the critical section for shear force is at a distance d_s from the intersection of the stem with the foot slab, defined as $d_s = X_3 - C_C$, where C_C is the concrete cover. The flexural moment and shear force demands of the stem at its critical section are computed by the following equations:

$$M_s = 1.6\left[\frac{1}{2}\left(\left(k_a q H^2\right) + \left(\frac{1}{3} k_a \gamma_{rs} H^3\right)\right)\right] \cos\beta, \tag{16}$$

$$V_s = 1.6\Big((k_a q(H - d_s) + \frac{1}{2} k_a \gamma_{rs} (H - d_s)^2\Big) \cos\beta. \quad (17)$$

2.2.2. Flexural Moment and Shear Force Demands of Toe Slab

The effective forces on the toe slab include the soil weight on the toe slab, the weight of the toe concrete slab, and the force caused by earth pressure under the toe slab. The foot of the front face of the stem is the critical section for flexural moment and the critical section for shear force is formed at a distance d_t from the front face of the stem ($d_t = X_5 - C_C$). The flexural moment and shear force demands of the toe slab at its critical section are computed by the following equations [6]:

$$M_t = \left[1.6\Big(\frac{q_2}{6} + \frac{q_{max}}{3}\Big) - 0.9(\gamma_c X_5 + \gamma_{bs}(D - X_5))\right] l_{toe}^2, \quad (18)$$

$$V_t = \left[1.6\Big(\frac{q_{dt} + q_{max}}{2}\Big) - 0.9(\gamma_c X_5 + \gamma_{bs}(D - X_5))\right] .(l_{toe} - dt), \quad (19)$$

where q_2 is the soil pressure intensity at the foot of the front face of the stem; γ_c is the unit weight of concrete; l_{toe} is the length of the toe slab; and q_{dt} is the soil pressure at a distance d_t from the foot of the stem front face.

2.2.3. Flexural Moment and Shear Force Demands of Heel Slab

The forces acting on the heel section include the soil weight of top of the heel slab, the weight of the heel concrete slab, the surcharge load, and the force resulting from earth pressure under the heel slab. The foot of the stem back face is the critical section for flexural moment and the critical section for shear force is at a distance d_h from the foot of the stem back face ($d_h = X_5 - C_C$). The flexural moment and shear force demands of the heel slab at its critical section are computed by the following equations [6]:

$$M_h = \left[\Big(\frac{1.6q + 1.2\gamma_c X_5 + 1.2\gamma_{rs} H}{2}\Big) + \frac{1.2W_{bs}}{3} - \Big(\frac{q_1 + 2q_{min}}{6}\Big)\right] l_{heel}^2, \quad (20)$$

$$V_h = \left[1.6q + 1.2\gamma_c X_5 + 1.2\gamma_{rs} H + 1.2\frac{W_{bs} + W_{bsdh}}{2} - 0.9\frac{q_{dh} + q_{min}}{2}\right] (l_{heel} - d_h), \quad (21)$$

in which W_{bs} is the maximum load due to triangular backfill soil weight at the top of the heel slab; q_1 is the soil pressure intensity at the foot of the stem back face; l_{heel} is the length of retaining wall heel; W_{bsdh} is the load resulting from triangular backfill soil weight; and q_{dh} is the intensity of soil pressure at a distance d_h from the stem back face.

3. Optimization Problem

3.1. *Formulation*

Optimization problems can be divided into two large groups in general—constrained and unconstrained problems. Because of the limitations required for the design of structures, the optimum design of RCRWs is a constrained optimization problem, which can be expressed as follows [25]:

$$\begin{gathered} Minimize\ F(x)\ subjected\ to\ g_i(x) \leq 0 \\ (i = 1, 2, \ldots, m;\ x_j \in R^d,\ j = 1, 2, \ldots, n) \end{gathered} \quad (22)$$

in which $F(x)$ is the objective function; $g_i(x)$ is the i-th constraint; m and n are the total number of constraints and design variables, respectively; R^d is a given set of discrete values from which the individual design variables x_j can take values. In this paper, an exterior penalty function method was used to transform the constrained structural optimization problem into an unconstrained one, as follows [26]:

$$\Phi(x, r_p) = F(x)\left[1 + r_p \sum_{r=1}^{m}\left(\max\left(\frac{g_r}{g_{r,all}} - 1, 0\right)\right)\right], \tag{23}$$

where ϕ is the pseudo (penalized) objective function; g_r and $g_{r,all}$ are the r-th constraint and its allowable value, respectively; and r_p is a positive penalty parameter, which in this study was assumed to be equal to 25.

3.2. Objective Function

In this work, the initial construction cost CC_{RCRW} of the RCRW was considered the single objective function of the optimization problem, to be minimized. The objective function was defined as follows:

$$CC_{RCRW} = C_c V_c + C_{st} W_{st}, \tag{24}$$

where C_c and C_{st} are the cost per unit volume of concrete and the cost per unit weight of steel reinforcement, respectively; V_c and W_{st} are the volume of concrete and weight of steel per unit length of the retaining wall, respectively. It is worth noting that the cost of the formwork, casting concrete, vibration, and generally all related labor costs were taken into consideration in the parameter C_c. In addition, the earthwork cost was not considered in the calculation of the total cost of the RCRWs.

3.3. Design Constraints

Two sets of constraints are considered for the optimum design of a RCRW. The first set (four constraints) is related to geotechnical requirements (wall stability). Of these, the first three are considered to provide safety factors against overturning, sliding, and bearing capacity failure modes, as shown below [21]:

$$g_1 = \frac{FS_O}{FS_{O,all}} - 1 \geq 0, \tag{25}$$

$$g_2 = \frac{FS_S}{FS_{S,all}} - 1 \geq 0, \tag{26}$$

$$g_3 = \frac{FS_B}{FS_{B,all}} - 1 \geq 0, \tag{27}$$

in which FS_O, FS_S, and FS_B are the safety factor demands against overturning, sliding, and bearing capacity failure modes, respectively, and $FS_{O,all}$, $FS_{S,all}$, and $FS_{B,all}$ are their allowable values.

The fourth constraint of the first set is used to avoid the presence of tensile stresses on the base soil, as follows:

$$g_4 = q_{min} \geq 0. \tag{28}$$

The second set of constraints is related to the structural requirements having to do with providing the required strength of wall components and reinforcement arrangements in their cross sections, in accordance with ACI318-14 [20]. To design the structural sections, the reinforcement area at each section of the retaining wall should satisfy the allowable amounts of the reinforcement area, as follows:

$$g_{5\text{–}8} = \left(\frac{A_s}{A_{s,min}}\right)_{st} - 1 \geq 0, \tag{29}$$

$$g_{9\text{–}12} = \left(\frac{A_{s,max}}{A_s}\right)_{st} - 1 \geq 0, \tag{30}$$

in which $A_{s,min}$ and $A_{s,max}$ are the minimum and maximum allowable area of steel reinforcement in accordance with the code, and A_s is defined as the cross-sectional area of steel reinforcement in each section. The subscript *st* refers to all sections of the RCRW including stem, heel, toe and shear key, and similarly subscripts 5 to 8 are used for the stem, toe, heel and shear key, respectively. The same is also the case for subscripts 9 to 12.

As mentioned earlier, the moment and shear capacities of all sections of the retaining wall should be greater than the corresponding demands, namely:

$$g_{13\text{–}16} = \left(\frac{M_n}{M_d}\right)_{st} - 1 \geq 0 \tag{31}$$

$$g_{17\text{–}20} = \left(\frac{V_n}{V_d}\right)_{st} - 1 \geq 0 \tag{32}$$

In the equations above, M_n and V_n are the moment and shear nominal capacity, and M_d and V_d are the moment and shear demands, respectively. M_n and V_n are the flexural moment and shear strength formulated before. The subscripts 13 to 16 are for stem, toe, heel, and shear key. The same is true for subscripts 17 to 20.

The following geometric constraints are also applied to avoid an impossible or impracticable shape of the wall:

$$g_{21} = \frac{X_1}{X_2 + X_3} - 1 \geq 0, \tag{33}$$

$$g_{22} = \frac{X_1}{X_6 + X_7} - 1 \geq 0. \tag{34}$$

The minimum development length of the steel reinforcement bars should be considered for all the structural components. At first, the minimum basic development length l_{db} against the allowable space is checked. If the available space is not enough, a hook is added to achieve the additional development length. In this case, a minimum hook development length l_{dh} and minimum hook length of $12d_{bh}$ (d_{bh} is the diameter of the hooked bar) should be satisfied. The following limitations are considered for stem, toe, heel, and shear key in the design, respectively:

$$g_{23} = \frac{l_{db,stem}}{X_5 - C_C} - 1 \geq 0 \quad or \quad g_{23} = \frac{l_{dh,stem}}{X_5 - C_C} - 1 \geq 0, \tag{35}$$

$$g_{24} = \frac{l_{db,toe}}{X_1 - X_2 - C_C} - 1 \geq 0 \quad or \quad g_{24} = \frac{12d_{b,toe}}{X_5 - C_C} - 1 \geq 0, \tag{36}$$

$$g_{25} = \frac{l_{db,heel}}{X_2 + X_3 - C_C} - 1 \geq 0 \quad or \quad g_{25} = \frac{12d_{b,heel}}{X_5 - C_C} - 1 \geq 0, \tag{37}$$

$$g_{26} = \frac{l_{db,key}}{X_5 - C_C} - 1 \geq 0 \quad or \quad g_{26} = \frac{l_{dh,key}}{X_5 - C_C} - 1 \geq 0, \tag{38}$$

It should be noted that the inequalities g_{23} to g_{26} as described in Equations (35)–(38) were actually not considered as design constraints in the present study. In fact, during the optimization process, the required development lengths were simply considered and their role in steel cost was computed and added to the construction cost. The cost of shrinkage reinforcement was also added to the total construction cost. Another important point is that the other constraints on the arrangements of steel bars in the wall sections, such as the number of allowable bars, bar size, and bar spacing, were all considered in the optimum design, as can be seen in Table 2.

3.4. PSO Algorithm

The successful application of the particle swarm optimization (PSO) algorithm in the optimum design of RC structures has been reported in the literature (e.g., [26–33]). In this paper, the PSO algorithm was used to optimize RCRWs. The algorithm was originally developed by Kennedy and

Eberhart [19] in the mid-1990s, and was first applied to simulate the social behavior of fish schooling and bird flocking as part of a socio-cognitive research. The PSO algorithm is considered a metaheuristic algorithm that optimizes a problem by iteratively trying to improve a candidate solution. It solves a problem by having a population of candidate solutions called particles. Each particle moves through the search space and its position is updated according to (a) its local best known position, and (b) the best known position for the whole swarm in the search space. Finally, the objective function is calculated for each particle and the fitness values of the candidate solutions are assessed to discover which position in the search space is the best. The algorithm searches for the optimum by adjusting the trajectory of each particle of the swarm in the multi-dimensional design space, in terms of paths created by positional vectors in a quasi-random manner. After finding the best values of position and velocity in each iteration k, these vectors are updated using the following equations:

$$\mathbf{X}_{\mathrm{i,k+1}} = \mathbf{X}_{\mathrm{i,k}} + \boldsymbol{V}_{\mathrm{i,k+1}}, \tag{39}$$

$$\boldsymbol{V}_{\mathrm{i,k+1}} = w_{\mathrm{k}} \boldsymbol{V}_{\mathrm{i,k}} + c_1 r_1 (\boldsymbol{P}_{\mathrm{i,k}} - \mathbf{X}_{\mathrm{i,k}}) + c_2 r_2 \left(\boldsymbol{P}_{\mathrm{g,k}} - \mathbf{X}_{\mathrm{i,k}}\right), \tag{40}$$

in which for the i-th particle, $\mathbf{X}_{\mathrm{i,k}}$ and $\boldsymbol{V}_{\mathrm{i,k}}$ are the current position vector and velocity vector at iteration k, respectively; $\boldsymbol{P}_{\mathrm{i,k}}$ is the best position that the particle has visited; $\boldsymbol{P}_{\mathrm{g,k}}$ is the global best position obtained so far by all the particles in the population; r_1 and r_2 are random numbers drawn from a uniform distribution in the range of [0, 1]; c_1 and c_2 are constants, called cognitive and social scaling parameters, and are usually in the range of [0, 2]; w_{k} known as the inertia weight, has a pivotal role in updating the position and the velocity vectors. In fact, this parameter is used to stabilize the motion of the particles, making the algorithm converge more quickly. In this paper, a linear weight-updating rule was implemented as follows:

$$w_{\mathrm{k}} = w_{\mathrm{max}} - \frac{w_{\mathrm{max}} - w_{\mathrm{min}}}{k_{\mathrm{max}}} k, \tag{41}$$

in which w_{max} and w_{min} are the upper and lower bounds of the current weight w_{k}; and k_{max} is the maximum number of iterations used. Figure 3 shows a schematic view for the PSO algorithm, i.e., how a particle's position is updated from one iteration to another toward finding the global optimum.

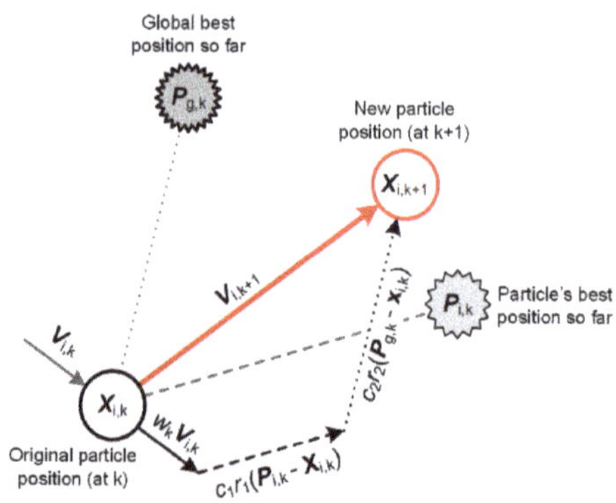

Figure 3. A schematic view for describing particle swarm optimization (PSO).

4. Methodology

As previously mentioned, although many studies have been implemented to optimize the retaining wall design, no studies have been conducted so far investigating the effect of various methods of determining the parameters of the soil ultimate bearing capacity on the design optimization of these structures. The current study deals with this issue and, in particular, the effect of using the methods developed by Meyerhof, Hansen, and Vesic (respectively abbreviated as MM, HM, and VM hereafter) on the cost of the retaining wall. For this purpose, a database containing discrete values of the mentioned variables (X_1 to X_8 and R_1 to R_4) was generated. Then, the PSO algorithm was used to solve the optimum design problem defined above. All the process of optimum design was implemented via a code written in MATLAB [18]. The flowchart of the design optimization process is shown in Figure 4. The next section investigates the effects of the above three methods on the optimum design of RCRWs.

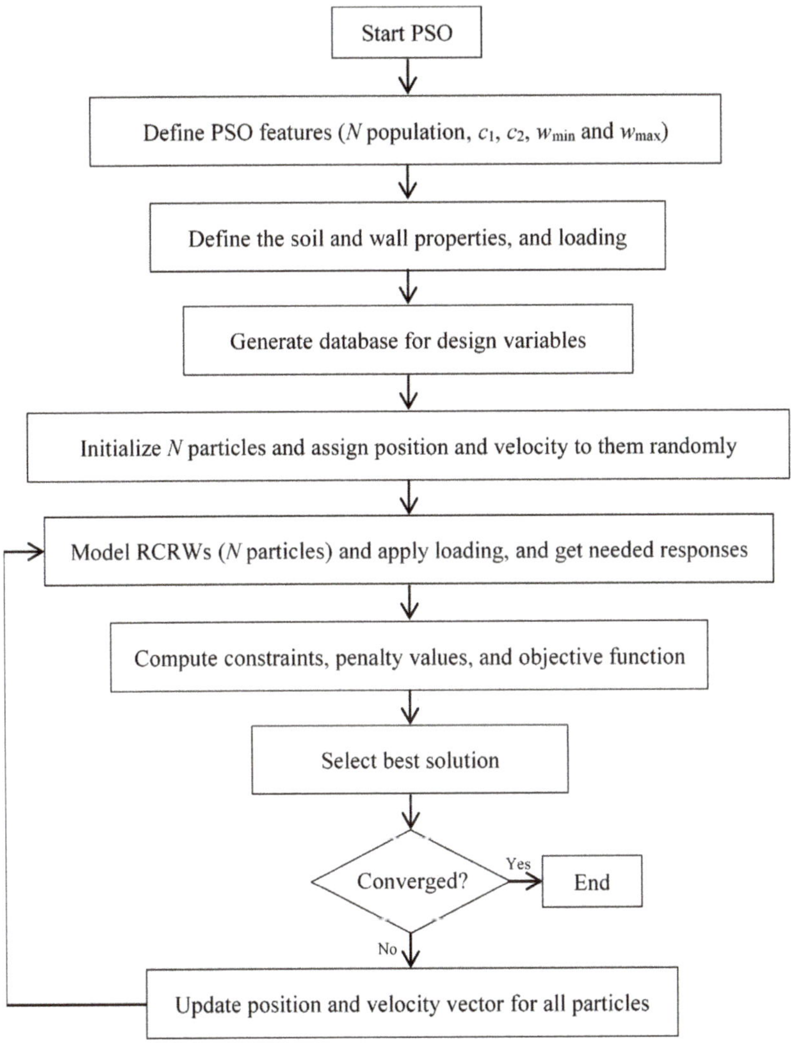

Figure 4. The flowchart of the design optimization of reinforced concrete retaining walls (RCRWs).

5. Design Examples

5.1. Assumptions

Three different RCRWs with different heights of 4.0, 5.5, and 7.0 m were considered for all three methods examined (MM, HM, and VM). A reasonable incremental step of 0.01 m was considered for the geometric variables. As shown in Table 3, a set of discrete values was considered for the steel reinforcement. It should be noted that the cross-sectional area of the steel reinforcement bars per unit length of the wall (1 m) was used during the optimization process, therefore, the number of used bars

(n_1 to n_4) needs to be obtained. All the values assumed for the modeling and design of the RCRWs are listed in Table 4.

Table 4. Values of input parameters for Examples 1, 2, and 3.

Parameter	Symbol	Value	Unit
		Examples 1–3	
Height of stem	H	4.0, 5.5, and 7.0	m
Steel yield strength	f_y	400	MPa
Shrinkage and temperature reinforcement ratio	ρ_{st}	0.002	-
Concrete compressive strength	f_c	21	MPa
Concrete cover	C_C	7	cm
Unit weight of concrete	γ_c	23.5	kN/m^3
Unit weight of steel bars	γ_s	78.5	kN/m^3
Surcharge load	q	15	kPa
Backfill slope	β	5	degrees
Internal friction angle of backfill	φ_{rs}	36	degrees
Unit weight of backfill	γ_{rs}	17.5	kN/m^3
Internal friction angle of base soil	φ_{bs}	39	degrees
Cohesion of base soil	c_{bs}	0	kPa
Unit weight of base soil	γ_{bs}	20	kN/m^3
Embedment depth of the toe	D	0.75	m
Cost of steel per unit of mass	C_{st}	0.4	\$/kg
Cost of concrete per unit of volume	C_c	40	\$/m^3

Because of the stochastic nature of the PSO algorithm, 20 independent runs were conducted for each case study corresponding to each method (MM, HM, and VM). In the PSO algorithm, the population was 20 and the maximum number of iterations was set to 6000; both c_1 and c_2 parameters were assumed to be equal to 2; w_{min} and w_{max} as the minimum and maximum value of the weight w_k, respectively, were considered to be 0.4 and 0.9.

5.2. Results and Discussion

After implementation of the optimization procedure, the best run among the 20 runs was chosen. The convergence history of all examples as Example 1 with H = 4.0 m, Example 2 with H = 5.5 m, and Example 3 with H = 7.0 m are shown in Figures 5–7, respectively. The convergence plots, which correspond to the best runs, start with a larger value of the cost, which is then minimized by the PSO algorithm at the final iteration. Note that since the optimization procedure was performed for a unit meter length of the wall, all the presented costs are per meter of length of the wall.

The optimum design variables, including the geometrical dimensions and the amounts of steel reinforcement for the three examples, are presented in Tables 5–7. As can be seen, all the results were within their allowable ranges defined in Tables 1 and 3. The last two columns of each table present the difference in the design variables for HM and VM with respect to MM as the chosen benchmark method. Based on the results, it can be noted that in general, the dimensions of X_1, X_4, X_6, X_7, and X_8 had relatively low sensitivity to the method used, while the other dimensions, namely X_2 and X_5, were quite sensitive. Concerning X_3, it seems to be sensitive to the method used merely for the case of wall height 4.0 m. The amounts of steel reinforcement (variables R_1, R_2, R_3, and R_4) were highly dependent on the method used.

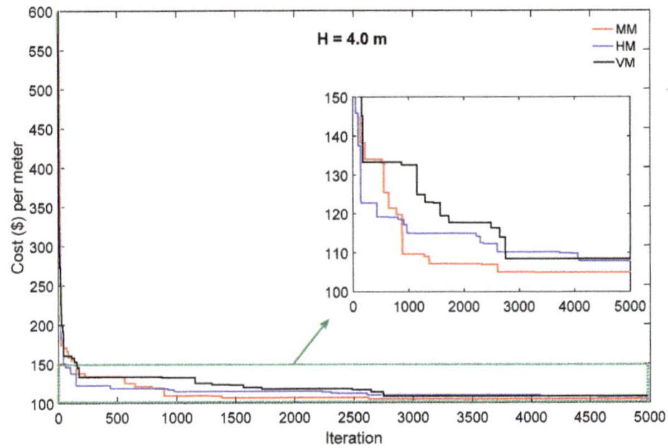

Figure 5. Convergence history of objective function—Example 1.

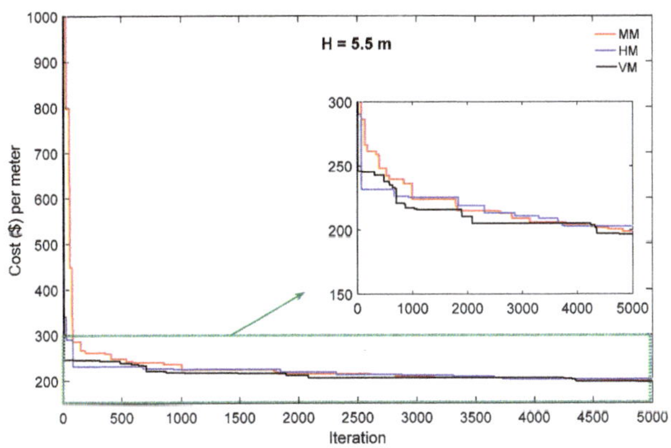

Figure 6. Convergence history of objective function—Example 2.

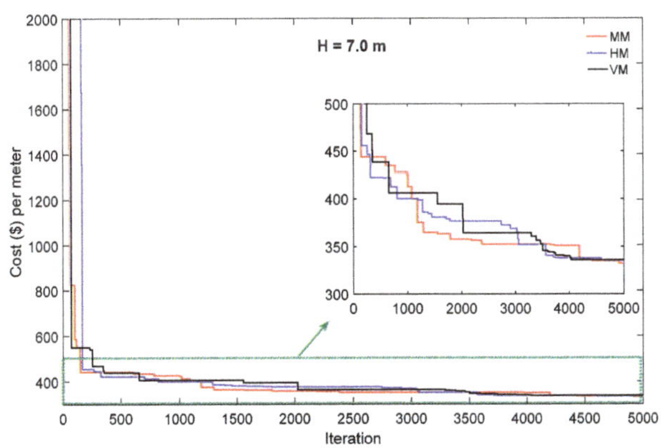

Figure 7. Convergence history of objective function—Example 3.

Table 5. Optimum design variables and comparison of the methods for Example 1—H = 4.0 m.

Variable	Method			Difference (%)	
	MM	HM	VM	100 × (HM − MM)/MM	100 × (VM − MM)/MM
X_1	2.33	2.47	2.38	6.01	2.15
X_2	0.88	1.03	1.05	17.05	19.32
X_3	0.37	0.34	0.32	−8.11	−13.51
X_4	0.20	0.20	0.20	0.00	0.00
X_5	0.24	0.25	0.27	4.17	12.50
X_6	2.01	2.01	2.05	0.00	1.99
X_7	0.26	0.27	0.28	3.85	7.69
X_8	0.23	0.22	0.24	−4.35	4.35
R_1	13ϕ12	20ϕ10	21ϕ10	6.84	12.18
R_2	13ϕ10	9ϕ12	14ϕ10	−0.31	7.69
R_3	13ϕ10	12ϕ10	9ϕ12	0.00	8.00
R_4	6ϕ12	14ϕ10	9ϕ10	62.04	4.17

Table 6. Optimum design variables and comparison of the methods for Example 2—H = 5.5 m.

Variable	Method			Difference (%)	
	MM	HM	VM	100 × (HM − MM)/MM	100 × (VM − MM)/MM
X_1	3.20	3.24	3.16	1.25	−1.25
X_2	0.99	1.19	1.33	20.20	34.34
X_3	0.47	0.48	0.46	2.13	−2.13
X_4	0.21	0.20	0.20	−4.76	−4.76
X_5	0.35	0.38	0.36	8.57	2.86
X_6	2.78	2.88	2.85	3.60	2.52
X_7	0.28	0.26	0.29	−7.14	3.57
X_8	0.23	0.20	0.25	−13.04	8.70
R_1	13ϕ16	22ϕ12	13ϕ16	−4.81	0.00
R_2	5ϕ16	5ϕ20	11ϕ14	56.25	68.44
R_3	11ϕ14	16ϕ10	7ϕ14	−25.79	−36.36
R_4	12ϕ10	7ϕ16	7ϕ12	49.33	−16.00

Table 7. Optimum design variables and comparison of the methods for Example 3—H = 7.0 m.

Variable	Method			Difference (%)	
	MM	HM	VM	100 × (HM − MM)/MM	100 × (VM − MM)/MM
X_1	4.04	3.90	3.79	−3.47	−6.19
X_2	1.34	1.51	1.76	12.69	31.34
X_3	0.49	0.48	0.48	−2.04	−2.04
X_4	0.21	0.21	0.20	0.00	−4.76
X_5	0.45	0.49	0.56	8.89	24.44
X_6	3.75	3.52	3.50	−6.13	−6.67
X_7	0.27	0.26	0.27	−3.70	0.00
X_8	0.21	0.21	0.21	0.00	0.00
R_1	19ϕ18	11ϕ24	19ϕ18	2.92	0.00
R_2	14ϕ12	16ϕ12	17ϕ12	14.29	21.43
R_3	16ϕ14	23ϕ10	22ϕ10	−26.66	−29.85
R_4	5ϕ18	22ϕ10	7ϕ12	35.80	−37.78

To show the convergence of the optimization process to the optimum solution, the best solutions in each iteration corresponding to Example 1 (with H = 4.0 m) considering the MM case are shown together in Figure 8. Based on this figure, PSO found the optimum solution in the design space very well. In addition, the optimum shapes of the three examples with consideration of the three methods of

MM, HM, and VM are shown in Figure 9. From this figure, it can be seen that the three methods resulted in different designs (geometrical dimensions) for all the optimum RCRWs with different heights.

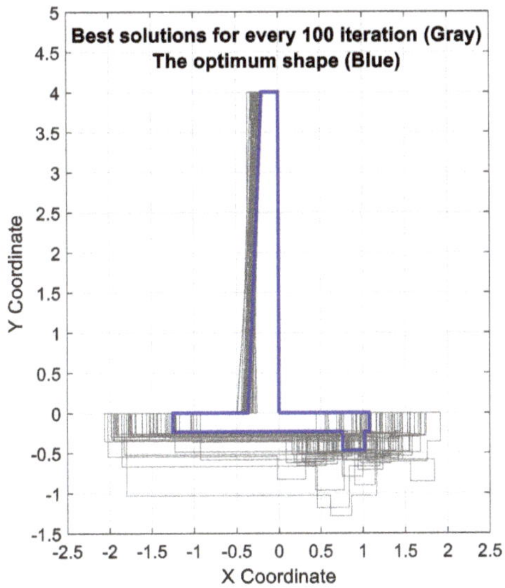

Figure 8. The candidate solutions and the final optimum shape (for H = 4.0 m, corresponding to the MM method).

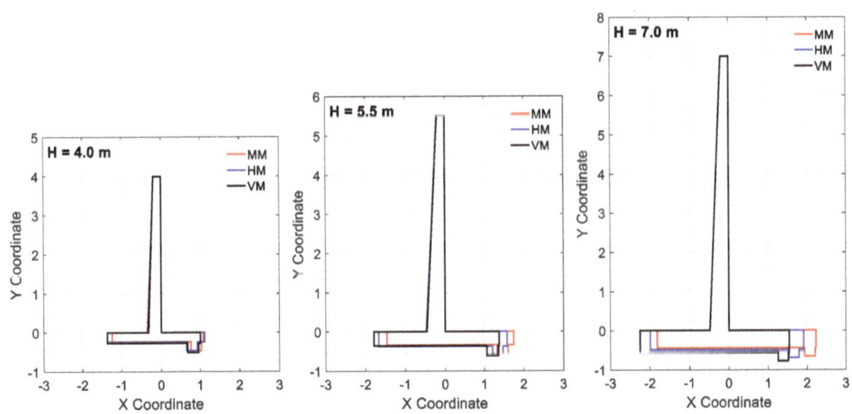

Figure 9. Optimal dimensions of the RCRWs for all examples with the three methods.

Figures 10–12 show the demand to capacity ratios of the constraints (all constraints $g_{1\text{–}3}$ and $g_{5\text{–}20}$, except for the constraint g_4 on $q_{\min}$) for the optimum solutions. As can be seen, all the corresponding values were less than 1.0, indicating that the optimum solutions have satisfied all the design constraints.

The constraint on the soil pressure intensity under the heel ($q_{\min}$), described as constraint g_4 in Equation (28), which is one of the fundamental constraints in the retaining walls design, has been excluded from the figures, yet it is examined in detail in Table 8. As described earlier, if $q_{\min}$ is negative, it means that some tensile stress is developed at the end of the heel, which is undesirable due to the negligible tensile strength of the soil. The final values for $q_{\min}$ corresponding to each example and method are listed in Table 8. As can be seen, all the values are greater than zero, indicating that no tensile stress appears beneath the heel slab of the RCRWs.

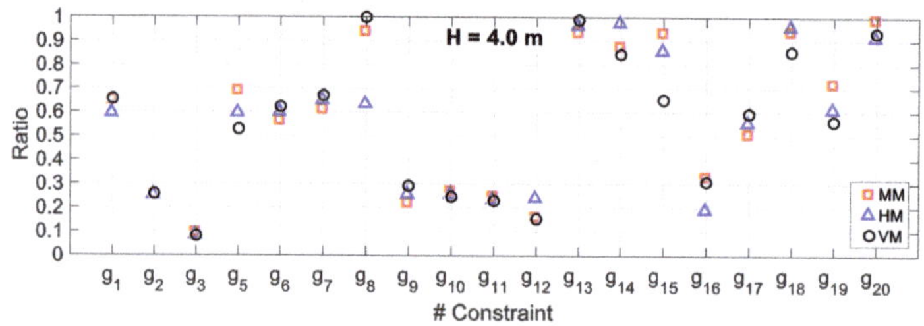

Figure 10. Demand to capacity ratio for Example 1.

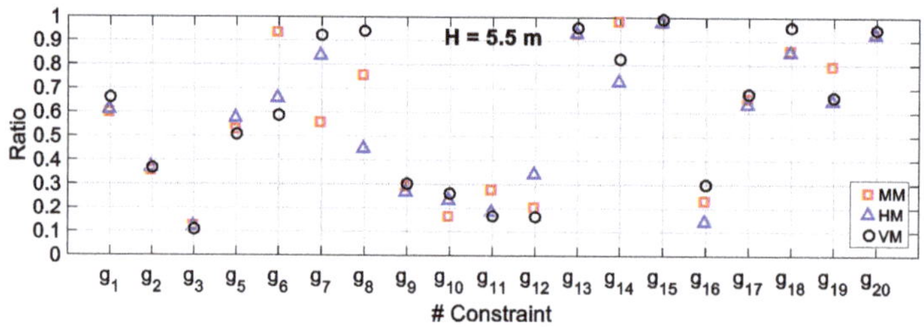

Figure 11. Demand to capacity ratio for Example 2.

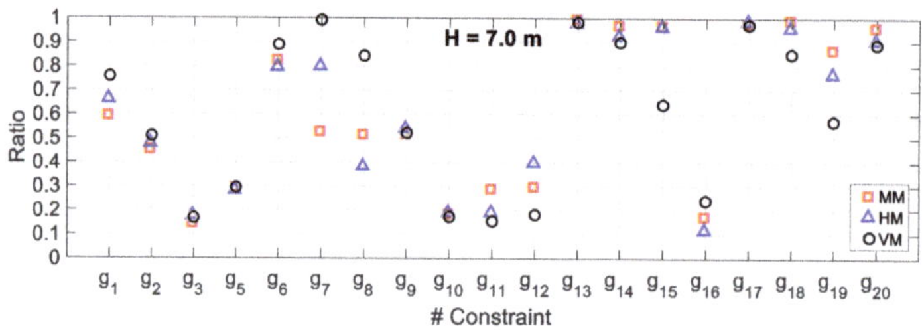

Figure 12. Demand to capacity ratio for Example 3.

Table 8. q_{min} (constraint g_4) values for all three examples and all three design methods.

Variable	Example 1—H = 4.0 m			Example 2—H = 5.5 m			Example 3—H = 7.0 m		
	MM	HM	VM	MM	HM	VM	MM	HM	VM
q_{min}	16.15	26.37	17.74	29.60	31.75	23.70	40.61	27.28	9.22

Next, the construction cost of the optimum designs was examined. The concrete, steel, and total cost for all of the optimum RCRWs with different heights, considering the three methods, are listed in Table 9. As can be seen, for H = 4.0 m and H = 7.0 m, the MM resulted in the minimum cost compared with the other two methods; also, the cost for HM was less than that for VM. For H = 5.5 m, the VM had the minimum cost and the cost for MM was less than that for HM. In addition, based on the results of Table 9 and as also shown in Tables 5–7, the provided amounts of reinforcement varied depending on the method used. Moreover, it was shown that the differences among the methods decreased with increasing the height of the RCRWs.

Figure 13 reveals the variation in cost components (concrete and steel materials) to total cost ratio with respect to the RCRW heights for each method. Based on this figure, it can be concluded that the concrete cost was reduced from 65.64% 52.93% on average with increasing the height, and the steel cost was increased from 34.36% to 47.07% on average with increasing the height.

Table 9. Cost of optimum RCRWs for different heights considering the three methods.

Variable	Example 1—H = 4.0 m			Example 2—H = 5.5 m			Example 3—H = 7.0 m		
	MM	HM	VM	MM	HM	VM	MM	HM	VM
Concrete ($)	70.07	70.60	70.27	122.00	126.39	120.80	172.15	175.37	184.04
Steel ($)	34.96	37.34	38.19	77.07	76.64	75.89	160.45	160.19	152.04
Total ($)	105.04	107.94	108.46	199.08	203.03	196.68	332.61	335.56	336.08
Diff. (%) to MM for total cost	-	2.76	3.26	-	1.99	−1.20	-	0.89	1.04

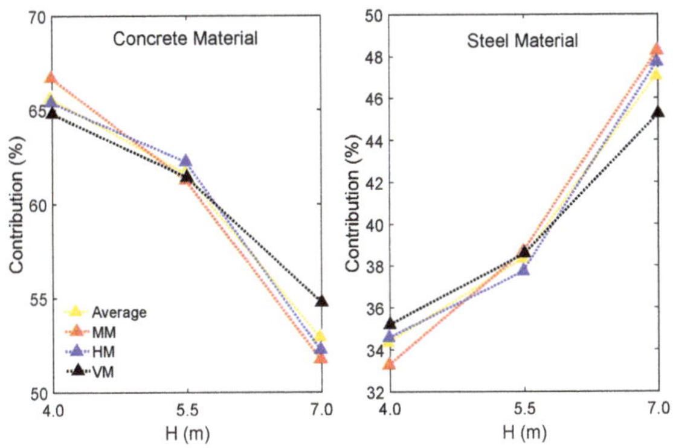

Figure 13. The contribution of concrete and steel materials in total cost for optimum RCRWs.

5.3. Comparative Study

In order to make a comparison with the literature, a specific example was chosen from the work conducted by Gandomi et al. [6] (Example 2, case 1 of the work [6] for $\beta = 0°$, see corresponding *Table 12* of *that* work). Using the methodology of the present work, all three methods of MM, HM, and VM were considered for the comparison. The results are listed in Table 10. As can be seen in this table, the results obtained for MM, HM, and VM were slightly different to those by Gandomi et al. [6], at least as far as the optimum cost is concerned. This can be attributed to the different analysis method and the different load combinations of ACI318-2014 used. Comparing the results also revealed the accuracy of the code programming used in this paper.

Table 10. Comparison of optimum cost obtained in this paper with the work by Gandomi et al. [6].

Work	Method	Cost ($/m)
Gandomi et al. [6]	-	162.37
This paper	MM	154.82
	HM	158.13
	VM	150.49

5.4. Effect of Backfill Slope

In this section, the effect of the parameter β (backfill slope) on the optimum cost of RCRWs was investigated, considering the three methods of MM, HM, and VM. Herein, this effect was assessed for Example 2—H = 5.5 m and the results are presented in Table 11. As can be seen, by increasing β, the cost was increased individually for each method. In addition, there were small differences (accounting for less than 3%) among the MM, HM, and VM methods, except for the case $\beta = 25°$, where the difference was slightly more, at 4.11%.

Table 11. Cost ($/m) of optimum RCRWs for Example 2—H = 5.5 m for different β for the three methods.

Method	β (degrees)					
	0	5	10	15	20	25
MM	193.52	199.08	200.78	215.98	224.86	231.04
HM	195.11	203.03	197.55	212.83	221.07	240.54
VM	191.05	196.68	202.96	209.59	220.57	234.34
HM to MM (Diff. %)	0.82	1.98	−1.61	−1.46	−1.69	4.11
VM to MM (Diff. %)	−1.28	−1.21	1.09	−2.96	−1.91	1.43

5.5. Effect of Surcharge Load

Herein, the effect of the parameter q (surcharge load) on the optimum cost of RCRWs was investigated, considering the three methods of MM, HM, and VM. Again, Example 2—H = 5.5 m was used to investigate the effect and the results are presented in Table 12. As shown, the optimum cost increased as q increased. In addition, there were small differences (accounting for less than 3%) among the MM, HM, and VM methods, except for the case that no surcharge load was applied (i.e., q = 0), which led to a difference of 4.38%.

Table 12. Cost ($/m) of optimum RCRWs for Example 2—H = 5.5 m with different q for the three methods.

Method	q (kPa)					
	0	5	10	15	20	25
MM	168.16	180.43	189.84	199.08	209.22	211.12
HM	172.47	180.88	186.93	203.03	203.69	213.82
VM	175.52	179.31	188.78	196.68	204.12	214.50
HM to MM (Diff. %)	2.56	0.25	−1.53	1.98	−2.64	1.28
VM to MM (Diff. %)	4.38	−0.62	−0.56	−1.21	−2.44	1.60

6. Conclusions

In the present study, three well-known methods of Meyerhof, Hansen, and Vesic were considered for the computation of the bearing capacity of RCRWs and, in particular, their influence on the optimum design of the wall. Three heights of walls were examined in the three test cases—4.0 m, 5.5 m, and 7.0 m. A code was developed in MATLAB where the PSO method was implemented for solving the constrained optimization problem. The design criteria were considered in accordance with ACI318-14 design code [20], where two sets of constraints were taken into account, the first on geotechnical requirements (wall stability) and the second on structural requirements (required strength of wall components and reinforcement arrangements). The PSO algorithm was successful in finding the optimum solutions fast and in a consistent way in all design cases, while the constraint handling mechanism was successful, managing to yield optimum solutions that satisfied the constraints in all cases.

The three methods resulted in slightly different designs for all the optimum RCRWs with different heights. In all three test examples, the MM (Meyerhof) method resulted in the design with the minimum total cost in comparison to the other two methods. It was also shown that the differences among the methods decreased with increasing the height of the RCRWs. When the height of the wall increases, the ratio of the cost of concrete to the total cost decreases, and the opposite happens with steel; the ratio of the cost of steel to the total cost increases. This observation was general and was made in all the three design methods examined. As regards the amounts of needed reinforcement, it can be concluded that the three methods give different results for the individual elements of the wall. On the other hand, as a general conclusion, it can be noted that the total cost of the wall corresponding to the three methods had only a small variation. In comparison to Meyerhof as the benchmark method, the maximum difference in the total cost was observed for the H = 4.0 m case and the VM method, and that accounted for only 3.26%. In all other cases the difference was even smaller.

In addition, the effect of the backfill slope β and the surcharge load q were examined. It was found that by increasing either the backfill slope or the surcharge load we obtained a higher total cost, but again in all cases the differences between the three methods were rather small, accounting for up to 4.5%. Finally, the results of the study were compared to results from an example of the work of Gandomi et al. [6] and were found only slightly different, which can be attributed to the different analysis method and the different load combinations used.

Author Contributions: Conceptualization, N.M. and S.G.; methodology, N.M. and S.G.; software, N.M. and S.G.; validation, V.P.; formal analysis, N.M., S.G. and V.P.; investigation, N.M., S.G. and V.P.; data curation, N.M. and S.G.; writing—original draft preparation, N.M. and S.G.; writing—review and editing, S.G. and V.P.; visualization, S.G., V.P.; supervision, S.G. and V.P.

Funding: This research received no external funding.

Conflicts of Interest: The authors declare no conflict of interest.

Appendix A

Table A1. The equations used for computing q_u based on each method.

Method	Equation	Condition
Meyerhof [22]	$N_{q,M} = e^{\pi \tan \phi_{bs}} \tan^2\left(45° + \frac{\phi_{bs}}{2}\right)$	ϕ_{bs} in degrees
	$N_{c,M} = \left(N_{q,M} - 1\right) \cot \phi_{bs}$	-
	$N_{\gamma,M} = \left(N_{q,M} - 1\right) \tan(1.4\phi_{bs})$	-
	$F_{cd,M} = 1 + 0.2\sqrt{k_p}\frac{D}{B-2e}$	
	$F_{qd,M} = F_{\gamma d,M} = 1 + 0.1\sqrt{k_p}\frac{D}{B-2e}$	$\phi_{bs} > 10°$
	$F_{cs,M} = F_{qs,M} = F_{\gamma s,M} = 1$	-
	$F_{ci} = F_{qi} = \left(1 - \frac{\theta}{90°}\right)^2$	θ in degrees
	$F_{\gamma i} = \left(1 - \frac{\theta}{\phi_{bs}}\right)^2$	
	$\theta° = \tan^{-1}\left(\frac{p_a \cos\beta}{\sum V}\right)$	for $\phi_{bs} > 0$
	θ is defined as angle of resultant measured from vertical direction	
Hansen [23]	$N_{q,H} = N_{q,M}$	-
	$N_{c,H} = N_{c,M}$	-
	$N_{\gamma,H} = 1.5\left(N_{q,H} - 1\right) \tan \phi_{bs}$	-
	$F_{cd,H} = 1 + 0.4\frac{D}{B-2e}$	$\frac{D}{B-2e} \le 1$
	$F_{cd,H} = 1 + 0.4 \tan^{-1}\left(\frac{D}{B-2e}\right)$	$\frac{D}{B-2e} > 1$
	$F_{qd,H} = 1 + 2\tan\phi_{bs}(1 - \sin\phi_{bs})^2 \frac{D}{B-2e}$	$\frac{D}{B-2e} \le 1$
	$F_{qd,H} = 1 + 2\tan\phi_{bs}(1 - \sin\phi_{bs})^2 \tan^{-1}\left(\frac{D}{B-2e}\right)$	$\frac{D}{B-2e} > 1$
	$F_{\gamma d,H} = 1$	-
	$F_{cs,H} = F_{qs,H} = F_{\gamma s,H} = 1$	-
	$F_{ci,H} = F_{qi,H} - \left(\frac{1-F_{qi,H}}{N_{q,H}-1}\right)$	$\phi_{bs} \neq 0$
	$F_{qi,H} = \left(1 - \frac{0.5p_a \cos\beta}{\sum V + (B-2e)c_{bs}\cot\phi_{bs}}\right)^5$	$\phi_{bs} \neq 0$
	$F_{\gamma i,H} = \left(1 - \frac{0.7p_a \cos\beta}{\sum V + (B-2e)c_{bs}\cot\phi_{bs}}\right)^5$	$\phi_{bs} \neq 0$
Vesic [24]	$N_{q,V} = N_{q,M}$	-
	$N_{c,V} = N_{c,M}$	-
	$N_{\gamma,V} = 2\left(N_{q,V} + 1\right) \tan \phi_{bs}$	-
	$F_{cd,V} = F_{cd,H}$	-
	$F_{qd,V} = F_{qd,H}$	-
	$F_{\gamma d,V} = F_{\gamma d,H}$	-
	$F_{ci,V} = F_{ci,H}$	-
	$F_{qi,V} = \left(1 - \frac{p_a \cos\beta}{\sum V + (B-2e)c_{bs}\cot\phi_{bs}}\right)^2$	$\phi_{bs} \neq 0$
	$F_{\gamma i,V} = \left(1 - \frac{p_a \cos\beta}{\sum V + (B-2e)c_{bs}\cot\phi_{bs}}\right)^3$	$\phi_{bs} \neq 0$

References

1. Sarıbaş, A.; Erbatur, F. Optimization and sensitivity of retaining structures. *J. Geotech. Eng.* **1996**, *122*, 649–656. [CrossRef]
2. Ceranic, B.; Fryer, C.; Baines, R. An application of simulated annealing to the optimum design of reinforced concrete retaining structures. *Comput. Struct.* **2001**, *79*, 1569–1581. [CrossRef]
3. Víctor, Y.; Julian, A.; Cristian, P.; Fernando, G.-V. A parametric study of optimum earth-retaining walls by simulated annealing. *Eng. Struct.* **2008**, *30*, 821–830.
4. Camp, C.V.; Akin, A. Design of retaining walls using big bang–big crunch optimization. *J. Struct. Eng.* **2011**, *138*, 438–448. [CrossRef]
5. Khajehzadeh, M.; Taha, M.R.; El-Shafie, A.; Eslami, M. Economic design of retaining wall using particle swarm optimization with passive congregation. *Aust. J. Basic Appl. Sci.* **2010**, *4*, 5500–5507.
6. Gandomi, A.H. Optimization of retaining wall design using recent swarm intelligence techniques. *Eng. Struct.* **2015**, *103*, 72–84. [CrossRef]
7. Kaveh, A.; Khayatazad, M. Optimal design of cantilever retaining walls using ray optimization method. Iranian Journal of Science and Technology. *Trans. Civ. Eng.* **2014**, *38*, 261.
8. Kaveh, A.; Soleimani, N. CBO and DPSO for optimum design of reinforced concrete cantilever retaining walls. *Asian J. Civ. Eng.* **2015**, *16*, 751–774.
9. Kaveh, A.; Farhoudi, N. Dolphin echolocation optimization for design of cantilever retaining walls. *Asian J. Civ. Eng.* **2016**, *17*, 193–211.
10. Kaveh, A.; Laien, D.J. Optimal design of reinforced concrete cantilever retaining walls using CBO, ECBO and VPS algorithms. *Asian J. Civ. Eng.* **2017**, *18*, 657–671.
11. Temur, R.; Bekdaş, G. Teaching learning-based optimization for design of cantilever retaining walls. *Struct. Eng. Mech.* **2016**, *57*, 763–783. [CrossRef]
12. Ukritchon, B.; Chea, S.; Keawsawasvong, S. Optimal design of Reinforced Concrete Cantilever Retaining Walls considering the requirement of slope stability. *KSCE J. Civ. Eng.* **2017**, *21*, 2673–2682. [CrossRef]
13. Aydogdu, I. Cost optimization of reinforced concrete cantilever retaining walls under seismic loading using a biogeography-based optimization algorithm with Levy flights. *Eng. Optim.* **2017**, *49*, 381–400. [CrossRef]
14. Kumar, V.N.; Suribabu, C. Optimal design of cantilever retaining wall using differential evolution algorithm. *Int. J. Optim. Civ. Eng.* **2017**, *7*, 433–449.
15. Gandomi, A.; Kashani, A.; Zeighami, F. Retaining wall optimization using interior search algorithm with different bound constraint handling. *Int. J. Numer. Anal. Methods Geomech.* **2017**, *41*, 1304–1331. [CrossRef]
16. Gandomi, A.H.; Kashani, A.R. Automating pseudo-static analysis of concrete cantilever retaining wall using evolutionary algorithms. *Measurement* **2018**, *115*, 104–124. [CrossRef]
17. Mergos, P.E.; Mantoglou, F. Optimum design of reinforced concrete retaining walls with the flower pollination algorithm. *Struct. Multidiscip. Optim.* **2019**, 1–11. [CrossRef]
18. MATLAB. *The Language of Technical Computing*; Math Works Inc.: Natick, MA, USA; 2005; Volume 2018a.
19. Kennedy, J.; Eberhart, R. Particle swarm optimization (PSO). In Proceedings of the IEEE International Conference on Neural Networks, Perth, Australia, 27 November–1 December 1995.
20. ACI. *American Concrete Institute: Building Code Requirements for Structural Concrete and Commentary*; ACI: Farmington Hills, MI, USA, 2014.
21. Das, B.M. *Principles of Foundation Engineering*; Cengage Learning: Boston, MA, USA, 2015.
22. Meyerhof, G.G. Some recent research on the bearing capacity of foundations. *Can. Geotech. J.* **1963**, *1*, 16–26. [CrossRef]
23. Hansen, J.B. *A Revised and Extended Formula for Bearing Capacity*; Danish Geotechnical Institute: Lyngby, Denmark, 1970.
24. Vesic, A.S. Analysis of ultimate loads of shallow foundations. *J. Soil Mech. Found. Div.* **1973**, *99*, 45–73. [CrossRef]
25. Arora, J.S. *Introduction to Optimum Design*; Elsevier: Amsterdam, The Netherlands, 2004.
26. Gharehbaghi, S.; Khatibinia, M. Optimal seismic design of reinforced concrete structures under time-history earthquake loads using an intelligent hybrid algorithm. *Earthq. Eng. Eng. Vib.* **2015**, *14*, 97–109. [CrossRef]
27. Gholizadeh, S.; Salajegheh, E. Optimal design of structures subjected to time history loading by swarm intelligence and an advanced metamodel. *Comput. Methods Appl. Mech. Eng.* **2009**, *198*, 2936–2949. [CrossRef]

28. Plevris, V.; Papadrakakis, M. A hybrid particle swarm-gradient algorithm for global structural optimization. *Comput.-Aided Civ. Infrastruct. Eng.* **2011**, *26*, 48–68. [CrossRef]
29. Gholizadeh, S. Layout optimization of truss structures by hybridizing cellular automata and particle swarm optimization. *Comput. Struct.* **2013**, *125*, 86–99. [CrossRef]
30. Yazdani, H. Probabilistic performance-based optimum seismic design of RC structures considering soil–structure interaction effects. *ASCE-ASME J. Risk Uncertain. Eng. Syst. Part A Civ. Eng.* **2016**, *3*, G4016004. [CrossRef]
31. Gharehbaghi, S.; Moustafa, A.; Salajegheh, E. Optimum seismic design of reinforced concrete frame structures. *Comput. Concr.* **2016**, *17*, 761–786. [CrossRef]
32. Gharehbaghi, S. Damage controlled optimum seismic design of reinforced concrete framed structures. *Struct. Eng. Mech.* **2018**, *65*, 53–68.
33. Khatibinia, M.; Jalali, M.; Gharehbaghi, S. Shape optimization of U-shaped steel dampers subjected to cyclic loading using an efficient hybrid approach. *Eng. Struct.* **2019**, *197*, 108874. [CrossRef]

MDPI
St. Alban-Anlage 66
4052 Basel
Switzerland
Tel. +41 61 683 77 34
Fax +41 61 302 89 18
www.mdpi.com

Mathematics Editorial Office
E-mail: mathematics@mdpi.com
www.mdpi.com/journal/mathematics

www.ingramcontent.com/pod-product-compliance
Lightning Source LLC
Chambersburg PA
CBHW061607210926
51850CB00009B/2244
9783039434954